Intestinal Tumorigenesis

Vincent W. Yang • Agnieszka B. Bialkowska
Editors

Intestinal Tumorigenesis

Mechanisms of Development & Progression

Springer

Editors
Vincent W. Yang
Department of Medicine
Stony Brook University
Stony Brook
USA

Agnieszka B. Bialkowska
Department of Medicine
Stony Brook University
Stony Brook
USA

ISBN 978-3-319-19985-6 ISBN 978-3-319-19986-3 (eBook)
DOI 10.1007/978-3-319-19986-3

Library of Congress Control Number: 2015947939

Springer Cham Heidelberg New York Dordrecht London

Printed on acid-free paper

Springer International Publishing AG Switzerland is part of Springer Science+Business Media (www.springer.com)

Preface

The intestinal epithelium is the largest mucosal surface in the body and functions as a barrier between the host and its external environment. Cells within the mucosal surface are renewed every four to five days and this renewal is subject to stringent regulation in order to maintain homeostasis [1–3, 6, 7]. Studies indicate that colorectal cancer is the consequence of stepwise perturbations to major signaling pathways important in controlling various cellular functions [4, 5]. Eventually these alterations lead to the development of disease, often with heterogeneous phenotype and clinical outcome. In the last decades considerable effort has been made towards understanding of the mechanisms underlying development and progression of intestinal tumorigenesis. Furthermore, new discoveries in the fields of stem cells research, epigenetics, non-coding RNA and epithelial-to-mesenchymal transition have greatly improved our understanding of this disease. Recent data regarding the function of the gut microbiota in influencing both normal intestinal homeostasis on the one hand and on the other chronic inflammation, which may lead to cancer formation, open new possibilities for preventive treatment. Finally, emerging proteomics and genomics approaches aimed at developing novel cancer biomarkers may lead to earlier detection and enhance treatment options.

Colorectal cancer (CRC) is the fourth most common type of the cancer in the world with the American Cancer Society estimating that 132,700 new cases and 49,700 deaths from cancers of the colon and rectum will occur in the United States in 2015 (see http://www.cancer.org/cancer/colonandrectumcancer/detailedguide/colorectal-cancer-key-statistics). Noteworthy, the death rate of the colorectal cancer has declined in the past two decades and this is mainly due to introduction of early screening and improvements in the treatment of this disease.

This volume includes in-depth reviews of many recent developments in the field of intestinal tumorigenesis with an emphasis on the latest discoveries that have resulted in identification of novel mechanisms underlying cancerous transformation. We hope this volume provides readers with the information necessary in understanding important recent work on the mechanisms of intestinal tumorigenesis and comprehending the complexity of the process. It is also our desire that this volume serves to entice new researchers to participate in future studies geared towards understanding the mechanisms underlying development and progression of colorectal cancer. Finally we hope that this volume opens discussions in the scientific community and shapes the future research in this field.

In Chap. 1, Gala and Chung introduce the fundamental basis of hereditary colorectal cancer syndromes. Chapter 2, by Prasad and Jung, comprehensively reviews the role of microsatellite instability in intestinal tumorigenesis; including mechanisms, mouse models, and finally the clinical implications. In Chap. 3, Bell, Lo and Shroyer review intestinal epithelial stem cell characteristics and the mechanisms of their function in epithelial homeostasis and touch upon the important role of these intestinal stem cells in tumor development. This is followed by a discussion by Harris, Gao and Evers on the role of the PI3K signaling pathway plays in intestinal tumorigenesis. In Chap. 5, Moreira, Balaguer and Goel review the role of epigenetics in intestinal tumorigenesis as well as the potential for using epigenetic modifications as CRC biomarkers. Chapter 6 by Nishida and Calin provides an overview of the functions of microRNAs in intestinal tumorigenesis. The effects of the gut microbiome, and associated inflammation, in homeostasis and tumor development are reviewed in Chap. 7 by Posocco, Dmitrieva and Grivennkov and Chap. 8 by Dheer, Davies and Abreu, respectively. Recently, studies have revealed an important role that lipids play in regulating intestinal homeostasis and tumorigenesis. In Chap. 9 Garcia-Barros, Coant and Snider discuss the involvement of shingolipids in intestinal tumorigenesis. There are many other important factors that lie at the basis of intestinal tumorigenesis, the effects of one that affects all—aging—are thoroughly reviewed by Nautiyal, Purvis and Majumdar in Chap. 10. In Chap. 11, Cao and colleagues presents an extensive overview of the role epithelial-to-mesenchymal transitioning plays in intestinal tumorigenesis. Koyuturk and Nibbe advocate the importance for developing robust biomarkers of CRC and discusses integrative networks in Chap. 12. Finally, Grady and colleagues summarize the latest targeted therapies being developed directed towards colorectal cancer development and progression in Chap. 13.

We would like to express our gratitude to all of the authors that kindly donated their time to make the publication of this monogram possible. We also thank the Springer publishing group and the Editorial Stuff for their assistance and professionalism.

1. Barker N (2014) Adult intestinal stem cells: critical drivers of epithelial homeostasis and regeneration. Nature reviews. Molecular cell biology 15:19–33
2. Barker N, Van De Wetering M, Clevers H (2008) The intestinal stem cell. Genes & development 22:1856–1864
3. Barker N, Van Es JH, Kuipers J et al. (2007) Identification of stem cells in small intestine and colon by marker gene *Lgr5*. Nature 449:1003–1007
4. Fearon ER (2011) Molecular genetics of colorectal cancer. Annual review of pathology 6:479–507
5. Fearon ER, Vogelstein B (1990) A genetic model for colorectal tumorigenesis. Cell 61:759–767
6. Snippert HJ, Van Der Flier LG, Sato T et al. (2010) Intestinal crypt homeostasis results from neutral competition between symmetrically dividing Lgr5 stem cells. Cell 143:134–144
7. Yan KS, Chia LA, Li X et al. (2012) The intestinal stem cell markers Bmi1 and Lgr5 identify two functionally distinct populations. Proceedings of the National Academy of Sciences of the United States of America 109:466–471

Contents

Contributors

Maria T. Abreu Division of Gastroenterology, Department of Medicine, University of Miami, Miller School of Medicine, Miami, FL, USA

Francesc Balaguer Department of Gastroenterology, Hospital Clínic, Centro de Investigación Biomédica en Red de Enfermedades Hepáticas y Digestivas (CIBERehd), Institut d'Investigacions Biomediques August Pi i Sunyer (IDIBAPS), University of Barcelona, Barcelona, Catalonia, Spain

Kristin Bell Division of Gastroenterology, Department of Medicine, Baylor College of Medicine, Houston, TX, USA

George A. Calin Department of Experimental Therapeutics and The Center for RNA Interference and Non-Coding RNAs, The University of Texas MD Anderson Cancer Center, Houston, TX, USA

Jian Cao Division of Cancer Prevention, Department of Medicine, Stony Brook University, Stony Brook, NY, USA

Jillian Cathcart Division of Cancer Prevention, Department of Medicine, Stony Brook University, Stony Brook, NY, USA

E. Gabriela Chiorean Department of Medicine, University of Washington School of Medicine, Seattle, WA, USA

Daniel C. Chung Department of Internal Medicine, Massachusetts General Hospital, Boston, MA, USA

Nicolas Coant Department of Medicine, Cancer Center, Stony Brook University, Stony Brook, NY, USA

Andrew Coveler Clinical Research Division, Fred Hutchinson Cancer Research Center, Seattle, WA, USA and Department of Medicine, University of Washington School of Medicine, Seattle, WA, USA

Julie M. Davies Division of Gastroenterology, Department of Medicine, University of Miami, Miller School of Medicine, Miami, FL, USA

Rishu Dheer Division of Gastroenterology, Department of Medicine, University of Miami, Miller School of Medicine, Miami, FL, USA

Oxana Dmitrieva Cancer Prevention and Control Program, Fox Chase Cancer Center, Philadelphia, PA, USA

B. Mark Evers Markey Cancer Center; Department of Surgery, University of Kentucky, Lexington, KY, USA

Manish Gala Department of Gastroenterology, Massachusetts General Hospital and Harvard Medical School, Boston, MA, USA

Tianyan Gao Molecular and Cellular Biochemistry, University of Kentucky, Lexington, KY, USA

Mónica García-Barros Department of Medicine, Cancer Center, Stony Brook University, Stony Brook, NY, USA

Ajay Goel Center for Gastrointestinal Cancer Research; Center for Epigenetics, Cancer Prevention and Cancer Genomics, Baylor Research Institute and Charles A Sammons Cancer Center, Baylor University Medical Center, Dallas, TX, USA

William M. Grady Clinical Research Division, Fred Hutchinson Cancer Research Center, Seattle, WA, USA and Department of Medicine, University of Washington School of Medicine, Seattle, WA, USA

Jon Grim Clinical Research Division, Fred Hutchinson Cancer Research Center, Seattle, WA, USA and Department of Medicine, University of Washington School of Medicine, Seattle, WA, USA and Research and Development Service, VA Puget Sound Health Care System, Seattle, WA, USA

Sergei Grivennikov Cancer Prevention and Control Program, Fox Chase Cancer Center, Philadelphia, PA, USA

Jennifer W. Harris Department of Surgery, University of Kentucky, Lexington, KY, USA

Barbara Jung Department of Medicine, Division of Gastroenterology and Hepatology, University of Illinois at Chicago, Chicago, IL, USA

Deborah Kim Department of Medicine, Division of Cancer Prevention, Stony Brook University, Stony Brook, NY, USA

Mehmet Koyuturk Department of Electrical Engineering & Computer Science Case Western Reserve University, Cleveland, OH, USA

Yuan-Hung Lo Division of Gastroenterology, Department of Medicine, Baylor College of Medicine, Houston, TX, USA

Adhip P. N. Majumdar Veterans Affairs Medical Center, Karmanos Cancer Institute, Departments of Internal Medicine, Wayne State University, Detroit, MI, USA

Leticia Moreira Department of Gastroenterology, Hospital Clínic, Centro de Investigación Biomédica en Red de Enfermedades Hepáticas y Digestivas (CIBERehd), Institut d'Investigacions Biomediques August Pi i Sunyer (IDIBAPS), University of Barcelona, Barcelona, Catalonia, Spain

Jyoti Nautiyal Department of Biological Sciences, College of Liberal Arts and Science, Wayne State University, Detroit, MI, USA

Rod K. Nibbe Systems Biology NeoProteomics, Inc. Cleveland, OH, USA

Naohiro Nishida Department of Experimental Therapeutics and The Center for RNA Interference and Non-Coding RNAs, The University of Texas MD Anderson Cancer Center, Houston, TX, USA

David Posocco Cancer Prevention and Control Program, Fox Chase Cancer Center, Philadelphia, PA, USA

Meena A. Prasad Northwestern University Feinberg School of Medicine, Division of Gastroenterology and Hepatology, Chicago, IL, USA

Ashleigh Pulkoski-Gross Department of Medicine, Division of Cancer Prevention, Stony Brook University, Stony Brook, NY, USA

Krystyn Purvis Department of Biological Sciences, College of Liberal Arts and Science, Wayne State University, Detroit, MI, USA

Noah F. Shroyer Division of Gastroenterology, Department of Medicine, Baylor College of Medicine, Houston, TX, USA

Ashley J. Snider Department of Medicine, Cancer Center, Stony Brook University, Stony Brook, NY, USA and Veterans Affairs Medical Center, Northport, NY, USA

Xi E. Zheng Department of Public Health, Weill Cornell Medical College, New York, NY, USA

Hereditary CRC Syndromes

Manish Gala and Daniel C. Chung

Abstract The genetic contribution to the development of colon cancer has been inferred to be approximately 35% from large Scandinavian twin registries. Independent studies indicate that 11–16% of those with a diagnosis of colon cancer also have a first-degree relative with colon cancer. A significant fraction of these familial cases are represented by well-characterized, high-risk familial cancer syndromes that display Mendelian or near-Mendelian patterns of inheritance. Accounting for nearly 5% of all colon cancer cases, these disorders are also associated with a diverse array of extracolonic malignancies. Careful integration of clinical features, family history, endoscopic findings, molecular pathology, and genetic testing are crucial components in establishing the diagnosis of a hereditary cancer syndrome. Recognition and diagnosis of these syndromes enable healthcare providers to pursue cancer risk-reducing measures, including chemoprevention, heightened surveillance, and prophylactic surgery, for these patients and affected family members. In addition, awareness of such syndromes has important ramifications for cancer management.

Keywords Lynch syndrome · Hereditary nonpolyposis colorectal cancer · Familial adenomatous polyposis · Attenuated familial adenomatous polyposis · *MUTYH*-associated polyposis · Juvenile polyposis · Peutz-Jeghers syndrome · Cowden syndrome · PTEN hamartoma syndrome · Serrated polyposis syndrome

D. C. Chung (✉)
Department of Internal Medicine, Massachusetts General Hospital, 55 Fruit Street, GRJ 704, Boston, MA 02114, USA
e-mail: dchung@partners.org

M. Gala
Department of Gastroenterology, Massachusetts General Hospital and Harvard Medical School, Boston, MA 02114, USA

V. W. Yang, A. B. Bialkowska (eds.), *Intestinal Tumorigenesis*,
DOI 10.1007/978-3-319-19986-3_1

1 Introduction

The genetic contribution to the development of colon cancer has been inferred to be approximately 35% from large Scandinavian twin registries [1]. Independent studies indicate that 11–16% of those with a diagnosis of colon cancer also have a first-degree relative with colon cancer [2, 3]. A significant fraction of these familial cases are represented by well-characterized, high-risk familial cancer syndromes that display Mendelian or near-Mendelian patterns of inheritance. Accounting for nearly 5% of all colon cancer cases, these disorders are also associated with a diverse array of extracolonic malignancies. Careful integration of clinical features, family history, endoscopic findings, molecular pathology, and genetic testing are crucial components in establishing the diagnosis of a hereditary cancer syndrome. Recognition and diagnosis of these syndromes enable healthcare providers to pursue cancer risk-reducing measures, including chemoprevention, heightened surveillance, and prophylactic surgery, for these patients and affected family members. In addition, awareness of such syndromes has important ramifications for cancer management.

Over the past few years, there has been an explosion in the amount of genetic data generated from next generation sequencing that promises to expand the current paradigms for hereditary colon cancer. Private mutations in several new genes have already been implicated as high-risk for the development of colon cancer through familial and case-control studies. It is also important to note that other hereditary cancer syndromes primarily affecting other organ systems, such as Li-Fraumeni, may occasionally manifest as hereditary colon cancer [4]. In this chapter, we will focus primarily on established, well-described hereditary colon cancer syndromes with moderate to high penetrance. Broadly characterized as either polyposis or nonpolyposis syndromes, we will describe their epidemiology, genetics and molecular mechanisms for carcinogenesis, diagnostic criteria and clinical presentation, surveillance and preventative strategies, and relevant cancer management approaches.

2 Nonpolyposis Syndromes

2.1 Lynch Syndrome (Hereditary Nonpolyposis Colon Cancer)

2.1.1 Genetics and Molecular Mechanisms

An autosomal dominant disorder, Lynch Syndrome is caused by deleterious germline mutations in DNA mismatch repair (MMR) genes. To date, a mutation in one of four MMR genes (*MLH1, MSH2, MSH6*, and *PMS2*) has been firmly implicated in the development of Lynch syndrome [5–7]. Mutations in a fifth gene (*EPCAM*) also result in Lynch Syndrome through its silencing effects on the adjacent *MSH2* gene [8]. Germline *MLH1* and *MSH2* mutations are responsible for approximately 70% of Lynch syndrome cases, and *MSH6* and *PMS2* mutations account for most of the remaining cases [7]. Case reports for mutations in other genes such as *PMS1* and *TGFBR2* have been described as causes for Lynch syndrome, but contribute to an exceedingly small fraction of the cases overall [9, 10].

Biallelic loss of these MMR genes results in DNA microsatellite instability, frequently abbreviated as MSI-H. DNA microsatellites (or simple sequence repeats) are repeating sequences of DNA usually 2–6 base pairs in length (Fig. 1). These sites are particularly vulnerable to base-pair mismatches, insertions, and deletions during DNA replication. In the genome, dinucleotide repeats of C-A occur most frequently, and these are often disrupted by defects in mismatch repair [11]. Microsatellite instability is evaluated through analysis of a set of five consensus microsatellite sites termed Bethesda markers (BAT25, BAT26, D2S123, D5S346, and D17S250). Instability in at least two markers is sufficient for a tumor from any primary site to be labeled as MSI-H. MMR enzymes function in pairs, and deficiency in one enzyme typically results in diminished or absent expression of its binding partner. MLH1 works in concert with PMS2, while MSH2 binds MSH6 to exert their DNA repair

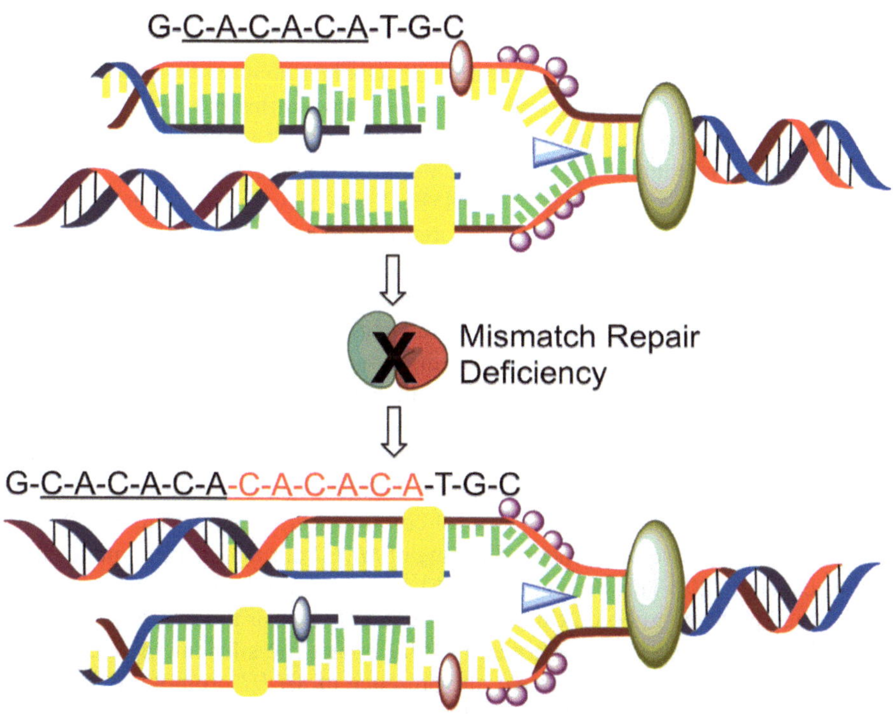

Fig. 1 Microsatellite instability in Lynch syndrome. In normal cell replication, mismatch repair proteins ensure the fidelity of multi-nucleotide repeats (*underlined*). In Lynch syndrome, these repeats can be expanded or deleted due to loss of mismatch repair proteins. The instability at these sites can result in frameshift mutations that can inactivate important tumor suppressor genes

function. This reciprocal relationship is utilized in immunohistochemical analysis of Lynch-associated cancers, as the staining of the inactivated protein as well as its binding partner is often diminished or missing [12]. The exception to this observation occurs with cases of Lynch Syndrome due to *PMS2* or *MSH6* mutations. Loss of PMS2 may still result in retained MLH1 expression, and *MSH6* mutations also lead to loss of staining for MSH6 with preserved MSH2 staining.

In addition to coding mutations and structural variations in these genes, heritable epigenetic changes may also cause Lynch Syndrome. Heritable gene silencing of *MLH1* through promoter CpG methylation has been infrequently observed in pedigrees of Lynch Syndrome [13]. Also, large deletions in the 3′ end of the gene *EPCAM* have recently been linked to Lynch Syndrome. The 3′ deletion in *EPCAM* results in epigenetic silencing of its distal locus on chromosome 2 which contains *MSH2*. Thus, Lynch syndrome patients with deletions in *EPCAM* will have deficiencies in MSH2 expression [8].

MMR deficiency shortens the duration of the canonical adenoma to carcinoma sequence from the prototypical 10 years to approximately 2–3 years [14]. The mechanism for this acceleration is twofold. First, many tumor suppressor genes contain microsatellite repeats, and deficiency in MMR facilitates their inactivation through accumulation of deleterious mutations in these regions. Second, MMR genes play an important role in activating the ATM-ATR DNA damage signaling network, which initiates cell growth arrest and thereby allows sufficient time for DNA repair prior to genome replication [15]. Distinct domains in MMR proteins have been identified that are responsible for these two different functions. The cumulative effects of these two impaired mechanisms result in tumors that display a higher mutation rate than most microsatellite stable colon cancers [16].

A fascinating observation regarding MMR deficient tumors has been their more favorable clinical outcomes when compared to microsatellite stable tumors [17]. MSI-H tumors are less likely to metastasize. In vitro studies have demonstrated that MSI-H tumor cells displayed impaired epithelial-mesenchymal transition (EMT), attributable to mutations in microsatellites of genes involved in the TGF-β signaling pathway [18]. In addition, tumor-infiltrating lymphocytes seen in MSI-H tumors may play an important role in immune surveillance that restrains tumor progression [19].

While cancers from individuals with Lynch Syndrome demonstrate MMR deficiency, it should be noted that most MMR deficient tumors are not due to Lynch Syndrome. Approximately 15–20 % of all colon cancers demonstrate MMR deficiency through acquired epigenetic silencing of *MLH1* during the adenoma to carcinoma sequence [20]. These "sporadic" MSI-H tumors are thought to arise through the serrated pathway to colon cancer, and frequently harbor the *BRAF* V600E mutation. Despite rare reports of heritable *MLH1* epigenetic silencing as a cause for Lynch syndrome as described above, the presence of *MLH1* promoter hypermethylation in tumor tissue generally excludes the diagnosis of Lynch Syndrome as the cause for the colon cancer [21]. A rare subset will have somatic, bialleic mutations in an MMR gene, resulting in a sporadic MSI-H tumor [22].

2.1.2 Epidemiology and Clinical Presentation

Lynch syndrome is the most prevalent hereditary colon cancer syndrome, responsible for approximately 2–3 % of all colon cancers [2, 23]. Liberal estimates can be as high as 5 % of all CRCs. Among individuals younger than the age of 50 years, Lynch Syndrome may be responsible for up to 10 % of colon cancers. The population prevalence of Lynch syndrome has been estimated to be 1 in 440 persons [24]. The lifetime risk of colon cancer is estimated to be approximately 50–80 %, with the lower estimates observed in more recent studies. While historically considered a disorder that causes early-onset colon cancer, more recent studies have demonstrated the median age of colon cancer diagnosis in Lynch Syndrome to be 61.2 years [25].

Individuals with Lynch syndrome may develop cancers in any location in the colon, but many studies have demonstrated a predilection for polyp and cancer development on the right side (70 % of cases) [26]. The reasons for this positional propensity are presently unknown. In addition, individuals with Lynch syndrome are at higher risk of developing metachronous and synchronous tumors. Overt polyposis is not a typical feature in individuals with Lynch syndrome.

MSI-H colon cancers, including those specifically associated with Lynch syndrome, often have histological features that can suggest mismatch repair deficiency. MSI-H tumors are more likely to be mucinous when compared to their MSS counterparts. In addition, MSI-H tumors may often have a Crohn's-like reaction surrounding the tumor, or high numbers of tumor-infiltrating lymphocytes. Such findings on pathology should increase the suspicion of MMR deficiency in the tumor.

Another important clinical feature in individuals and families with Lynch syndrome is the high rate of extracolonic cancers. Mismatch repair deficiency and subsequent microsatellite instability can occur in a diverse set of extracolonic tissues including endometrium, stomach, small bowel, hepatobiliary tract, pancreas, urinary tract, ovaries, and the central nervous system. More recently, an increased risk of breast cancer has been suggested in Lynch Syndrome [27].

Despite the broad diversity of extracolonic tumors identified in individuals with Lynch syndrome, special consideration should be given to female patients. In women with Lynch syndrome, the lifetime risk of endometrial cancer can be as high as or exceed that of colon cancer, ranging from 27 to 71 % [28, 29]. Lynch syndrome may be responsible for up to 5 % of all endometrial cancers. In addition, the lifetime risk of ovarian cancer in such patients may be as high as 14 %.

Additional terms have been designated for those subsets of Lynch families who develop specific extracolonic cancers. Muir-Torre syndrome, a subset of Lynch Syndrome, occurs when individuals with Lynch syndrome develop a sebaceous gland tumor (adenoma, epithelioma, or carcinoma) in combination with colon cancer. Muir-Torre occurs in approximately 28 % of Lynch families and 9.2 % of individuals with Lynch syndrome [30]. Approximately 70 % of the extracolonic tumors associated with Muir-Torre syndrome display microsatellite instability. The majority of cases occur with germline mutations in *MLH1* or *MSH2*. Turcot syndrome represents a hereditary colon cancer syndrome with the presence of a glioblastoma or medulloblastoma. Turcot syndrome may occur in individuals and families with either Lynch Syndrome or Familial Adenomatous Polyposis (see section below). In Lynch-associated Turcot syndrome, glioblastomas are more common than medulloblastomas.

2.1.3 Diagnosis

A positive genetic test is considered the diagnostic test of choice in Lynch syndrome. Several clinical criteria have been formulated to determine when genetic testing is appropriate for Lynch Syndrome. While supplanted by other screening modalities, these criteria still remain important to recognize individuals who have a higher pretest probability of having the syndrome. Developed in 1990, the first widely used criteria were the Amsterdam I Criteria. Colloquially referred to the "3-2-1 rule", families required three individuals with colon cancer (two of which are first-degree relatives) that spanned two successive generations, and at least one individual developed CRC at age less than 50. Polyposis syndromes such as FAP should be excluded. Families satisfying such criteria were recommended to undergo germline testing of MMR genes. The sensitivity and specificity of Amersterdam I criteria have been reported to be 61 and 67 %, respectively [31]. These criteria were subsequently revised and expanded in 1999 to include Lynch-associated extracolonic cancers (Table 1). Known as the Amsterdam II Criteria, they have a sensitivity and specificity of 72 and 78 %, respectively. In some cases, patients will satisfy Amsterdam criteria but will not harbor germline mutations in MMR genes or exhibit MSI in their tumors. This constellation is called familial cancer syndrome X. Individuals and families with syndrome X tend to have lower risks for the extracolonic tumors associated with Lynch syndrome, but are placed on similar colonoscopy surveillance regimens.

The Bethesda criteria were developed in part to determine which individuals and families should proceed to MSI testing of tumors prior to germline testing of MMR genes. Intentionally designed to capture those missed by the Amsterdam criteria, these criteria have a sensitivity of 94 % and specificity of 25 % [31]. Modified Bethesda criteria have also been developed to further enhance the sensitivity (Table 1).

In addition to these criteria, several computational models incorporating family history and age of presentation have been developed to help guide testing for Lynch Syndrome. Nevertheless, these strategies have some limitations. More recently, studies have demonstrated the cost-effectiveness of population-based screening of all newly diagnosed cases of CRC [32]. In particular, many large academic centers have implemented this strategy of universal IHC staining of MMR protein and somatic *BRAF* mutation testing. The presence of the *BRAF* V600E mutation rules out Lynch syndrome as the cause of the colon cancer. Absence of IHC staining for any MMR protein coupled with the absence of an activating *BRAF* mutation should be followed up by specific germline testing for those genes with absent expression. In addition, universal screening for Lynch syndrome has also been recommended for individuals who develop endometrial cancer before the age of 50. An important challenge for clinicians and geneticists is that many eligible for germline testing refuse to proceed further.

Table 1 Clinical criteria utilized for Lynch syndrome screening

Amsterdam II criteria	Each of the following criteria must be satisfied:
	Three or more relatives with an associated cancer (colorectal cancer, or cancer of the endometrium, small intestine, ureter or renal pelvis)
	Two or more successive generations affected
	One or more relatives diagnosed before the age of 50 years
	One should be a first-degree relative of the other two
	Familial adenomatous polyposis (FAP) should be excluded in cases of colorectal carcinoma
	Tumors should be verified by pathologic examination
Revised Bethesda guidelines	One of the following criteria must be satisfied:
	Colorectal carcinoma diagnosed in a patient who is less than 50 years old
	Presence of synchronous or metachronous CRC or other Lynch syndrome-associated tumors regardless of age
	CRC with high microsatellite instability histology diagnosed in a patient less than 60 years old
	CRC diagnosed in one or more first-degree relatives with a Lynch syndrome-associated tumor, with one of the cancers being diagnosed at less than 50 years of age
	CRC diagnosed in two or more first-degree or second-degree relatives with Lynch syndrome-associated tumors, regardless of age

2.1.4 Surveillance and Cancer Prevention

Once a diagnosis of Lynch syndrome has been established in a proband, testing for the specific mutation should be offered to all at-risk relatives. Genetic testing in at-risk individuals should commence between the age of 20 and 25 years, or ten years prior to the earliest cancer, whichever comes first. Individuals and affected family members should be placed on a careful cancer surveillance program to reduce mortality. Colonoscopies should begin at age 20–25 years, or 10 years prior to the earliest colon cancer in the family, whichever comes first. In addition, they should be repeated at 1–2 year intervals. Regular, frequent colonoscopies can reduce the rate of colon cancer by 63 % in affected individuals in controlled trials [33].

Given the high risk of endometrial cancer, women with Lynch syndrome frequently undergo annual endometrial sampling starting at age 30–35 years. While no evidence presently exists that an endometrial cancer surveillance regimen improves mortality, it has been regularly employed by many high-risk genetics groups. Women are also encouraged to undergo prophylactic total abdominal hysterectomy and bilateral salpingo-oophorectomy (TAH-BSO) after child-bearing, or potentially at the time of colectomy for colon cancer.

While these surveillance measures for colon cancer reduce the morbidity and mortality of those with Lynch syndrome, it should be noted that approximately 61 % of cancer deaths are now attributable to malignancies other than colon or endometrial [34]. Many centers routinely perform urine cytology, transvaginal ultrasounds, and serum CA-125 levels to screen for these other Lynch-associated malignancies.

However, no data exist to support the efficacy of these other cancer screening modalities. Many experts recommend that such screening should be confined to the research setting, while others tailor approaches ad-hoc based on the family history.

Despite the lack of data for these cancer prevention efforts, clinicians can still have an impact on the development of cancers in Lynch syndrome. Two modifiable risk factors that have been observed to influence cancer-risk are smoking and obesity. Individuals with Lynch syndrome who smoke should be offered cessation measures, as the hazard ratio for adenoma development has been observed to be sixfold compared to non-smokers [35]. In addition, former smokers have a threefold increased hazard ratio for the development of polyps. Certain dietary patterns have also been linked to increased adenoma formation in individuals with Lynch syndrome [36].

In addition to lifestyle modifications, chemoprevention may offer some benefit in reducing the risk of Lynch-associated cancers. Although a randomized placebo-controlled trial of high dose aspirin (600 mg/day) did not demonstrate any reduction in cancer risk in Lynch syndrome after 4 years, an extended post-trial analysis revealed a benefit after 6 years [37, 38]. Presently, additional studies are underway to determine the dose-response relationship at which such chemopreventative effects can be sustained.

2.1.5 Cancer Management

The management of colon cancer differs when there is a concurrent diagnosis of Lynch syndrome. Whereas a segmental colectomy remains the mainstay for individuals with sporadic cancers, a subtotal colectomy is recommended in individuals with Lynch syndrome due to the high risk of metachronous lesions. Women who have finished child-bearing may also be offered a TAH-BSO at the time of colectomy to minimize endometrial and ovarian cancer risks. Concordant with better outcomes in individuals with Lynch compared to stage-matched MSS cancers, no benefit with adjuvant 5-FU treatment has been observed with stage II CRCs in Lynch syndrome [39].

3 Polyposis Syndromes

3.1 Familial Adenomatous Polyposis (FAP) and Attenuated Familial Adenomatous Polyposis (AFAP)

3.1.1 Genetics and Molecular Mechanisms

An autosomal dominant disorder, Familial Adenomatous Polyposis (FAP) results from deleterious germline mutations in the *APC* gene, and the hallmark feature is the carpeting of the colon by adenomatous polyps [40]. APC is a key component of the β-catenin destruction complex that regulates Wnt signaling (Fig. 2). By facilitating the phosphorylation of β-catenin, APC promotes the destruction of β-catenin in

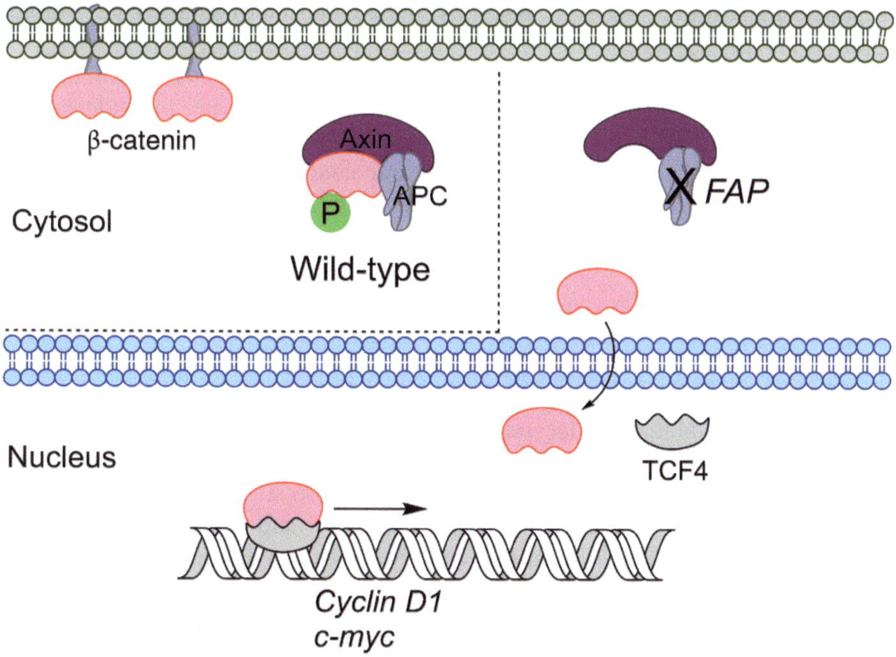

Fig. 2 Aberrant Wnt signaling in familial adenomatous polyposis. In individuals with functional APC, β-catenin typically serves as a protein involved in cell-cell adhesions. Cytosolic β-catenin is targeted for destruction by phosphorylation by a large protein complex that contains APC and Axin. In individuals with FAP, cytosolic destruction of β-catenin is impaired due to mutations in *APC*. β-catenin translocates into the nucleus, where it binds to TCF4 to initiate gene transcription of a stem-cell like program

the cytosol. Mutations in *APC* prevent this destruction, resulting in nuclear translocation of β-catenin. In the nucleus, β-catenin binds to a transcription factor, TCF4, to activate a transcriptional program of cell growth and stem-cell like characteristics. For example, active Wnt signaling upregulates important oncogenic proteins such as Cyclin D1 and c-Myc. More than 60 % of *APC* mutations in FAP occur between codons 1286 and 1583 in exon 15, a region referred to as the mutation cluster region (MCR) [41].

A subset of individuals with *APC* mutations develop a milder form of the disease, known as Attenuated Familial Adenomatous Polyposis (AFAP). Individuals with AFAP develop fewer colonic polyps (<100), and develop CRC at a later age compared to FAP patients. In AFAP, mutations in *APC* are more commonly located in the proximal 5′ end, exon 9, or the distal 3′ end of the gene. Unlike the mutations observed in classic FAP patients, these mutations typically produce a partially functional protein. Mutations toward the proximal 5′ end result in the activation of an internal translational start site [42]. Mutations in the distal 3′ end result in a minimally truncated protein since the mutation is located distal to the boundary required to induce nonsense-mediated decay (typically 50 base pairs prior to the last exon-exon junction). Exon 9 mutations observed in AFAP are found typically in the

5′ splice-site region of the exon, resulting in the inclusion of an alternate in-frame exon 9a into the sequence [43].

In addition to the attenuated phenotype in AFAP, there are other clinical correlations between *APC* germline mutations and clinical phenotype. Individuals with FAP may develop desmoid tumors and congenital hypertrophy of the retinal pigment epithelium (CHRPE), among other extracolonic manifestations. A desmoid cluster region comprised of mutations occurring between codons 1445 and 1580 has been observed [44]. *APC* mutations in FAP patients with CHRPE commonly occur between codons 457 and 1444 [45].

A second hit is usually required to induce the biallelic inactivation required for polyp development in the colon, as well as neoplasia in other organs. However, observations of large cohorts of FAP patients reveal that the location of the second, somatic hit can be predicted by the specific germline mutation in the host [46]. As such, it appears that the location of the second mutation is often nonrandom, resulting in an optimized level of Wnt activation for neoplasia. This optimized level of Wnt activity correlates with observations in in vitro studies that too much Wnt activation can precipitate DNA damage, senescence, and cell growth arrest, and too little Wnt is insufficient to sustain cell growth and stemness. This relationship has been observed in both desmoids and tubular adenomas from FAP patients. If germline mutations occur between codons 1194 and 1392, strong selection exists in tubular adenomas for allelic loss of *APC* as the second hit. Otherwise, the second hit is more likely to be a truncating mutation in the MCR.

3.1.2 Epidemiology and Clinical Presentation

The incidence of FAP is estimated to be 1:10,000 and the disease accounts for fewer than 1 % of all colon cancer cases. As a truly Mendelian autosomal dominant disease, FAP has nearly 100 % penetrance. However, it is important to recognize that 20 % of cases represent de novo cases that demonstrate no family history of polyposis or colon cancer [47]. Individuals typically develop hundreds to thousands of polyps during the second decade of life, and will invariably develop colon cancer if left untreated (Fig. 3). The mean age at which colon cancer presents in FAP is 39 years. Synchronous cancers may also be found in such individuals.

Given the importance of Wnt signaling in a variety of other organ systems, it is unsurprising to find extracolonic manifestations in individuals with FAP. Fundic gland and duodenal polyps, ampullary adenomas, desmoid tumors, osteomas, CHRPE, papillary thyroid cancers, and hepatoblastomas can develop in individuals with FAP. Desmoid tumors (occurring in 8–14 %) represent the second leading cause of cancer-related morbidity and mortality in FAP patients due to their local effects on neighboring organs and high rate of recurrence [48]. Papillary thyroid cancers occur in 1–2 % of FAP patients, typically by the age of 28 [49]. Duodenal, periampullary, or ampullary adenomas occur in over 90 % of patients, and the lifetime risk of adenocarcinoma from periampullary and duodenal cancers ranges from 3 to 5 % [50]. Gardner syndrome refers to a subset of FAP patients with osteomas,

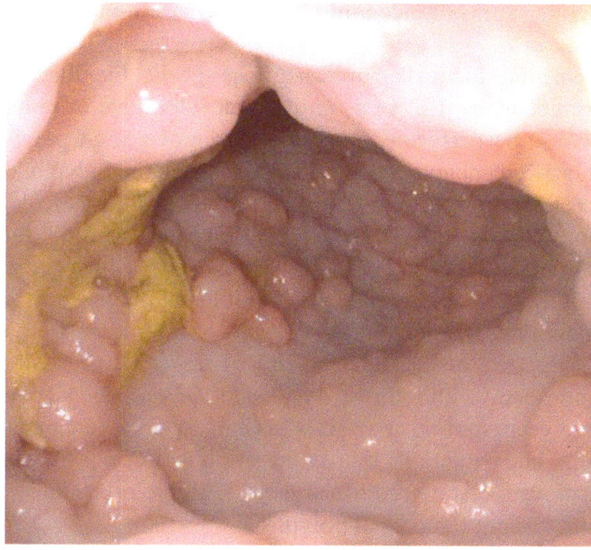

Fig. 3 Endoscopic appearance of familial adenomatous polyposis. The colon has an extensive burden of adenomatous polyps

thyroid cancer, desmoids, or epidermoid cysts. Turcot syndrome, as it relates to FAP, primarily manifests as medulloblastomas.

Attenuated Familial Adenomatous Polyposis (AFAP) constitutes a milder clinical phenotype, with typically fewer than 100 polyps seen in such individuals (average age of onset ~30 years). Given the lower number of polyps and the delayed age of onset of polyposis, the mean age at which colon cancer develops in affected individuals (50 years) is much later than classical FAP [51]. Individuals with AFAP have a similar risk of upper gastrointestinal cancers, but display a considerably lower risk of desmoids and CHRPE compared to their classical FAP counterparts.

3.1.3 Diagnosis

Individuals with as few as ten cumulative polyps should be referred for genetic testing for polyposis syndromes. Individuals with as few as 10–19 polyps have a 5 % chance of harboring pathogenic germline *APC* mutations [52]. Individuals who harbor more than 1000 polyps have an 80 % chance. Family history of polyposis or CRC is not required for a diagnosis of FAP since 20 % of cases represent de novo mutational events. Familial cases often exhibit an autosomal dominant family history of colon cancer if not previously diagnosed. Individuals who test negative for *APC* mutations should be tested for *MUTYH* (see next section) given the phenotypic overlap. First-degree relatives should be screened for *APC* mutations starting at age 12 years.

3.1.4 Surveillance and Cancer Prevention

Screening for FAP should begin at age 12. Flexible sigmoidoscopy should be performed every 1–2 years until the development of polyps, after which colonoscopy should be performed annually to remove large polyps. Individuals who develop advanced adenomas, or in those in which the polyp burden becomes unmanageable to endoscopists, should be referred for total proctocolectomy with ileal pouch-anal anastomosis (IPAA). In classical FAP, the need for surgery typically becomes apparent in the second or third decade of life. Given the risk of adenoma development in the pouch and rectal cuff, flexible sigmoidoscopy should be performed at 6–12 month intervals thereafter.

Because of the elevated risk of upper GI tract malignancies in FAP, an upper endoscopy with a forward and side viewing endoscope should be performed every 1–3 years starting at age 20–25 years. A side-viewing endoscope is required to visualize and potentially biopsy the ampulla. Lesions may be monitored, endoscopically removed, or referred to surgery (Whipple procedure) depending on the pathology and extent of duodenal polyposis.

Screening for other extracolonic cancers in FAP is not standardized. Given the higher lifetime risk of thyroid cancers, some providers elect to perform thyroid ultrasounds in addition to careful physical palpation for nodules. In addition, some centers perform abdominal ultrasounds with AFP serum measurements for hepatoblastomas during infancy. Such practices vary considerably between institutions, and more data are required before broadly recommending them. However, it should be noted that thyroid ultrasounds are successful in detecting cancers in FAP patient at early stages, and notably when the physical exam has been unremarkable.

Recommendations for AFAP differ from classical FAP because of the attenuated phenotype. Colonoscopy should commence at age 18 years, and then should be repeated every 1–2 years depending on polyp burden. The decision for colectomy should be made on the basis of the polyp burden, since the polyp burden can be quite heterogeneous. Patients may demonstrate as few as ten polyps or as many as 100. For those individuals in whom surgery is required, a subtotal colectomy with ileorectal anastomosis may also be considered in lieu of an IPAA if the rectal burden of polyps is low. Surveillance of upper GI neoplasms should be performed similarly to those with classical FAP.

Chemoprevention has been explored in detail with FAP. In particular, NSAIDs have been used to reduce polyp burden. Sulindac was the first drug shown to reduce polyp burden by almost 50 %, and this was particularly important for those who had undergone subtotal colectomy with a retained rectum [53]. The drug was less efficacious for upper GI neoplasms. Despite these results, sulindac does not obviate the need for colectomy, and side effects of gastrointestinal irritation are noted. Subsequently, COX-2 selective inhibitors were tested in FAP patients. Drugs, such as celecoxib, reduced polyp formation by 28 % and also reduced duodenal adenoma formation [54]. However, long-term use of another drug in the same class, rofecoxib, was noted to carry an increased risk of cardiovascular deaths. The use of aspirin is being explored in FAP. Currently, NSAIDs cannot be considered primary therapy for FAP but can play an adjunctive role to reduce postoperative polyp recurrence rates.

3.1.5 Cancer Management

Individuals with FAP who have a resectable colon cancer discovered at the time of the diagnosis should have a total proctocolectomy performed given the high risk of synchronous and metachronous lesions. Desmoid tumors are the second leading cause of cancer-related morbidity and mortality in FAP patients and typically develop after colectomy. Their morbidity is due to their location; in FAP they are typically intra-abdominal or in the abdominal wall. First-line therapy for large or growing desmoids includes sulindac and tamoxifen [55]. If the lesion is focal, surgical resection can be considered. Desmoid tumors that do not respond to this regimen or are unresectable are subsequently treated with chemotherapy. Radiation therapy can be considered, but the risk of injury to adjacent bowel is significant. For duodenal or ampullary malignancies, a Whipple procedure is generally indicated.

3.2 MUTYH-*associated Polyposis*

3.2.1 Genetics and Molecular Mechanisms

MUTYH-associated polyposis (MAP) is another hereditary syndrome that is characterized by the development of 3–100s of tubular adenomas in the colon. Caused by mutations in the gene encoding the DNA glycosylase, *MUTYH*, the disease is inherited in an autosomal recessive fashion [56]. MUTYH, a base excision repair enzyme, repairs nucleotides that are damaged by oxidation. In particular, oxidation of guanine nucleotides causes the formation of 7,8-dihydroxy-8-oxoguanine (8-oxo-G). Deficiency in this enzyme results in G:C to T:A transversions during the next round of DNA replication [57]. These transversions occur in many established tumor suppressor genes and oncogenes, including *APC* and *KRAS*, to promote polyp development [58, 59]. In addition to its base excision repair function, MUTYH has also been implicated in DNA damage signaling. Loss of MUTYH in vitro has been linked to diminished activation of ATR and CHK1—key components of the DNA damage signaling cascade [60].

The majority of pathogenic mutations occur from two founder missense mutations (Y165C and G382D). These mutations constitute 90 % of the deleterious mutations observed in the gene from Northern European populations. Pathogenic *MUTYH* mutations have also been found in most populations around the world. While the polyposis phenotype is observed in homozygotes or compound heterozygotes, recent studies have suggested that heterozygotes may also have an elevated colon cancer risk, particularly if there is also a family history of CRC [61].

3.2.2 Epidemiology and Clinical Presentation

MAP often presents with 10s–100s of tubular adenomas by age 50, and may frequently mimic the clinical presentations of de novo forms of FAP/AFAP. The penetrance of the disease is presently unknown. A small subset of individuals with MAP may also present with large numbers of serrated polyps and satisfy criteria for serrated polyposis syndrome (see section below) [62]. Similarly to FAP, individuals are also at higher risk of upper gastrointestinal malignancies; the lifetime risk of duodenal cancer is 4 %. Duodenal adenomas are present in 17–25 % of individuals [63].

3.2.3 Diagnosis

Germline *MUTYH* testing should be considered in all individuals who develop greater than ten polyps, particularly in the setting of pedigrees that demonstrate an autosomal recessive inheritance or negative germline *APC* mutation testing. In addition, siblings of those affected with MAP should be offered testing. Before testing children, the spouse may be tested to see if he/she harbors a pathogenic allele. Given the increased risk of CRC observed in heterozygotes with family histories of colon cancer as described above, testing may be considered for all first-degree relatives when there is a family history of colon cancer.

In addition, MAP should be suspected in individuals who develop 1–10 adenomas at a young age, particularly before the age of 40. Also, consideration for testing should be made in individuals who display an autosomal recessive pattern of inheritance for colon cancer when polyposis information is unavailable. Finally, the high rate of somatic tranversions seen in CRCs of MAP patients results in the distinct *KRAS* mutation c.34G>T in codon 12 in approximately 64 % of these individuals [59]. The presence of this mutation may serve to identify additional individuals who harbor germline *MUTYH* mutations.

3.2.4 Cancer Surveillance

Guidelines on cancer surveillance in MAP differ between the United States and Europe. In the United States, colonoscopy is recommended starting at age 25–30 and then repeated every 1–2 years. Upper endoscopies with forward and side-viewing capabilities are recommended every 3–5 years starting at age 30–35. In Europe, colonoscopy is recommended at age 18–20 years with follow-up determined by polyp burden. Upper endoscopies are recommended to start at age 25–30.

3.3 Juvenile Polyposis Syndrome

3.3.1 Genetics and Molecular Mechanisms

Juvenile Polyposis Syndrome (JPS) is a rare autosomal dominant disorder that predisposes affected individuals to the development of juvenile polyps throughout the gastrointestinal tract. Juvenile polyps are hamartomatous polyps with cystically dilated glands and inflammatory stroma. Mucinous cuboidal or columnar cells line the epithelium, and in contrast to Peutz-Jeghers syndrome (see next section), there is no smooth muscle proliferation. While sporadic juvenile polyps pose little cancer risk, the juvenile polyps in JPS can undergo malignant transformation.

While the genetic basis of many cases of JPS remains unknown, mutations in the TGF-β pathway have been firmly implicated in its pathogenesis. Approximately 20 % of cases harbor a pathogenic mutation in *SMAD4*, a nuclear transcription factor that responds to TGF-β stimuli [64]. In response to TGF-β, SMAD2 and SMAD3 become phosphorylated and form complexes with SMAD4 in the nucleus to regulate gene transcription. Deleterious mutations in *SMAD4* render cells unresponsive to TGF-β stimulation. Interestingly, conditional deletion of *SMAD4* in the epithelial tissues of mice fails to promote carcinogenesis in mouse models, whereas T-cell specific deletion is sufficient to generate epithelial cancers [65]. T-cell deletion of *SMAD4* encourages a Th2 cytokine response that promotes stromal proliferation and an inflammatory infiltrate. These findings stress the importance of the role of dysregulation between the stroma and the epithelial compartment in transformation and carcinogenesis in JPS.

Another 20 % of cases are attributable to deleterious mutations in *BMPR1A*, a receptor in the TGF-β superfamily [66]. Loss of BMPR1A results in an expansion in the stem cell compartment of the intestine, and the subsequent development of intestinal polyps and JPS in animal models. In addition, BMPR1A can negatively regulate Wnt signaling, thus controlling the number of stem cells [67].

3.3.2 Epidemiology and Clinical Presentation

JPS is a rare syndrome, occurring in less than 1 in 100,000 persons. Approximately 75 % of individuals with JPS will have an affected parent, whereas the remaining 25 % of cases may reflect de novo mutations. Due to the large size and vascularity of juvenile polyps, individuals may be symptomatic due to the ability of polyps to cause abdominal pain through obstruction, anemia from bleeding, and protein-losing enteropathy. The lifetime risk of colon cancer is 39 %, and the mean age at diagnosis is 46 years [68]. Individuals are also at increased risk for duodenal, gastric, and pancreatic cancers. In JPS, only 20 % of affected individuals develop polyps in the stomach and small intestine, and the majority of the polyps occur in the large intestine. The diagnosis of JPS is typically made before the age of 20.

Approximately 15–22 % of individuals who harbor *SMAD4* mutations will also have overalp symptoms of hereditary hemorrhagic telangiectasia (otherwise known as Osler–Weber–Rendu syndrome), a genetic disorder that predisposes individuals toward arteriovenous malformations and bleeding [69].

3.3.3 Diagnosis

Clinical diagnostic criteria for JPS include any of the following: (1) at least five juvenile polyps in the large intestine, (2) at least one juvenile polyp in the upper GI tract and one in the colon, or (3) one juvenile polyp and a family history of JPS. Germline testing for mutations in *SMAD4* or *BMPR1A*, if positive, can reinforce the diagnosis and clarify risk for first degree family members.

3.3.4 Cancer Surveillance

Guidelines have been issued by the British Society of Gastroenterology [70]. Affected individuals should undergo colonoscopy every 1–2 years starting at 15–18 years of age, or earlier if symptoms are present. Surveillance regimens should be continued until the age of 70. Examination of the upper GI system should commence starting at age 25 with upper endoscopy or enteroscopy, upper GI series, or small bowel follow-through. If there is no known familial mutation, at-risk relatives should begin screening at the same age as those affected but can be discontinued by age 35 if no lesions have been discovered. Prophylactic surgery may be indicated when endoscopic surveillance is not feasible due to extensive polyp burden, excessive bleeding or other symptoms, or if there are multiple polyps with adenomatous changes or high-grade dysplasia.

3.4 Peutz-Jeghers Syndrome

3.4.1 Genetics and Molecular Mechanisms

Peutz-Jeghers syndrome (PJS) is another autosomal dominant, hamartomatous polyposis syndrome. Mutations in *STK11* (formerly referred to *LKB1*), located on chromosome 19, predispose individuals toward the disease [71]. When cellular energy levels are scarce, STK11 activates AMPK and AMPK-related kinases to restrain cell growth and maintain polarity. Thus, STK11 is important for energy homeostasis as a negative regulator of the mTOR growth pathway [72]. STK11 itself forms a complex and becomes allosterically activated when bound to the pseudokinase, STRAD, and adaptor protein, MO25 [73].

3.4.2 Epidemiology and Clinical Presentation

PJS is a rare disorder, thought to occur in approximately 1 in 300,000 individuals. Polyps from PJS are hamartomas, characterized by an arborizing pattern of smooth muscle hyperplasia. The epithelium may appear as islands within the smooth muscle hyperplasia. Individuals also develop mucocutaneous pigmentation and melanin spots that most commonly occur across the mouth and cross the vermilion border. In addition, these skin lesions may be found in the palms, soles, nostrils, fingers, toes, and perianal regions. They can fade with age, particularly after adolescence. Lesions may persist in the buccal mucosa.

Individuals with PJS may also present with anemia, abdominal pain, obstruction, and gastrointestinal bleeding due to polyp burden. In addition, intussusception occurs frequently in PJS, often a cause of obstruction and first presentation of the disease. Gynecomastia may also be present in prepubertal males due to the predilection toward large-cell calcifying Sertoli cell tumors [74]. The bulk of polyps occur in the small intestine, followed by the colon.

The lifetime risk of cancer is high in patients with PJS. Individuals have a 85–93 % risk of any malignancy by age 70 [75]. The lifetime risk for gastrointestinal cancers is 57 %, 45 % for breast cancer, 18 % for gynecological tumors, and 11 % for pancreatic cancer. The lifetime risk of colon cancer is 39 %, and constitutes the most frequent type of luminal gastrointestinal cancer.

3.4.3 Diagnosis

Clinical suspicion can often be aroused in infancy due to the mucocutaneous findings observed in those with PJS. Clinical diagnostic criteria are presented in Table 2. Approximately 94 % of individuals who satisfy clinical criteria possess a germline mutation or deletion in *STK11* [76]. Genetic testing starting at age eight may be helpful in establishing a diagnosis and then clarifying risk in first-degree relatives who do not have obvious physical exam findings. Genetic testing at an early age may reduce the morbidity from complications such as obstruction or intussusception.

Table 2 Diagnostic clinical criteria for Peutz-Jeghers syndrome

Scenario	Critieria
Histologically confirmed PJS polyp	Two of three required: 1. Family history consistent with autosomal dominant inheritance 2. Mucocutaneous hyperpigmentation 3. Small Bowel Polyposis
Without histologically confirmed PJS polyp	Probable diagnosis can be made from at least two of the above criteria
No family history of PJS	At least two histologically confirmed PJS polyps required
Positive family history of PJS	Mucocutaneous hyperpigmentation is sufficient for presumptive diagnosis

3.4.4 Cancer Surveillance

No clear consensus guidelines exist for cancer surveillance in individuals with PJS. However, recommendations have been offered by European experts after careful consideration of the literature [77]. Affected individuals should have annual physical exams with complete blood counts and liver function tests. Males should also have careful testicular exams for Sertoli cell tumors during this time until age 12. An index upper and lower endoscopy should be performed at age 8. If there are any polyps, a repeat exam should be performed in 3 years. If negative, the next exam should occur at 18 years of age or the development of symptoms, whichever comes first. Until age 50, individuals should have surveillance endoscopies every 3 years. After 50 years, they should be performed every 1–2 years. Upper endoscopy should also performed with small bowel imaging (video capsule, barium SBFT, or CT enterography).

For women, cervical cancer screening should begin by age 21, and annual breast MRIs should be performed starting at age 25. No clear recommendations exist for pancreatic or ovarian cancer screening, but protocols have been implemented by many centers.

3.5 PTEN Hamartoma Syndromes

3.5.1 Genetics and Molecular Mechanisms

The PTEN hamartoma syndromes (PHTS) consist of three different syndromes: Cowden syndrome, Bannayan-Riley-Ruvalcaba syndrome, and Proteus syndrome. Despite the phenotypic differences, all three syndromes have been linked to deleterious germline mutations in *PTEN*. *PTEN* is a tumor suppressor gene frequently mutated in a variety of tumors, and is a negative regular of the PI3K-Akt-mTOR signaling pathway that is critical to cell growth, metabolism and cell-cycle progression [78]. PTEN also interfaces with other important tumor suppressor pathways including the ATR-ATM DNA damage pathway [79]. While initially thought to be responsible for over 80 % of cases of Cowden syndrome, more recent studies have implicated *PTEN* mutations in only 34 % of cases [80]. In addition, other genes have been recently implicated. Additional genes involved in metabolism and the PI3K-Akt-mTOR pathway have been identified in Cowden syndrome; mutations in *SDHB, SDHD, AKT1*, and *PIK3CA* have also been implicated [81, 82]. Finally, epigenetic silencing of *KLLN* has also been linked to Cowden Syndrome [83].

3.5.2 Epidemiology and Clinical Presentation

While difficult to assess, the prevalence of Cowden disease is 1 in 200,000 cases and felt to be an underestimate. The penetrance of the disease is approximately 90 % by age 20, and is transmitted in an autosomal dominant order. Clinical characteristics of CS include multiple hamartomas of many tissue types, but skin involvement in nearly 100 %. Mucocutaneous lesions include trichilemmomas, acral keratoses, facial papules, and oral papillomas. Individuals with Cowden syndrome are at elevated risk of breast, thyroid, endometrial, and renal cancers. Subsets of Cowden disease and those with *PTEN* germline mutations may develop a hamartoma in the cerebellum, dysplastic gangliocytoma, and this is referred to as adult Lhermitte-Duclos disease. Intracranial vascular malformations, macrocephaly, and mental retardation can also be seen with the disorder.

In the gastrointestinal tract, individuals with Cowden syndrome frequently develop esophageal acanthosis, as well as gastric and duodenal polyps. Polyps may be adenomatous, hamartomatous, ganglioneuromas, inflammatory, or hyperplastic. Such polyps also develop in the colon. Recently, the risk of early onset of colon cancer has been better appreciated. The lifetime risk of colon cancer by age 60 has been reported to be 18 % [84].

The prevalence of BRRS and Proteus syndrome are too rare to be accurately estimated. Many of the clinical manifestations are similar to CS. Individuals with BRRS typically exhibit macrocephaly, mental retardation, penile lentigines, myopathy, thyroiditis, and subcutaneous lipomas.

3.5.3 Diagnosis

Recently, revised diagnostic criteria for PTEN hamartoma syndrome have been proposed after careful review of the existing literature (Table 3) [85]. Less stringent criteria apply when a family member already has the clinical diagnosis or carries a pathogenic *PTEN* mutation.

3.5.4 Cancer Surveillance

Despite the perceived lower risk of malignancy observed in BRRS, all individuals with PTEN hamartoma syndromes should be placed on a cancer surveillance program. The NCCN recommends that affected individuals should have annual physical exams with careful examination of the thyroid, breasts, and the skin. Thyroid ultrasounds should be performed annually starting at age 18. In addition, breast MRIs and mammography should be performed in women starting at age 30–35, or 5–10 years prior to the earliest breast cancer in the family, whichever comes first. Colonoscopy is recommended starting at age 35, with follow-up every 5 years, or more frequently depending on findings.

3.6 *Serrated Polyposis Syndrome*

3.6.1 Genetics and Molecular Mechanisms

Serrated polyps are colonic premalignant polyps that typically arise from activating mutations in *BRAF*, or less commonly *KRAS*, in the colonic epithelium. In contrast to tubular adenomas, these polyps lack activating Wnt pathway mutations in *APC* or

Table 3 Clinical diagnostic criteria for PTEN Harmatoma syndrome (Ohio State University)

Major criteria	Endometrial cancer
	Thyroid cancer (follicular)
	Breast cancer
	Gastrointestinal hamartomas (≥ 3)
	Lhermitte-Duclos disease (adult)
	Macrocephaly
	Macular pigmentation of the glans penis
	Mucocutaneous lesions: Multiple trichilemmomas (≥ 3, biopsy proven), acral keratoses (≥ 3), Mucocutaneous neuromas (≥ 3), Oral papillomas (≥ 3)
Minor criteria	Autism (any spectrum)
	Colon cancer
	Esophageal glycogenic acanthosis (≥ 3)
	Lipoma (≥ 3)
	Mental retardation (IQ < 75)
	Renal cell carcinoma
	Testicular lipomatosis
	Thyroid cancer (papillary)
	Thyroid structural abnormalities
	Vascular abnormalities (including congential CNS AVMs)

Diagnosis is made when an individual satisfies at least three major criteria (one must include gastrointestinal hamartomas, Lhermitte-Duclos, or macrocephaly) or two major and three minor criteria. In individuals with a family member with PTEN hamartoma syndrome or a pathogenic *PTEN* mutation, any two major criteria or one major and two minor or three minor criteria are required

CTNNB1 as initiating steps. Mouse models of serrated neoplasia demonstrate that *BRAF* or *KRAS* mutations are sufficient to recapitulate the serrated morphology, as well as the sporadic MSI-H tumors observed in humans [86, 87]. Individuals have been identified who develop multiple serrated polyps, and this condition is designated serrated polyposis syndrome.

In contrast to the other hereditary syndromes, the genetics of this disorder are extremely heterogeneous and complex. First-degree relatives have a 35 % chance of developing multiple serrated polyps, and have a fivefold increased risk of colon cancer [88]. Attempts to identify a monogenic cause for this disorder have been unsuccessful. A minority of patients with *MUTYH*-associated polyposis can develop numerous serrated polyps; however, mutations in this gene comprise less than 1 % of cases of SPS [62, 89].

A recent exome study of individuals with SPS has revealed that many harbor germline, loss-of-function mutations in key oncogene-induced senescence pathways [90]. Oncogene-induced senescence is a tumor suppressive mechanism that is triggered by the cellular stress from large bursts of cells division and replication. The ATR-ATM mediated DNA damage response and p16-Rb pathways play key roles in oncogene-induced senescence. In mouse models, bypass of these mechanisms increases polyp multiplicity and progression to cancer. Deleterious mutations in genes such as *ATM* have been identified in patients with SPS. Also, germline mutations in *RNF43*, a negative regulator of Wnt signaling and critical enzyme for

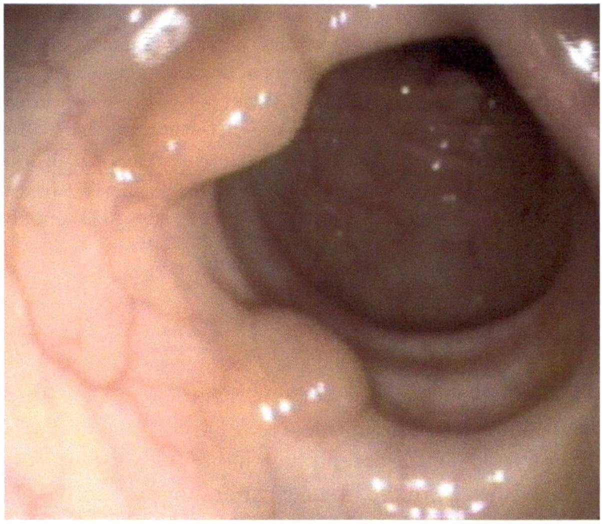

Fig. 4 Endoscopic appearance of serrated polyposis. Two large sessile serrated adenomas are present in the ascending colon. It is not uncommon to find adherent mucin attached to these polyps

the DNA damage response, have been identified as high-risk variants, validated in independent studies.

3.6.2 Epidemiology and Clinical Presentation

The prevalence of SPS has been reported to be as frequent as 1:3000 [91]. Individuals with SPS have as high as a 54 % lifetime risk of colon cancer [92]. In addition, many studies describe strong family histories of extracolonic malignancies. A recent single-center study has demonstrated an increased risk of pancreatic cancer among family members with SPS [88]. The term "serrated polyp" is used broadly and includes multiple histologic subtypes including goblet-cell rich hyperplastic polyps, microvesicular hyperplastic polyps, sessile serrated polyps, and traditional serrated adenomas. These polyps tend to occur predominantly on the right side, have a flat morphology, and may be covered with mucin (Fig. 4). Tumors arising from patients with SPS include both MSI-H and MSS CRCs. In addition, it appears that many colon cancers found in SPS patients may arise from an accelerated traditional pathway. This observation may be explained by the host genetics. Defects in oncogene-induced senescence will also accelerate the progression of tubular adenomas to carcinomas.

3.6.3 Diagnosis

At present, no genetic testing is yet recommended for SPS. In those with SPS who also demonstrate large numbers of tubular adenomas, *MUTYH* testing may not be unreasonable. The diagnosis is primarily a clinical one based upon the revised WHO guidelines (Table 4).

Table 4 Revised World Health Organization criteria for serrated polyposis

Criterion A	At least five serrated polyps proximal to the sigmoid colon, two of which are greater than 10 mm in diameter
Criterion B	Any number of serrated polyps occurring proximal to the sigmoid colon in an individual who has a first-degree relative with serrated polyposis
Criterion C	More than 20 serrated polyps of any size distributed throughout the colon

At least one criterion must be satisfied for the diagnosis to be established. Serrated polyps include sessile serrated adenomas, traditional serrated adenomas, microvesicular hyperplastic polyps, and goblet-cell rich hyperplastic polyps.

3.6.4 Cancer Surveillance

Upon diagnosis, colonoscopy is recommended every 1–2 years. Presently, no clear guidelines exist for first-degree relatives, despite the elevated epidemiological risk of colon cancer observed. Such individuals should undergo a colonoscopy 10 years prior to the age of the first cancer or polyps reported in the family, or age 40, whichever comes first.

4 Conclusion

Hereditary colon cancer syndromes are a diverse group of conditions that provide insights into the heterogeneous molecular mechanisms underlying the development of colon cancer. Recognizing and managing these families requires successful integration of genetics, molecular biology, and diverse clinical clues. In this chapter, we reviewed the syndromes that have been best characterized. As genomic sequencing becomes more widespread, clinicians and researchers will be challenged to translate the ever-increasing number of newly identified high-risk genes into clinical care and management. Individuals and families may harbor private, pathogenic mutations in important tumor suppressor genes not observed otherwise in the general population. Developing precise management and testing strategies will require a deep understanding of individual clinical phenotypes and molecular mechanisms of tumor pathogenesis.

References

1. Lichtenstein P, Holm NV, Verkasalo PK, Iliadou A, Kaprio J, Koskenvuo M, Pukkala E, Skytthe A, Hemminki K (2000) Environmental and heritable factors in the causation of cancer—analyses of cohorts of twins from Sweden, Denmark, and Finland. N Engl J Med 343(2):78–85. doi:10.1056/NEJM200007133430201
2. Salovaara R, Loukola A, Kristo P, Kaariainen H, Ahtola H, Eskelinen M, Harkonen N, Julkunen R, Kangas E, Ojala S, Tulikoura J, Valkamo E, Jarvinen H, Mecklin JP, Aaltonen LA, de la Chapelle A (2000) Population-based molecular detection of hereditary nonpolyposis colorectal cancer. J Clin Oncol 18(11):2193–2200

3. St John DJ, McDermott FT, Hopper JL, Debney EA, Johnson WR, Hughes ES (1993) Cancer risk in relatives of patients with common colorectal cancer. Ann Intern Med 118(10):785–790
4. Wong P, Verselis SJ, Garber JE, Schneider K, DiGianni L, Stockwell DH, Li FP, Syngal S (2006) Prevalence of early onset colorectal cancer in 397 patients with classic Li–Fraumeni syndrome. Gastroenterology 130(1):73–79. doi:10.1053/j.gastro.2005.10.014
5. Fishel R, Lescoe MK, Rao MR, Copeland NG, Jenkins NA, Garber J, Kane M, Kolodner R (1993) The human mutator gene homolog MSH2 and its association with hereditary nonpolyposis colon cancer. Cell 75(5):1027–1038
6. Papadopoulos N, Nicolaides NC, Wei YF, Ruben SM, Carter KC, Rosen CA, Haseltine WA, Fleischmann RD, Fraser CM, Adams MD et al (1994) Mutation of a mutL homolog in hereditary colon cancer. Science 263(5153):1625–1629
7. Thompson E, Meldrum CJ, Crooks R, McPhillips M, Thomas L, Spigelman AD, Scott RJ (2004) Hereditary non-polyposis colorectal cancer and the role of hPMS2 and hEXO1 mutations. Clin Genet 65(3):215–225
8. Ligtenberg MJ, Kuiper RP, Chan TL, Goossens M, Hebeda KM, Voorendt M, Lee TY, Bodmer D, Hoenselaar E, Hendriks-Cornelissen SJ, Tsui WY, Kong CK, Brunner HG, van Kessel AG, Yuen ST, van Krieken JH, Leung SY, Hoogerbrugge N (2009) Heritable somatic methylation and inactivation of MSH2 in families with Lynch syndrome due to deletion of the 3′ exons of TACSTD1. Nat Genet 41(1):112–117. doi:10.1038/ng.283
9. Lu SL, Kawabata M, Imamura T, Akiyama Y, Nomizu T, Miyazono K, Yuasa Y (1998) HNPCC associated with germline mutation in the TGF-beta type II receptor gene. Nat Genet 19(1):17–18. doi:10.1038/ng0598-17
10. Nicolaides NC, Papadopoulos N, Liu B, Wei YF, Carter KC, Ruben SM, Rosen CA, Haseltine WA, Fleischmann RD, Fraser CM et al (1994) Mutations of two PMS homologues in hereditary nonpolyposis colon cancer. Nature 371(6492):75–80. doi:10.1038/371075a0
11. Oki E, Oda S, Maehara Y, Sugimachi K (1999) Mutated gene-specific phenotypes of dinucleotide repeat instability in human colorectal carcinoma cell lines deficient in DNA mismatch repair. Oncogene 18(12):2143–2147. doi:10.1038/sj.onc.1202583
12. Baudhuin LM, Burgart LJ, Leontovich O, Thibodeau SN (2005) Use of microsatellite instability and immunohistochemistry testing for the identification of individuals at risk for Lynch syndrome. Fam Cancer 4(3):255–265. doi:10.1007/s10689-004-1447-6
13. Gazzoli I, Loda M, Garber J, Syngal S, Kolodner RD (2002) A hereditary nonpolyposis colorectal carcinoma case associated with hypermethylation of the MLH1 gene in normal tissue and loss of heterozygosity of the unmethylated allele in the resulting microsatellite instability-high tumor. Cancer Res 62(14):3925–3928
14. Markowitz S (2000) DNA repair defects inactivate tumor suppressor genes and induce hereditary and sporadic colon cancers. J Clin Oncol 18(21 Suppl):75S–80S
15. Liu Y, Fang Y, Shao H, Lindsey-Boltz L, Sancar A, Modrich P (2010) Interactions of human mismatch repair proteins MutSalpha and MutLalpha with proteins of the ATR-Chk1 pathway. J Biol Chem 285(8):5974–5982. doi:10.1074/jbc.M109.076109
16. Cancer Genome Atlas Network (2012) Comprehensive molecular characterization of human colon and rectal cancer. Nature 487(7407):330–337. doi:10.1038/nature11252
17. Lochhead P, Kuchiba A, Imamura Y, Liao X, Yamauchi M, Nishihara R, Qian ZR, Morikawa T, Shen J, Meyerhardt JA, Fuchs CS, Ogino S (2013) Microsatellite instability and BRAF mutation testing in colorectal cancer prognostication. J Natl Cancer Inst 105(15):1151–1156. doi:10.1093/jnci/djt173
18. Pino MS, Kikuchi H, Zeng M, Herraiz MT, Sperduti I, Berger D, Park DY, Iafrate AJ, Zukerberg LR, Chung DC (2010) Epithelial to mesenchymal transition is impaired in colon cancer cells with microsatellite instability. Gastroenterology 138(4):1406–1417. doi:10.1053/j.gastro.2009.12.010
19. Zlobec I, Terracciano LM, Lugli A (2008) Local recurrence in mismatch repair-proficient colon cancer predicted by an infiltrative tumor border and lack of CD8+ tumor-infiltrating lymphocytes. Clin Cancer Res 14(12):3792–3797. doi:10.1158/1078-0432.CCR-08-0048

20. Boland CR (2000) Molecular genetics of hereditary nonpolyposis colorectal cancer. Ann N Y Acad Sci 910:50–59; discussion 59–61
21. Wang L, Cunningham JM, Winters JL, Guenther JC, French AJ, Boardman LA, Burgart LJ, McDonnell SK, Schaid DJ, Thibodeau SN (2003) BRAF mutations in colon cancer are not likely attributable to defective DNA mismatch repair. Cancer Res 63(17):5209–5212
22. Mensenkamp AR, Vogelaar IP, van Zelst-Stams WA, Goossens M, Ouchene H, Hendriks-Cornelissen SJ, Kwint MP, Hoogerbrugge N, Nagtegaal ID, Ligtenberg MJ (2014) Somatic mutations in MLH1 and MSH2 are a frequent cause of mismatch-repair deficiency in Lynch syndrome-like tumors. Gastroenterology 146(3):643–646, e648. doi:10.1053/j.gastro.2013.12.002
23. Cunningham JM, Kim CY, Christensen ER, Tester DJ, Parc Y, Burgart LJ, Halling KC, McDonnell SK, Schaid DJ, Walsh Vockley C, Kubly V, Nelson H, Michels VV, Thibodeau SN (2001) The frequency of hereditary defective mismatch repair in a prospective series of unselected colorectal carcinomas. Am J Hum Genet 69(4):780–790. doi:10.1086/323658
24. Chen S, Wang W, Lee S, Nafa K, Lee J, Romans K, Watson P, Gruber SB, Euhus D, Kinzler KW, Jass J, Gallinger S, Lindor NM, Casey G, Ellis N, Giardiello FM, Offit K, Parmigiani G, Colon Cancer Family Registry (2006) Prediction of germline mutations and cancer risk in the Lynch syndrome. JAMA 296(12):1479–1487. doi:10.1001/jama.296.12.1479
25. Hampel H, Stephens JA, Pukkala E, Sankila R, Aaltonen LA, Mecklin JP, de la Chapelle A (2005) Cancer risk in hereditary nonpolyposis colorectal cancer syndrome: later age of onset. Gastroenterology 129(2):415–421. doi:10.1016/j.gastro.2005.05.011
26. Lynch HT, Lynch J (2000) Lynch syndrome: genetics, natural history, genetic counseling, and prevention. J Clin Oncol 18(21 Suppl):19S–31S
27. Win AK, Lindor NM, Jenkins MA (2013) Risk of breast cancer in Lynch syndrome: a systematic review. Breast Cancer Res 15(2):R27. doi:10.1186/bcr3405
28. Koornstra JJ, Mourits MJ, Sijmons RH, Leliveld AM, Hollema H, Kleibeuker JH (2009) Management of extracolonic tumours in patients with Lynch syndrome. Lancet Oncol 10(4):400–408. doi:10.1016/S1470-2045(09)70041-5
29. Barrow E, Robinson L, Alduaij W, Shenton A, Clancy T, Lalloo F, Hill J, Evans DG (2009) Cumulative lifetime incidence of extracolonic cancers in Lynch syndrome: a report of 121 families with proven mutations. Clin Genet 75(2):141–149. doi:10.1111/j.1399-0004.2008.01125.x
30. South CD, Hampel H, Comeras I, Westman JA, Frankel WL, de la Chapelle A (2008) The frequency of Muir–Torre syndrome among Lynch syndrome families. J Natl Cancer Inst 100(4):277–281. doi:10.1093/jnci/djm291
31. Syngal S, Fox EA, Eng C, Kolodner RD, Garber JE (2000) Sensitivity and specificity of clinical criteria for hereditary non-polyposis colorectal cancer associated mutations in MSH2 and MLH1. J Med Genet 37(9):641–645
32. Ward RL, Hicks S, Hawkins NJ (2013) Population-based molecular screening for Lynch syndrome: implications for personalized medicine. J Clin Oncol 31(20):2554–2562. doi:10.1200/JCO.2012.46.8454
33. Jarvinen HJ, Aarnio M, Mustonen H, Aktan-Collan K, Aaltonen LA, Peltomaki P, De La Chapelle A, Mecklin JP (2000) Controlled 15-year trial on screening for colorectal cancer in families with hereditary nonpolyposis colorectal cancer. Gastroenterology 118(5):829–834
34. Pylvanainen K, Lehtinen T, Kellokumpu I, Jarvinen H, Mecklin JP (2012) Causes of death of mutation carriers in Finnish Lynch syndrome families. Fam Cancer 11(3):467–471. doi:10.1007/s10689-012-9537-3
35. Winkels RM, Botma A, van Duijnhoven FJ, Nagengast FM, Kleibeuker JH, Vasen HF, Kampman E (2012) Smoking increases the risk for colorectal adenomas in patients with Lynch syndrome. Gastroenterology 142(2):241–247. doi:10.1053/j.gastro.2011.10.033
36. Botma A, Vasen HF, van Duijnhoven FJ, Kleibeuker JH, Nagengast FM, Kampman E (2013) Dietary patterns and colorectal adenomas in Lynch syndrome: the GEOLynch cohort study. Cancer 119(3):512–521. doi:10.1002/cncr.27726

37. Burn J, Bishop DT, Mecklin JP, Macrae F, Moslein G, Olschwang S, Bisgaard ML, Ramesar R, Eccles D, Maher ER, Bertario L, Jarvinen HJ, Lindblom A, Evans DG, Lubinski J, Morrison PJ, Ho JW, Vasen HF, Side L, Thomas HJ, Scott RJ, Dunlop M, Barker G, Elliott F, Jass JR, Fodde R, Lynch HT, Mathers JC, Investigators C (2008) Effect of aspirin or resistant starch on colorectal neoplasia in the Lynch syndrome. N Engl J Med 359(24):2567–2578. doi:10.1056/NEJMoa0801297
38. Burn J, Gerdes AM, Macrae F, Mecklin JP, Moeslein G, Olschwang S, Eccles D, Evans DG, Maher ER, Bertario L, Bisgaard ML, Dunlop MG, Ho JW, Hodgson SV, Lindblom A, Lubinski J, Morrison PJ, Murday V, Ramesar R, Side L, Scott RJ, Thomas HJ, Vasen HF, Barker G, Crawford G, Elliott F, Movahedi M, Pylvanainen K, Wijnen JT, Fodde R, Lynch HT, Mathers JC, Bishop DT, CAPP2 Investigators (2011) Long-term effect of aspirin on cancer risk in carriers of hereditary colorectal cancer: an analysis from the CAPP2 randomised controlled trial. Lancet 378(9809):2081–2087. doi:10.1016/S0140-6736(11)61049-0
39. Sargent DJ, Marsoni S, Monges G, Thibodeau SN, Labianca R, Hamilton SR, French AJ, Kabat B, Foster NR, Torri V, Ribic C, Grothey A, Moore M, Zaniboni A, Seitz JF, Sinicrope F, Gallinger S (2010) Defective mismatch repair as a predictive marker for lack of efficacy of fluorouracil-based adjuvant therapy in colon cancer. J Clin Oncol 28(20):3219–3226. doi:10.1200/JCO.2009.27.1825
40. Nishisho I, Nakamura Y, Miyoshi Y, Miki Y, Ando H, Horii A, Koyama K, Utsunomiya J, Baba S, Hedge P (1991) Mutations of chromosome 5q21 genes in FAP and colorectal cancer patients. Science 253(5020):665–669
41. Miyoshi Y, Nagase H, Ando H, Horii A, Ichii S, Nakatsuru S, Aoki T, Miki Y, Mori T, Nakamura Y (1992) Somatic mutations of the APC gene in colorectal tumors: mutation cluster region in the APC gene. Hum Mol Genet 1(4):229–233
42. Heppner Goss K, Trzepacz C, Tuohy TM, Groden J (2002) Attenuated APC alleles produce functional protein from internal translation initiation. Proc Natl Acad Sci U S A 99(12):8161–8166. doi:10.1073/pnas.112072199
43. Cheah PY, Wong YH, Koh PK, Loi C, Chew MH, Tang CL (2013) A novel indel in exon 9 of APC upregulates a 'skip exon 9' isoform and causes very severe familial adenomatous polyposis. Eur J Hum Genet. doi:10.1038/ejhg.2013.245
44. Friedl W, Caspari R, Sengteller M, Uhlhaas S, Lamberti C, Jungck M, Kadmon M, Wolf M, Fahnenstich J, Gebert J, Moslein G, Mangold E, Propping P (2001) Can APC mutation analysis contribute to therapeutic decisions in familial adenomatous polyposis? Experience from 680 FAP families. Gut 48(4):515–521
45. Nieuwenhuis MH, Vasen HF (2007) Correlations between mutation site in APC and phenotype of familial adenomatous polyposis (FAP): a review of the literature. Crit Rev Oncol Hematol 61(2):153–161. doi:10.1016/j.critrevonc.2006.07.004
46. Crabtree M, Sieber OM, Lipton L, Hodgson SV, Lamlum H, Thomas HJ, Neale K, Phillips RK, Heinimann K, Tomlinson IP (2003) Refining the relation between 'first hits' and 'second hits' at the APC locus: the 'loose fit' model and evidence for differences in somatic mutation spectra among patients. Oncogene 22(27):4257–4265. doi:10.1038/sj.onc.1206471
47. Hes FJ, Nielsen M, Bik EC, Konvalinka D, Wijnen JT, Bakker E, Vasen HF, Breuning MH, Tops CM (2008) Somatic APC mosaicism: an underestimated cause of polyposis coli. Gut 57(1):71–76. doi:10.1136/gut.2006.117796
48. Quintini C, Ward G, Shatnawei A, Xhaja X, Hashimoto K, Steiger E, Hammel J, Diago Uso T, Burke CA, Church JM (2012) Mortality of intra-abdominal desmoid tumors in patients with familial adenomatous polyposis: a single center review of 154 patients. Ann Surg 255(3):511–516. doi:10.1097/SLA.0b013e31824682d4
49. Perrier ND, van Heerden JA, Goellner JR, Williams ED, Gharib H, Marchesa P, Church JM, Fazio VW, Larson DR (1998) Thyroid cancer in patients with familial adenomatous polyposis. World J Surg 22(7):738–742; discussion 743
50. Kadmon M, Tandara A, Herfarth C (2001) Duodenal adenomatosis in familial adenomatous polyposis coli. A review of the literature and results from the Heidelberg Polyposis Register. Int J Colorectal Dis 16(2):63–75

51. Knudsen AL, Bisgaard ML, Bulow S (2003) Attenuated familial adenomatous polyposis (AFAP). A review of the literature. Fam Cancer 2(1):43–55
52. Grover S, Kastrinos F, Steyerberg EW, Cook EF, Dewanwala A, Burbidge LA, Wenstrup RJ, Syngal S (2012) Prevalence and phenotypes of APC and MUTYH mutations in patients with multiple colorectal adenomas. JAMA 308(5):485–492. doi:10.1001/jama.2012.8780
53. Cruz-Correa M, Hylind LM, Romans KE, Booker SV, Giardiello FM (2002) Long-term treatment with sulindac in familial adenomatous polyposis: a prospective cohort study. Gastroenterology 122(3):641–645
54. Steinbach G, Lynch PM, Phillips RK, Wallace MH, Hawk E, Gordon GB, Wakabayashi N, Saunders B, Shen Y, Fujimura T, Su LK, Levin B, Godio L, Patterson S, Rodriguez-Bigas MA, Jester SL, King KL, Schumacher M, Abbruzzese J, DuBois RN, Hittelman WN, Zimmerman S, Sherman JW, Kelloff G (2000) The effect of celecoxib, a cyclooxygenase-2 inhibitor, in familial adenomatous polyposis. N Engl J Med 342(26):1946–1952. doi:10.1056/NEJM200006293422603
55. Hansmann A, Adolph C, Vogel T, Unger A, Moeslein G (2004) High-dose tamoxifen and sulindac as first-line treatment for desmoid tumors. Cancer 100(3):612–620. doi:10.1002/cncr.11937
56. Sieber OM, Lipton L, Crabtree M, Heinimann K, Fidalgo P, Phillips RK, Bisgaard ML, Orntoft TF, Aaltonen LA, Hodgson SV, Thomas HJ, Tomlinson IP (2003) Multiple colorectal adenomas, classic adenomatous polyposis, and germ-line mutations in MYH. N Engl J Med 348(9):791–799. doi:10.1056/NEJMoa025283
57. Al-Tassan N, Chmiel NH, Maynard J, Fleming N, Livingston AL, Williams GT, Hodges AK, Davies DR, David SS, Sampson JR, Cheadle JP (2002) Inherited variants of MYH associated with somatic G:C→T:A mutations in colorectal tumors. Nat Genet 30(2):227–232. doi:10.1038/ng828
58. Jones S, Emmerson P, Maynard J, Best JM, Jordan S, Williams GT, Sampson JR, Cheadle JP (2002) Biallelic germline mutations in MYH predispose to multiple colorectal adenoma and somatic G:C→T:A mutations. Hum Mol Genet 11(23):2961–2967
59. Jones S, Lambert S, Williams GT, Best JM, Sampson JR, Cheadle JP (2004) Increased frequency of the k-ras G12C mutation in MYH polyposis colorectal adenomas. Br J Cancer 90(8):1591–1593. doi:10.1038/sj.bjc.6601747
60. Hahm SH, Park JH, Ko SI, Lee YR, Chung IS, Chung JH, Kang LW, Han YS (2011) Knockdown of human MutY homolog (hMYH) decreases phosphorylation of checkpoint kinase 1 (Chk1) induced by hydroxyurea and UV treatment. BMB Rep 44(5):352–357. doi:10.5483/BMBRep.2011.44.5.352
61. Win AK, Dowty JG, Cleary SP, Kim H, Buchanan DD, Young JP, Clendenning M, Rosty C, Macinnis RJ, Giles GG, Boussioutas A, Macrae FA, Parry S, Goldblatt J, Baron JA, Burnett T, Le Marchand L, Newcomb PA, Haile RW, Hopper JL, Cotterchio M, Gallinger S, Lindor NM, Tucker KM, Winship IM, Jenkins MA (2014) Risk of colorectal cancer for carriers of mutations in MUTYH, with and without a family history of cancer. Gastroenterology. doi:10.1053/j.gastro.2014.01.022
62. Boparai KS, Dekker E, van Eeden S, Polak MM, Bartelsman JF, Mathus-Vliegen EM, Keller JJ, van Noesel CJ (2008) Hyperplastic polyps and sessile serrated adenomas as a phenotypic expression of MYH-associated polyposis. Gastroenterology 135(6):2014–2018. doi:10.1053/j.gastro.2008.09.020
63. Vogt S, Jones N, Christian D, Engel C, Nielsen M, Kaufmann A, Steinke V, Vasen HF, Propping P, Sampson JR, Hes FJ, Aretz S (2009) Expanded extracolonic tumor spectrum in MUTYH-associated polyposis. Gastroenterology 137(6):1976–1985, e1–10. doi:10.1053/j.gastro.2009.08.052
64. Sayed MG, Ahmed AF, Ringold JR, Anderson ME, Bair JL, Mitros FA, Lynch HT, Tinley ST, Petersen GM, Giardiello FM, Vogelstein B, Howe JR (2002) Germline SMAD4 or BMPR1A mutations and phenotype of juvenile polyposis. Ann Surg Oncol 9(9):901–906
65. Kim BG, Li C, Qiao W, Mamura M, Kasprzak B, Anver M, Wolfraim L, Hong S, Mushinski E, Potter M, Kim SJ, Fu XY, Deng C, Letterio JJ (2006) Smad4 signalling in T cells is required

for suppression of gastrointestinal cancer. Nature 441(7096):1015–1019. doi:10.1038/nature04846
66. Howe JR, Sayed MG, Ahmed AF, Ringold J, Larsen-Haidle J, Merg A, Mitros FA, Vaccaro CA, Petersen GM, Giardiello FM, Tinley ST, Aaltonen LA, Lynch HT (2004) The prevalence of MADH4 and BMPR1A mutations in juvenile polyposis and absence of BMPR2, BMPR1B, and ACVR1 mutations. J Med Genet 41(7):484–491
67. He XC, Zhang J, Tong WG, Tawfik O, Ross J, Scoville DH, Tian Q, Zeng X, He X, Wiedemann LM, Mishina Y, Li L (2004) BMP signaling inhibits intestinal stem cell self-renewal through suppression of Wnt-beta-catenin signaling. Nat Genet 36(10):1117–1121. doi:10.1038/ng1430
68. Brosens LA, van Hattem A, Hylind LM, Iacobuzio-Donahue C, Romans KE, Axilbund J, Cruz-Correa M, Tersmette AC, Offerhaus GJ, Giardiello FM (2007) Risk of colorectal cancer in juvenile polyposis. Gut 56(7):965–967. doi:10.1136/gut.2006.116913
69. Gallione CJ, Repetto GM, Legius E, Rustgi AK, Schelley SL, Tejpar S, Mitchell G, Drouin E, Westermann CJ, Marchuk DA (2004) A combined syndrome of juvenile polyposis and hereditary haemorrhagic telangiectasia associated with mutations in MADH4 (SMAD4). Lancet 363(9412):852–859. doi:10.1016/S0140-6736(04)15732-2
70. Cairns SR, Scholefield JH, Steele RJ, Dunlop MG, Thomas HJ, Evans GD, Eaden JA, Rutter MD, Atkin WP, Saunders BP, Lucassen A, Jenkins P, Fairclough PD, Woodhouse CR, British Society of Gastroenterology, Association of Coloproctology for Great Britain and Ireland (2010) Guidelines for colorectal cancer screening and surveillance in moderate and high risk groups (update from 2002). Gut 59(5):666–689. doi:10.1136/gut.2009.179804
71. Hemminki A, Markie D, Tomlinson I, Avizienyte E, Roth S, Loukola A, Bignell G, Warren W, Aminoff M, Hoglund P, Jarvinen H, Kristo P, Pelin K, Ridanpaa M, Salovaara R, Toro T, Bodmer W, Olschwang S, Olsen AS, Stratton MR, de la Chapelle A, Aaltonen LA (1998) A serine/threonine kinase gene defective in Peutz–Jeghers syndrome. Nature 391(6663):184–187. doi:10.1038/34432
72. Shaw RJ, Bardeesy N, Manning BD, Lopez L, Kosmatka M, DePinho RA, Cantley LC (2004) The LKB1 tumor suppressor negatively regulates mTOR signaling. Cancer Cell 6(1):91–99. doi:10.1016/j.ccr.2004.06.007
73. Hawley SA, Boudeau J, Reid JL, Mustard KJ, Udd L, Makela TP, Alessi DR, Hardie DG (2003) Complexes between the LKB1 tumor suppressor, STRAD alpha/beta and MO25 alpha/beta are upstream kinases in the AMP-activated protein kinase cascade. J Biol 2(4):28. doi:10.1186/1475-4924-2-28
74. Cantu JM, Rivera H, Ocampo-Campos R, Bedolla N, Cortes-Gallegos V, Gonzalez-Mendoza A, Diaz M, Hernandez A (1980) Peutz–Jeghers syndrome with feminizing sertoli cell tumor. Cancer 46(1):223–228
75. Giardiello FM, Brensinger JD, Tersmette AC, Goodman SN, Petersen GM, Booker SV, Cruz-Correa M, Offerhaus JA (2000) Very high risk of cancer in familial Peutz–Jeghers syndrome. Gastroenterology 119(6):1447–1453
76. Aretz S, Stienen D, Uhlhaas S, Loff S, Back W, Pagenstecher C, McLeod DR, Graham GE, Mangold E, Santer R, Propping P, Friedl W (2005) High proportion of large genomic STK11 deletions in Peutz–Jeghers syndrome. Hum Mutat 26(6):513–519. doi:10.1002/humu.20253
77. Beggs AD, Latchford AR, Vasen HF, Moslein G, Alonso A, Aretz S, Bertario L, Blanco I, Bulow S, Burn J, Capella G, Colas C, Friedl W, Moller P, Hes FJ, Jarvinen H, Mecklin JP, Nagengast FM, Parc Y, Phillips RK, Hyer W, Ponz de Leon M, Renkonen-Sinisalo L, Sampson JR, Stormorken A, Tejpar S, Thomas HJ, Wijnen JT, Clark SK, Hodgson SV (2010) Peutz–Jeghers syndrome: a systematic review and recommendations for management. Gut 59(7):975–986. doi:10.1136/gut.2009.198499
78. Song MS, Salmena L, Pandolfi PP (2012) The functions and regulation of the PTEN tumour suppressor. Nat Rev Mol Cell Biol 13(5):283–296. doi:10.1038/nrm3330
79. Puc J, Keniry M, Li HS, Pandita TK, Choudhury AD, Memeo L, Mansukhani M, Murty VV, Gaciong Z, Meek SE, Piwnica-Worms H, Hibshoosh H, Parsons R (2005) Lack of PTEN sequesters CHK1 and initiates genetic instability. Cancer Cell 7(2):193–204. doi:10.1016/j.ccr.2005.01.009

80. Pilarski R, Stephens JA, Noss R, Fisher JL, Prior TW (2011) Predicting PTEN mutations: an evaluation of Cowden syndrome and Bannayan–Riley–Ruvalcaba syndrome clinical features. J Med Genet 48(8):505–512. doi:10.1136/jmg.2011.088807
81. Hobert JA, Mester JL, Moline J, Eng C (2012) Elevated plasma succinate in PTEN, SDHB, and SDHD mutation-positive individuals. Genet Med 14(6):616–619. doi:10.1038/gim.2011.63
82. Orloff MS, He X, Peterson C, Chen F, Chen JL, Mester JL, Eng C (2013) Germline PIK3CA and AKT1 mutations in Cowden and Cowden-like syndromes. Am J Hum Genet 92(1):76–80. doi:10.1016/j.ajhg.2012.10.021
83. Bennett KL, Mester J, Eng C (2010) Germline epigenetic regulation of KILLIN in Cowden and Cowden-like syndrome. JAMA 304(24):2724–2731. doi:10.1001/jama.2010.1877
84. Nieuwenhuis MH, Kets CM, Murphy-Ryan M, Colas C, Moller P, Hes FJ, Hodgson SV, Olderode-Berends MJ, Aretz S, Heinimann K, Gomez Garcia EB, Douglas F, Spigelman A, Timshel S, Lindor NM, Vasen HF (2012) Is colorectal surveillance indicated in patients with PTEN mutations? Colorectal Dis 14(9):e562–566. doi:10.1111/j.1463-1318.2012.03121.x
85. Pilarski R, Burt R, Kohlman W, Pho L, Shannon KM, Swisher E (2013) Cowden syndrome and the PTEN hamartoma tumor syndrome: systematic review and revised diagnostic criteria. J Natl Cancer Inst 105(21):1607–1616. doi:10.1093/jnci/djt277
86. Rad R, Cadinanos J, Rad L, Varela I, Strong A, Kriegl L, Constantino-Casas F, Eser S, Hieber M, Seidler B, Price S, Fraga MF, Calvanese V, Hoffman G, Ponstingl H, Schneider G, Yusa K, Grove C, Schmid RM, Wang W, Vassiliou G, Kirchner T, McDermott U, Liu P, Saur D, Bradley A (2013) A genetic progression model of Braf(V600E)-induced intestinal tumorigenesis reveals targets for therapeutic intervention. Cancer Cell 24(1):15–29. doi:10.1016/j.ccr.2013.05.014
87. Bennecke M, Kriegl L, Bajbouj M, Retzlaff K, Robine S, Jung A, Arkan MC, Kirchner T, Greten FR (2010) Ink4a/Arf and oncogene-induced senescence prevent tumor progression during alternative colorectal tumorigenesis. Cancer Cell 18(2):135–146. doi:10.1016/j.ccr.2010.06.013
88. Win AK, Walters RJ, Buchanan DD, Jenkins MA, Sweet K, Frankel WL, de la Chapelle A, McKeone DM, Walsh MD, Clendenning M, Pearson SA, Pavluk E, Nagler B, Hopper JL, Gattas MR, Goldblatt J, George J, Suthers GK, Phillips KD, Woodall S, Arnold J, Tucker K, Field M, Greening S, Gallinger S, Aronson M, Perrier R, Woods MO, Green JS, Walker N, Rosty C, Parry S, Young JP (2012) Cancer risks for relatives of patients with serrated polyposis. Am J Gastroenterol 107(5):770–778. doi:10.1038/ajg.2012.52
89. Clendenning M, Young JP, Walsh MD, Woodall S, Arnold J, Jenkins M, Win AK, Hopper JL, Sweet K, Gallinger S, Rosty C, Parry S, Buchanan DD (2013) Germline mutations in the polyposis-associated genes BMPR1A, SMAD4, PTEN, MUTYH and GREM1 are not common in individuals with serrated polyposis syndrome. PLoS ONE 8(6):e66705. doi:10.1371/journal.pone.0066705
90. Gala MK, Mizukami Y, Le LP, Moriichi K, Austin T, Yamamoto M, Lauwers GY, Bardeesy N, Chung DC (2014) Germline mutations in oncogene-induced senescence pathways are associated with multiple sessile serrated adenomas. Gastroenterology 146(2):520–529. doi:10.1053/j.gastro.2013.10.045
91. Rosty C, Parry S, Young JP (2011) Serrated polyposis: an enigmatic model of colorectal cancer predisposition. Pathol Res Int 2011:157073. doi:10.4061/2011/157073
92. Hyman NH, Anderson P, Blasyk H (2004) Hyperplastic polyposis and the risk of colorectal cancer. Dis Colon Rectum 47(12):2101–2104. doi:10.1007/s10350-004-0709-6

Microsatellite Instability and Intestinal Tumorigenesis

Meena A. Prasad and Barbara Jung

Abstract Microsatellites are simple, repetitive sequences of DNA present throughout the genome prone to DNA replication errors. The mismatch repair machinery is crucial to maintaining genetic integrity. When mismatch repair is defective, microsatellite instability occurs, resulting in deregulated tissue growth and tumorigenesis. Although defunct mismatch repair proteins can occur sporadically, these defects are the hallmark of Lynch syndrome, a hereditary colon cancer syndrome due to germline mutations in the mismatch repair genes *MLH1, MSH2, MSH6* or *PMS2*. Loss of mismatch repair protein function results in an intestinal and extra-intestinal tumor phenotype. This chapter reviews the mechanisms involved in the mismatch repair process, mouse models of defective mismatch repair, genetic versus sporadic microsatellite instability, methods of testing for microsatellite instability, and the clinical implications of microsatellite instability in the large bowel, small bowel and stomach.

Keywords Colorectal cancer · Gastrointestinal cancer · Intestinal tumorigenesis · Microsatellite instability · Mismatch repair · Lynch syndrome

1 Microsatellite Instability and Mismatch Repair

1.1 Microsatellites

Microsatellites are simple, repetitive sequences of DNA that are found throughout the genome and are prone to DNA replication errors [1]. They are typically composed of 1–6 base pair units that may be repeated up to 100 times. Given their

---

B. Jung (✉)
Department of Medicine, Division of Gastroenterology and Hepatology, University of Illinois at Chicago, 840 South Wood Street, 738A CSB, Chicago, IL 60612, USA
e-mail: bjung@uic.edu

M. A. Prasad
Division of Gastroenterology and Hepatology, Northwestern University Feinberg School of Medicine, 676 N. St Clair Street, Suite 1400, Chicago, IL 60611, USA
e-mail: MeenaAPrasad@gmail.com

V. W. Yang, A. B. Bialkowska (eds.), *Intestinal Tumorigenesis,*
DOI 10.1007/978-3-319-19986-3_2

repetitive nature, they are prone to errors that can occur during DNA replication [2]. The DNA mismatch repair (MMR) system recognizes and repairs base-pair mismatches that occur during DNA replication.

1.2 Mismatch Repair

MMR is a complex process dependent upon MMR proteins and multiple other proteins that interact with DNA. During DNA replication, DNA polymerase is prone to errors in incorporating the correct number of bases in long repetitive DNA sequences, such as microsatellites. These errors occur due to slippage during replication of such a repetitive sequence [3]. MMR was first recognized in yeast with the key genes involved termed *Mut S homologue* (*MSH*), *Mut L homologue* (*MLH*) and *post-meiotic segregation (PMS)* and numbered *MSH1* through *MSH6*, *MLH1* through *MLH3* and *PMS1* [3]. Some of the key proteins involved in human MMR include MutL homolog 1 (MLH1), MutS homolog 2 (MSH2), MutS homolog 6 (MSH6), and postmeiotic segregation increased 2 (PMS2).

In mammals, there are three different MutS homologs (MSH2, MSH3 and MSH6) and four different MutL homologs (MLH1, PMS1, PMS2, MLH3) [4]. MMR involves formation of heterodimers of key MMR proteins. The MSH2 and MSH6 proteins associate as a heterodimer (termed MutSα), which then constitutes a sliding clamp on the newly synthesized strand of DNA and recognizes base-base mispairs as well as 1 base pair insertion/deletion mutations (see Fig. 1) [5]. MSH2 also forms a heterodimer with MSH3 (termed MutSβ), which recognizes 1–4 base pair insertion/deletion mutations. When MSH2 heterodimer recognizes a DNA base pair mismatch, it recruits the MLH1-PMS2 heterodimer (termed MutLα) to assist in repairing the mismatch [4]. The interactions of the MLH1-PMS2 heterodimer with MutSα (MSH2-MSH6) and MutSβ (MSH2-MSH3) are critical for activation of subsequent MMR steps, including excision of the DNA strand with the mismatched base(s) and its resynthesis [4]. The precise mechanism of repair remains unknown [6]. Of note, MutL homolog 3 (MLH3) also forms a heterodimer with MLH1 (termed MutLγ) and participates in mismatch repair. In cases where a loss of PMS2 is present, function can be partially compensated by MLH3 [3]. However, loss of MLH3 has not been implicated in microsatellite instability and clinical pathology. MLH1 also forms a heterodimer with PMS1 (termed MutLβ), but has little or no role in mismatch repair [7].

If MMR proteins (specifically MSH2, MSH6, MLH1 and PMS2) do not correct errors from slippage during replication, a temporary insertion-deletion loop is created. This may result in microsatellite instability manifested by frame shift mutation and possible downstream nonsense mutation with production of a truncated, nonfunctional protein if the mutation is located in a coding region [3].

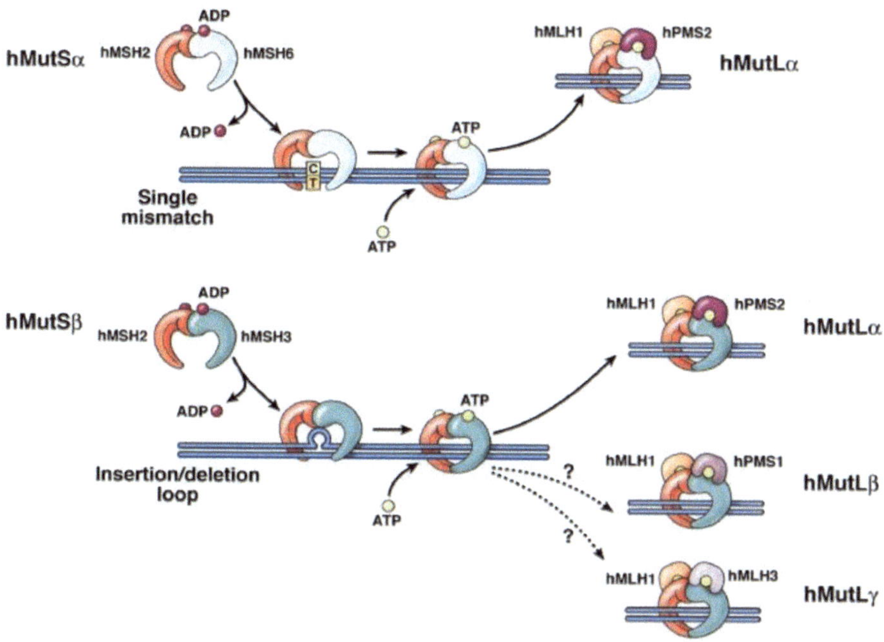

Fig. 1 The DNA MMR system

1.3 Microsatellite Instability

Microsatellite instability (MSI) is defined as a change of any length of DNA due to either insertion or deletion of repeating units in a microsatellite within a tumor when compared to normal tissue [2]. Typically MSI manifests itself in frame shifts that the defective MMR system is unable to correct [8]. MSI tumors rarely have large-scale genomic alterations but rather harbor mutations in coding microsatellite sequences in *TGFBR2, BAX, BRAF* and *ACVR2* [9]. Mutations in *APC, KRAS, P53* and *PIK3CA* are less common in MSI tumors [10].

2 Mouse Models of Mismatch Repair

Mouse models with inactivating mutations in MMR genes have allowed the methodical study of the individual function of each gene and their respective contribution to cancer susceptibility. In addition, specific loss of function of a gene may approximate the clinical phenotype. However, unlike Lynch syndrome (LS) patients with genetic defects in MMR, heterozygous MMR mutant mice do not develop early onset tumors. This may be due to the shorter life span of mice associated with a lower likelihood of a second hit induced somatic loss of the wild-type

MMR allele leading to tumor development. Only homozygous MMR mutant mice are prone to developing early tumors including gastrointestinal cancer, skin cancer and lymphoma. One limitation of mouse models to study MMR is that due to the development of aggressive lymphomas, the study of spontaneous GI tumors is difficult. Patients with LS (and monoallelic mutations) do not typically develop early lymphoma, but the rare patients with biallelic mutations in *MSH2, MSH6, MLH1* and *PMS2* have a severely reduced life span due to hematologic, brain and gastrointestinal malignancies similar to the knockout mice [11]. Interestingly, MMR deficient mice do not specifically develop colorectal cancer (CRC), but predominantly small bowel tumors [4].

2.1 MutS Homolog Deficient Mice

Of the three MutS homologs (MSH2, MSH6, MHS3), only deficiency in MSH2 and MSH6 demonstrated a strong tumor predisposition phenotype (summarized in Table 1).

In particular, the phenotypes are as follows:

2.2 MSH2

Msh2 deficient mice display severely reduced survival with 50 % mortality by 6 months and 100 % mortality by 11–12 months. Phenotypically, these mice have a strong predisposition to malignancy with lymphomas observed in 80 % and intestinal neoplasms observed in 70 % of knockout mice 6 months or older and skin neoplasms (sebaceous gland tumors and squamous cell carcinoma) in 7 % [12]. MSI is more common in carcinomas than in adenomas, but uncommon in normal tissue. Cell extracts from these mice are unable to repair single-base mismatches or insertion/deletion loops of one to four nucleotides, supporting the

Table 1 Mouse lines with MutS MMR gene mutations

Genotype	Fertility M/F	Mononucleotide MSI panel	Dinucleotide MSI panel	Tumor incidence	Tumor spectrum
Msh2$^{-/-}$	+/+	High	High	High	Lymphoma, GI, skin, other
Msh3$^{-/-}$	+/+	Moderate	High	High	GI tumors
Msh6$^{-/-}$	+/+	None	Low	High	Lymphoma, GI, other
Msh3$^{-/-}$ *Msh6*$^{-/-}$	+/+	High	High	High	Lymphoma, GI, skin, other
Msh4$^{-/-}$	–/–	N/A	N/A	None	None
Msh5$^{-/-}$	–/–	N/A	N/A	None	None

concept that loss of Msh2 in mice results in complete MMR deficiency. Clinically, LS patients with MSH2 deficiency have earlier onset and more aggressive malignancy phenotype.

Due to the aggressive nature of lymphomas affecting *Msh2* deficient mice, the study of GI tumors is challenging as they often do not live long enough to develop GI tumors. To avoid early death caused by tumorigenesis in extraintestinal organs, conditional *Msh2* knockout mice with intestine specific gene inactivation have been generated. These mice carry an $Msh2^{loxP}$ allele that results in somatic deletion of *Msh2* in the intestinal epithelium by expression of the *Villin-Cre* recombinase transgene. Tumorigenesis is isolated to the small intestine and formation of 1–2 adenomas and/or adenocarcinomas is observed within the first year of life. This *Msh2*-deficient model, while slower, provides a more useful means of studying intestinal tumorigenesis [13].

2.3 *MSH6*

Knockout of *Msh6* in mice led to the search for and later implication of *Msh6* as a late-onset cancer susceptibility gene. Phenotypically, these mice have a tumor spectrum phenotype similar to *Msh2* deficient mice, but the onset of tumor development is delayed and the mice survive up to 16 months. Molecular analysis of these mice revealed a partial repair defect with defective repair of single nucleotide mismatches but intact repair of 1, 2 and 4 nucleotide insertion/deletion mismatches [14]. As a result, these mice accumulate base substitutions rather than the frame shift mutations that occur in the absence of *Msh2*. Due to preservation of insertion/deletion mismatches, the tumors in *Msh6* deficient mice do not display MSI phenotype. In concordance with the *Msh6* deficient mouse model, LS patients with germ line mutation in *MSH6* develop GI cancers at age 60 and older with variable MSI phenotypes [15]. Interestingly, both the female carriers of *MSH6* germ line mutation and mice with global *Msh6* deficiency develop endometrial cancer [16].

2.4 *MSH3*

Msh3 deficient mice neither have a reduced survival or cancer phenotype, likely due to only moderate defects in MMR. These mice do develop tumors later in life, but the incidence is comparable to wild-type mice. The cell extracts from these mice show impaired repair of insertion/deletion mutations of 1–4 bases, but efficient repair of single-base substitution mutations, likely because of functional Msh2-Msh6 complexes. However, when *Msh3* deficiency is combined with *Msh6* deficiency via crosses, there is a complete loss of repair activity in cell extracts and the tumor predisposition phenotype is similar to that in *Msh2* deficiency. This suggests a functional redundancy between MSH2-MSH6 and MSH2-MSH3 complexes and indicates that MSH3 is not implicated in LS [4, 17].

2.5 MutL Homolog Deficient Mice

Of the four different MutL homologs (MLH1, PMS1, PMS2, MLH3), only mice deficient in *Mlh1* or *Pms2*, which together form the MutLα heterodimer, demonstrate a strong tumor predisposition phenotype (summarized in Table 2).

2.6 MLH1

Mice deficient in *Mlh1* have reduced survival, a strong cancer phenotype similar to *Msh2* deficient mice and a tumor spectrum that includes T-cell lymphoma, intestinal adenoma/adenocarcinoma and skin neoplasia [4]. Typically these mice live up to 12 months. Extracts from these mice display complete MMR deficiency with MSI high tumors. This suggests that *Mlh1* is essential for MMR.

2.7 PMS2

Pms2 deficient mice harbor a cancer predisposition phenotype that is less severe and different in its tumor spectrum compared to *Mlh1, Msh2* or *Msh6* mutant mice. These mice develop lymphomas and sarcomas, but not intestinal tumors [18]. They also exhibit increased mutation frequencies at mononucleotide repeat tracts, with an increase in mutation frequency up to three times lower when compared to *Mlh1* deficient mice [13, 19].

2.8 MLH3

Although MLH3 deficiency in humans is not known to result in a predisposition to malignancy, *Mlh3*-deficient mice develop adenomas/adenocarcinomas in their small intestines as well as lymphomas, basal cell carcinoma of the skin and other

Table 2 Mouse lines with exonuclease and MutL MMR gene mutations

Genotype	Fertility M/F	Mononucleotide MSI panel	Dinucleotide MSI panel	Tumor incidence	Tumor spectrum
$Mlh1^{-/-}$	–/–	High	High	High	Lymphoma, GI, skin, other
$Pms1^{-/-}$	+/+	Low	Low	None	None
$Pms2^{-/-}$	–/+	High	High	High	Lymphoma and sarcoma
$Mlh3^{-/-}$	–/–	Low	Low	N/A	N/A
$Exo1^{-/-}$	–/–	High	Low	Moderate	Lymphoma

tumors. These mice also display MSI at mononucleotide repeat sequences in their genomic DNA, but to a lesser extent than *Pms2* [13].

2.9 Exonuclease 1

Exonuclease 1 (Exo1)-deficient mice do display a significant increase in GI tumors and therefore it was concluded that this gene might only rarely be involved in LS. Loss of Exo1 function causes defects in the repair of single base mismatches and 1 base insertion/deletion loops but does not affect repair of 2 base insertion/deletion loops. This mouse line has an increased predisposition to lymphoma late in life. Tumor tissues in these mice display MSI at mononucleotide repeat markers. When compared with MMR defects in *Msh2* or *Mlh1* deficient mice, Exo1 deficiency results in only a partial repair defect, suggesting that other redundant exonucleases with unknown identity are present [13] (Table 2).

2.10 Homozygous MMR Mutant Mice with Heterozygous Adenomatous polyposis coli (APC) Mutation

Many colorectal cancers with MSI also carry somatic mutations in the *APC* gene, which implies that loss of APC function is critical for tumorigenesis. Mice with homozygous mutation of *Msh2, Msh6, Mlh1* or *Pms2* and heterozygous germ-line mutation of *Apc* develop tumors predominantly in the intestinal tract. Phenotypically, in *Apc*-mutant mice, loss of Msh2 or Mlh1 dramatically increases intestinal tumor burden while loss of Msh6 or Pms2 causes a milder intestinal tumor. In contrast, loss of Msh3 does not result in any significant increase in tumors. Overall, the level of tumor burden correlates with the type of MMR defects in MMR deficient mice [20–23].

3 Microsatellite Instability in Human Cancers

3.1 Lynch Syndrome or germline mutations in MMR genes

LS is an autosomal dominant disease characterized by MSI high tumors of the colon and other organs. Previously patients with LS were identified by the Amsterdam I criteria, which required fulfillment of the following diagnostic criteria: 3 or more relatives with CRC, 2 or more successive generations affected, 1 or more relatives diagnosed before age 50 [24]. In 1998, this criterion was broadened to include the extracolonic tumors, known as Amterdam II. Amsterdam II criteria requires fulfillment of the following diagnostic criteria: 3 or more relatives with a LS associated cancer (CRC, endometrial, small bowel, ureter or renal pelvis), 2 or more successive generations affected, 1 or more relatives diagnosed before age 50, 1 should be a first-degree relative of the other two, familial adenomatous polyposis should be excluded, and tumors should be verified by pathologic examination [25]. In 1997, the National Cancer Institute (NCI) developed the Bethesda guidelines in order to identify individuals to undergo MSI testing and genetic testing for LS, which were subsequently revised in 2004 [26]. The revised Bethesda guidelines recommend genetic testing for MSI high tumors in the following situations: CRC diagnosed in a patient age less than 50, presence of synchronous or metachronous CRC or other LS associated tumor regardless of age, CRC with high MSI histology diagnosed in a patient less than 60 years old, CRC diagnosed in one or more first degree relatives with a LS associated tumor with one cancer diagnosed at age less than 50, CRC diagnosed in two or more first-degree or second-degree relatives with LS associated tumors regardless of age (summarized in Table 3).

LS is now defined by the presence of a germline mutation in a DNA MMR gene. Approximately 15–20 % of colorectal cancers are related to a familial cancer syndrome with 3 % of all CRC associated with LS and a germline mutation in a MMR protein [6]. LS is characterized by an increased risk of CRC (lifetimes risk: 54–74 % males and 30–52 % females) and earlier age of onset (mean age 42–61) [27]. Females with LS have a 28–68 % lifetime risk for endometrial cancer and modestly increased risks for other cancers including gastric, small bowel, ovarian, urinary tract, pancreatic, biliary, brain and sebaceous gland tumors (summarized in Table 4) [27].

Table 3 Clinical criteria to identify patients with Lynch syndrome

Amsterdam I Criteria
Amsterdam II Criteria
Revised Bethesda Guidelines
3 or more relatives with CRC
2 or more successive generations affected
1 or more relatives diagnosed before age 50 years
3 or more relatives with a LS associated cancer (CRC, endometrial, small bowel, ureter or renal pelvis)
2 or more successive generations affected
1 or more relatives diagnosed before age 50
1 should be a first-degree relative of the other two
Familial adenomatous polyposis should be excluded
Tumors should be verified by pathologic examination
CRC diagnosed in a patient age less than 50
Presence of synchronous or metachronous CRC or other LS associated tumor regardless of age
CRC with high MSI histology diagnosed in a patient less than 60 years old
CRC diagnosed in one or more first degree relatives with a LS associated tumor with one cancer diagnosed at age less than 50
CRC diagnosed in two or more first-degree or second-degree relatives with LS associated tumors regardless of age

Table 4 MMR gene mutation and clinical human phenotype

Gene affected	Age onset	Tumor spectrum/risk
MLH1+PMS2	Early	CRC Endometrial cancer Any LS-associated cancer
MSH2+MSH6	Early	Endometrial cancer CRC Any LS-associated cancer
MSH6	Late	Endometrial cancer CRC Any LS-associated cancer
PMS2	Late	CRC 15–20 % Endometrial cancer 15 % Any LS-associated cancer 25–32 % [27]

The MMR genes implicated in LS include *MLH1, MSH2, MSH6,* and *PMS2*; however, not all causative gene mutations are known. More recently, deletions in the 3′ region of the gene *EPCAM* have been identified as a cause of constitutional epigenetic inactivation of *MSH2*. Of note, cancer risks and age of onset vary depending on which gene is affected. Lower risk of malignancy and later age of onset has been noted for individuals with *MSH6* and *PMS2* mutations [28, 29]. LS patients with endometrial cancer tend to have *MSH2* and *MSH6* gene mutations [30]. The contribution of each MMR protein to LS is 32 % for MLH1, 38 % for MSH2, 14 % for MSH6, and 15 % for PMS2 [31].

As discussed previously, monoallelic inactivation of a MMR gene results in a normal phenotype but with an increased risk of developing cancer. Although LS is inherited in an autosomal dominant manner with a single mutated allele present in the germline of an affected individual, a second hit on the other allele is required before the defect in MMR occurs. Thus, MSI is not found in normal tissue unless biallelic germline mutation is present, but this is exceedingly rare. When a second alteration occurs that inactivates the wild-type allele in a cell of a target tissue, there is loss of DNA MMR activity, resulting in MSI [5]. Inactivation of the remaining wild-type allele can occur by a variety of mechanisms including deletion, gene conversion, methylation or point mutation.

Biallelic germline mutation of MMR genes has been documented in a rare childhood cancer syndrome [32]. The clinical phenotype of these children includes development of brain tumors, leukemias and lymphomas in the first decade of life. Those children who survive these initial malignancies may later develop gastrointestinal tumors, typically multiple adenomatous polyps that evolve into CRC and small bowel cancer, as well as café-au-lait macules. In patients with biallelic MMR gene mutations with gastrointestinal cancers, only 14 % harbor *MLH1* or *MSH2* gene mutations and more than half biallelic *PMS2* mutations. Accordingly, parents of these children are obligate carriers of a heterozygous MMR gene mutation and require LS cancer screening. Siblings have a 25 % chance of also having biallelic MMR gene mutations and a 50 % chance of being heterozygous for either MMR gene mutation [32].

MMR gene mutations associated with LS include point mutations, large genomic deletions or rearrangements. Large genomic deletions can be challenging to detect since they are missed by qualitative standard sequencing technology given the potential for loss of the area's primer binding for amplification in the affected allele, resulting in amplification of the intact wild type allele only. Next generation sequencing, however, which allows for much more overlap and even assessment of mosaics, may detect such lesions.

3.2 Sporadic MSI

MSI is the hallmark feature of LS, but can occur in sporadic CRC as well. The most common etiology of sporadic MSI is acquired promoter hypermethylation of *MLH1* that occurs in a background of global hypermethylation of gene promoters known as CpG island methylator phenotype (CIMP). Methylation of the CpG site on the 3′ end of the *MLH1* promoter, close to the start codon, mediates somatic gene silencing of *MLH1*. When CpG sites on the promoter regions of both copies of *MLH1* are hypermethylated, transcription factor binding is blocked and MLH1 expression is lost in epithelial cells, resulting in genomic MSI [33]. *MLH1* promoter hypermethylation is highly associated with the V600E mutation (c.1799T>A) in the BRAF oncogene that occurs in 40–50% of sporadic MSI CRC. *BRAF* mutations lead to constitutive activity of a kinase involved in response to growth signals. *BRAF* mutation is typically absent in LS associated tumors and helps to distinguish between sporadic and germline MSI [6]. Hypermethylation of the *MLH1* gene promoter is a somatic change associated with advanced age, female sex and proximal colon location. Sessile serrated adenomatous polyps may be the precursor lesions for the development of colon cancer in this pathway [33].

3.3 Constitutional Epimutations

Recently, cases of constitutional *MLH1* and *MSH2* epimutations have been reported in patients with normal genetic testing for MMR proteins, but meeting clinical criteria for LS.

In cases of constitutional *MLH1* epimutations, the promoter of one of the *MLH1* alleles is methylated in the germline, resulting in transcriptional silencing of this allele in non-neoplastic tissue [34]. Typically, an *MLH1* epimutation is identified when MLH1 and PMS2 are proteins that are not expressed, so these proteins are absent on immunohistochemistry (IHC) but methylation testing is positive for tumor tissue and adjacent normal tissue, no *BRAF* mutation is identified and no germline mutation identified with genetic testing. *MLH1* epimutations are reversible between generations and thus display non-Mendelian inheritance [34]. However, dominant transmission of *MLH1* epimutation has been documented [35].

In cases of constitutional *MSH2* epimutations, *MSH2* methylation occurs when a deletion in the 3′ end of the gene upstream of *MSH2* (*TACSTD1*, which encodes Ep-CAM) is present. This results in the loss of the stop codon, which leads to methylation of the CpG island downstream and constitutional silencing of *MSH2* in tissue that is Ep-CAM positive, but not tissue that is EpCAM negative [36]. Constitutional epimutations of *MSH2* are inherited in a classic autosomal dominant manner [3].

3.4 Lynch-Like Syndrome

Although biallelic *MLH1* promoter hypermethylation is the most common etiology of sporadic MSI, a significant number of patients have a Lynch-like syndrome, defined by patients with tumors that display MSI, have no germline DNA MMR gene mutation detected and show no hypermethylation of *MLH1* as the etiology for MSI [10]. The molecular mechanism of tumor development in Lynch-like tumors is unknown. Hypotheses include the presence of unknown germline mutations in MMR genes, the presence of germline mutations in MMR genes not identified by current detection methods, or the presence of a genetic process within the tumors other than germline mutation coupled with a second hit.

A recent cohort study of 232 MSI-positive colorectal and endometrial cancers further evaluated 25 patients with tumor MSI but no underlying germline mutation or *MLH1* promoter methylation. The authors identified that two somatic mutations in either *MLH1* or *MSH2* were present in 52 % of these tumors [37]. However, with our current knowledge, LS cannot be ruled out with certainty in these Lynch-like patients given our incomplete understanding of all its genetic causes and therefore these patients are treated with the same surveillance recommendations as LS patients.

4 Testing for MSI

The primary reason for testing of MSI and defects in MMR protein expression and later identification of patients with LS is the long-term clinical and preventive implications for patients and families. For patients with LS, the option of subtotal colectomy should be discussed when a new CRC is diagnosed. Screening colonoscopy for affected family members starts earlier and intervals are shortened due to the more rapid progression of polyps to malignancy. Intensive surveillance with colonoscopy has been shown to reduce CRC incidence and related mortality [38–40]. Current National Comprehensive Cancer Network (NCCN) guidelines recommend colonoscopy every 1–2 years beginning at age 20–25 years (www.NCCN.org). Patients with LS also need to be screened for extra-colonic LS-related malignancies [41].

There has been a recent push for universal screening in any CRC patient for either MSI or loss of MMR protein expression by the Evaluation of Genomic Applications in Practice and Prevention Working Group with the goal of reducing morbidity and mortality of family members [42]. Of the 2.8 % of subjects identified with LS by universal screening (MSI and IHC), 50 % were over age 50 years old, and 25 % did not meet either the Amsterdam II diagnostic criteria or the revised Bethesda guidelines for LS [43]. Solely relying on age at diagnosis, family history and tumor histology as defined by these clinical criteria would miss a significant number of affected individuals. The current NCCN clinical practice guidelines for CRC screening endorse MSI/MMR protein expression screening in individuals with a strong family history of colon cancer (www.NCCN.org).

Several algorithms have been proposed to methodically evaluate all patients with newly diagnosed CRC for LS. Sequential testing beginning with IHC and/or MSI testing by polymerase chain reaction (PCR) followed by targeted direct mutation analysis could be a cost-effective approach, eliminating the expense of full gene sequencing in individuals unlikely to have LS (summarized in Table 5).

4.1 Microsatellite instability PCR testing

PCR-based testing for MSI detects alterations in size of microsatellite repeats in tumor versus healthy tissue from the same patient. The presence of new microsatellite allele sizes in the tumor due to errors that have not been corrected indicates a MMR deficiency. In 1997, a National Cancer Institute sponsored workshop developed the Bethesda panel of 5 microsatellite loci, which consists of 2 mononucleotide repeats (BAT25 and BAT 26) and 3 dinucleotide repeats (D2S123, D5S346 and D17S250), to evaluate for MSI [44]. Subsequently in 2002, a panel of 5 mononucleotide repeats (BAT25, BAT26, NR-21, NR-22 and NR-24) was developed. A third panel (BAT25, BAT26, MONO-27, NR-21 and NR-24) is also available. The advantage of the mononucleotide panel is that it allows for detection of MSI without the need for healthy tissue for comparison. Mononucleotide markers are also more reliable than dinucleotide markers in the identification of MSI high tumors [26]. MSI testing is performed at molecular facilities.

Two categories of MSI have been defined, based on results from PCR testing for MSI. MSI high (*MSI-H*) is defined as instability in 2 or more out of 5 markers in the panels currently available [45]. MSI-H tumors share the phenotype of LS with a characteristic pathology, proximal colonic predominance and a better prognosis. About 20–30 % of these tumors come from patients with LS and the other 75–80 % arise from acquired hypermethylation of the promoter of the *MLH1* gene, which does not involve germline mutation [46, 47]. MSI low (*MSI-L*) is defined as instability of 1 of 5 markers in the current panels. These tumors behave clinically like microsatellite stable tumors (*MSS*), in which none of the five microsatellites tested is mutated. The mechanism of MSI-L may be multifactorial, but non-genetic down-regulation of the *MSH3* gene has been shown to produce MSI-L results. Loss of MSH3 activity by IHC correlates with the presence of somatic mutations in tri- and

Table 5 Clinical testing algorithm for MSI-H tumors

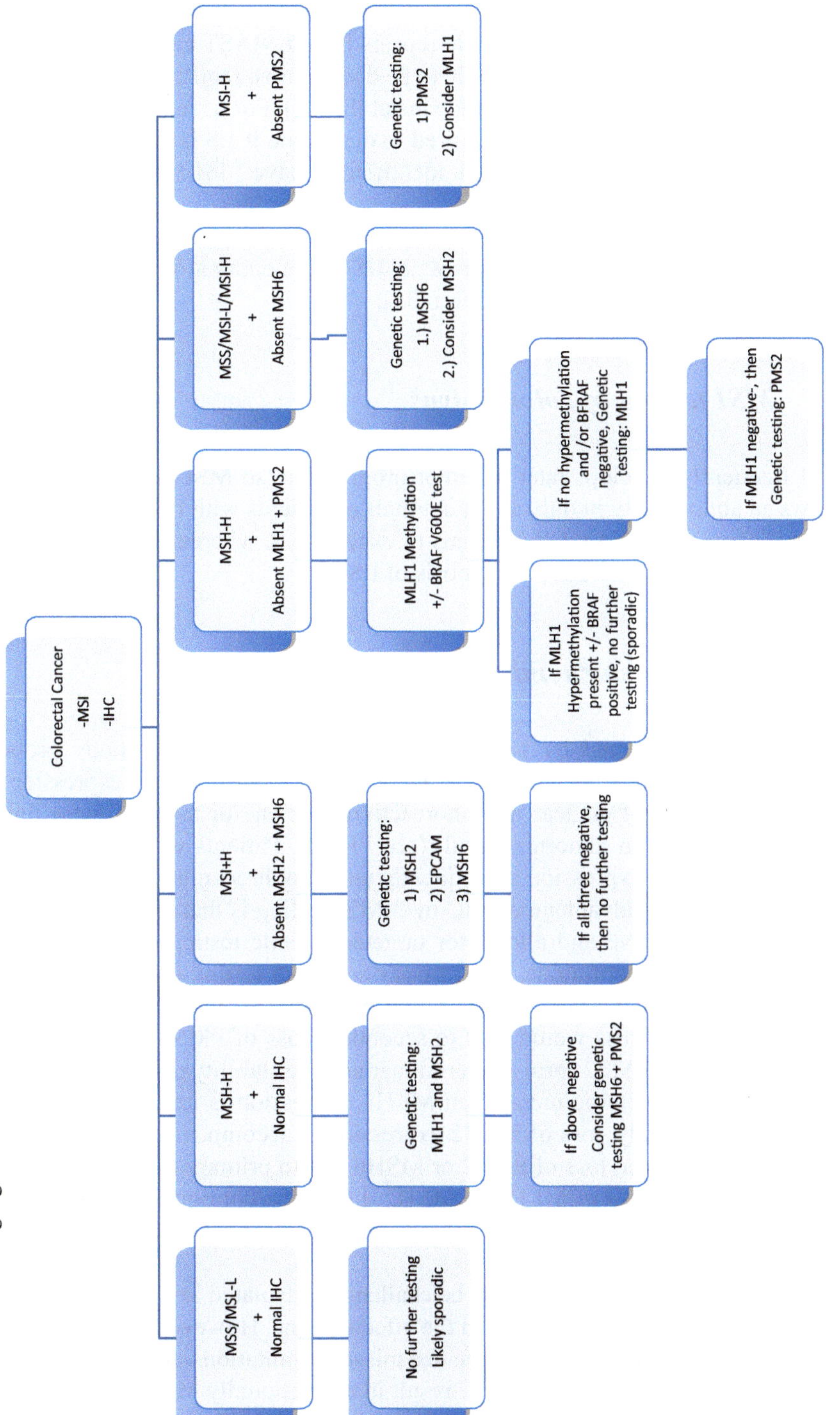

tetra-nucleotide repeats, which has been termed EMAST, for elevated microsatellite alterations at selected tetranucleotide repeats [48]. EMAST can be found in ~60 % of CRC that are MSI-L and is thought to be due to down regulation of MSH3 resulting in dinucleotide and tetranucleotide instability [8]. For patients identified to have MSI-H tumor, further testing is required to determine if LS is present or if this is a sporadic MSI-H tumor. For patients identified to have MSI-L or MSS tumors, no additional testing is warranted.

MSI testing has been found to be 80 % sensitive for patients with *MLH1* gene mutation, 84 % sensitive for patients with *MSH2* gene mutation and 55 % sensitive for patients with *MSH6* gene mutation [31].

4.2 MSI testing of colon polyps

MSI frequently develops later in tumor progression, so MSI testing on adenomas shows an abnormality in only about one-half of patients with LS [49]. Lack of MSI in an adenoma is not a reliable means to exclude LS, whereas identification of an MSS CRC reliably excludes a diagnosis of LS.

4.3 Immunohistochemistry

IHC testing, performed by pathology departments, utilizes antibody probes to MMR proteins MLH1, MSH2, MSH6 and PMS2 to identify loss of expression. The complete absence of nuclear immunoreactivity for one or more of these proteins in tumor tissue is an abnormal result (see Fig. 2). Patients who are identified to loss MMR protein expression subsequently undergo direct mutation analysis of the relevant gene. The advantage of IHC over MSI testing is that it is easily available, relatively inexpensive and allows for targeted genetic testing of a gene in question [50]. In some institutions, a 2-antibody approach testing for PMS2 and MSH6 takes advantage of the heterodimerization properties of the MMR proteins since MLH1 and MSH2 loss would lead to secondary loss of PMS2 or MSH6, respectively. MLH1 and PMS2 form a heterodimer and the stability of PMS2 relies on the formation of this heterodimer. When MLH1 expression is lost, PMS2 expression is also lost. Similarly, loss of MSH2 expression is accompanied by loss of MSH6 expression. However, loss of PMS2 or MSH6 due to primary defect in these genes does not result in loss of MLH1 or MSH2 due to the configuration of the protein complex. The combined sensitivity of IHC on CRC is thought to be 93 % for all four MMR proteins [30].

Interpretation of IHC results can be challenging. Isolated loss of MSH6 or PMS2 is likely due to germline mutation in the affected gene. However, expression of proteins that have lost their function due to a missense mutation but not their antigenicity may result in a false normal IHC result in a functionally MMR-deficient tumor. Thus missense mutation of *MLH1* could result in isolated loss of PMS2 staining on

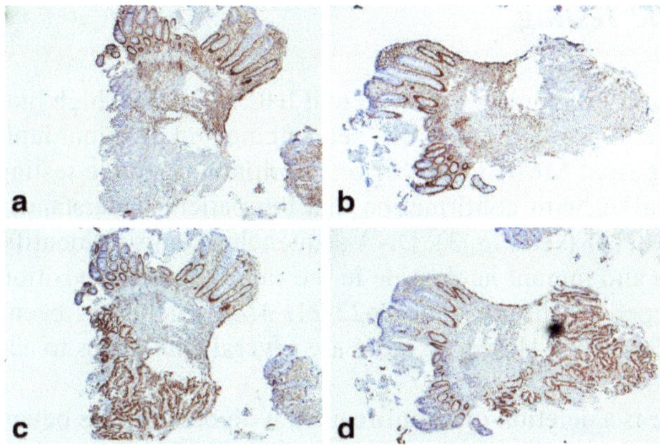

Fig. 2 **a** MHL1 staining is positive in normal colon epithelium and inflammatory cells but absent in invasive carcinoma cells, indicating loss of protein in tumor. **b** PMS2 shows same pattern of staining/deletion as MLHl (tumor is at *right* portion of tissue). **c** MSH2 is present in normal colonic epithelium, inflammatory cells, and tumor cells. **d** MSH6 shCNVS same pattern of staining as MSH2

IHC when the true germline mutation is actually present in *MLH1* [50]. When IHC shows loss of expression of MLH1, then additional testing is warranted to determine if this result is due to LS or is sporadic.

4.4 Hypermethylation Testing

The *BRAF* p.V600E (c.1799T>A) mutation and *MLH1* promoter methylation are commonly identified in sporadic MSI tumors and rarely identified in LS associated tumors. Only 1 in 5 MSI-H tumors are due to LS. When IHC implies a defect in MLH1, reflex testing for *BRAF* p.V600E gene mutation and *MLH1* promoter methylation is a cost-effective method for eliminating unnecessary genetic testing [6]. *MLH1* methylation testing requires PCR amplification using primers specific for CpG islands in the proximal promoter region because other regions do not correlate as well with loss of MLH1 expression [51].

Clinically, if a *BRAF* mutation is identified, then the tumor is likely sporadic and no further testing is required. If BRAF testing is negative, then *MLH1* promoter methylation analysis may be performed and if positive, tumor is likely sporadic and no further testing is required. However, there has been one case identified of coexisting monoallelic germline *MLH1* mutation and somatic *MLH1* promoter hypermethylation [52]. If *MLH1* promoter methylation testing is negative, genetic testing for *MLH1* would be reasonable.

4.5 Genetic Testing

When preliminary screening with MSI and IHC suggests a high likelihood of LS, it is reasonable to proceed with DNA sequencing and deletion/duplication analysis of the suggested MMR genes. Prior to initiating genetic testing, appropriate genetic counseling with confirmation that the patient understands the risks and benefits must occur (see Fig. 3). DNA sequencing looks to identify the presence of a wild-type and mutant nucleotide in the same sequence position. *MSH2* gene has been mapped to chromosome 2p22-21. *MLH1* gene has been identified on chromosome 3p21 [3]. However, there are several challenges to accurate genetic testing.

When there is a deletion of an entire exon with breakpoints beyond the primers for that exon, the analysis gives a normal result because only the wild-type strand is sequenced and the deletion is not detected [3]. Large deletions account for >33 % of all *MSH2* mutations associated with LS. Because the 5′ end of the *MSH2* gene is located within a cluster of Alu sequences that are prone to internal recombination followed by excision of the DNA between the hybrid sequences, *MSH2* gene is prone to large deletions. Identification of these large deletions often requires utilization of multiplex ligation-dependent probe amplification to quantify whether one or two alleles at each exon have been sequenced [53].

Molecular genetic analysis will miss 15 % of individuals with clinical LS [33]. Alternative mechanisms of LS such as *EPCAM* deletions and constitutional *MLH1* epimutations (see above) will result in normal testing of MMR genes with MMR deficient phenotype. Germline deletions in the *TACSTD1* gene, which is directly upstream of *MSH2* and encodes the epithelial cell adhesion molecule EPCAM, has been identified as the causative mutation in some families with LS [36]. Deletion of *EPCAM* results in epigenetic inactivation of *MSH-2*. *EPCAM* deletions are implicated in 1–3 % of LS families [54].

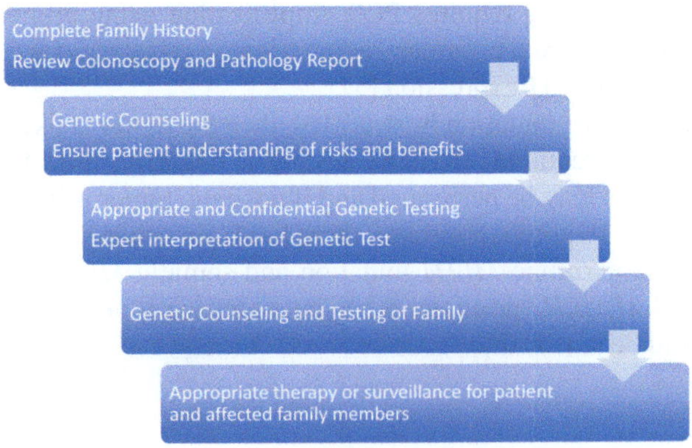

Fig. 3 Approach to Genetic Testing

Identification of *PMS2* germline mutation is challenging due to the presence of many pseudogenes in the human genome with sequences homologous with *PMS2* [55]. This results in technical difficulties identifying large deletions in *PMS2*.

One of the biggest challenges in the analysis for mutations in DNA MMR genes is the identification of variants of unknown significance (VUS). These variants may be either disease causing yet undescribed, or, more commonly, benign polymorphisms. Determination of this distinction is an ongoing challenge. Functional assays of the cloned mutant allele and its product may aid and bioinformatic algorithms have been developed to classify *MLH1* and *MSH2* missense variants as pathogenic or non-pathogenic based on evolutionary sequence conservation data and predicted protein structural change to assist with interpretation of aberrations in genetic sequencing [56].

Once the putative gene is identified, family members can be tested with 100% accuracy. The probability of a positive genetic test is 50% in first-degree relatives and 25% probability in second-degree relatives given autosomal dominant inheritance [27].

4.6 Classifying CRC by molecular features

One method of classifying CRC based on the presence of MSI and CIMP was suggested with the following molecular and histopathologic features [57]:

1. Type 1 (CIMP high/MSI-H/*BRAF* mutation): Accounts for 12% of CRC. Originates in serrated adenomas and is characterized by *BRAF* mutation and *MLH1* methylation.
2. Type 2 (CIMP high/MSI –L or MSS): Accounts for 8% of CRC. Originates in serrated adenomas and is characterized by *BRAF* mutation and methylation of multiple genes.
3. Type 3 (CIMP low/MSI-L or MSS): Accounts for 20% of CRC. Originates in tubular, tubulovillous, or serrated adenomas and is characterized by chromosomal instability (CIN), *K-RAS* mutation, and methylguanine methyltransferase (*MGMT*) methylation.
4. Type 4 (CIMP negative/MSS): Accounts for 57% of CRC. Originates in traditional adenoma and is characterized by CIN.
5. Type 5 or Lynch Syndrome (CIMP negative/MSI-H): Accounts for 3% of CRC. Genetic testing needed to confirm diagnosis.

5 Clinical Implications of MSI in the Large Bowel, Small Bowel and Stomach

The life time risks for LS associated cancers are highest for CRC with a risk of 52–82%, followed by a 25–60% risk of endometrial, and a 6–13% risk of gastric cancer [58]. Although the majority of extra-colonic gastrointestinal cancers are sporadic, the role of MSI testing in identifying LS patients has not been as clearly defined as it has in CRC.

5.1 Microsatellite Instability in Colorectal Cancer

MSI-H colorectal tumors, both inherited and sporadic, share several phenotypic features. Clinically, these tumors tend to occur in the right side of the colon. Histologically, they have medullary carcinoma phenotype, the presence of mucinous or signet-ring component, the presence of tumor-infiltrating and peritumoral lymphocytes, and a Crohn's disease-like lymphocytic infiltration [2, 59].

CRC with MSI-H portend a significantly better prognosis than MSS tumors with intact MMR. Specifically, patients with MSI-H CRC were found to have a better 5-year survival rate and overall survival rate than patients with MSS or MSI-L tumors [60]. MSI-H tumors that are stage II have lower rates of tumor recurrence than MSS tumors [61]. MSI status also affects the benefit of chemotherapy. Patients with MSS stage II and III benefited from adjuvant therapy with 5-FU with an increase in overall survival whereas patients with MSI-H stage II or III CRC treated with adjuvant 5-FU had no overall survival difference when compared with patients without adjuvant 5-FU [62]. For patients with stage II and III CRC, postoperative adjuvant FU-based chemotherapy significantly improved disease free survival (DFS) in patients with MSS tumors, but no improvement in DFS in patients with tumors deficient in MMR protein [63]. However, there are other studies that have reported patients with MSI-H tumors to have similar or better outcomes than patients with MSS tumors with adjuvant chemotherapy [64, 65]. Because of the conflicting data on this topic, MSI status is currently not a factor in adjuvant chemotherapy decisions.

5.2 Microsatellite Stable Colorectal Cancer: Familial Colorectal Cancer Type X

Patients meeting Amsterdam criteria with MSS tumors have an overall lower CRC risk and later average age at diagnosis than patients with LS. They also have no significantly elevated extracolonic cancer risks and do not demonstrate the rapid progression to CRC seen in MMR mutation carriers. The molecular origin of this entity has not yet been identified.

5.3 MSI in Small Bowel Cancer

The published prevalence of MSI-H small bowel carcinoma in LS ranges from 19.6 to 33.3% [1]. The risk for small bowel cancer is significantly increased in patients with LS with a relative risk greater than 100 when compared to the general population. The development of small bowel cancer also occurs 20 years earlier in LS when compared to the general population. The lifetime risk of small bowel cancer in patients with LS is about 1–4 % and has been reported to be higher in *MLH1* mutation carriers when compared to *MSH2* mutation carriers [66]. Small bowel cancer may be the initial presentation of LS in up to 45–57 % of cases [67, 68].

In addition to LS, there have been reports of MSI-H small bowel carcinoma associated with celiac disease and Crohn's disease [69–72]. Patients with celiac disease with small bowel adenocarcinoma had higher rates MSI-H tumor (73 vs 9 %), indicating increased MMR deficiency, as well as better survival than stage-matched controls [69]. In one study, the mechanism for three MSI-H small bowel cancer in celiac disease patients was identified to be loss of MLH1 expression due to hypermethylation of the *MLH1* promoter [71]. Thus celiac disease must remain on the differential diagnosis for patients with MSI-H small bowel cancer. One case report described a Crohns disease patient with MSI-H small bowel cancer likely due to hypermethylation since no mutation was identified on genetic testing [72].

Few studies have assessed the characteristics of intestinal tumorigenesis in the small bowel. One retrospective review of the German HNPCC Consortium database of 1986 families identified 31 patients with 32 small bowel cancers [67]. Of these patients 69 % were men and the median age of diagnosis was 39 years. Small bowel cancer was the first presentation of LS in 45 % and 55 % had a previous LS-related malignancy prior to diagnosis of small bowel cancer. Forty seven percent of these small bowel cancers were located in the duodenum with decreasing frequency from duodenum to ileum. High MSI was detected in 95 % of tumors, loss of MMR protein expression in 89 % and pathogenic germline mutations were identified in 81 %. The overall 10-year survival rate was 87 %.

Similar to MSI-H CRC, MSI-H adenocarcinoma of the small bowel has been shown to portend a better prognosis. One study of 35 patients with a curative resection of an adenocarcinoma of the small bowel showed that patients with MSI-H tumors had a cancer-specific survival of 49.6 months in contrast to patients with MSS tumors, who had a cancer-specific survival of 23.3 months [73].

5.4 *MSI in Gastric Cancer*

It is estimated that 15 % of all gastric tumors exhibit MSI-H histology, but the majority of these have acquired this phenotype through sporadic mutations, so further genetic testing is necessary to confirm a molecular diagnosis of LS [58]. MSI-H gastric carcinomas have been associated with increased prevalence in older patients, in female patients and in patients from European countries [1]. Gastric carcinoma is one of the most common extracolonic tumor sites in LS. The cumulative incidence of gastric carcinoma in LS has been reported to be between 2.1 and 18.9 % [1]. MSI-H tumors are associated with distal/antral location, larger size and lower overall stage and nodal status. In an analysis of 15 gastric carcinomas in family members of LS patients, MSI-H status was identified using the NCI panel in 13 out of 15 cases, of which all 13 showed germline MMR mutation [74]. The two other cases were MSS and did not show a germline mutation. Other studies have shown that mononucleotide markers, especially BAT-25, BAT-26 and BAT-40 are highly sensitive and specific for detection of MSI-H status in identifying MSI-H gastric carcinoma with absent expression of mismatch repair proteins by IHC [75–77]. For patients meeting the revised Bethesda clinical criteria for LS, it is reasonable to recommend screening of gastric carcinomas with molecular techniques and/or IHC for MMR proteins [58].

In summary, the detailed knowledge about the mismatch repair process and the development of tools to detect MSI and MMR deficiency have much improved our understanding of intestinal tumorigenesis in general and LS specifically. However, much still remains to be learned regarding the variable phenotypes, alternative causes for LS as well as Lynch-like syndrome and familial colorectal cancer X. This knowledge is crucial when diagnosing or managing patients with colon cancers and their family members.

References

1. Williams AS, Huang WY (2013) The analysis of microsatellite instability in extracolonic gastrointestinal malignancy. Pathology 45(6):540–552. doi:10.1097/PAT.0b013e3283653307
2. Bedeir A, Krasinskas AM (2011) Molecular diagnostics of colorectal cancer. Arch Pathol Lab Med 135(5):578–587. doi:10.1043/2010-0613-RAIR.1
3. Boland CR, Goel A (2010) Microsatellite instability in colorectal cancer. Gastroenterology 138(6):2073–2087e2073. doi:10.1053/j.gastro.2009.12.064
4. Edelmann L, Edelmann W (2004) Loss of DNA mismatch repair function and cancer predisposition in the mouse: animal models for human hereditary nonpolyposis colorectal cancer. Am J Med Genet C Semin Med Genet 129C(1):91–99. doi:10.1002/ajmg.c.30021
5. Boland CR, Jung B, Carethers JM (2013) Cancer of the colon and gastrointestinal tract. In: Rimoin DL, Pyeritz RE, Korf BR (eds) Emery and Rimoin's principles and practice of medical genetics, 6th edn. Elsevier/Academic, San Diego
6. Geiersbach KB, Samowitz WS (2011) Microsatellite instability and colorectal cancer. Arch Pathol Lab Med 135(10):1269–1277. doi:10.5858/arpa.2011-0035–RA
7. Shah SN, Hile SE, Eckert KA (2010) Defective mismatch repair, microsatellite mutation bias, and variability in clinical cancer phenotypes. Cancer Res 70(2):431–435. doi:10.1158/0008-5472.CAN-09-3049
8. Al-Sohaily S, Biankin A, Leong R, Kohonen-Corish M, Warusavitarne J (2012) Molecular pathways in colorectal cancer. J Gastroenterol Hepatol 27(9):1423–1431. doi:10.1111/j.1440-1746.2012.07200.x
9. Jung BH, Beck SE, Cabral J, Chau E, Cabrera BL, Fiorino A, Smith EJ, Bocanegra M, Carethers JM (2007) Activin type 2 receptor restoration in MSI-H colon cancer suppresses growth and enhances migration with activin. Gastroenterology 132(2):633–644. doi:10.1053/j.gastro.2006.11.018
10. Carethers JM (2014) Differentiating lynch-like from lynch syndrome. Gastroenterology 146(3):602–604. doi:10.1053/j.gastro.2014.01.041
11. Felton KE, Gilchrist DM, Andrew SE (2007) Constitutive deficiency in DNA mismatch repair. Clin Genet 71(6):483–498. doi:10.1111/j.1399-0004.2007.00803.x

12. Reitmair AH, Redston M, Cai JC, Chuang TC, Bjerknes M, Cheng H, Hay K, Gallinger S, Bapat B, Mak TW (1996) Spontaneous intestinal carcinomas and skin neoplasms in Msh2-deficient mice. Cancer Res 56(16):3842–3849
13. Taketo MM, Edelmann W (2009) Mouse models of colon cancer. Gastroenterology 136(3):780–798
14. Edelmann W, Yang K, Umar A, Heyer J, Lau K, Fan K, Liedtke W, Cohen PE, Kane MF, Lipford JR, Yu N, Crouse GF, Pollard JW, Kunkel T, Lipkin M, Kolodner R, Kucherlapati R (1997) Mutation in the mismatch repair gene Msh6 causes cancer susceptibility. Cell 91 (4):467–477
15. Kolodner RD, Tytell JD, Schmeits JL, Kane MF, Gupta RD, Weger J, Wahlberg S, Fox EA, Peel D, Ziogas A, Garber JE, Syngal S, Anton-Culver H, Li FP (1999) Germ-line msh6 mutations in colorectal cancer families. Cancer Res 59(20):5068–5074
16. Wijnen J, de Leeuw W, Vasen H, van der Klift H, Moller P, Stormorken A, Meijers-Heijboer H, Lindhout D, Menko F, Vossen S, Moslein G, Tops C, Brocker-Vriends A, Wu Y, Hofstra R, Sijmons R, Cornelisse C, Morreau H, Fodde R (1999) Familial endometrial cancer in female carriers of MSH6 germline mutations. Nat Genet 23(2):142–144. doi:10.1038/13773
17. Edelmann W, Umar A, Yang K, Heyer J, Kucherlapati M, Lia M, Kneitz B, Avdievich E, Fan K, Wong E, Crouse G, Kunkel T, Lipkin M, Kolodner RD, Kucherlapati R (2000) The DNA mismatch repair genes Msh3 and Msh6 cooperate in intestinal tumor suppression. Cancer Res 60(4):803–807
18. Baker SM, Bronner CE, Zhang L, Plug AW, Robatzek M, Warren G, Elliott EA, Yu J, Ashley T, Arnheim N, Flavell RA, Liskay RM (1995) Male mice defective in the DNA mismatch repair gene PMS2 exhibit abnormal chromosome synapsis in meiosis. Cell 82(2):309–319
19. Prolla TA, Baker SM, Harris AC, Tsao JL, Yao X, Bronner CE, Zheng B, Gordon M, Reneker J, Arnheim N, Shibata D, Bradley A, Liskay RM (1998) Tumour susceptibility and spontaneous mutation in mice deficient in Mlh1, Pms1 and Pms2 DNA mismatch repair. Nat Genet 18(3):276–279. doi:10.1038/ng0398-276
20. Edelmann W, Yang K, Kuraguchi M, Heyer J, Lia M, Kneitz B, Fan K, Brown AM, Lipkin M, Kucherlapati R (1999) Tumorigenesis in Mlh1 and Mlh1/Apc1638N mutant mice. Cancer Res 59(6):1301–1307
21. Reitmair AH, Cai JC, Bjerknes M, Redston M, Cheng H, Pind MT, Hay K, Mitri A, Bapat BV, Mak TW, Gallinger S (1996) MSH2 deficiency contributes to accelerated APC-mediated intestinal tumorigenesis. Cancer Res 56(13):2922–2926
22. Baker SM, Harris AC, Tsao JL, Flath TJ, Bronner CE, Gordon M, Shibata D, Liskay RM (1998) Enhanced intestinal adenomatous polyp formation in Pms2-/-;Min mice. Cancer Res 58(6):1087–1089
23. Kuraguchi M, Yang K, Wong E, Avdievich E, Fan K, Kolodner RD, Lipkin M, Brown AM, Kucherlapati R, Edelmann W (2001) The distinct spectra of tumor-associated Apc mutations in mismatch repair-deficient Apc1638N mice define the roles of MSH3 and MSH6 in DNA repair and intestinal tumorigenesis. Cancer Res 61(21):7934–7942
24. Vasen HF, Mecklin JP, Khan PM, Lynch HT (1991) The International Collaborative Group on Hereditary Non-Polyposis Colorectal Cancer (ICG-HNPCC). Dis Colon Rectum 34(5):424–425
25. Vasen HF, Watson P, Mecklin JP, Lynch HT (1999) New clinical criteria for hereditary non-polyposis colorectal cancer (HNPCC, Lynch syndrome) proposed by the International Collaborative group on HNPCC. Gastroenterology 116(6):1453–1456
26. Umar A, Boland CR, Terdiman JP, Syngal S, de la Chapelle A, Ruschoff J, Fishel R, Lindor NM, Burgart LJ, Hamelin R, Hamilton SR, Hiatt RA, Jass J, Lindblom A, Lynch HT, Peltomaki P, Ramsey SD, Rodriguez-Bigas MA, Vasen HF, Hawk ET, Barrett JC, Freedman AN, Srivastava S (2004) Revised Bethesda Guidelines for hereditary nonpolyposis colorectal cancer (Lynch syndrome) and microsatellite instability. Je Natl Cancer Inst 96(4):261–268
27. Bellcross CA, Bedrosian SR, Daniels E, Duquette D, Hampel H, Jasperson K, Joseph DA, Kaye C, Lubin I, Meyer LJ, Reyes M, Scheuner MT, Schully SD, Senter L, Stewart SL,

St Pierre J, Westman J, Wise P, Yang VW, Khoury MJ (2012) Implementing screening for Lynch syndrome among patients with newly diagnosed colorectal cancer: summary of a public health/clinical collaborative meeting. Genet Med 14(1):152–162. doi:10.1038/gim.0b013e31823375ea

28. Bonadona V, Bonaiti B, Olschwang S, Grandjouan S, Huiart L, Longy M, Guimbaud R, Buecher B, Bignon YJ, Caron O, Colas C, Nogues C, Lejeune-Dumoulin S, Olivier-Faivre L, Polycarpe-Osaer F, Nguyen TD, Desseigne F, Saurin JC, Berthet P, Leroux D, Duffour J, Manouvrier S, Frebourg T, Sobol H, Lasset C, Bonaiti-Pellie C, French Cancer Genetics N (2011) Cancer risks associated with germline mutations in MLH1, MSH2, and MSH6 genes in Lynch syndrome. JAMA 305(22):2304–2310. doi:10.1001/jama.2011.743
29. Senter L, Clendenning M, Sotamaa K, Hampel H, Green J, Potter JD, Lindblom A, Lagerstedt K, Thibodeau SN, Lindor NM, Young J, Winship I, Dowty JG, White DM, Hopper JL, Baglietto L, Jenkins MA, de la Chapelle A (2008) The clinical phenotype of Lynch syndrome due to germ-line PMS2 mutations. Gastroenterology 135(2):419–428. doi:10.1053/j.gastro.2008.04.026
30. Weissman SM, Burt R, Church J, Erdman S, Hampel H, Holter S, Jasperson K, Kalady MF, Haidle JL, Lynch HT, Palaniappan S, Wise PE, Senter L (2012) Identification of individuals at risk for Lynch syndrome using targeted evaluations and genetic testing: National Society of Genetic Counselors and the Collaborative Group of the Americas on Inherited Colorectal Cancer joint practice guideline. J Genet Couns 21(4):484–493. doi:10.1007/s10897-011-9465–7
31. Palomaki GE, McClain MR, Melillo S, Hampel HL, Thibodeau SN (2009) EGAPP supplementary evidence review: DNA testing strategies aimed at reducing morbidity and mortality from Lynch syndrome. Genet Med 11(1):42–65. doi:10.1097/GIM.0b013e31818fa2db
32. Durno CA, Holter S, Sherman PM, Gallinger S (2010) The gastrointestinal phenotype of germline biallelic mismatch repair gene mutations. Am J Gastroenterol 105(11):2449–2456. doi:10.1038/ajg.2010.215
33. Rybak C, Hall MJ (2011) Interpretation of genetic testing for lynch syndrome in patients with putative familial colorectal cancer. J Natl Compr Canc Netw 9(11):1311–1320
34. Hitchins MP, Ward RL (2009) Constitutional (germline) MLH1 epimutation as an aetiological mechanism for hereditary non-polyposis colorectal cancer. J Med Genet 46(12):793–802. doi:10.1136/jmg.2009.068122
35. Hitchins MP, Rapkins RW, Kwok CT, Srivastava S, Wong JJ, Khachigian LM, Polly P, Goldblatt J, Ward RL (2011) Dominantly inherited constitutional epigenetic silencing of MLH1 in a cancer-affected family is linked to a single nucleotide variant within the 5′UTR. Cancer Cell 20(2):200–213. doi:10.1016/j.ccr.2011.07.003
36. Ligtenberg MJ, Kuiper RP, Chan TL, Goossens M, Hebeda KM, Voorendt M, Lee TY, Bodmer D, Hoenselaar E, Hendriks-Cornelissen SJ, Tsui WY, Kong CK, Brunner HG, van Kessel AG, Yuen ST, van Krieken JH, Leung SY, Hoogerbrugge N (2009) Heritable somatic methylation and inactivation of MSH2 in families with Lynch syndrome due to deletion of the 3′ exons of TACSTD1. Nat Genet 41(1):112–117. doi:10.1038/ng.283
37. Mensenkamp AR, Vogelaar IP, van Zelst-Stams WA, Goossens M, Ouchene H, Hendriks-Cornelissen SJ, Kwint MP, Hoogerbrugge N, Nagtegaal ID, Ligtenberg MJ (2014) Somatic mutations in MLH1 and MSH2 are a frequent cause of mismatch-repair deficiency in lynch syndrome-like tumors. Gastroenterology 146(3):643–646 e648. doi:10.1053/j.gastro.2013.12.002
38. Jarvinen HJ, Aarnio M, Mustonen H, Aktan-Collan K, Aaltonen LA, Peltomaki P, De La Chapelle A, Mecklin JP (2000) Controlled 15-year trial on screening for colorectal cancer in families with hereditary nonpolyposis colorectal cancer. Gastroenterology 118(5):829–834
39. Jarvinen HJ, Renkonen-Sinisalo L, Aktan-Collan K, Peltomaki P, Aaltonen LA, Mecklin JP (2009) Ten years after mutation testing for Lynch syndrome: cancer incidence and outcome in mutation-positive and mutation-negative family members. J Clin Oncol 27(28):4793–4797. doi:10.1200/JCO.2009.23.7784

40. Dove-Edwin I, de Jong AE, Adams J, Mesher D, Lipton L, Sasieni P, Vasen HF, Thomas HJ (2006) Prospective results of surveillance colonoscopy in dominant familial colorectal cancer with and without Lynch syndrome. Gastroenterology 130(7):1995–2000. doi:10.1053/j.gastro.2006.03.018
41. Vasen HF, Moslein G, Alonso A, Bernstein I, Bertario L, Blanco I, Burn J, Capella G, Engel C, Frayling I, Friedl W, Hes FJ, Hodgson S, Mecklin JP, Moller P, Nagengast F, Parc Y, Renkonen-Sinisalo L, Sampson JR, Stormorken A, Wijnen J (2007) Guidelines for the clinical management of Lynch syndrome (hereditary non-polyposis cancer). J Med Genet 44(6):353–362. doi:10.1136/jmg.2007.048991
42. Evaluation of Genomic Applications in P, Prevention Working G (2009) Recommendations from the EGAPP Working Group: genetic testing strategies in newly diagnosed individuals with colorectal cancer aimed at reducing morbidity and mortality from Lynch syndrome in relatives. Genet Med 11(1):35–41. doi:10.1097/GIM.0b013e31818fa2ff
43. Hampel H, Frankel WL, Martin E, Arnold M, Khanduja K, Kuebler P, Nakagawa H, Sotamaa K, Prior TW, Westman J, Panescu J, Fix D, Lockman J, Comeras I, de la Chapelle A (2005) Screening for the Lynch syndrome (hereditary nonpolyposis colorectal cancer). New Engl J Med 352(18):1851–1860. doi:10.1056/NEJMoa043146
44. Boland CR, Thibodeau SN, Hamilton SR, Sidransky D, Eshleman JR, Burt RW, Meltzer SJ, Rodriguez-Bigas MA, Fodde R, Ranzani GN, Srivastava S (1998) A National Cancer Institute Workshop on Microsatellite Instability for cancer detection and familial predisposition: development of international criteria for the determination of microsatellite instability in colorectal cancer. Cancer Res 58(22):5248–5257
45. Dietmaier W, Wallinger S, Bocker T, Kullmann F, Fishel R, Ruschoff J (1997) Diagnostic microsatellite instability: definition and correlation with mismatch repair protein expression. Cancer Res 57(21):4749–4756
46. Veigl ML, Kasturi L, Olechnowicz J, Ma AH, Lutterbaugh JD, Periyasamy S, Li GM, Drummond J, Modrich PL, Sedwick WD, Markowitz SD (1998) Biallelic inactivation of hMLH1 by epigenetic gene silencing, a novel mechanism causing human MSI cancers. Proc Natl Acad Sci U S A 95(15):8698–8702
47. Kane MF, Loda M, Gaida GM, Lipman J, Mishra R, Goldman H, Jessup JM, Kolodner R (1997) Methylation of the hMLH1 promoter correlates with lack of expression of hMLH1 in sporadic colon tumors and mismatch repair-defective human tumor cell lines. Cancer Res 57(5):808–811
48. Haugen AC, Goel A, Yamada K, Marra G, Nguyen TP, Nagasaka T, Kanazawa S, Koike J, Kikuchi Y, Zhong X, Arita M, Shibuya K, Oshimura M, Hemmi H, Boland CR, Koi M (2008) Genetic instability caused by loss of MutS homologue 3 in human colorectal cancer. Cancer Res 68(20):8465–8472. doi:10.1158/0008-5472.CAN-08-0002
49. Stoffel EM, Syngal S (2005) Adenomas in young patients: what is the optimal evaluation? Am J Gastroenterol 100(5):1150–1153. doi:10.1111/j.1572-0241.2005.41967.x
50. Shia J (2008) Immunohistochemistry versus microsatellite instability testing for screening colorectal cancer patients at risk for hereditary nonpolyposis colorectal cancer syndrome. Part I. The utility of immunohistochemistry. J Mol Diagn 10(4):293–300. doi:10.2353/jmoldx.2008.080031
51. Capel E, Flejou JF, Hamelin R (2007) Assessment of MLH1 promoter methylation in relation to gene expression requires specific analysis. Oncogene 26(54):7596–7600. doi:10.1038/sj.onc.1210581
52. Rahner N, Friedrichs N, Steinke V, Aretz S, Friedl W, Buettner R, Mangold E, Propping P, Walldorf C (2008) Coexisting somatic promoter hypermethylation and pathogenic MLH1 germline mutation in Lynch syndrome. J Pathol 214(1):10–16. doi:10.1002/path.2263
53. Wijnen J, van der Klift H, Vasen H, Khan PM, Menko F, Tops C, Meijers Heijboer H, Lindhout D, Moller P, Fodde R (1998) MSH2 genomic deletions are a frequent cause of HNPCC. Nat Genet 20(4):326–328. doi:10.1038/3795
54. Kuiper RP, Vissers LE, Venkatachalam R, Bodmer D, Hoenselaar E, Goossens M, Haufe A, Kamping E, Niessen RC, Hogervorst FB, Gille JJ, Redeker B, Tops CM, van Gijn ME,

van den Ouweland AM, Rahner N, Steinke V, Kahl P, Holinski-Feder E, Morak M, Kloor M, Stemmler S, Betz B, Hutter P, Bunyan DJ, Syngal S, Culver JO, Graham T, Chan TL, Nagtegaal ID, van Krieken JH, Schackert HK, Hoogerbrugge N, van Kessel AG, Ligtenberg MJ (2011) Recurrence and variability of germline EPCAM deletions in Lynch syndrome. Hum Mutat 32(4):407–414. doi:10.1002/humu.21446

55. Nicolaides NC, Carter KC, Shell BK, Papadopoulos N, Vogelstein B, Kinzler KW (1995) Genomic organization of the human PMS2 gene family. Genomics 30(2):195–206. doi:10.1006/geno.1995.9885
56. Chao EC, Velasquez JL, Witherspoon MS, Rozek LS, Peel D, Ng P, Gruber SB, Watson P, Rennert G, Anton-Culver H, Lynch H, Lipkin SM (2008) Accurate classification of MLH1/MSH2 missense variants with multivariate analysis of protein polymorphisms-mismatch repair (MAPP-MMR). Hum Mutat 29(6):852–860. doi:10.1002/humu.20735
57. Jass JR (2007) Classification of colorectal cancer based on correlation of clinical, morphological and molecular features. Histopathology 50(1):113–130. doi:10.1111/j.1365-2559.2006.02549.x
58. Chun N, Ford JM (2012) Genetic testing by cancer site: stomach. Cancer J 18(4):355–363. doi:10.1097/PPO.0b013e31826246dc
59. Rodriguez-Bigas MA, Boland CR, Hamilton SR, Henson DE, Jass JR, Khan PM, Lynch H, Perucho M, Smyrk T, Sobin L, Srivastava S (1997) A National Cancer Institute Workshop on Hereditary Nonpolyposis Colorectal Cancer Syndrome: meeting highlights and Bethesda guidelines. J Natl Cancer Inst 89(23):1758–1762
60. Popat S, Hubner R, Houlston RS (2005) Systematic review of microsatellite instability and colorectal cancer prognosis. J Clin Oncol 23(3):609–618. doi:10.1200/JCO.2005.01.086
61. Sinicrope FA, Foster NR, Thibodeau SN, Marsoni S, Monges G, Labianca R, Kim GP, Yothers G, Allegra C, Moore MJ, Gallinger S, Sargent DJ (2011) DNA mismatch repair status and colon cancer recurrence and survival in clinical trials of 5-fluorouracil-based adjuvant therapy. J Natl Cancer Inst 103(11):863–875. doi:10.1093/jnci/djr153
62. Ribic CM, Sargent DJ, Moore MJ, Thibodeau SN, French AJ, Goldberg RM, Hamilton SR, Laurent-Puig P, Gryfe R, Shepherd LE, Tu D, Redston M, Gallinger S (2003) Tumor microsatellite-instability status as a predictor of benefit from fluorouracil-based adjuvant chemotherapy for colon cancer. New Engl J Med 349(3):247–257. doi:10.1056/NEJMoa022289
63. Sargent DJ, Marsoni S, Monges G, Thibodeau SN, Labianca R, Hamilton SR, French AJ, Kabat B, Foster NR, Torri V, Ribic C, Grothey A, Moore M, Zaniboni A, Seitz JF, Sinicrope F, Gallinger S (2010) Defective mismatch repair as a predictive marker for lack of efficacy of fluorouracil-based adjuvant therapy in colon cancer. J Clin Oncol 28(20):3219–3226. doi:10.1200/JCO.2009.27.1825
64. Elsaleh H, Iacopetta B (2001) Microsatellite instability is a predictive marker for survival benefit from adjuvant chemotherapy in a population-based series of stage III colorectal carcinoma. Clin Colorectal Cancer 1(2):104–109. doi:10.3816/CCC.2001.n.010
65. Hemminki A, Mecklin JP, Jarvinen H, Aaltonen LA, Joensuu H (2000) Microsatellite instability is a favorable prognostic indicator in patients with colorectal cancer receiving chemotherapy. Gastroenterology 119(4):921–928
66. Vasen HF, Wijnen JT, Menko FH, Kleibeuker JH, Taal BG, Griffioen G, Nagengast FM, Meijers-Heijboer EH, Bertario L, Varesco L, Bisgaard ML, Mohr J, Fodde R, Khan PM (1996) Cancer risk in families with hereditary nonpolyposis colorectal cancer diagnosed by mutation analysis. Gastroenterology 110(4):1020–1027
67. Schulmann K, Brasch FE, Kunstmann E, Engel C, Pagenstecher C, Vogelsang H, Kruger S, Vogel T, Knaebel HP, Ruschoff J, Hahn SA, Knebel-Doeberitz MV, Moeslein G, Meltzer SJ, Schackert HK, Tympner C, Mangold E, Schmiegel W, German HC (2005) HNPCC-associated small bowel cancer: clinical and molecular characteristics. Gastroenterology 128(3):590–599
68. Rodriguez-Bigas MA, Vasen HF, Lynch HT, Watson P, Myrhoj T, Jarvinen HJ, Mecklin JP, Macrae F, St John DJ, Bertario L, Fidalgo P, Madlensky L, Rozen P (1998) Characteristics of

small bowel carcinoma in hereditary nonpolyposis colorectal carcinoma. International Collaborative Group on HNPCC. Cancer 83(2):240–244
69. Potter DD, Murray JA, Donohue JH, Burgart LJ, Nagorney DM, van Heerden JA, Plevak MF, Zinsmeister AR, Thibodeau SN (2004) The role of defective mismatch repair in small bowel adenocarcinoma in celiac disease. Cancer Res 64(19):7073–7077. doi:10.1158/0008-5472.CAN-04-1096
70. Diosdado B, Buffart TE, Watkins R, Carvalho B, Ylstra B, Tijssen M, Bolijn AS, Lewis F, Maude K, Verbeke C, Nagtegaal ID, Grabsch H, Mulder CJ, Quirke P, Howdle P, Meijer GA (2010) High-resolution array comparative genomic hybridization in sporadic and celiac disease-related small bowel adenocarcinomas. Clin Cancer Res 16(5):1391–1401. doi:10.1158/1078-0432.CCR-09-1773
71. Bergmann F, Singh S, Michel S, Kahlert C, Schirmacher P, Helmke B, Von Knebel Doeberitz M, Kloor M, Blaker H (2010) Small bowel adenocarcinomas in celiac disease follow the CIM-MSI pathway. Oncol Rep 24(6):1535–1539
72. Chan RC, Katelaris PH, Stewart P, Lin BP (2006) Small bowel adenocarcinoma with high levels of microsatellite instability in Crohn's disease. Hum Pathol 37(5):631–634. doi:10.1016/j.humpath.2005.12.013
73. Brueckl WM, Heinze E, Milsmann C, Wein A, Koebnick C, Jung A, Croner RS, Brabletz T, Gunther K, Kirchner T, Hahn EG, Hohenberger W, Becker H, Reingruber B (2004) Prognostic significance of microsatellite instability in curatively resected adenocarcinoma of the small intestine. Cancer Lett 203(2):181–190
74. Gylling A, Abdel-Rahman WM, Juhola M, Nuorva K, Hautala E, Jarvinen HJ, Mecklin JP, Aarnio M, Peltomaki P (2007) Is gastric cancer part of the tumour spectrum of hereditary nonpolyposis colorectal cancer? A molecular genetic study. Gut 56(7):926–933. doi:10.1136/gut.2006.114876
75. Bacani J, Zwingerman R, Di Nicola N, Spencer S, Wegrynowski T, Mitchell K, Hay K, Redston M, Holowaty E, Huntsman D, Pollett A, Riddell R, Gallinger S (2005) Tumor microsatellite instability in early onset gastric cancer. J Mol Diagn 7(4):465–477. doi:10.1016/S1525-1578(10)60577-6
76. Halling KC, Harper J, Moskaluk CA, Thibodeau SN, Petroni GR, Yustein AS, Tosi P, Minacci C, Roviello F, Piva P, Hamilton SR, Jackson CE, Powell SM (1999) Origin of microsatellite instability in gastric cancer. Am J Pathol 155(1):205–211. doi:10.1016/S0002-9440(10)65114-0
77. Kim HS, Lee BL, Woo DK, Bae SI, Kim WH (2001) Assessment of markers for the identification of microsatellite instability phenotype in gastric neoplasms. Cancer Lett 164(1):61–68

Biology of Intestinal Epithelial Stem Cells

Noah F. Shroyer, Kristin Bell and Yuan-Hung Lo

Abstract The small and large intestines' central function is water, electrolyte, and nutrient digestion and absorption. Simultaneously, the intestines must provide a selective barrier to the noxious lumenal contents, and provide feedback to regulate digestive functions. To achieve these roles, the epithelium is composed of several distinct cell types with different capabilities, including absorptive enterocytes/colonocytes, microfold, goblet, enteroendocrine, tuft, and Paneth cells. These intestinal epithelial cells are relatively short-lived, and must be continuously replaced by intestinal stem cells (ISCs), which produce rapidly dividing progenitors, termed transit amplifying cells, that subsequently differentiate into the various distinct epithelial cells. In this chapter, we focus on the biology of ISCs. First, we will summarize our current knowledge of adult intestinal stem cell types and markers. Second, we will review cell specification and differentiation of each cell type in the adult intestinal epithelium. Third, we will examine the important signaling pathways (Wnt/β-catenin, Notch, BMP, Hedgehog, RTK/MAPK, and Hippo/YAP) involved in stem cell maintenance. Finally, we highlight examples where stem cells impact major diseases of the intestines.

Keywords Small intestine · Large intestine · Intestinal stem cells · Crypt base columnar cells · Label retaining cells · Cellular differentiation · Absorptive cells · Secretory cells · Wnt · Notch · Bone morphogenetic proteins · Hedgehog · Receptor tyrosine kinases · Hippo/Yes-associated protein

N. F. Shroyer (✉) · K. Bell · Y.-H. Lo
Department of Medicine, Division of Gastroenterology, Baylor College of Medicine, one Baylor Plaza, Alkek Building, Room N1317-08, Houston, TX 77030, USA.
e-mail: noah.shroyer@bcm.edu

N. F. Shroyer
Division of Developmental Biology, Cincinnati Childrens Hospital Medical Center, Cincinnati, OH, USA

N. F. Shroyer · K. Bell
Developmental Biology, University of Cincinnati, Cincinnati, OH, USA

N. F. Shroyer · Y.-H. Lo
Cancer and Cell Biology, University of Cincinnati, Cincinnati, OH, USA

V. W. Yang, A. B. Bialkowska (eds.), *Intestinal Tumorigenesis*,
DOI 10.1007/978-3-319-19986-3_3

1 Structure of the Adult Intestines

The small and large intestines' central function is water, electrolyte, and nutrient digestion and absorption [1]. Simultaneously, the intestines must provide a selective barrier to the noxious lumenal contents, and provide feedback to regulate digestive functions such as satiety and pancreatic function. To achieve these functions, the epithelium is composed of several distinct cell types with different functions, including absorptive enterocytes/colonocytes, microfold, goblet, enteroendocrine, tuft, and Paneth cells. These intestinal epithelial cells are relatively short-lived, and must be continuously replaced by intestinal stem cells (ISCs), which produce rapidly dividing progenitor cells, termed transient amplifying cells, that subsequently differentiate into the various distinct epithelial cells. ISCs are located at the base of crypts of Lieberkühn, pocket structures that project from the lumen into the wall of the intestine (Fig. 1). Daughter cells derived from ISCs migrate lumenally toward the mouth of the crypts as they divide, eventually exiting the cell cycle and terminally differentiating near the top of the crypts. In the colon, differentiated cells migrate out onto the flat surface where they reside for a few days before being shed into the lumen. In the small intestine, villi extend into the lumen and increase the functional surface area for digestion and absorption, with cells migrating up the villi prior to being shed from the villus tips.

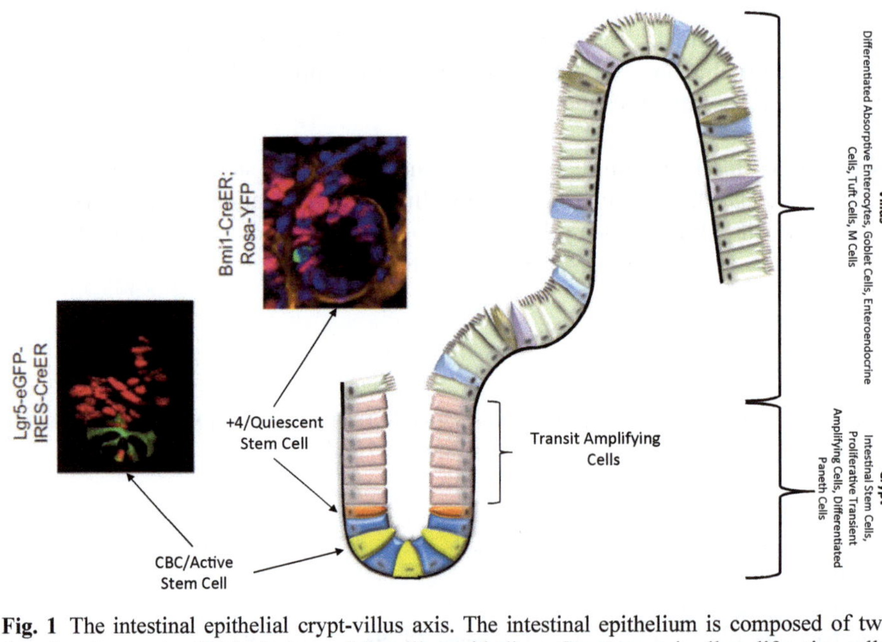

Fig. 1 The intestinal epithelial crypt-villus axis. The intestinal epithelium is composed of two domains: the crypts of Lieberkuhn, and the villus epithelium. Crypts contain all proliferating cells. Quiescent stem cells, here shown as a Bmi1 positive cell, are thought to be located at position + 4 in the crypt. Active stem cells, or crypt based columnar cells, here shown as Lgr5+ cells, are interspersed between Paneth cells at the crypt base. Transit amplifying progenitor cells are rapidly dividing progenitors that reside in the upper crypt. Images used with permission [2]

2 Stem Cells in the Intestinal Epithelium

The adult intestinal epithelium is one of the most proliferative organs in the body, necessitated by the rapid turnover of epithelial cells, which have average lifespans of about 5–7 days. In order to maintain epithelial integrity, continuous proliferation of stem/progenitor cells is required to produce all cell types. In this chapter, we focus on the biology of intestinal epithelial stem cells. First, we will summarize our current knowledge of adult intestinal stem cell types and markers. Second, we will review cell specification and differentiation of each cell type in the adult intestinal epithelium. Third, we will highlight the important signaling pathways involved in stem cell maintenance. Finally, we will highlight examples where stem cells impact major diseases of the intestines.

3 Locating the Intestinal Stem Cell (ISC)

Identification of intestinal stem cells with specific molecular markers is crucial to the study of the intestines. This knowledge provides the tools to examine effects on ISCs, to purify them for study, and to create animal models necessary to study the physiology of individual cell lineages, mechanisms of disease, and origins of cancer. The location and activity of ISCs has been the subject of intense study and debate for almost 50 years. An early paper in 1965, using 3H-thymidine incorporation into nascent DNA to study cell proliferation within the crypt, surmised that ISCs reside above the Paneth cells [3]. Nearly a decade later this idea was further expanded by Potten and colleagues who identified ISCs based on their "cryptogenic" capacity, that is, their ability to regenerate a crypt after injury. Potten and colleagues localized these cryptogenic cells at four cell positions above the base of the crypt (+4 cells), directly above the Paneth cell. This radiosensitive cell had a division rate that was about half that of its more proliferative progeny, and maintained DNA label retention, meaning that these ISCs retained the original template DNA following replication and passed on the "new" DNA to daughter cells. This ability was defined in the +4 cells by their capability to retain 3H-thymidine after irradiation and subsequent crypt regeneration. This label retention capacity was proposed to limit replication mutations that arise during the S-phase of the cell cycle, and therefore these label retaining cells (LRCs) were proposed to be guardians of the genome in the intestine [4–6]. Meanwhile, around the same time, Cheng and Leblond utilized radioactive labeled phagosomes as a rudimentary cell lineage tracing technique to identify the location of ISCs nestled between Paneth cells at the crypt base; they termed these putative ISCs "crypt base columnar cells" or CBCs [7]. Later, Cheng and Bjerknes would expand their stem cell hypothesis by suggesting that a stem cell remains multipotent provided it resides within the permissive environment, the stem cell zone or niche, located below position five, or the origin of differentiation [8]. Since the publication of these studies, many investigators have tried to reconcile the nature of these stem cells and their exact locations. Current views predict that the intestine may harbor a mixed pool of quiescent/+4 and active/CBC ISCs.

4 Marking the Intestinal Stem Cell

While the position of the ISC still remains a topic of interest, it is agreed that all cells in the intestinal epithelium arise from a common ISC, an idea known as the Unitarian Theory [7]. In an effort to refine the earliest lineage tracing studies that utilized marked phagosomes and radiation injury, elegantly engineered mouse models, typically using Cre/loxP genetic recombination technology, became a new way of studying the stem cell (reviewed in [9]). Arguably the most important, and most widely accepted, stem cell marker is Lgr5 (Leucine-rich repeat containing G-protein-coupled Receptor 5). Lgr5 is a component of the Wnt receptor complex; when liganded by R-spondin this amplifies Wnt signaling to enhance proliferation signals to the stem cells (See section below on Wnt signaling in ISCs). In 2007 Barker and colleagues found that Lgr5 is a specific marker for the CBCs originally described by Cheng and Leblond in 1974. Using Cre/loxP lineage tracing to mark Lgr5-expressing cells and their progeny, Barker et al. were able to visualize the placement of the CBCs between the Paneth cells, and follow its progeny, consisting of all intestinal epithelial cell types within the small intestine and colon [10]. Later it was found that these Lgr5-positive stem cells efficiently gave rise to tumors in the small bowel upon deletion of *Apc* [11] making Lgr5 an attractive marker for those studying cancer within the gastrointestinal tract. More recently, these cells were isolated and cultured in vitro. The resulting organoid (now termed "enteroid" to reflect their small intestinal origin [12]), consisted of a 3-dimensional cystic structure composed of interconnected crypt-like buds and all differentiated cell types, further confirming Lgr5 as a marker of ISCs [13, 14].

Lgr5 is the best known marker of actively cycling ISCs, or CBCs; however, its mRNA and protein expression levels are low and the transgenic reporter mice sometimes do not provide accurate representation of the CBCs. This created the need to locate other markers that can be used in conjunction with Lgr5 to further substantiate studies pertaining to the relevance and mechanisms of stem cell biology. A glycoprotein, Olfactomedin 4 (OLFM4), was found to be coexpressed with Lgr5 in the murine and human small intestine. In the human, expression extended into the colon, and was also found to be enriched in colonic adenocarcinoma, thus providing some evidence that this may be a putative marker for human cancer studies [15]. The mRNA of the basic helix-loop-helix transcription factor, Achaete scute-like 2 (Ascl2), was discovered in isolated Lgr5-expressing crypt cells. It was confirmed to regulate the expression of Lgr5 as conditional ablation of this gene caused a disappearance of the receptor and, ultimately, the Lgr5+ stem cells [16]. PW1/Peg3 (a gene with an unknown function) [17], Smoc2 (a BMP signaling inhibitor), Igfpb4 (an inhibitor of the Igf pathway), and Pcdh8 (protocadherin 8) [18, 19] have been identified as putative markers of the active stem cells (See Table 1). Other proteins such as Musashi-1,

Table 1 Enzymatic catalytic subunit of mouse telomerase

Gene name	Aliases	Name	Functional description	Active vs. Quiescent	Reference
Alcam	*CD166, MEMD*	Tctivated leukocyte cell adhesion molecule	Transmembrane glycoprotein	Active	20826154
Ascl2	*ASH2, HASH2, MASH2, bHLHa45*	Achaete-scute family bHLH transcription factor 2	Basic helix loop helix transcription factor	Active	19269367
Bmi1	*RP11-573G6.1, FLVI2/BMI1, PCGF4, RNF51*	Bmi1 polycomb ring finger oncogene	Polycomb transcription repressor complex	Quiescent	18536716, 22190486, 21927002
Dckl1	*RP11-113P14.1, CL1, CLICK1, DCAMKL1, DCDC3A, DCLK*	Doublecortin and CaM kinase-like-1	Microtubule-associated protein kinase	Active, Quiescent	18055444, 24487592
EphB2	*CAPB, DRT, EK5, EPHT3, ERK, Hek5, PCBC, Tyro5*	Ephrin type-B receptor 2	Ephrin receptor	Active	21419747
Hopx	*CAMEO, HOD, HOP, LAGY, NECC1, OB1, SMAP31, TOTO*	Homeodomain-only protein	An atypical homeobox protein	Quiescent	22075725
Igfbp4	*BP-4, HT29-IGFBP, IBP4, IGFBP-4*	Insulin-like growth factor binding protein 4	Inhibitor of the Igf pathway	Active	21419747
Itgb1	*RP11-479G22.2, CD29, FNRB, GPIIA, MDF2, MSK12, VLA-BETA, VLAB*	β1 integrin	Fibronectin receptor beta	Active	16285956
Lgr5	*FEX, GPR49, GPR67, GRP49, HG38*	Luciene-rich repeat containing G-protein-coupled receptor	R-spondin receptor	Active	17934449, 22473993
Lrig1	*LIG-1, LIG1*	Leucine-rich repeats and immunoglobulin-like domains protein 1	ErbB inhibitor	Active, Quiescent	22464327, 22388892
mTert	*CMM9, DKCA2, DKCB4, EST2, PFBMFT1, TCS1, TP2, TRT, hEST2, hTRT*	Mouse telomerase reverse transcriptase	Enzymatic catalytic subunit of mouse telomerase	Quiescent	21173232
Musashi-1	*MSI1*	Musashi RNA-binding protein 1	Eranslational repressor and regulator Notch signaling	Active	17122772

Table 1 (continued)

Gene name	Aliases	Name	Functional description	Active vs. Quiescent	Reference
Olfm4	*UNQ362/PRO698, GC1, GW112, OLM4, OlfD, UNQ362, bA209J19.1, hGC-1, hOLfD*	Olfactomedin 4	Glycoprotein	Active	19450592
Phlda1	*DT1P1B11, PHRIP, TDAG51*	Protocadherin8	Cadherin	Active, Quiescent	21558389
Prom1	*MSTP061, AC133, CD133, CORD12, MCDR2, PROML1, RP41, STGD4*	Prominin1	Glycoprotein	Active	19092805
Peg3	*hCG_1685807, PW1, ZKSCAN22, ZNF904, ZSCAN24*	Paternally expressed gene 3	Unknown	Active	21709251
Smoc2	*RP11-270C4__A.1, DTDP1, MST117, MSTP117, MSTP140, SMAP2, bA270C4A.1, bA37D8.1, dJ421D16.1*	SPARC-related modular calcium-binding protein 2	BMP signaling inhibitor	Active	21419747, 22692129
Sox9	*CMD1, CMPD1, SRA1*	SRY (sex determining region Y)-box 9	Transcription factor	Active	19228882
Tnfrsf19	*UNQ1888/PRO4333, TAJ, TAJ-alpha, TRADE, TROY*		Tumor necrosis factor receptor family member	Active	23142137

an enhancer of Notch signaling [20], Prom1/CD133 [21, 22], and the receptor EphB2 [18] have been found to be coexpressed with Lgr5 positive CBCs and intestinal tumors, but their expression extends further up the crypt into the proliferative transit amplifying zone thus rendering them ambiguous stem cell markers.

Studies on +4 position stem cell markers fueled the notion that there exists a stem cell hierarchy, with a quiescent or reserve stem cell population giving rise to a more active stem cell. These +4 cells were identified by their DNA label retention ability based on the hypothesis that these cells asymmetrically divide giving rise to one stem cell that will retain the original template DNA strand and one daughter cell that receives the newly synthesized DNA. Label retention was determined by durable labeling of cells marked by incorporation of BrdU or tritiated thymidine into DNA, as well as labeled chromatin marked by Histone-GFP fusion proteins [23, 24]. Additionally, these cells, unlike the CBCs previously mentioned, have been shown to be resistant to inactivation or perturbations of the cell cycle suggesting a more slowly dividing cell [2, 25]. Under high Wnt signaling, a strong driver of intestinal stem cell proliferation (see the section below), these cells possess the ability to maintain high cytoplasmic and transient nuclear expression of β-catenin due to its interaction with protein 14-3-3ζ after Akt phosphorylation [26]. These proteins were highly expressed in the +4 cells and may infer a mechanism for label retention under stringent regulation of canonical Wnt signaling. Bmi1, a component of the polycomb transcription repressor complex, was found to be expressed in +4 cells. These Bmi1 positive cells possessed the capability of producing all cell lineages within the intestine and forming adenomas in the context of aberrant canonical Wnt signaling. It was further found that ablation of Bmi1-expressing cells with diphtheria toxin lead to the loss of crypts [27]. Simultaneous lineage tracing experiments utilizing Lgr5-GFP and Bmi1-YPF indicated that Bmi1+ cells appeared to be functionally distinct from the active Lgr5+ CBC pool. Lgr5+ cells produced a distinct striping pattern as progeny containing GFP differentiated and migrated up the villi as a coherent stripe. In contrast, YFP labeled Bmi1+ cells failed to produce the same efficient striping pattern suggesting a difference in their proliferative profiles. Furthermore, only Lgr5+ cells were affected by manipulation of Wnt signaling, a pathway known to affect crypt maintenance and architecture (discussed below), as either increased or decreased signaling concomitantly resulted in increased or decreased Lgr5+ expansion. Bmi1+ cells remained unaffected by these perturbations. Irradiation treatment resulted in loss of Lgr5+ cells; however crypts were able to regenerate within a few days after treatment. In contrast Bmi1+ cells, post radiation, showed enhanced proliferation and their production of daughter cells leading to villus stripes expanded profoundly [2]. In fact, it was shown that the Bmi1+ cell population could sustain epithelial homeostasis in the absence of Lgr5 marked cells. Loss of Lgr5 for 10 consecutive days did not overtly affect tissue. Cellular production from Bmi1+ stem cells was enough to enable proper cellular proliferation and expansion [28]. This provides evidence of more than one stem cell population. In further support of this concept, Bmi1+ cells have been shown to give rise to the Lgr5+ cell population. After Lgr5 cells were ablated with diphtheria toxin for 6 days, lineage tracing indicated Bmi1+ cells were able to give rise to the active stem cell population 3 days later [28]. Similar outcomes were found in vitro, when Bmi1+ cells were isolated and allowed to grow in culture to examine their stem cell functionality. Not only did proper enteroids form from Bmi1+ cells, Lgr5+ cells were also found within these Bmi1+ cell derived enteroids in the CBC position [2]. These results imply the existence of a non-CBC pool of stem cells, marked by Bmi1, capable of regenerating lost epithelium in response to injury and replenishing the active stem cell pool.

While Bmi1 has been the subject of intensive recent study, its expression is limited to the proximal small intestine, prompting the investigation of other +4 stem cell markers. Hopx, an atypical homeodomain-containing protein, was found to mark the +4 position along the entire length of the small intestine, and co-localize with Bmi1. Furthermore, isolated Hopx cells remained quiescent in culture and, upon stimulation, produced Lgr5 and Olfm4 positive cells within cultured organoids [29]. mTert, or mouse telomerase reverse transcriptase, was found to mark a distinct population of slowly cycling, radiation resistant stem cells throughout the small intestine and colon that can give rise to Lgr5 positive cells [30]. The different expression patterns of Bmi1, Hopx, and mTert suggest the existence of some regional specificity in the nature of ISCs within the gastrointestinal tract.

A few other putative +4 ISC markers are ambiguous in their specificity: Dclk1, which was first identified to be enriched in ISCs at the crypt base [31], has been shown to mark tuft cells [32, 33]. Lrig1, an ErbB inhibitor, was reported to be a specific marker of ISCs [34], but also to have a broad pattern of expression throughout the proliferating cells in the crypt [35]. Despite the overlap in expression patterns, Lrig1-expressing cells exhibited some of the same attributes such as quiescence, injury induced proliferation, and production of Lgr5-positive ISCs. *Phlda1*, a gene thought to have an anti-proliferative effect, was found to be expressed in the CBCs of all crypts throughout the intestine; however, some crypts contained Phlda1 expression in the +4 stem cells [36]. It is of interest to note that the +4 cells discussed here, and the characteristics they possess, are not the same as Potten's original description of this cell. His kinetics studies indicated a cell cycle time of 24 h, appropriate for maintenance of high epithelial turnover [6] while the +4 cells discussed in this paragraph do not contribute to the homeostatic regeneration of tissue.

5 A Stem Cell Hierarchy

It has been proposed that tissues in need of a continuous cellular supply retain both active and quiescent stem cell pools to provide a defense mechanism against injury and mutation while simultaneously allowing for expansion and rapid turnover necessary to maintain organ homeostasis (reviewed by [37]). In the intestine, the active stem cell (CBC) is not a label retaining cell and is sensitive to both mutation and injury [38, 39]. In contrast, the quiescent stem cell (+4 ISC or LRC) may provide a "back-up" to render the intestine less vulnerable to damage as they are a means of replacement for the active stem cell pool, as previously mentioned. This concept of a hierarchy among intestinal stem cells originated at least as early as Potten's work as he and his colleagues suggested that multiple types of stem cells existed due to their variable resistances to radiation [40].

However, other theories exist about the nature of the quiescent stem cell and whether there is a stringent hierarchy wherein quiescent stem cells' primary function is to give rise to active stem cells in response to specific stimuli, or if quiescent stem cells have a separate function. Analysis of the mRNA and protein expression in Lgr5+ stem cells found many quiescent stem cell markers [19]. This information conflicts with the idea of a definitive +4 quiescent signature. Stem cells have been shown to revert back and forth in different contexts as described above for Bmi1-to-Lgr5 and Hopx-to-Lgr5 "conversion" [29, 2]. This ability to produce active stem cells does not rest solely with the quiescent and active stem cell populations as described above. Recent studies suggest that progenitors that are normally restricted to produce only non-proliferative progeny can be stimulated to produce stem cells in response to injury. This was shown for Dll1-expressing secretory progenitor cells (see below: "Absorptive versus secretory cell fate"), a daughter of the Lgr5 positive cells, which upon injury revert back to a stem cell fate to populate the depleted CBC pool and give rise to all cell types [41]. Another study shows that following injury, Paneth cells can begin to express Bmi1 and proliferate to promote wound repair [42]. A separate study examined the fate of quiescent stem cells based on their property of label retention. Utilizing YFP marked histones, they found that these label retaining cells were destined, under physiological conditions, to become Paneth or enteroendocrine cells rather than an active stem cell or transit amplifying cell. However, after injury, these specific quiescent, Lgr5-positive, label-retaining cells, could repopulate the crypt [43]. These data indicate that quiescent stem cells may not be a distinct reserve population of stem cells that give rise to a more active stem cells following injury, but instead may be long-lived crypt constituent cells (e.g., Paneth or enteroendocrine cells) that retain the capacity for reversion to active stem cells when necessary to repair the tissue after injury. Furthermore, these findings indicate, that rather than a hierarchical stem cell model, there exists a dynamic relationship with neighboring cells and signaling cues. These recent studies propose new ways of thinking about the stem cells dynamics and how they function under homeostatic conditions.

6 Progeny of Intestinal Stem Cells

6.1 Absorptive Versus Secretory Cell Fate

The proliferative ISCs give rise to the numerous transit amplifying cells, highly proliferative progenitor cells with limited renewal capacity. Under the influence of an assortment of molecular signals, these cells withdraw from the cell cycle and differentiate to become either absorptive enterocytes (colonocytes in the colon) or one of several secretory cell types that collectively constitute the functional intes-

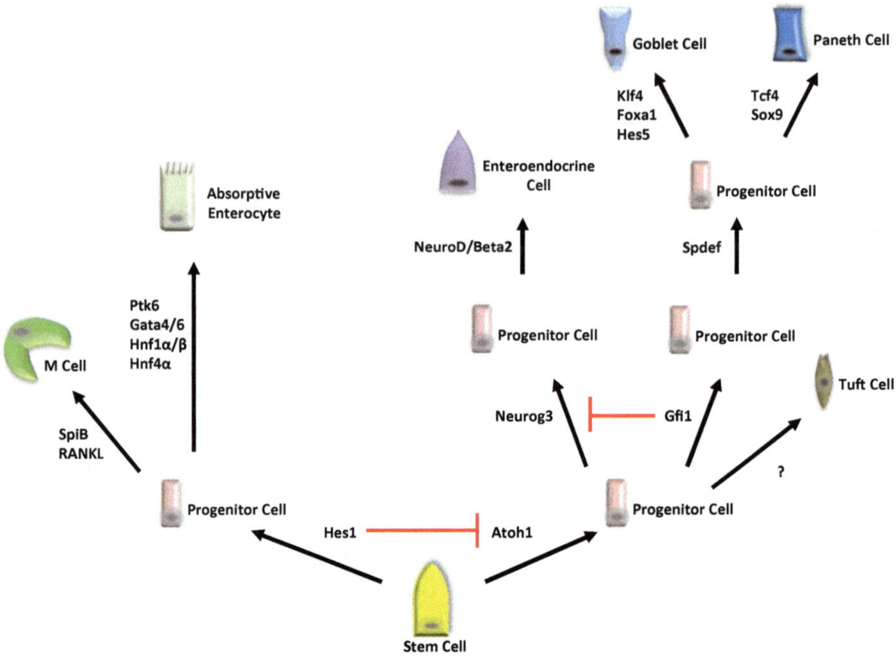

Fig. 2 Intestinal stem cells and their progeny. Stem cells divide to become an absorptive or secretory progenitor based on Atoh1 expression. Notch-directed Hes1 repression of Atoh1 results in an absorptive progenitor that will divide to become either an absorptive enterocyte or M cell. Expression of Ptk6, Gata4/6, Hnf1α/β, and Hnf4α have been shown to affect absorptive enterocyte differentiation. Both SpiB and RANKL are required for M cell differentiation. Atoh1 expression in stem cell progeny produces a secretory progenitor. Gfi1 repression of Neurog3 leads to a Goblet/Paneth cell progenitor that later expresses the Spdef transcription factor. Klf4, Foxa1, and Hes5 have been shown to be necessary requirements for goblet cell differentiation, while Tcf4 and Sox9 are involved in Paneth cell differentiation. Little information is known about tuft cell origin and differentiation, but Atoh1 seems to be a requirement

tinal epithelium (Fig. 2). Notch signaling (see Notch signaling, below) is critical for determining the types of cells produced by differentiating ISCs, directing cells toward either the absorptive of secretory fate by controlling the expression of *HES* genes (promoting absorptive fate) or Atoh1 (promoting secretory fate) (Reviewed in [44]). In addition to Notch regulation, the secretory versus absorptive fate decision can be influenced by other pathways such as Wnt/β-catenin. Atoh1 was found to be directly regulated and targeted for proteosomal degradation by GSK3β, a component of the canonical Wnt pathway that targets β-catenin for degradation. Inhibition of GSK3β led to stabilization of Atoh1 with consequent upregulation of the goblet cell mucin Muc2 [45, 46]. Mice with ectopic expression of Dickkopf1 (Dkk1), a secreted Wnt inhibitor, lacked Atoh1 expression, suggesting that canonical Wnt signaling may be necessary for *Atoh1* gene expression [47]. In addition to Wnt/GSK3β regulation of Atoh1, Wnt/β-catenin has been shown to regulate Hes1 expression [48]. Thus, Atoh1 and Hes1 are likely to be regulated by both Notch and

Wnt signaling pathways. Additional transcription factors such as Neurog3, Gfi1, Mtgr1, and Spdef are thought to function in Atoh1-specified secretory cell progenitors to determine which of several possible fates are adopted by these progenitor cells. Detailed reviews of the factors that specify individual intestinal cell fates are available [44, 49, 50, 51].

6.2 The Absorptive Enterocyte

The majority of cells in the small and large intestines are absorptive enterocytes/colonocytes. These cells are necessary for food digestion and nutrient absorption necessary to maintain overall health of an organism (reviewed in [52, 53]). Following Notch-directed Hes1 repression of Atoh1 to commit to the absorptive lineage (see Notch signaling, below), further molecular regulation is required for maturation of these cells to occur. Protein tyrosine kinase 6 (Ptk6), was shown to have no effect on the secretory cell population upon its depletion; however, an increase in proliferation and decreased expression of the enterocyte marker I-FABP suggesting a role in enterocyte differentiation [54]. Gata transcription factors play a role in regulation of regional identity and function of enterocytes. Gata4 is required to maintain the proximal identity of the small intestine including its function in cholesterol and fat absorption [55, 56]. Loss of Gata6 interfered with Notch signaling and altered expression of some absorptive genes involved in lipid metabolism [57]. Members of the hepatocyte nuclear family also seem to be involved in absorptive cell fate. Deletion of both *Hnf1α* and *Hnf1β* resulted in reduced water absorption and decreased mRNA expression of *DPPIV*, a brush border enzyme found in enterocytes, within the mouse intestine. Hnf1α was further found to directly regulate *Cdx2*, a gene involved in development and commitment of the intestinal epithelium [58]. Another member of the family, Hnf4α, was able to force intestinal cell differentiation in non-intestinal cell culture experiments. Furthermore, it was shown to increase secretory cell numbers upon deletion in mice [59, 60]. Bioinformatics analyses indicate that Hnf4α may play a role in binding to the enterocyte specific genes such as alkaline phosphatase, another enterocyte brush border enzyme [61, 62]. Despite being the most abundant cellular component of the intestinal epithelium, much remains to be learned about the molecular regulation of absorptive enterocytes.

6.2.1 M Cells

M cells, or Microfold cells, are specialized epithelial cells normally found in the follicle-associated epithelium overlying the Peyer's patches, but can also be found occasionally on the villi intermingled with differentiated intestinal cells. Their primary function is endocytosis and delivery of antigens from the gut lumen to immune cells [63, 64]. It is not clear how these M cells are differentiated, but it likely involves Notch signaling. Mice with a Delta3 ligand mutation produced more of an intermediate type M cell (characterized by their shape) rather than the fully mature M cells. Furthermore, these studies showed that contact with B-cells is responsible for maintaining the mature M cell phenotype [65]. In another study, RANKL (also known as TNF-related activation-induced cytokine) was shown to be required for M cell differentiation. Depletion of RANKL or interrupting interaction with its receptor, RANK, resulted in fewer intestinal M cells; rescue experiments with exogenous RANKL were able to restore the M cell population [66]. Most recently, the Ets transcription factor SpiB, was shown to be the mediator of M cell differentiation as mice lacking this transcription factor failed to form M cells. In vitro assays further defined the role of RANKL as a key inducer of SpiB [67, 68].

6.2.2 Cup Cells

Cup cells, first identified in rabbits, guinea pigs, and monkeys, comprise approximately 3–6 % of the cell population within the ilieum. While electron microscopy depicts a unique ultrastructure for these cells, they are difficult to distinguish from typical intestinal enterocytes [69]. They are defined by a short brush border, weak alkaline phosphatase staining, and vimentin expression, a protein not normally expressed in epithelial cells [70, 71]. Little is known about their function.

6.3 Secretory Cells

6.3.1 Enteroendocrine Cells

The enteroendocrine (EE) cells are hormone-producing cells within the intestine. With approximately 16 different types, these cells are responsible for regulating satiation, immunity, and inflammation. These actions have all been reviewed in depth [72]. Following Atoh1-mediated commitment of differentiating ISCs to the secretory lineage, EE differentiation is specified by Neurogenin3 (Neurog3), a transcriptional regulator that is localized to proliferating secretory progenitor cells [73, 74]. Lineage tracing of cells expressing Neurog3 found that these cells eventually become positive for ChromograninA (ChgA), a marker of mature EE cells. Furthermore, deletion of *Neurog3* resulted in EE cell deficiency [74], while transgenic expression lead to increased EE type hormones and a decrease in the number of goblet cells [75]. Downstream of Neurog3 is Beta2/NeuroD [74], a transcription factor found in mature, post-mitotic EE cells that is required for development of hormones, such as secretin [76]. Beta2 may activate Pax6 expression which is required for EE hormones Glp1 and 2 in the distal intestine and GIP in the duodenum [77–79]). Other genes important for maturation of EE cells include *Foxa1/a2*, *Insm1*, *Nkx2.2*, and *Arx* [80–83]. Both *Pdx1* and *Pax4* are regional-specific genes that control EE differentiation within the small intestine [78, 84, 85].

6.3.2 Goblet Cells

Goblet cells produce mucous and additional proteins required for physical protection of the epithelium and lubrication of the luminal surface to facilitate fecal expulsion from the gut (reviewed in [86]). Differentiation of these cells from Atoh1-specified secretory progenitors relies on expression of Gfi1, a negative regulator of Neurog3. Depletion of this zinc-finger transcriptional repressor results in intestines with fewer Paneth and goblet cells. Additionally, loss of Gfi1 leads to increased expression of Neurog3 and a subsequent increase in EE cells [87, 88]. Downstream of Atoh1 and Gfi1, Spdef, an Ets-transcription factor, is also required for the differentiation of goblet cells. Transgenic expression of this gene results in increased goblet cell number, while depletion in vitro was shown to have the opposite effect [89]. While *Spdef* null mice did retain expression of the goblet cell marker Tff3, these cells were not fully mature as visualized through poor PAS staining [90]. Notch signaling may also play a role in goblet cell differentiation through Hes5. When Notch signaling was increased in mice by overexpression of NICD in post mitotic cells, an increase of Hes5 and goblet cells were found specifically above the crypt region [91]. The placement of this regulation in the post mitotic cells suggests that Notch signaling exerts its effects on specified secretory progenitors (i.e. after expression of Atoh1). Further evidence of Notch effects on goblet cell differentiation were found when increased Hes5 expression was found in mice lacking Lkb1 kinase activity, which also have increased Muc2, a well-known goblet cell marker, levels [92]. Other goblet cell differentiation protein include Foxa1 that binds to the goblet cell specific *Muc2* promoter [83]; and *Klf4 gene*, which when mutated caused a goblet cell deficiency in the embryonic colon [93], and decreased acidic mucin production in the adult small intestine [94].

6.3.3 Paneth Cells

Unlike the other secretory cells, Paneth cells reside in the crypt bottom of the small intestine only and have a longer turnover time of about 30 days. Their granulocytic products include lysozyme and defensins which provide antimicrobial defense for the intestine, further reviewed by Clevers, 2013 [95]. These cells also provide a source of Wnt3a, EGF, TGFα, and Dll4 required for stem cell maintenance [96]. Additionally, mTOR expression enables them to be nutritional sensors that can augment stem cell renewal dependent upon calorie intake [97]. Similar to goblet cells, Paneth cell secretory progenitor fate arises through expression of Atoh1, Gfi1, and Spdef [88, 90]. Subsequently, Wnt signaling through Frizzled is then required for differentiation of the Paneth cell lineage. While conditional loss of APC induced proliferation and decreased villus cell differentiation, it resulted in accumulation of β-catenin and Tcf4 which also led to increased expression of Paneth cell markers lysozyme and cryptdin. It was then shown that Tcf4 mediates the transcriptional control of the cryptdin genes [98, 99]. Furthermore, Sox9, transcriptional target of Wnt signaling, was shown to be a requirement for Paneth cell differentiation [100, 101].

6.3.4 Tuft Cells

Tuft cells, (reviewed by [102]) also known as brush cells or caveolated cells, represent a lesser known group of secretory cells that were originally discovered in 1956 [103, 104]. Thought to play a role in the taste receptor pathway and opioid secretion, their function within the intestine remains a mystery [105, 106]. While there is limited knowledge of these special cells, they are known to express DCAMKL1 [32]. It was also found that Atoh1 is required for their differentiation, while Ngn3, Sox9, Gfi1, and Spdef proved to be dispensable in this process. These cells were also present in mouse and human adenomas suggesting they may play a role in tumorigeneis [107]. Later studies indicate Atoh1 is not required for the adult tuft cell. In fact, depletion of Atoh1 led to an increase in this lineage [108]. Further studies are needed to elucidate the mechanism of tuft cell differentiation and function.

7 Signaling Pathways That Regulate Intestinal Stem Cells

7.1 *Wnt/β-catenin Signaling is the Major Driver of ISC Activity*

The Wnt signaling pathway plays a pivotal role in the proliferation, stem cell maintenance, differentiation and oncogenesis of the intestinal epithelium [109–112]. Wnt ligands and receptors are highly evolutionarily conserved proteins. In humans, there are nineteen cysteine-rich Wnt ligands and more than fifteen different Wnt receptors and/or co-receptors in seven protein families [113, 114]. Interaction between Wnt ligands with their Wnt receptor complexes triggers an activation of several intracellular signaling cascades, generally classified into the canonical Wnt/β-catenin pathway and the non-canonical Wnt pathway, which is independent of β-catenin signaling. In this section, we will focus on the impact of canonical Wnt/β-catenin signaling pathway due to its central role in supporting the proliferation, maintenance and differentiation of adult ISCs. Excellent reviews cover other aspects of the Wnt/β-catenin-independent signaling pathway, including Wnt/Planar Cell Polarity (PCP), Wnt/protein kinase C (PKC)-dependent pathways, Wnt/JNK and Wnt/calcium [115, 116].

7.1.1 Mechanism of Canonical Wnt/β-catenin Signaling

In the absence of Wnt ligands, cytoplasmic β-catenin is captured by the destruction complex, which is composed of Axin, adenomatous polyposis coli (APC), glycogen synthase kinase 3β (GSK3β) and casein kinase 1 (CK1). In the destruction complex, Axin and APC function as scaffolds and interact with β-catenin directly [112]. Sequential phosphorylation of β-catenin by CK1 and GSK3β at Ser45, Thr41, Ser37,

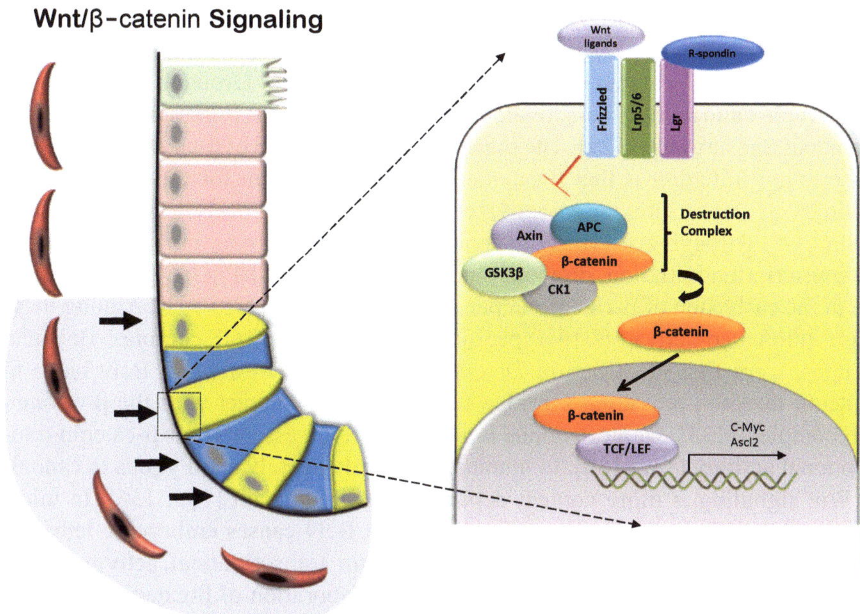

Fig. 3 Wnt/β-catenin signaling. Binding of Wnt ligands to Frizzled/Lrp receptor complexes triggers downstream Wnt/β-catenin signaling. The R-spondin/Lgr complexes can associate with the Wnt/Frizzled/Lrp-receptor complex to amplify Wnt/β-catenin signaling. Transduction of Wnt/β-catenin signaling results in dissociation of β-catenin from the destruction complex. Accumulated cytoplasmic β-catenin then translocates to the nucleus. In the nucleus, β-catenin interacts with TCF/LEF transcription factors and activates its target genes, such as c-Myc and Ascl2, which are critical for ISC maintenance

and Ser33 triggers the degradation of β-catenin through a proteasome-dependent mechanism [117–120]. Ser33 and Ser37 phosphorylated β-catenin is recognized by β-TrCP, a component of E3 ubiquitin ligase complex [117]. As a consequence, ubiquitinated β-catenin is degraded rapidly by the 26S proteasome [121]. Thus, the destruction complex plays a critical role in the stability of β-catenin. Activation of Wnt/β-catenin signaling is triggered by the binding of Wnt ligands with Frizzled receptors and low-density lipoprotein receptor-related protein (Lrp) co-receptors. Upon the activation of Wnt signaling, earlier studies suggested that β-catenin is released from the destruction complex, resulting in disassembly of the destruction complex [122]. However, recent studies suggested that the complex does not disassemble upon Wnt signaling. In this model, Wnt signaling inhibits β-catenin ubiquitination within the destruction complex, resulting the saturation of the destruction complex by phosphorylated β-catenin and thereby newly synthesized β-catenin could accumulate [123]. In both models, β-catenin accumulates in the cytoplasm followed by translocation into the nucleus through an unknown mechanism (Fig. 3).

In the nucleus, β-catenin interacts with TCF/LEF family transcription factors, including TCF1, TCF3, TCF4 and LEF1 in mammals, which guide β-catenin to specific DNA loci (reviewed in [124]. Without β-catenin, TCF/LEF proteins function as transcriptional co-repressors by interacting with Groucho/TLE co-repressors [125–128]. When β-catenin enters the nucleus, it displaces Groucho/TLE from the TCF complex and recruits co-activators, thereby inducing the initiation of β-catenin-dependent transcription [129]. The most fundamental model of β-catenin-dependent transcription initiation is that β-catenin provides a platform for the sequential recruitment of transcriptional co-activators, such as histone acetyltransferase CBP/p300 [129]. Through acetylating histones and opening chromatin structures to allow transcription initiation, previous studies showed that CBP/p300 plays a critical role in the early step of β-catenin-dependent transcription [130, 131]. Among these transcriptional co-activators, two specific for the β-catenin/TCF complex, Bcl9 and Pygopus, were first identified in *Drosophila* [132]. In *Drosophila*, Bcl9 binds to β-catenin directly, providing a bridge for Pygopus to interact with the β-catenin/TCF complex [133], while Pygopus is suggested to be essential for β-catenin transcriptional activity [132, 134]. In mammals, the role of Bcl9 and Pygous in canonical Wnt signaling is more context-dependent and redundant [135, 136]. In mice, disrupting the interaction between β-catenin and Bcl9 causes embryonic lethality, accompanied by the downregulation of β-catenin transcriptional activity [137]. Reviews summarizing more details about the collaboration of β-catenin and other co-factors are available [129, 138].

7.1.2 Expression Pattern of Wnt Signaling Components in the Adult Intestine

Canonical Wnt/β-catenin signaling is confined to the crypt epithelial cells, with prominent nuclear β-catenin apparent in Paneth cells. The expression patterns of other Wnt signaling components in the adult intestine have been examined by using specific RNA probes [139]. High expression of Wnt-3, Wnt-6, Wnt-9b, Frizzled 4, Frizzled 6, Frizzled 7, and low-density lipoprotein receptor-related protein 5 (Lrp5) were observed in the intestinal crypt. Other Wnts and Frizzleds were detected in differentiated epithelial and mesenchymal cells. Previous studies showed Paneth cells provide secreted Wnt ligands, which are crucially required for the maintenance of ISCs [96], although subsequent reports indicate additional sources of Wnt ligands can support the stem cell niche in vivo [140, 141]. Besides Paneth cells, stromal myofibroblasts were suggested to induce Wnt/β-catenin signaling by secreting hepatocyte growth factor (HGF), suggesting additional regulation of Wnt/β-catenin signaling by extrinsic cues provided by the microenvironment of ISCs [142].

7.1.3 Functions and Phenotypes Associated with Canonical Wnt/β-catenin Signaling in Intestinal Epithelium

Activation of canonical Wnt/β-catenin signaling in adult intestinal epithelium by deletion of *APC* induced a significant expansion of the crypt compartment and cell proliferation associated with a "crypt progenitor-like" phenotype, as well as aberrant cell migration along the crypt-villus axis [98, 143]. In addition, loss of APC inhibited terminal differentiation of the goblet, enteroendocrine and enterocyte lineages, while simultaneously driving expression of Paneth cell markers [98, 143]. Similarly, an earlier study showed that forced expression of gain-of-function N-terminal truncated β-catenin in intestinal epithelium stimulates cell proliferation restricted to crypt regions [144]. On the other hand, previous research showed that the proliferating compartments are completely lost between crypts regions and villus in neonatal *Tcf4* null mice, causing these mice to die quickly after birth [145, 146]. In adult small intestines, Tcf4 is expressed throughout the entire crypt-villus axis, while in adult colons, Tcf4 is expressed higher in the non-cycling cells in the upper crypts, but lower at the base of the crypts [139, 145]. In contrast with Tcf4, Tcf1 and Tcf3 are expressed in the proliferative zone of small intestinal crypt and colonic crypt, respectively [139]. Moreover, *Lef1* transcripts are not detectable in normal intestinal epithelium [139]. Conditional knockout of TCFs in adult intestinal epithelium indicated that Tcf4, but not Tcf1 or Tcf3, plays an essential role in maintaining intestinal stem cell homeostasis [147]. However, whether Tcf4 directly mediates intestinal differentiation and whether TCF/LEF members play distinct functions in the adult intestine remain unclear. Together, these studies provided solid evidence for the importance of canonical Wnt signaling in intestinal stem cell maintenance, proliferation and differentiation.

7.1.4 R-spondin/Lgr5 and Its Signaling

Of note, the ISC marker Lgr5 is also a target of Wnt/β-catenin signaling, and is thought to function as a co-activator of the Fzd/Lrp Wnt receptor complex [148, 149]. In human colon cancer cell lines, inhibition of WNT signaling activity results in a decrease of Lgr5 expression [149]. *Lgr5* null mice die at birth characterized by gastrointestinal distension [150], and also display premature Paneth cell differentiation and concomitant upregulation of Wnt signaling in developing intestine, suggesting a possible negative feedback loop on Wnt signaling through Lgr5 [151]. In contrast, conditional deletion of *Lgr5* in the adult intestinal epithelium had no significant phenotype [152]. Interestingly, deletion of *Lgr4*, a close homolog *Lgr5*, from the adult intestine results in loss of cell proliferation and Paneth cell differentiation, without affecting stem cell markers and Wnt target gene expression [152, 153]. In the same study, simultaneous deletion of both *Lgr5* and *Lgr4* displayed the most significant phenotype: a dramatic downregulation of canonical Wnt target genes and completely disappearance of intestinal crypts. These results suggested an essential, but redundant function of the homologues Lgr5 and Lgr4 in maintaining canonical Wnt signaling in the stem cells and progenitor cells. Recent biochemical studies linking Lgr proteins to canonical Wnt signaling have shed light on the mechanisms underlying these phenotypes.

In 2011, R-spondins were first identified as endogenous ligands of Lgr5 homologous receptors (Lgr4-6) [152, 154, 155]. Overexpression of human R-spondin in mouse intestine induced rapid crypt epithelial hyperplasia and a significant expansion of proliferating cells [156]. In vitro, R-spondin is required for the growth of single Lgr5+ active ISC-driven organoids (enteroids), which contain an intestinal architecture and self-renewing capacity that are remarkably similar to physiological crypt-villus units [14]. The R-spondin/Lgr complexes can associate with the Wnt/Frizzled/Lrp-receptor complex to enhance canonical Wnt/β-catenin signaling, suggesting the crucial role of Lrg5 homologous receptors in regulating Wnt-dependent intestinal stem cell homeostasis. Notably, although Lgr receptors are homologous to the glycoprotein hormone receptor class of G protein-coupled receptors (GPCRs) [157–159], Wnt-related activity of Lgr proteins was found to be independent of G-protein coupling, suggesting that Lgr receptors mediate R-spondin signals through a unique mechanism [154, 160]. Furthermore, the Lrg5/Wnt-coreceptors super-complexes were suggested to be internalized through a clathrin-dependent mechanism to mediate Wnt/β-catenin signaling [161]. R-spondin/Lgr complexes have been shown to associate with the transmembrane E3 ubiquitin ligase ZNRF3 or its functional homolog, RNF43 [162, 163]. Both ZNRF3 and RNF43 were identified as target genes of Wnt/β-catenin signaling specifically enriched in Lgr5+ active ISCs [16, 164]. Simultaneous deletion of *Znrf3* and *Rnf43* in the intestine induced strong activation of Wnt/β-catenin target gene expression and increased both progenitors and Paneth cells [163]. Mechanistically, *Znrf3* and *Rnf43* were identified to be negative regulators of Wnt signaling by selectively ubiquitinating Frizzled/Lrp-receptor complex, resulting in Wnt receptor degradation on the cell surface [162, 163]. R-spondin/Lgr complexes can inhibit Znrf3/Rnf43 activities through inducing endocytosis of Znrf3/Rnf43, resulting in increase level of Frizzled/Lrp for Wnt signaling [162]. The X-ray structures of R-spondin bound to the ectodomains of Lgr4-6 and RNF43 were reported recently, providing structure insights into the role of R-spondin/Lgr4-6/RNF43 complex in Wnt signaling pathway [165–167].

7.1.5 Important Wnt/β-catenin Targets in the Adult Intestine Crypts

More and more target genes of the Wnt/β-catenin signaling have been identified during the last few years. Among these genes, some have been shown to be required for appropriate differentiation of the intestinal epithelium. For example, the HMG-box transcription factor SOX9 is expressed in the epithelial cells of the intestinal crypts and is a target of Wnt/β-catenin signaling [168]. Conditional inactivation of *Sox9* in the intestinal epithelium results in disappearance of Paneth cells, fewer goblet cells and increased proliferating cells throughout the crypts, [100, 101]. Using *Sox9*^{EGFP} reporter mice, previous studies showed that *Sox9*^{EGFPlo} cells, but not *Sox9*^{EGFPhi} cells, exhibit ISC properties [169–171]. The expression of transcription factor achaete scute-like 2 (*Ascl2*), one of the genes within Lgr5^{+} active ISCs signature, is also Wnt/β-catenin-dependent [143, 164, 172]. Gain-of-function and loss-of-function studies suggested that Ascl2 plays a central role in controlling Lgr5^{+} active ISCs [16]. Telomerase activity, which controls telomere length, has been reported to be regulated by Wnt/β-catenin signaling in the intestine [173]. In this study, the binding of β-catenin to the *Tert* promoter was detected only in crypt but not villus nuclei. $Lgr5^{+}$ active ISCs containing high telomerase activity have also been reported [39]. This link between β-catenin activity in Lgr5+ ISCs and Tert expression somewhat confounds the finding that the *Tert* promoter marks quiescent ISCs; this quandary remains unresolved. EphB2/EphB3 tyrosine kinase receptors, which are positively controlled by Wnt/β-catenin signaling, are highly expressed at the bottom of crypts [149, 174]. Conversely, the expression of ephrin-B1 and ephrin-B2 ligands of EphB receptors are negatively regulated by Wnt/β-catenin signaling [174]. In *EphB2/EphB3* double-knockout mice, the boundary between the proliferative and the differentiated cells was observably disrupted, implicating EphB/ephrin signaling in controlling cell position in the crypt-villus axis [174]. *EphB3* null mice, but not *EphB2* null mice, exhibited mislocalized Paneth cells which had migrated onto the villus [174]. Interestingly, *Frizzled-5* mutant mice displayed similar mislocalization of Paneth cells, providing a genetic link between EphB3-mediated crypt-villus polarity and Frizzled-5-dependent Wnt/β-catenin signaling [99]. A subsequent study showed that EphB/ephrin signaling, primarily mediated by EphB2, is also necessary for maintenance of crypt cell proliferation, suggesting an important role in supporting both ISC renewal and the stem cell niche [175]. Further studies showed that high expression of EphB2 characterizes both human and mouse ISCs, providing a cell surface molecule which can allow affinity purification of ISCs without genetic manipulation [18, 176]. The proto-oncogene c-MYC was identified as a target of Wnt/β-catenin signaling both in colorectal cancer cells and in intestinal crypts [177, 143]. Overexpression of c-MYC in intestinal epithelium is sufficient to induce cell proliferation and apoptosis, which is comparable to activation of Wnt/β-catenin signaling through *Apc* knockout [178]. Conversely, conditional deletion of *c-MYC* specifically in the adult intestinal epithelium led to rapid disappearance of c-MYC deficient crypts and replacement by c-MYC expressing crypts through crypt fission [179]. More importantly, *c-MYC* deletion completely rescued the progenitor cell expansion and loss of differentiation upon deletion of *Apc* in the adult intestine, indicating that c-MYC functions as a critical mediator of Wnt/β-catenin signaling [180]. Previous studies demonstrated that Cyclin D1 is induced by Wnt/β-catenin signaling through β-catenin/LEF1 binding to the *Cyclin D1* promoter [181, 182]. Mutation of *Cyclin D1* or *Cyclin D2* conferred protection from hyperproliferation and tumor formation upon APC loss, suggesting that CyclinD/CDK4 activity is essential for β-catenin mediated hyperproliferation and transformation [183, 184]. Taken together, these results support the important role of Wnt/β-catenin signaling in maintaining adult ISCs homeostasis.

7.2 *Notch Signaling Controls the Fate of ISCs*

Notch signaling has a critical function controlling the fate of ISCs. Mammals have four Notch receptors (NOTCH 1-4) and five Notch ligands, which are divided into two subgroups: Delta-like (DLL1, DLL3, and DLL4) and Serrate-like (JAGGED1

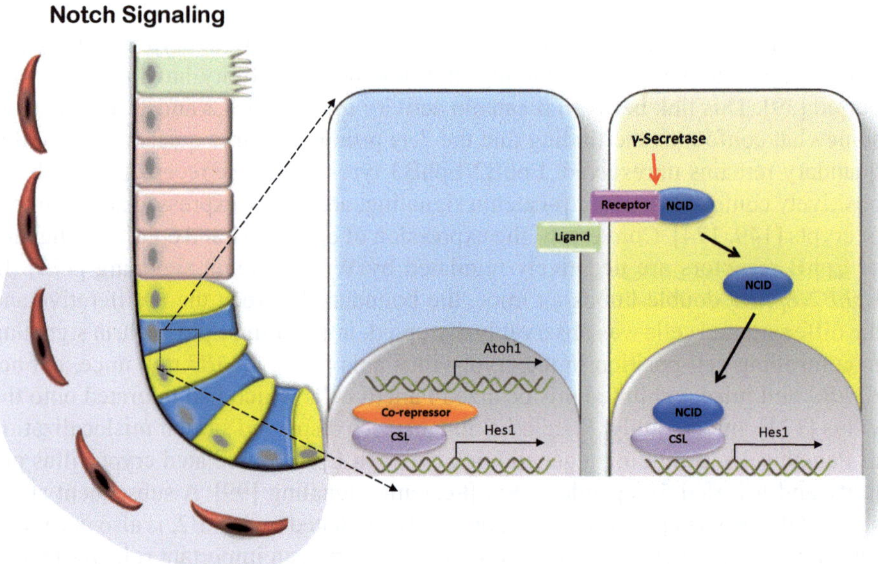

Fig. 4 Notch signaling. Canonical Notch signaling is dependent on direct cell-cell contact. Notch ligands and receptors are transmembrane proteins. Upon binding, γ-secretase cleaves the Notch receptor in its transmembrane domain, releasing the Notch intracellular domain (*NICD*), which subsequently translocates into the nucleus. In the nucleus, NICD binds to the CSL complex, displaces transcriptional co-repressors, and recruits transcriptional co-activator, resulting in the activation of Notch target genes such as Hes1

and JAGGED2) [44, 185]. Canonical Notch signaling is dependent on direct cell-cell contact. Upon binding of the transmembrane Notch ligand to the transmembrane Notch receptor expressed on adjacent cells, a conformational change of Notch receptor triggers unfolding of its juxtamembrane domain and allows an ADAM metalloprotease to cleave and release the extracellular domain of Notch receptor. Subsequently, γ-secretase cleaves Notch receptor in its transmembrane domain, resulting in the release of the Notch intracellular domain (NICD) which subsequently translocates into the nucleus. In the nucleus, NICD forms a complex with the DNA-binding transcription factor CSL (Rbp-Jκ in mice), displacing co-repressor Groucho proteins and recruiting co-activators including master-mind-like (MAML). Thus, ligand binding to Notch results in NICD converting the CSL-complex from a transcriptional repressor into a transcriptional activator of Notch target genes such as members of the HES and HEY family (Fig. 4). More details of canonical Notch signaling are referred in the following excellent reviews [44, 51, 186, 187]. In this section, we will focus on the importance of canonical Notch signaling pathway in cell fate decisions in ISCs.

7.2.1 Expression Pattern of Notch Signaling in the Adult Intestine

The expression pattern of Notch receptors, ligands, and canonical targets in the embryonic and adult intestinal epithelium has been reported [188, 189]. Notch1 is prominently expressed throughout the intestinal crypts, as well as at lower levels in the differentiated villus, muscle, mesenchyme and mesothelium [189, 190]. In contrast, Notch2 is only detected diffusely within the crypts of small intestine and in the muscle [189, 190]. Notch3 and Notch4 are restricted to the mesenchyme, endothelium and the villus [189, 190]. Notch ligands Dll1, Dll4, Jagged1 and Jagged2 are expressed in the crypts [189, 191, 192]. The expression of Jagged1 and Jagged2 shows extensive overlap with the expression pattern of Notch1 [188]. Of note, the expression of Dll1 and Dll4 have also been localized to intestinal secretory cells, such as Paneth cells in the small intestine and cKit$^+$ crypt base goblet cells in the colon, adjacent to ISCs. Other components of Notch pathway, such as Hes genes, are also expressed in the epithelium of the crypt [189]. Taken together, these expression studies suggested that Notch signaling is active in the ISCs and crypt progenitors.

7.2.2 Genetic and Pharmacological Evidence of Notch Function in ISCs

Transgenic mouse models provide genetic evidence that Notch signaling as an important player in the self-renewal of ISCs. Inhibition of canonical Notch signaling by conditional deletion of *Rbp-Jκ*, pharmacological inhibition of Notch activation using γ-secretase inhibitors, or antibodies that block the Notch receptors, resulted in profound loss of proliferative stem/progenitor cells, premature differentiation, and secretory cell hyperplasia [193–196]. Conditional deletion of *Notch1* or *Notch2* alone does not lead to significant intestinal phenotype [190, 197]. However, combinatorial deletion of *Notch1* and *Notch2* resulted in loss of crypt progenitor proliferation and secretory cell hyperplasia, nearly identical to the effect of global Notch inhibition, indicating that Notch1 and Notch2 are key receptors maintaining cell proliferation and terminal differentiation in the adult intestine [190]. Consistent with these observations, activation of Notch signaling by overexpression of NICD results in increased cell proliferation in the intestinal crypts as well as a decrease of secretory lineages, including goblet cells, Paneth cells and enteroendocrine cells [198, 199]. These results show that hyperactivation of Notch signaling perturbs intestinal homeostasis and may cause uncontrolled cell proliferation and contribute to tumorigenesis. Partial functional redundancy was also observed in Notch ligands Dll1 and Dll4. Deletion of *Dll1* in intestinal epithelium, but not *Dll4* or *Jagged1*, resulted in a moderate increase in number of goblet cells without significant change to cell proliferation in the crypts [200]. However, simultaneous deletion of *Dll1* and *Dll4*, but not *Jagged1*, caused the complete conversion of proliferating progenitors into postmitotic secretory cells concomitant with loss of the expression of active ISCs markers Lgr5, Ascl2 and Olmf4 [200]. These results indicate that Dll1 and Dll4 are the key Notch ligands required to maintain ISCs homeostasis.

Posttranslational modifications of Notch receptors, such as glycosylation, are essential for Notch activity [201]. Deletion of O-fucosyl transferase1 (*Pofut1*), which is important for Notch glycosylation and protein folding, was embryonic lethal and phenocopied the Rbp-Jκ mutant phenotype [202]. Moreover, consistent with Notch loss in the intestinal epithelium, mice with a disrupted fucose synthesis pathway exhibited reduced cell proliferation and secretory cell hyperplasia similar to Notch loss of function [203]. Similarly, conditional knockout of *Pofut1* in the intestinal epithelium results in reduced cell proliferation and increased production of secretory lineages and resembles mice intestines lack Rbp-Jκ or both Notch1 and Notch2 [204]. Endocytosis of Notch ligands mediated by Mind bomb1 (Mib1), an E3 ubiquitin ligase, is required for the activation of Notch signaling in mammals [205, 206]. Specific deletion of *Mib1* in the intestinal epithelium results in a Notch mutant phenotype as well as the mislocalization of Paneth cells [207]. Moreover, these phenotypes could be restored by overexpression of constitutively active Notch1 [207]. In addition, studies of Numb and Musashi1 indicated these components could modulate Notch activity in the intestine. For example, deletion of *Numb*, a negative regulator of Notch signaling, in human intestinal epithelium cell lines led to loss of goblet cell phenotype [208]. On the other hand, Musashi1, an RNA-binding protein, enhances Notch signaling through translational repression of *Numb* mRNA [209, 210].

Hes proteins, basic helix-loop-helix transcriptional repressors, are direct targets of NICD and expressed in the crypt base [211]. *Hes1* null mice die at birth with a modest increase in goblet cell and all endocrine lineages, but fewer intestinal enterocytes [211]. Conditional deletion of *Hes1*, *Hes3* and *Hes5*, but not *Hes1* alone, results in reduced cell proliferation, increase differentiation of secretory lineages as well as altered intestinal architecture in adult mice, suggesting other Hes family members may partially compensate for loss of Hes1 [212]. However, the Lgr5$^+$ ISCs are still maintained in both *Hes1* and *Hes1/Hes3/Hes5* mutants, indicating that these *Hes* genes are important mediators of Notch signaling, but not entirely sufficient for self-renewal of Notch-dependent ISCs [212]. The expression of Olfm4, a specific marker of active ISCs, is directly regulated by Notch signaling at the transcription level, providing a separate potential mechanism for Notch-medicated ISCs maintenance [195].

Atoh1, a basic helix-loop-helix transcriptional activator, is required for the differentiation of all secretory cell lineages [213]. *Atoh1* null mice die shortly after birth and fail to form any secretory lineages without affecting enterocytes [213]. Consistent with these observations, specific deletion of *Atoh1* in the adult intestine confirmed that Atoh1 is essential for all secretory cell lineages commitment as well as required for the normal cell cycle exit of crypt progenitors [214]. Conversely, in the embryo, overexpression of *Atoh1* directs intestinal progenitors to secretory cell fate at the expanse of absorptive enterocytes and progenitor cell self-renewal [215]. Inhibition of Notch signaling activity either by γ-secretase inhibitor or by knocking out *Rbp-Jκ* in adult Atoh1-deficient intestine failed to induce secretory cell differentiation or decrease proliferation of crypt stem/progenitor cells, suggesting that the primary function of Notch signaling is regulation of Atoh1 in uncommitted crypt progenitors, to determine commitment to secretory or absorptive lineages [216–218]. In addition, previous studies showed that the expression of Atoh1 is directly repressed by Hes1 [219]. Interestingly, inhibition of Atoh1 induced the expression of Hes1, suggesting Atoh1 might inhibit Hes1 indirectly [216, 217]. Taken together, these studies indicated that Notch/Hes signaling and Atoh1 play critical, but opposing roles, in the determination of secretory versus absorptive cell commitment.

7.3 *Mesenchymal BMPs Provide Niche Signals for the Intestinal Stem Cell Compartment*

Bone morphogenetic proteins (BMP) are members of the TGF-β superfamily. These multifunctional growth factors are well known for their roles in regulation of cell proliferation, differentiation, and tissue pattern formation in several systems. BMP ligands and their receptor are found in both the epithelium and the stroma of the intestine, and play various roles in intestinal development, homeostasis, and disease. Extracellular BMP ligands bind to transmembrane type 2 receptors that subsequently recruit one of three type 1 receptors, such as Bmpr1a, to form a heterodimer. Intracellular signal transduction by the newly formed ligand/receptor complex results in the phosphorylation of receptor-associated Smad proteins, Smad1/5/8, within the cytoplasm. This post-translational modification enables Smad proteins to dimerize with Smad4 and translocate to the nucleus to regulate transcription by binding to the appropriate promoter region of specific genes [220] (Fig. 5).

The most prominent player of the bone morphogenetic proteins in intestinal stem cell regulation is BMP4. Under physiological conditions, this ligand is restricted to the mesenchyme of the intervillus region and surrounding area at the base of the

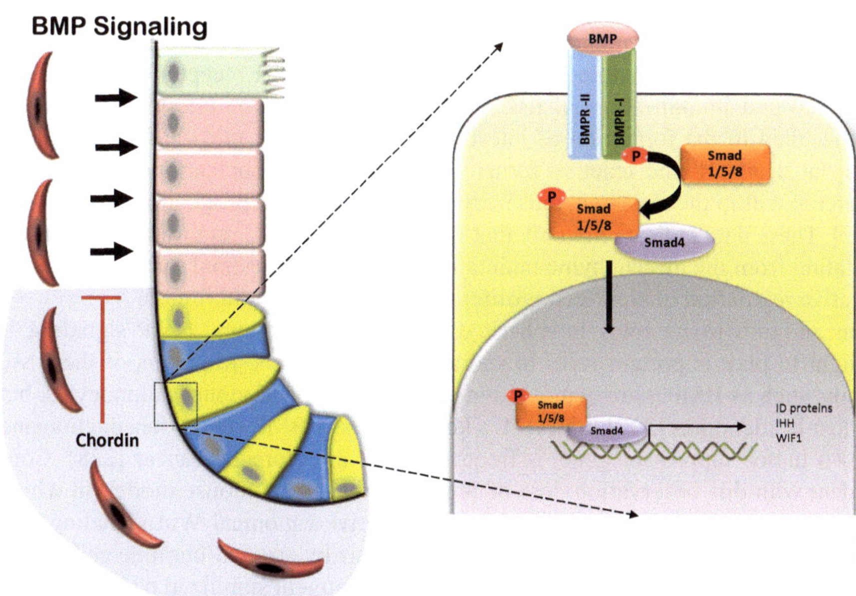

Fig. 5 Bone Morphogenetic Protein (BMP) signaling. BMP signals generally originate from the mesenchyme and activate Bmp receptors located on the epithelial cells. BMP antagonists, such as Chordin, surround the crypt base to allow only transient activation of BMP signaling within the stem cell compartment to balance the proliferative signals. Upon BMP binding to the type 2 BMP receptor, a heterodimer forms with the BMP type 1 receptor. This allows for phosphorylation of Smads1/5/8, which then bind to Smad4 in the cytoplasm. SMAD dimers then translocate to the nucleus and mediate transcription of downstream factors that regulate proliferative signals

crypt [221, 222]. Defects of BMP signaling in vivo can induce proliferation and cause diseases such as juvenile polyposis syndrome, an autosomal dominant inherited disorder has been characterized by germline mutations in the *Bmpr1a* receptor and/or *Smad4* genes. Inheritance of this disease puts patients at a 39–68 % increased risk of gastrointestinal cancer [223]. In mice, transgenic epithelial expression of Noggin, a Bmp antagonist, effectively inhibited Bmp signaling and drove ectopic crypt formation within the villi, induced polyp formation, and increased stromal and epithelial proliferation [224, 221]. These studies examined the effect of Bmp signaling loss simultaneously in both the epithelium and mesenchyme. When isolated to the epithelium, deletion of the *Bmpr1a* receptor showed increased proliferation and Cyclin D1 expression, but lacked de novo crypt formation and changes in Wnt signaling seen in previous studies. Additionally, terminal differentiation of the secretory cell lineages was impaired as evidenced by lack of mature genes expression [225]. These sets of experiments distinguished tissue specific roles for BMP signaling in the intestine. It appears that BMP signaling through mesenchymal cells is responsible for the origin of crypt formation and juvenile polyposis while epithelial BMP signaling is responsible for cellular proliferation and stem cell renewal. In addition to these positive signals, BMP antagonists including Gremlin 1, Gremlin 2, and Chordin-like 1 in surrounding mesenchymal cells help mediate the balance between a proliferative and differentiating states [226].

Mechanistically, BMP signaling was found to exert its effect as a suppressor of Wnt signaling through enhanced activation of PTEN, a negative regulator of PI3K activity. Interferon-inducible deletion of *Bmpr1a*, a type I receptor found in both epithelial and immune/stromal tissues, caused expansion of intestinal stem cells and resulted in the formation of intestinal polyps. Mutant mice had higher phosphorylated PTEN (the inactive form) which resulted in nuclear accumulation of β-catenin within the stem cell, thus increasing stem cell and progenitor proliferation [222]. These data indicate not only that PI3K modulates ISC renewal, but that BMP signaling from the mesenchyme maintains a balance of intestinal turn over through negative regulation of stem-cell proliferation via PI3K-Akt inhibition.

In addition to its roles in inherited polyposis syndromes, BMP signaling is thought to play important roles in sporadic tumorigenesis. Members of the BMP family, such as BMP2, are expressed in normal human and mouse colonocytes, but are lost in adenomas [227]. SMAD4, a key effector in BMP signal transduction and known tumor suppressor gene, is frequently lost in colorectal cancer [228]. Concordant with this observation, loss of Smad4 in the Apc^{min} mouse model (in which benign intestinal adenomas result from hyperactive canonical Wnt signaling) led to tumor invasion and expansion, mediated in part by stromal/immune cells [229, 230]. In *KRAS* mutated cancer cell lines, BMP2 treatment stabilized p21WAF1 protein expression and suppressed growth [231]. Other BMPs have been implicated in colorectal cancer suppression such as BMP3 [232] while some have been found to play a role in development and invasion such as BMP7 and BMP4 [233, 234]. Overall the role of BMP is to regulate the balance of proliferative signals within the intestinal crypt, however, this role is highly regulated and context dependent.

7.4 Epithelial Hedgehog Signaling to the Mesenchyme Regulates Stem Cell Niche Signals

Hedgehog signaling plays a role in stem cell maintenance indirectly as it affects the surrounding mesenchymal niche signals needed for ISC homeostasis. In the absence of Hedgehog family ligands, transmembrane Patched receptors suppress Smoothened, the intracellular signal transducer, promoting Gli phosphorylation and proteolytic processing to a repressor state, resulting intranscriptional inhibition of Gli target genes. In the presence of Hedgehog, Patched releases Smoothened which is subsequently phosphorylated by several kinases allowing for interaction with proteins in the primary cilium including Sufu. Accumulation of the phospho-Smoothened-ciliary protein complex prevents phosphorylation of Gli proteins which then dissociate from Sufu and translocate to the nucleus to activate transcriptional targets [235] (Fig. 6).

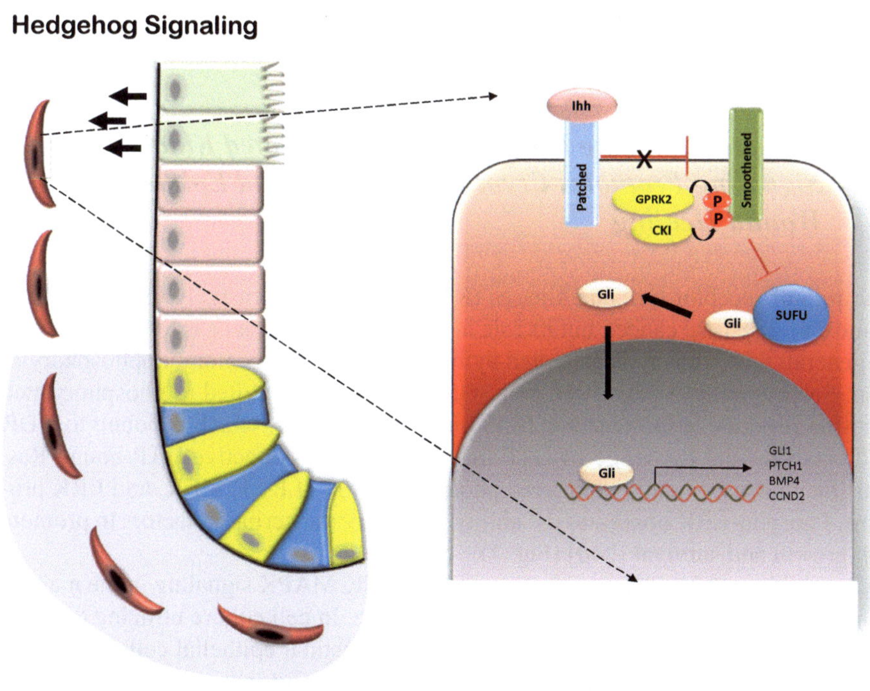

Fig. 6 Hedgehog signaling. Indian Hedgehog (Ihh) is the most prominent Hedgehog ligand in the mammalian intestine, where it originates from differentiated epithelial cells. Hedgehog ligands bind to Patched receptors on mesenchymal cells, which relieves Patched suppression of Smoothened, allowing phosphorylation by GPRK2 and CKI. Phosphorylated Smoothened binds to proteins localized on the primary cilium (not shown here). This complex prevents Gli phosphorylation and allows dissociation of Gli from Sufu and subsequent translocation into the nucleus to activate downstream transcriptional targets. Downstream effects of the Hedgehog pathway include altered expression of BMP ligand that originates from the mesenchyme thereby affecting proliferative signals in the stem cell

In the intestine, Indian hedgehog (Ihh) ligand is expressed by mature, differentiated epithelial cells and, via a mesenchymal intermediate, limits proliferation in the crypts [236–238]. Deletion of *Ihh* in the adult intestine resulted in increased nuclear β-catenin and downstream targets of Wnt signaling, including stem cell markers Lgr5 and Olfm4. Consequently, increased proliferation, crypt fission, crypt depth, and villus length were also found. Bmp4 signaling was reduced as evidenced by loss of phosphorylated Smad1/5/8, possibly owing their absence to the reduced number of mesenchymal cells found in the villus core [236, 237, 239]. Conversely conditional activation of Ihh signaling in the adult mouse colon led to decreased proliferation, reduced nuclear Tcf/β-catenin accumulation within the stem cell, crypt hypoplasia, and premature cell differentiation. Additionally, there was a significant accrual of colonic myofibroblasts as well as Bmp4 expression [238]. Together, these results suggest that, under physiological conditions, Ihh functions as part of a negative feedback mechanism to limit production of progenitor cells by ISCs. Ihh exerts its effects directly on the mesenchyme, the source of BMP signals. A surface-to-crypt Bmp4 gradient then, in turn, limits canonical Wnt signaling to the base of the crypt where it limits proliferation of the ISCs as previously discussed above.

7.5 Receptor Tyrosine Kinase/Mitogen-Activated Kinase Signaling Promotes Crypt Proliferation and Epithelial Wound Healing

The role of receptor tyrosine kinases (RTK) has been extensively explored in the context of intestinal cancer, but its role in intestinal maintenance is less well known. In general, this pathway is activated by ligand-mediated receptor autophosphorylation. Subsequently, SCH, GRB, and SOS proteins are recruited to phosphorylated tyrosine residues of the receptor. SOS binds inactive Ras, which is bound to GDP, and catalyzes the exchange of GTP for GDP resulting in active GTP-bound Ras. This facilitates a cascade of ordered phosphorylation of RAF, MEK, and ERK proteins. Phospho-ERK possesses the ability to modify transcription factors to promote cell growth and survival [240] (Fig. 7).

Several lines of evidence point to a role for RTK/MAPK signaling in the maintenance of proliferative capacity within the intestine. In cell culture utilizing colorectal cancer cells and nontransformed rat jejunal intestinal epithelial cells, inhibition of MAPK signaling correlated with cell cycle arrest and increased differentiation [241]. EGFR is an ERBB family transmembrane receptor tyrosine kinase that is expressed highly proliferative cells of the crypts [242], while its corresponding ligand, EGF, is secreted from the stroma, salivary glands, pancreas, differentiated intestinal cells, and the Brunner's glands of the duodenum [243]). Both the receptor and ligand were found to be highly upregulated in human colorectal cancer [244]. In mice, inhibition of Egfr kinase (the active form of the receptor) resulted in decreased

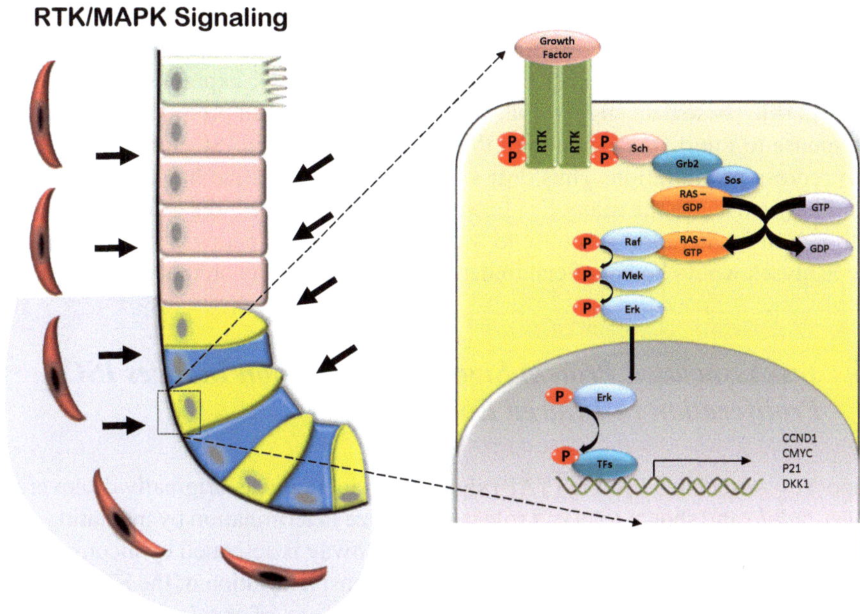

Fig. 7 Receptor Tyrosine Kinase/Mitogen Activated Protein Kinase (RTK/MAPK) signaling. RTK/MAPK signaling originates from growth factors found in the stroma and lumen of the intestine. When bound to the receptor, homodimer autophosphorylation results in complex formation with Sch, Grb2, and Sos. Sos is a guanine nucleotide exchange protein that converts inactive GDP-Ras to active GTP-Ras. This triggers a downstream activation of the MAP kinases that induces transcription factor binding in the nucleus. Roles of this pathway involve pro-proliferative effects include proliferative maintenance and wound repair

proliferation followed by increased cell death and loss of differentiated epithelial cells suggesting the inability for cellular repopulation of the epithelium [242]. The requirement for EGF signaling specifically in stem cell proliferation has been found in *Drosophila* intestines [245] however, this signaling through Egfr may be limited to the transit amplifying population of the normal mammalian intestine. ErbB is another RTK that is activated in response Egf and Egf-like ligand bonding. The stem cell marker Lrig1 was shown to interact with the ErbB receptor to maintain proper ISC proliferation. Experimentally, Lrig1-specific deletion of *ErbB* resulted in stem cell expansion as well as increased expression of activated MAPK signaling [35]. These data indicate that Lrig1 negatively regulates MAPK signaling and thus prevents unwarranted ISC expansion.

Fibroblast growth factors promote growth within the intestine. Fgf signals originate in the mesenchyme and bind to Fgf receptors found on the epithelial cells. Fgf signaling may have a direct effect on ISC survival. After irradiation, expression of Fgf2 mRNA and protein increased in the mesenchyme surrounding the crypts of the small intestine. Systemic administration of this growth factor enhanced stem cell survival when given prior to radiation [246]. Similar effects had been seen with

Fgf7 treatment under these conditions [247]. Alternatively, knock out of the Fgf receptor, *FgfR3*, increased proliferation in intestinal crypts; complementary in vitro experiments indicating this may be due to increased pERK expression and loss of pAKT [248]. These data suggest that while Fgf signaling can promote ISC survival in response to injury, the function of this pathway in normal ISC homeostasis may be to repress proliferation. Thus, Fgf signaling in the intestine has differential effects likely owing to the many Fgf ligand and receptor types; regardless, epithelial proliferation is altered in response to manipulation of this pathway. Stem cell specific analyses would be beneficial to determine direct effects on this population.

7.6 Yes-Associated Protein Signal Transduction Induces ISC Proliferation in Injured Epithelium

Hippo/Yes-Associated Protein (YAP) signal transduction was originally discovered in *Drosophila* and shown to play a role in organ size determination by mediating cell cycle and anti-apoptosis genes. This signaling pathway is activated by incompletely characterized receptor interaction that results in phosphorylation of the STE20 family kinases MST1 and MST2, mammalian orthologues of the *Drosophila* Hippo protein. These kinases then, in turn, phosphorylate NRD family proteins LATS1 and LATS2 as well as MOB1. MOB1 interacts with LATS1/2 to phosphorylate YAP and TAZ proteins. This modification facilitates interaction with 14-3-3 proteins in the cytoplasm and prevents nuclear translocation. In the absence of YAP and TAZ phosphorylation, these proteins act as coactivators for various transcription factors that regulate cellular proliferation and organ size [249, 250] (Fig. 8).

Within the proliferative compartment of the intestine, strong expression of MST1 is found in the cytoplasm, whereas YAP accumulates in the nucleus inducing gene transcription. In villus cells, nuclear YAP is absent whereas phosphorylated YAP is seen in the cytoplasm, suggesting activation of MST-LATS-MOB1 signal transduction as cells terminally differentiate [251]. YAP does not seem to be required for normal mammalian intestinal homeostasis [252, 253], however ablation of both Mst1 and Mst2 leads to mucosal hypertrophy, stem cell expansion, and tumorigenesis, suggesting an important role for the Hippo/YAP pathway in ISCs [254]. This role may be especially important in the context of wound healing: YAP1 was required for regeneration after chemically-induced colonic ulceration, and overexpression of YAP can facilitate colitis-associated polyp formation [253]. In contrast, another group reported that ablation of YAP1 after irradiation resulted in crypt hyperplasia and epithelial overgrowth during regeneration [252]. These conflicting reports may suggest differential requirements for the YAP pathway in different stem cell pools or in distinct phases of regeneration. Manipulation of the Hippo/YAP pathway affects other important signaling pathways in the intestine. Overexpression of YAP throughout the intestinal epithelium reduced β-catenin activity and downstream Wnt target genes and resulted in crypt loss in the intestine [252]. Consistent with this, loss of Mst1/2 resulted in increased β-catenin activity [254]. Regulation of Notch signaling by the

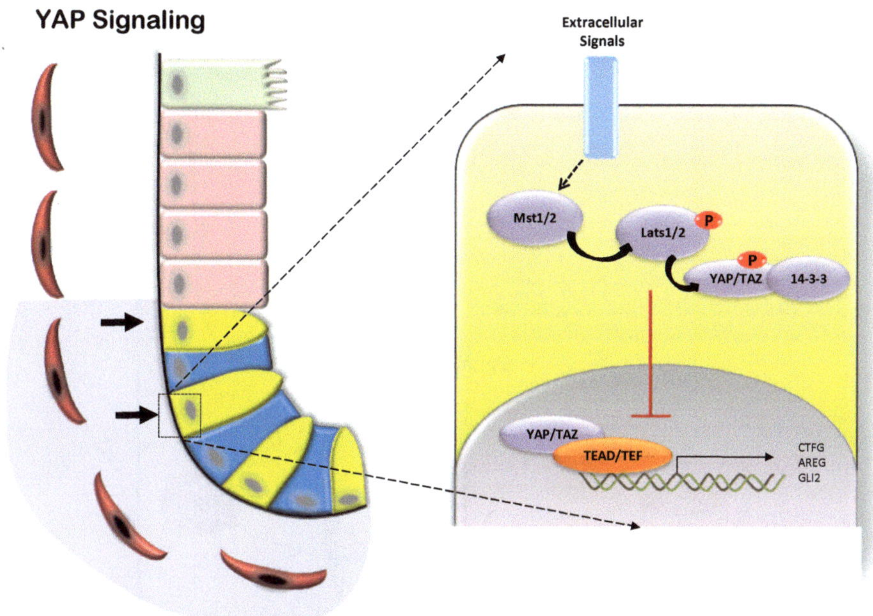

Fig. 8 Yes-associated Protein (YAP) Signaling. YAP signal transduction begins when extracellular signals, via an uncharacterized receptor, result in a series of downstream phosphorylation events beginning with Mst1 and Mst2 kinases. Lats1 and Lats2 phosphorylation of YAP facilitates its interaction with 14-3-3 proteins thereby preventing translocation of YAP into the nucleus. In the absence of phosphorylation, YAP enters the nucleus where it acts as a coactivator for the TEAD family of transcription factors, resulting in production of various proteins involved in cellular proliferation

YAP pathway has also been reported [254]. Despite the duality in reported YAP expression and downstream proliferative events, the Hippo/YAP pathway appears to have a prominent role in both wound healing and epithelial tumor initiation prompting the need for further investigation into the nature of this pathway in the intestine.

8 Intestinal Stem Cells in Digestive Diseases

In response to intestinal injury, the ISCs are essential for the regeneration of the entire intestinal epithelium through appropriate proliferation and differentiation. After injury, it is likely ISCs undergo symmetrical division to expand the ISC pool and replace damaged ISCs. To achieve these functions, ISCs must integrate input from multiple sources using the signaling pathways described above as well as others (Fig. 9). Misregulation of one or more signaling pathways can result in uncontrolled ISC proliferation or aberrant differentiation that is undesirable and can contribute to disease. Thus, the biological function of ISCs is highly relevant to gastrointestinal diseases such as inflammatory bowel disease and colorectal cancer.

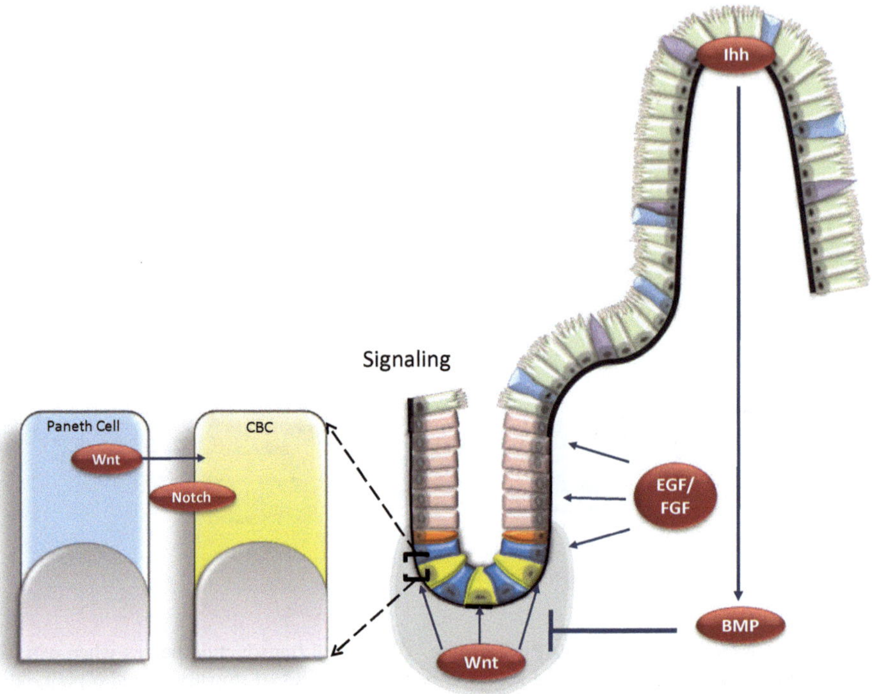

Fig. 9 Integration of signaling pathways on Intestinal stem cell regulation. A gradient of Wnt signaling is highest at the bottom of crypts, providing important signals for ISC maintenance and differentiation. Notch signaling, which is dependent on direct cell-cell contact, has a critical function controlling the fate of ISC. BMP signaling is found in both the epithelium and the stroma of the intestine, and plays various roles in mediating the balance between proliferative and differentiating states. Hedgehog signaling plays a role in stem cell maintenance indirectly as it affects the surrounding mesenchymal niche signals needed for ISC homeostasis. EGF/FGF-mediated receptor tyrosine kinase signaling promotes crypt proliferation and epithelial wound healing. YAP signaling, not shown, is thought to interact with both Wnt and Notch pathways

8.1 Inflammatory Bowel Disease (IBD)

The pathogenesis of two main entities of inflammatory bowel disease (IBD), Crohn's disease (CD) and ulcerative colitis (UC), is complex and still largely unknown. In healthy intestinal mucosa, the constituency of luminal microbes is controlled by the secretion of antimicrobial peptides and by the mucus layer. However, high concentrations of intestinal bacteria are found in diseased tissues from CD and UC, suggesting a critical role of luminal microbes in the development of IBD. Recent studies suggested that defective differentiation from ISC into Paneth and goblet cells results in a defective antimicrobial shield in CD and mucus layer in UC, respectively. Canonical Wnt/β-catenin signaling is essential for Paneth cell differentiation [99]. Importantly, β-catenin/TCF4-mediated transcription regulates the expression level of Paneth cell specific protein defensins, which protects the intestinal epithelium against a large number of microbes [255]. In ileal CD, diminished expression of TCF4 causes deficiency of Paneth cell defensins, resulting in the disruption of the mucosal barrier, which allows invasion of luminal microbes and causes inflammation [256, 257]. Besides *TCF4*, other genes such as *NOD2*, *ATG16L1*, *XBP1*, *HD5*, *TLR9* and *KCNN4*, which are necessary for functional Paneth cells, have allelic variants that confer increased risk for ileal CD (reviewed in [258, 259]. Goblet cells release mucins into the intestinal lumen to form the mucus layer that functions as a barrier between luminal components and the intestinal epithelium. Previous studies showed that a defective colonic mucus layer and accompanying paucity of goblet cells is associated with UC [260, 261]. Further studies suggest that an insufficient induction of mucins in UC, but not CD, is due to defective differentiation of ISCs into goblet cells [262]. Taken together, understanding the role of ISC differentiation towards Paneth and goblet cells could improve our current knowledge and provide novel avenues for therapy of CD and UC, respectively.

8.2 Cancer

Colorectal cancer (CRC) is one of the most common and lethal cancers worldwide [263]. Several questions concerning CRC carcinogenesis are still largely unknown, such as the cell of the origin. The traditional stochastic model of tumorigenesis suggests that every single cell within a tumor is capable of initiating and propagating cancer development. However, the cancer stem cell (CSC) model proposes that only a small subpopulation of cells has the exclusive ability to regenerate entire tumors. Like normal stem cells, CSCs are responsible for self-renewal and differentiation to generate heterogeneous progeny of the entire tumor [264]. Most likely, in the intestine, CSC might be derived from self-renewing ISC (reviewed in [265]). Previous studies showed that over activation of Wnt/β-catenin signaling in ISCs is sufficient to initiate tumorigenesis [11, 22, 266]. In this case, Wnt/β-catenin-driven tumor initiation is more efficient in stem cells than in more differentiated cells, suggesting that stem cells may preferentially act as the cell of origin in CRC. These results suggested that CRC could be a consequence of ISC dysregulation, at least in some cases. Of note, it is also possible that differentiated intestinal epithelial cells might give rise to CSCs through dedifferentiation and acquisition of stem-cell-like properties [267]. Emerging evidence supported the existence of CSCs in CRC. However, most of the CSC markers, such as LGR5 and CD133, are shared by ISC in the intestines [11, 268–270] Thus, identifying CSC specific markers will be an important challenge. Recently, Nakanishi and colleagues beautifully showed that Dclk1 specifically marked CSC, but not ISC, providing a potential therapeutic strategy based on targeting Dclk1 positive cells in CRC [271]. Taken together, understanding the molecular mechanisms that regulate ISC self-renewal and differentiation becomes more and more important for further therapies because the machinery involved in these pathways likely contributes to tumor development.

References

1. Shroyer NF, Kocoshis SA (2011) Anatomy and physiology of the small and large intestines. In: Wyllie R, Hyams J, Kay M (eds) Pediatric gastrointestinal and liver disease. Elsevier, Philadelphia, pp. 324–336
2. Yan KS, Chia LA, Li X, Ootani A, Su J, Lee JY, Su N, Luo Y, Heilshorn SC, Amieva MR et al (2012) The intestinal stem cell markers Bmi1 and Lgr5 identify two functionally distinct populations. Proc Natl Acad Sci U S A 109:466–471
3. Cairnie AB, Lamerton LF, Steel GG (1965) Cell proliferation studies in the intestinal epithelium of the rat. II. Theoretical aspects. Exp Cell Res 39:539–553
4. Potten CS (1975) Kinetics and possible regulation of crypt cell populations under normal and stress conditions. Bull Cancer 62:419–430
5. Potten CS (1977) Extreme sensitivity of some intestinal crypt cells to X and gamma irradiation. Nature 269:518–521
6. Potten CS, Kovacs L, Hamilton E (1974) Continuous labelling studies on mouse skin and intestine. Cell Tissue Kinet 7:271–283
7. Cheng H, Leblond CP (1974) Origin, differentiation and renewal of the four main epithelial cell types in the mouse small intestine. V. Unitarian theory of the origin of the four epithelial cell types. Am J Anat 141:537–561
8. Bjerknes M, Cheng H (1981) The stem-cell zone of the small intestinal epithelium. I. Evidence from Paneth cells in the adult mouse. Am J Anat 160:51–63
9. Blanpain C, Simons BD (2013) Unravelling stem cell dynamics by lineage tracing. Nat Rev Mol Cell Biol 14:489–502
10. Barker N, van Es JH, Kuipers J, Kujala P, van den Born M, Cozijnsen M, Haegebarth A, Korving J, Begthel H, Peters PJ et al (2007) Identification of stem cells in small intestine and colon by marker gene Lgr5. Nature 449:1003–1007
11. Barker N, Ridgway RA, van Es JH, van de Wetering M, Begthel H, van den Born M, Danenberg E, Clarke AR, Sansom OJ, Clevers H (2009) Crypt stem cells as the cells-of-origin of intestinal cancer. Nature 457:608–611
12. Stelzner M, Helmrath M, Dunn JC, Henning SJ, Houchen CW, Kuo C, Lynch J, Li L, Magness ST, Martin MG et al (2012) A nomenclature for intestinal in vitro cultures. Am J Physiol Gastrointest Liver Physiol 302:G1359–G1363
13. Mahe MM, Aihara E, Schumacher MA, Zavros Y, Montrose MH, Helmrath MA, Sato T, Shroyer NF (2013) Establishment of gastrointestinal epithelial organoids. In Moore D (ed) Current protocols in mouse biology. Wiley, pp. 217–240
14. Sato T, Vries RG, Snippert HJ, van de Wetering M, Barker N, Stange DE, van Es JH, Abo A, Kujala P, Peters PJ et al (2009) Single Lgr5 stem cells build crypt-villus structures in vitro without a mesenchymal niche. Nature 459:262–265
15. van der Flier LG, Haegebarth A, Stange DE, van de Wetering M, Clevers H (2009a) OLFM4 is a robust marker for stem cells in human intestine and marks a subset of colorectal cancer cells. Gastroenterology 137:15–17
16. van der Flier LG, van Gijn ME, Hatzis P, Kujala P, Haegebarth A, Stange DE, Begthel H, van den Born M, Guryev V, Oving I et al (2009b) Transcription factor achaete scute-like 2 controls intestinal stem cell fate. Cell 136:903–912
17. Besson V, Smeriglio P, Wegener A, Relaix F, Nait Oumesmar B, Sassoon DA, Marazzi G (2011) PW1 gene/paternally expressed gene 3 (PW1/Peg3) identifies multiple adult stem and progenitor cell populations. Proc Natl Acad Sci U S A 108:11470–11475
18. Merlos-Suarez A, Barriga FM, Jung P, Iglesias M, Cespedes MV, Rossell D, Sevillano M, Hernando-Momblona X, da Silva-Diz V, Munoz P et al (2011) The intestinal stem cell signature identifies colorectal cancer stem cells and predicts disease relapse. Cell Stem Cell 8:511–524
19. Munoz J, Stange DE, Schepers AG, van de Wetering M, Koo BK, Itzkovitz S, Volckmann R, Kung KS, Koster J, Radulescu S et al (2012) The Lgr5 intestinal stem cell signature: robust expression of proposed quiescent '+4' cell markers. EMBO J 31:3079–3091

20. Potten CS, Booth C, Tudor GL, Booth D, Brady G, Hurley P, Ashton G, Clarke R, Sakakibara S, Okano H (2003) Identification of a putative intestinal stem cell and early lineage marker; musashi-1. Differentiation 71:28–41
21. Snippert HJ, van Es JH, van den Born M, Begthel H, Stange DE, Barker N, Clevers H (2009) Prominin-1/CD133 marks stem cells and early progenitors in mouse small intestine. Gastroenterology 136:2187–2194, e2181
22. Zhu L, Gibson P, Currle DS, Tong Y, Richardson RJ, Bayazitov IT, Poppleton H, Zakharenko S, Ellison DW, Gilbertson RJ (2009) Prominin 1 marks intestinal stem cells that are susceptible to neoplastic transformation. Nature 457:603–607
23. Hughes KR, Gandara RM, Javkar T, Sablitzky F, Hock H, Potten CS, Mahida YR (2012) Heterogeneity in histone 2B-green fluorescent protein-retaining putative small intestinal stem cells at cell position 4 and their absence in the colon. Am J Physiol Gastrointest Liver Physiol 303:G1188–G1201
24. Potten CS, Owen G, Booth D (2002) Intestinal stem cells protect their genome by selective segregation of template DNA strands. J Cell Sci 115:2381–2388
25. Lee G, White LS, Hurov KE, Stappenbeck TS, Piwnica-Worms H (2009) Response of small intestinal epithelial cells to acute disruption of cell division through CDC25 deletion. Proc Natl Acad Sci U S A 106:4701–4706
26. Tian Q, Feetham MC, Tao WA, He XC, Li L, Aebersold R, Hood L (2004) Proteomic analysis identifies that 14-3-3zeta interacts with beta-catenin and facilitates its activation by Akt. Proc Natl Acad Sci U S A 101:15370–15375
27. Sangiorgi E, Capecchi MR (2008) Bmi1 is expressed in vivo in intestinal stem cells. Nat Genet 40:915–920
28. Tian H, Biehs B, Warming S, Leong KG, Rangell L, Klein OD, de Sauvage FJ (2011) A reserve stem cell population in small intestine renders Lgr5-positive cells dispensable. Nature 478:255–259
29. Takeda N, Jain R, LeBoeuf MR, Wang Q, Lu MM, Epstein JA (2011) Interconversion between intestinal stem cell populations in distinct niches. Science 334:1420–1424
30. Montgomery RK, Carlone DL, Richmond CA, Farilla L, Kranendonk ME, Henderson DE, Baffour-Awuah NY, Ambruzs DM, Fogli LK, Algra S et al (2011) Mouse telomerase reverse transcriptase (mTert) expression marks slowly cycling intestinal stem cells. Proc Natl Acad Sci U S A 108:179–184
31. Giannakis M, Stappenbeck TS, Mills JC, Leip DG, Lovett M, Clifton SW, Ippolito JE, Glasscock JI, Arumugam M, Brent MR et al (2006) Molecular properties of adult mouse gastric and intestinal epithelial progenitors in their niches. J Biol Chem 281:11292–11300
32. Gerbe F, Brulin B, Makrini L, Legraverend C, Jay P (2009) DCAMKL-1 expression identifies Tuft cells rather than stem cells in the adult mouse intestinal epithelium. Gastroenterology 137:2179–2180; author reply 2180–2171
33. May R, Riehl TE, Hunt C, Sureban SM, Anant S, Houchen CW (2008) Identification of a novel putative gastrointestinal stem cell and adenoma stem cell marker, doublecortin and CaM kinase-like-1, following radiation injury and in adenomatous polyposis coli/multiple intestinal neoplasia mice. Stem Cells 26:630–637
34. Powell AE, Wang Y, Li Y, Poulin EJ, Means AL, Washington MK, Higginbotham JN, Juchheim A, Prasad N, Levy SE et al (2012) The pan-ErbB negative regulator Lrig1 is an intestinal stem cell marker that functions as a tumor suppressor. Cell 149:146–158
35. Wong VW, Stange DE, Page ME, Buczacki S, Wabik A, Itami S, van de Wetering M, Poulsom R, Wright NA, Trotter MW et al (2012) Lrig1 controls intestinal stem-cell homeostasis by negative regulation of ErbB signalling. Nat Cell Biol 14:401–408
36. Sakthianandeswaren A, Christie M, D'Andreti C, Tsui C, Jorissen RN, Li S, Fleming NI, Gibbs P, Lipton L, Malaterre J et al (2011) PHLDA1 expression marks the putative epithelial stem cells and contributes to intestinal tumorigenesis. Cancer Res 71:3709–3719
37. Li L, Clevers H (2010) Coexistence of quiescent and active adult stem cells in mammals. Science 327:542–545

38. Escobar M, Nicolas P, Sangar F, Laurent-Chabalier S, Clair P, Joubert D, Jay P, Legraverend C (2011) Intestinal epithelial stem cells do not protect their genome by asymmetric chromosome segregation. Nat Commun 2:258
39. Schepers AG, Vries R, van den Born M, van de Wetering M, Clevers H (2011) Lgr5 intestinal stem cells have high telomerase activity and randomly segregate their chromosomes. EMBO J 30:1104–1109
40. Potten CS, Chadwick C, Ijiri K, Tsubouchi S, Hanson WR (1984) The recruitability and cell-cycle state of intestinal stem cells. Int J Cell Cloning 2:126–140
41. van Es JH, Sato T, van de Wetering M, Lyubimova A, Nee AN, Gregorieff A, Sasaki N, Zeinstra L, van den Born M, Korving J et al (2012b) Dll1+ secretory progenitor cells revert to stem cells upon crypt damage. Nat Cell Biol 14:1099–1104
42. Roth S, Franken P, Sacchetti A, Kremer A, Anderson K, Sansom O, Fodde R (2012) Paneth cells in intestinal homeostasis and tissue injury. PLoS One 7:e38965
43. Buczacki SJ, Zecchini HI, Nicholson AM, Russell R, Vermeulen L, Kemp R, Winton DJ (2013) Intestinal label-retaining cells are secretory precursors expressing Lgr5. Nature 495:65–69
44. Noah TK, Shroyer NF (2013) Notch in the intestine: regulation of homeostasis and pathogenesis. Annu Rev Physiol 75:263–288
45. Aragaki M, Tsuchiya K, Okamoto R, Yoshioka S, Nakamura T, Sakamoto N, Kanai T, Watanabe M (2008) Proteasomal degradation of Atoh1 by aberrant Wnt signaling maintains the undifferentiated state of colon cancer. Biochem Biophys Res Commun 368:923–929
46. Tsuchiya K, Nakamura T, Okamoto R, Kanai T, Watanabe M (2007) Reciprocal targeting of Hath1 and beta-catenin by Wnt glycogen synthase kinase 3beta in human colon cancer. Gastroenterology 132:208–220
47. Pinto D, Gregorieff A, Begthel H, Clevers H (2003) Canonical Wnt signals are essential for homeostasis of the intestinal epithelium. Genes Dev 17:1709–1713
48. Peignon G, Durand A, Cacheux W, Ayrault O, Terris B, Laurent-Puig P, Shroyer NF, Van Seuningen I, Honjo T, Perret C et al (2011) Complex interplay between beta-catenin signalling and Notch effectors in intestinal tumorigenesis. Gut 60 166–176
49. Li HJ, Ray SK, Singh NK, Johnston B, Leiter AB (2011) Basic helix-loop-helix transcription factors and enteroendocrine cell differentiation. Diabetes Obes Metab 13 (Suppl 1):5–12
50. Noah TK, Donahue B, Shroyer NF (2011) Intestinal development and differentiation. Exp Cell Res 317:2702–2710
51. Vanuytsel T, Senger S, Fasano A, Shea-Donohue T (2013) Major signaling pathways in intestinal stem cells. Biochim Biophys Acta 1830:2410–2426
52. Snoeck V, Goddeeris B, Cox E (2005) The role of enterocytes in the intestinal barrier function and antigen uptake. Microbes Infect 7:997–1004
53. Ziv E, Bendayan M (2000) Intestinal absorption of peptides through the enterocytes. Microsc Res Tech 49:346–352
54. Haegebarth A, Bie W, Yang R, Crawford SE, Vasioukhin V, Fuchs E, Tyner AL (2006) Protein tyrosine kinase 6 negatively regulates growth and promotes enterocyte differentiation in the small intestine. Mol Cell Biol 26:4949–4957
55. Battle MA, Bondow BJ, Iverson MA, Adams SJ, Jandacek RJ, Tso P, Duncan SA (2008) GATA4 is essential for jejunal function in mice. Gastroenterology 135:1676–1686, e1671
56. Bosse T, Piaseckyj CM, Burghard E, Fialkovich JJ, Rajagopal S, Pu WT, Krasinski SD (2006) Gata4 is essential for the maintenance of jejunal-ileal identities in the adult mouse small intestine. Mol Cell Biol 26:9060–9070
57. Beuling E, Baffour-Awuah NY, Stapleton KA, Aronson BE, Noah TK, Shroyer NF, Duncan SA, Fleet JC, Krasinski SD (2011) GATA factors regulate proliferation, differentiation, and gene expression in small intestine of mature mice. Gastroenterology 140:1219–1229, e1211–e1212
58. D'Angelo A, Bluteau O, Garcia-Gonzalez MA, Gresh L, Doyen A, Garbay S, Robine S, Pontoglio M (2010) Hepatocyte nuclear factor 1alpha and beta control terminal differentiation and cell fate commitment in the gut epithelium. Development 137:1573–1582

59. Babeu JP, Darsigny M, Lussier CR, Boudreau F (2009) Hepatocyte nuclear factor 4alpha contributes to an intestinal epithelial phenotype in vitro and plays a partial role in mouse intestinal epithelium differentiation. Am J Physiol Gastrointest Liver Physiol 297:G124–G134
60. Cattin AL, Le Beyec J, Barreau F, Saint-Just S, Houllier A, Gonzalez FJ, Robine S, Pincon-Raymond M, Cardot P, Lacasa M et al (2009) Hepatocyte nuclear factor 4alpha, a key factor for homeostasis, cell architecture, and barrier function of the adult intestinal epithelium. Mol Cell Biol 29:6294–6308
61. Olsen L, Bressendorff S, Troelsen JT, Olsen J (2005) Differentiation-dependent activation of the human intestinal alkaline phosphatase promoter by HNF-4 in intestinal cells. Am J Physiol Gastrointest Liver Physiol 289:G220–G226
62. Stegmann A, Hansen M, Wang Y, Larsen JB, Lund LR, Ritie L, Nicholson JK, Quistorff B, Simon-Assmann P, Troelsen JT et al (2006) Metabolome, transcriptome, and bioinformatic cis-element analyses point to HNF-4 as a central regulator of gene expression during enterocyte differentiation. Physiol Genom 27:141–155
63. Jang MH, Kweon MN, Iwatani K, Yamamoto M, Terahara K, Sasakawa C, Suzuki T, Nochi T, Yokota Y, Rennert PD et al (2004) Intestinal villous M cells: an antigen entry site in the mucosal epithelium. Proc Natl Acad Sci U S A 101:6110–6115
64. Kucharzik T, Lugering N, Rautenberg K, Lugering A, Schmidt MA, Stoll R, Domschke W (2000) Role of M cells in intestinal barrier function. Ann N Y Acad Sci 915:171–183
65. Mach J, Hshieh T, Hsieh D, Grubbs N, Chervonsky A (2005) Development of intestinal M cells. Immunol Rev 206:177–189
66. Knoop KA, Kumar N, Butler BR, Sakthivel SK, Taylor RT, Nochi T, Akiba H, Yagita H, Kiyono H, Williams IR (2009) RANKL is necessary and sufficient to initiate development of antigen-sampling M cells in the intestinal epithelium. J Immunol 183:5738–5747
67. de Lau W, Kujala P, Schneeberger K, Middendorp S, Li VS, Barker N, Martens A, Hofhuis F, DeKoter RP, Peters PJ et al (2012) Peyer's patch M cells derived from Lgr5(+) stem cells require SpiB and are induced by RankL in cultured "miniguts". Mol Cell Biol 32:3639–3647
68. Kanaya T, Hase K, Takahashi D, Fukuda S, Hoshino K, Sasaki I, Hemmi H, Knoop KA, Kumar N, Sato M et al (2012) The Ets transcription factor Spi-B is essential for the differentiation of intestinal microfold cells. Nat Immunol 13:729–736
69. Madara JL, Carlson SL (1985) Cup cells: further structural characterization of the brush border and the suggestion that they may serve as an attachment site for an unidentified bacillus in guinea pig ileum. Gastroenterology 89:1374–1386
70. Fujimura Y, Iida M (2001) A new marker for cup cells in the rabbit small intestine: expression of vimentin intermediate filament protein. Med Electron Microsc 34:223–229
71. Ramirez C, Gebert A (2003) Vimentin-positive cells in the epithelium of rabbit ileal villi represent cup cells but not M-cells. J Histochem Cytochem 51:1533–1544
72. Schonhoff SE, Giel-Moloney M, Leiter AB (2004) Minireview: development and differentiation of gut endocrine cells. Endocrinology 145:2639–2644
73. Bjerknes M, Cheng H (2006) Neurogenin 3 and the enteroendocrine cell lineage in the adult mouse small intestinal epithelium. Dev Biol 300:722–735
74. Jenny M, Uhl C, Roche C, Duluc I, Guillermin V, Guillemot F, Jensen J, Kedinger M, Gradwohl G (2002) Neurogenin3 is differentially required for endocrine cell fate specification in the intestinal and gastric epithelium. EMBO J 21:6338–6347
75. Lopez-Diaz L, Jain RN, Keeley TM, VanDussen KL, Brunkan CS, Gumucio DL, Samuelson LC (2007) Intestinal Neurogenin 3 directs differentiation of a bipotential secretory progenitor to endocrine cell rather than goblet cell fate. Dev Biol 309:298–305
76. Naya FJ, Huang HP, Qiu Y, Mutoh H, DeMayo FJ, Leiter AB, Tsai MJ (1997) Diabetes, defective pancreatic morphogenesis, abnormal enteroendocrine differentiation in BETA2/neuroD-deficient mice. Genes Dev 11:2323–2334
77. Hill ME, Asa SL, Drucker DJ (1999) Essential requirement for Pax6 in control of enteroendocrine proglucagon gene transcription. Mol Endocrinol 13:1474–1486
78. Larsson LI, St-Onge L, Hougaard DM, Sosa-Pineda B, Gruss P (1998) Pax 4 and 6 regulate gastrointestinal endocrine cell development. Mech Dev 79:153–159

79. Marsich E, Vetere A, Di Piazza M, Tell G, Paoletti S (2003) The PAX6 gene is activated by the basic helix-loop-helix transcription factor NeuroD/BETA2. Biochem J 376:707–715
80. Desai S, Loomis Z, Pugh-Bernard A, Schrunk J, Doyle MJ, Minic A, McCoy E, Sussel L (2008) Nkx2.2 regulates cell fate choice in the enteroendocrine cell lineages of the intestine. Dev Biol 313:58–66
81. Du A, McCracken KW, Walp ER, Terry NA, Klein TJ, Han A, Wells JM, May CL (2012) Arx is required for normal enteroendocrine cell development in mice and humans. Dev Biol 365:175–188
82. Gierl MS, Karoulias N, Wende H, Strehle M, Birchmeier C (2006) The zinc-finger factor Insm1 (IA-1) is essential for the development of pancreatic beta cells and intestinal endocrine cells. Genes Dev 20:2465–2478
83. Ye DZ, Kaestner KH (2009) Foxa1 and Foxa2 control the differentiation of goblet and enteroendocrine L- and D-cells in mice. Gastroenterology 137:2052–2062
84. Offield MF, Jetton TL, Labosky PA, Ray M, Stein RW, Magnuson MA, Hogan BL, Wright CV (1996) PDX-1 is required for pancreatic outgrowth and differentiation of the rostral duodenum. Development 122:983–995
85. Yamada S, Kojima H, Fujimiya M, Nakamura T, Kashiwagi A, Kikkawa R (2001) Differentiation of immature enterocytes into enteroendocrine cells by Pdx1 overexpression. Am J Physiol Gastrointest Liver Physiol 281:G229–236
86. Specian RD, Oliver MG (1991) Functional biology of intestinal goblet cells. Am J Physiol 260:C183–C193
87. Bjerknes M, Cheng H (2010) Cell Lineage metastability in Gfi1-deficient mouse intestinal epithelium. Dev Biol 345:49–63
88. Shroyer NF, Wallis D, Venken KJ, Bellen HJ, Zoghbi HY (2005) Gfi1 functions downstream of Math1 to control intestinal secretory cell subtype allocation and differentiation. Genes Dev 19:2412–2417
89. Noah TK, Kazanjian A, Whitsett J, Shroyer NF (2010) SAM pointed domain ETS factor (SP-DEF) regulates terminal differentiation and maturation of intestinal goblet cells. Exp Cell Res 316:452–465
90. Gregorieff A, Stange DE, Kujala P, Begthel H, van den Born M, Korving J, Peters PJ, Clevers H (2009) The ets-domain transcription factor Spdef promotes maturation of goblet and paneth cells in the intestinal epithelium. Gastroenterology 137:1333–1345, e1331–e1333
91. Zecchini V, Domaschenz R, Winton D, Jones P (2005) Notch signaling regulates the differentiation of post-mitotic intestinal epithelial cells. Genes Dev 19:1686–1691
92. Shorning BY, Zabkiewicz J, McCarthy A, Pearson HB, Winton DJ, Sansom OJ, Ashworth A, Clarke AR (2009) Lkb1 deficiency alters goblet and paneth cell differentiation in the small intestine. PLoS One 4:e4264
93. Katz JP, Perreault N, Goldstein BG, Lee CS, Labosky PA, Yang VW, Kaestner KH (2002) The zinc-finger transcription factor Klf4 is required for terminal differentiation of goblet cells in the colon. Development 129:2619–2628
94. Ghaleb AM, McConnell BB, Kaestner KH, Yang VW (2011) Altered intestinal epithelial homeostasis in mice with intestine-specific deletion of the Kruppel-like factor 4 gene. Dev Biol 349:310–320
95. Clevers HC, Bevins CL (2013) Paneth cells: maestros of the small intestinal crypts. Annu Rev Physiol 75:289–311
96. Sato T, van Es JH, Snippert HJ, Stange DE, Vries RG, van den Born M, Barker N, Shroyer NF, van de Wetering M, Clevers H (2011) Paneth cells constitute the niche for Lgr5 stem cells in intestinal crypts. Nature 469:415–418
97. Yilmaz OH, Katajisto P, Lamming DW, Gultekin Y, Bauer-Rowe KE, Sengupta S, Birsoy K, Dursun A, Yilmaz VO, Selig M et al (2012) mTORC1 in the Paneth cell niche couples intestinal stem-cell function to calorie intake. Nature 486:490–495
98. Andreu P, Colnot S, Godard C, Gad S, Chafey P, Niwa-Kawakita M, Laurent-Puig P, Kahn A, Robine S, Perret C et al (2005) Crypt-restricted proliferation and commitment to the Paneth cell lineage following Apc loss in the mouse intestine. Development 132:1443–1451

99. van Es JH, Jay P, Gregorieff A, van Gijn ME, Jonkheer S, Hatzis P, Thiele A, van den Born M, Begthel H, Brabletz T et al (2005a) Wnt signalling induces maturation of Paneth cells in intestinal crypts. Nat Cell Biol 7:381–386
100. Bastide P, Darido C, Pannequin J, Kist R, Robine S, Marty-Double C, Bibeau F, Scherer G, Joubert D, Hollande F et al (2007) Sox9 regulates cell proliferation and is required for Paneth cell differentiation in the intestinal epithelium. J Cell Biol 178:635–648
101. Mori-Akiyama Y, van den Born M, van Es JH, Hamilton SR, Adams HP, Zhang J, Clevers H, de Crombrugghe B (2007) SOX9 is required for the differentiation of paneth cells in the intestinal epithelium. Gastroenterology 133:539–546
102. Gerbe F, Legraverend C, Jay P (2012) The intestinal epithelium tuft cells: specification and function. Cell Mol Life Sci 69:2907–2917
103. Jarvi O, Keyrilainen O (1956) On the cellular structures of the epithelial invasions in the glandular stomach of mice caused by intramural application of 20-methylcholantren. Acta Pathol Microbiol Scand 39:72–73
104. Rhodin J, Dalhamn T (1956) Electron microscopy of the tracheal ciliated mucosa in rat. Z Zellforsch Mikrosk Anat 44:345–412
105. Hofer D, Drenckhahn D (1998) Identification of the taste cell G-protein, alpha-gustducin, in brush cells of the rat pancreatic duct system. Histochem Cell Biol 110:303–309
106. Kokrashvili Z, Mosinger B, Margolskee RF (2009) Taste signaling elements expressed in gut enteroendocrine cells regulate nutrient-responsive secretion of gut hormones. Am J Clin Nutr 90:822S–825S
107. Gerbe F, van Es JH, Makrini L, Brulin B, Mellitzer G, Robine S, Romagnolo B, Shroyer NF, Bourgaux JF, Pignodel C et al (2011) Distinct ATOH1 and Neurog3 requirements define tuft cells as a new secretory cell type in the intestinal epithelium. J Cell Biol 192:767–780
108. Bjerknes M, Khandanpour C, Moroy T, Fujiyama T, Hoshino M, Klisch TJ, Ding Q, Gan L, Wang J, Martin MG et al (2012) Origin of the brush cell lineage in the mouse intestinal epithelium. Dev Biol 362:194–218
109. Clevers H, Nusse R (2012) Wnt/beta-catenin signaling and disease. Cell 149:1192–1205
110. Holland JD, Klaus A, Garratt AN, Birchmeier W (2013) Wnt signaling in stem and cancer stem cells. Curr Opin Cell Biol 25:254–264
111. Kuhl SJ, Kuhl M (2013) On the role of Wnt/beta-catenin signaling in stem cells. Biochim Biophys Acta 1830:2297–2306
112. Nusse R, Varmus H (2012) Three decades of Wnts: a personal perspective on how a scientific field developed. EMBO J 31:2670–2684
113. Niehrs C (2012) The complex world of WNT receptor signalling. Nat Rev Mol Cell Biol 13:767–779
114. Willert K, Nusse R (2012) Wnt proteins. Cold Spring Harbor Perspect Biol 4:a007864
115. Angers S, Moon RT (2009) Proximal events in Wnt signal transduction. Nat Rev Mol Cell Biol 10:468–477
116. van Amerongen R, Mikels A, Nusse R (2008) Alternative wnt signaling is initiated by distinct receptors. Sci Signal 1:re9
117. Liu C, Kato Y, Zhang Z, Do VM, Yankner BA, He X (1999) beta-Trcp couples beta-catenin phosphorylation-degradation and regulates Xenopus axis formation. Proc Natl Acad Sci U S A 96:6273–6278
118. Liu C, Li Y, Semenov M, Han C, Baeg GH, Tan Y, Zhang Z, Lin X, He X (2002) Control of beta-catenin phosphorylation/degradation by a dual-kinase mechanism. Cell 108:837–847
119. Winston JT, Strack P, Beer-Romero P, Chu CY, Elledge SJ, Harper JW (1999) The SCFbeta-TRCP-ubiquitin ligase complex associates specifically with phosphorylated destruction motifs in IkappaBalpha and beta-catenin and stimulates IkappaBalpha ubiquitination in vitro. Genes Dev 13:270–283
120. Wu G, He X (2006) Threonine 41 in beta-catenin serves as a key phosphorylation relay residue in beta-catenin degradation. Biochemistry 45:5319–5323
121. Hart M, Concordet JP, Lassot I, Albert I, del los Santos R, Durand H, Perret C, Rubinfeld B, Margottin F, Benarous R et al (1999) The F-box protein beta-TrCP associates with phosphorylated beta-catenin and regulates its activity in the cell. Curr Biol 9:207–210

122. Valenta T, Hausmann G, Basler K (2012) The many faces and functions of beta-catenin. EMBO J 31:2714–2736
123. Li VS, Ng SS, Boersema PJ, Low TY, Karthaus WR, Gerlach JP, Mohammed S, Heck AJ, Maurice MM, Mahmoudi T et al (2012) Wnt signaling through inhibition of beta-catenin degradation in an intact Axin1 complex. Cell 149:1245–1256
124. Cadigan KM, Waterman ML (2012) TCF/LEFs and Wnt signaling in the nucleus. Cold Spring Harbor Perspect Biol 4
125. Cavallo RA, Cox RT, Moline MM, Roose J, Polevoy GA, Clevers H, Peifer M, Bejsovec A (1998) Drosophila Tcf and Groucho interact to repress Wingless signalling activity. Nature 395:604–608
126. Daniels DL, Weis WI (2005) Beta-catenin directly displaces Groucho/TLE repressors from Tcf/Lef in Wnt-mediated transcription activation. Nat Struct Mol Biol 12:364–371
127. Levanon D, Goldstein RE, Bernstein Y, Tang H, Goldenberg D, Stifani S, Paroush Z, Groner Y (1998) Transcriptional repression by AML1 and LEF-1 is mediated by the TLE/Groucho corepressors. Proc Natl Acad Sci U S A 95:11590–11595
128. Roose J, Molenaar M, Peterson J, Hurenkamp J, Brantjes H, Moerer P, van de Wetering M, Destree O, Clevers H (1998) The Xenopus Wnt effector XTcf-3 interacts with Groucho-related transcriptional repressors. Nature 395:608–612
129. Mosimann C, Hausmann G, Basler K (2009) Beta-catenin hits chromatin: regulation of Wnt target gene activation. Nat Rev Mol Cell Biol 10:276–286
130. Hecht A, Vleminckx K, Stemmler MP, van Roy F, Kemler R (2000) The p300/CBP acetyltransferases function as transcriptional coactivators of beta-catenin in vertebrates. EMBO J 19:1839–1850
131. Parker DS, Ni YY, Chang JL, Li J, Cadigan KM (2008) Wingless signaling induces widespread chromatin remodeling of target loci. Mol Cell Biol 28:1815–1828
132. Kramps T, Peter O, Brunner E, Nellen D, Froesch B, Chatterjee S, Murone M, Zullig S, Basler K (2002) Wnt/wingless signaling requires BCL9/legless-mediated recruitment of pygopus to the nuclear beta-catenin-TCF complex. Cell 109:47–60
133. Sampietro J, Dahlberg CL, Cho US, Hinds TR, Kimelman D, Xu W (2006) Crystal structure of a beta-catenin/BCL9/Tcf4 complex. Mol Cell 24:293–300
134. Thompson B, Townsley F, Rosin-Arbesfeld R, Musisi H, Bienz M (2002) A new nuclear component of the Wnt signalling pathway. Nat Cell Biol 4:367–373
135. de la Roche M, Worm J, Bienz M (2008) The function of BCL9 in Wnt/beta-catenin signaling and colorectal cancer cells. BMC cancer 8:199
136. Deka J, Wiedemann N, Anderle P, Murphy-Seiler F, Bultinck J, Eyckerman S, Stehle JC, Andre S, Vilain N, Zilian O et al (2010) Bcl9/Bcl9l are critical for Wnt-mediated regulation of stem cell traits in colon epithelium and adenocarcinomas. Cancer Res 70:6619–6628
137. Valenta T, Gay M, Steiner S, Draganova K, Zemke M, Hoffmans R, Cinelli P, Aguet M, Sommer L, Basler K (2011) Probing transcription-specific outputs of beta-catenin in vivo. Genes Dev 25:2631–2643
138. Shitashige M, Hirohashi S, Yamada T (2008) Wnt signaling inside the nucleus. Cancer Sci 99:631–637
139. Gregorieff A, Pinto D, Begthel H, Destree O, Kielman M, Clevers H (2005) Expression pattern of Wnt signaling components in the adult intestine. Gastroenterology 129:626–638
140. Durand A, Donahue B, Peignon G, Letourneur F, Cagnard N, Slomianny C, Perret C, Shroyer NF, Romagnolo B (2012) Functional intestinal stem cells after Paneth cell ablation induced by the loss of transcription factor Math1 (Atoh1). Proc Natl Acad Sci U S A 109:8965–8970
141. Kim TH, Escudero S, Shivdasani RA (2012) Intact function of Lgr5 receptor-expressing intestinal stem cells in the absence of Paneth cells. Proc Natl Acad Sci U S A 109:3932–3937
142. Vermeulen L, De Sousa EMF, van der Heijden M, Cameron K, de Jong JH, Borovski T, Tuynman JB, Todaro M, Merz C, Rodermond H et al (2010) Wnt activity defines colon cancer stem cells and is regulated by the microenvironment. Nat Cell Biol 12:468–476
143. Sansom OJ, Reed KR, Hayes AJ, Ireland H, Brinkmann H, Newton IP, Batlle E, Simon-Assmann P, Clevers H, Nathke IS et al (2004) Loss of Apc in vivo immediately perturbs Wnt signaling, differentiation, migration. Genes Dev 18:1385–1390

144. Wong MH, Rubinfeld B, Gordon JI (1998) Effects of forced expression of an NH2-terminal truncated beta-Catenin on mouse intestinal epithelial homeostasis. J Cell Biol 141:765–777
145. Angus-Hill ML, Elbert KM, Hidalgo J, Capecchi MR (2011) T-cell factor 4 functions as a tumor suppressor whose disruption modulates colon cell proliferation and tumorigenesis. Proc Natl Acad Sci U S A 108:4914–4919
146. Korinek V, Barker N, Moerer P, van Donselaar E, Huls G, Peters PJ, Clevers H (1998) Depletion of epithelial stem-cell compartments in the small intestine of mice lacking Tcf-4. Nat Genet 19:379–383
147. van Es JH, Haegebarth A, Kujala P, Itzkovitz S, Koo BK, Boj SF, Korving J, van den Born M, van Oudenaarden A, Robine S et al (2012a) A critical role for the Wnt effector Tcf4 in adult intestinal homeostatic self-renewal. Mol Cell Biol 32:1918–1927
148. Schuijers J, Clevers H (2012) Adult mammalian stem cells: the role of Wnt, Lgr5 and R-spondins. EMBO J 31:2685–2696
149. van de Wetering M, Sancho E, Verweij C, de Lau W, Oving I, Hurlstone A, van der Horn K, Batlle E, Coudreuse D, Haramis AP et al (2002) The beta-catenin/TCF-4 complex imposes a crypt progenitor phenotype on colorectal cancer cells. Cell 111:241–250
150. Morita H, Mazerbourg S, Bouley DM, Luo CW, Kawamura K, Kuwabara Y, Baribault H, Tian H, Hsueh AJ (2004) Neonatal lethality of LGR5 null mice is associated with ankyloglossia and gastrointestinal distension. Mol Cell Biol 24:9736–9743
151. Garcia MI, Ghiani M, Lefort A, Libert F, Strollo S, Vassart G (2009) LGR5 deficiency deregulates Wnt signaling and leads to precocious Paneth cell differentiation in the fetal intestine. Dev Biol 331:58–67
152. de Lau W, Barker N, Low TY, Koo BK, Li VS, Teunissen H, Kujala P, Haegebarth A, Peters PJ, van de Wetering M et al (2011) Lgr5 homologues associate with Wnt receptors and mediate R-spondin signalling. Nature 476:293–297
153. Mustata RC, Van Loy T, Lefort A, Libert F, Strollo S, Vassart G, Garcia MI (2011) Lgr4 is required for Paneth cell differentiation and maintenance of intestinal stem cells ex vivo. EMBO Rep 12:558–564
154. Carmon KS, Gong X, Lin Q, Thomas A, Liu Q (2011) R-spondins function as ligands of the orphan receptors LGR4 and LGR5 to regulate Wnt/beta-catenin signaling. Proc Natl Acad Sci U S A 108:11452–11457
155. Glinka A, Dolde C, Kirsch N, Huang YL, Kazanskaya O, Ingelfinger D, Boutros M, Cruciat CM, Niehrs C (2011) LGR4 and LGR5 are R-spondin receptors mediating Wnt/beta-catenin and Wnt/PCP signalling. EMBO Rep 12:1055–1061
156. Kim KA, Kakitani M, Zhao J, Oshima T, Tang T, Binnerts M, Liu Y, Boyle B, Park E, Emtage P et al (2005) Mitogenic influence of human R-spondin1 on the intestinal epithelium. Science 309:1256–1259
157. Hsu SY, Kudo M, Chen T, Nakabayashi K, Bhalla A, van der Spek PJ, van Duin M, Hsueh AJ (2000) The three subfamilies of leucine-rich repeat-containing G protein-coupled receptors (LGR): identification of LGR6 and LGR7 and the signaling mechanism for LGR7. Mol Endocrinol 14:1257–1271
158. Hsu SY, Liang SG, Hsueh AJ (1998) Characterization of two LGR genes homologous to gonadotropin and thyrotropin receptors with extracellular leucine-rich repeats and a G protein-coupled, seven-transmembrane region. Mol Endocrinol 12:1830–1845
159. McDonald T, Wang R, Bailey W, Xie G, Chen F, Caskey CT, Liu Q (1998) Identification and cloning of an orphan G protein-coupled receptor of the glycoprotein hormone receptor subfamily. Biochem Biophys Res Commun 247:266–270
160. Ruffner H, Sprunger J, Charlat O, Leighton-Davies J, Grosshans B, Salathe A, Zietzling S, Beck V, Therier M, Isken A et al (2012) R-Spondin potentiates Wnt/beta-catenin signaling through orphan receptors LGR4 and LGR5. PLoS One 7:e40976
161. Carmon KS, Lin Q, Gong X, Thomas A, Liu Q (2012) LGR5 interacts and cointernalizes with Wnt receptors to modulate Wnt/beta-catenin signaling. Mol Cell Biol 32:2054–2064
162. Hao HX, Xie Y, Zhang Y, Charlat O, Oster E, Avello M, Lei H, Mickanin C, Liu D, Ruffner H et al (2012) ZNRF3 promotes Wnt receptor turnover in an R-spondin-sensitive manner. Nature 485:195–200

163. Koo BK, Spit M, Jordens I, Low TY, Stange DE, van de Wetering M, van Es JH, Mohammed S, Heck AJ, Maurice MM et al (2012) Tumour suppressor RNF43 is a stem-cell E3 ligase that induces endocytosis of Wnt receptors. Nature 488:665–669
164. Van der Flier LG, Sabates-Bellver J, Oving I, Haegebarth A, De Palo M, Anti M, Van Gijn ME, Suijkerbuijk S, Van de Wetering M, Marra G et al (2007) The intestinal Wnt/TCF signature. Gastroenterology 132:628–632
165. Chen PH, Chen X, Lin Z, Fang D, He X (2013) The structural basis of R-spondin recognition by LGR5 and RNF43. Genes Dev 27:1345–1350
166. Peng WC, de Lau W, Forneris F, Granneman JC, Huch M, Clevers H, Gros P (2013) Structure of stem cell growth factor R-spondin 1 in complex with the ectodomain of its receptor LGR5. Cell Reports 3:1885–1892
167. Wang D, Huang B, Zhang S, Yu X, Wu W, Wang X (2013) Structural basis for R-spondin recognition by LGR4/5/6 receptors. Genes Dev 27:1339–1344
168. Blache P, van de Wetering M, Duluc I, Domon C, Berta P, Freund JN, Clevers H, Jay P (2004) SOX9 is an intestine crypt transcription factor, is regulated by the Wnt pathway, and represses the CDX2 and MUC2 genes. J Cell Biol 166:37–47
169. Formeister EJ, Sionas AL, Lorance DK, Barkley CL, Lee GH, Magness ST (2009) Distinct SOX9 levels differentially mark stem/progenitor populations and enteroendocrine cells of the small intestine epithelium. Am J Physiol Gastrointest Liver Physiol 296:G1108–G1118
170. Gracz AD, Ramalingam S, Magness ST (2010) Sox9 expression marks a subset of CD24-expressing small intestine epithelial stem cells that form organoids in vitro. Am J Physiol Gastrointest Liver Physiol 298:G590–G600
171. Van Landeghem L, Santoro MA, Krebs AE, Mah AT, Dehmer JJ, Gracz AD, Scull BP, McNaughton K, Magness ST, Lund PK (2012) Activation of two distinct Sox9-EGFP-expressing intestinal stem cell populations during crypt regeneration after irradiation. Am J Physiol Gastrointest Liver Physiol 302:G1111–G1132
172. Jubb AM, Chalasani S, Frantz GD, Smits R, Grabsch HI, Kavi V, Maughan NJ, Hillan KJ, Quirke P, Koeppen H (2006) Achaete-scute like 2 (ascl2) is a target of Wnt signalling and is upregulated in intestinal neoplasia. Oncogene 25:3445–3457
173. Hoffmeyer K, Raggioli A, Rudloff S, Anton R, Hierholzer A, Del Valle I, Hein K, Vogt R, Kemler R (2012) Wnt/beta-catenin signaling regulates telomerase in stem cells and cancer cells. Science 336:1549–1554
174. Batlle E, Henderson JT, Beghtel H, van den Born MM, Sancho E, Huls G, Meeldijk J, Robertson J, van de Wetering M, Pawson T et al (2002) Beta-catenin and TCF mediate cell positioning in the intestinal epithelium by controlling the expression of EphB/ephrinB. Cell 111:251–263
175. Holmberg J, Genander M, Halford MM, Anneren C, Sondell M, Chumley MJ, Silvany RE, Henkemeyer M, Frisen J (2006) EphB receptors coordinate migration and proliferation in the intestinal stem cell niche. Cell 125:1151–1163
176. Jung P, Sato T, Merlos-Suarez A, Barriga FM, Iglesias M, Rossell D, Auer H, Gallardo M, Blasco MA, Sancho E et al (2011) Isolation and in vitro expansion of human colonic stem cells. Nat Med 17:1225–1227
177. He TC, Sparks AB, Rago C, Hermeking H, Zawel L, da Costa LT, Morin PJ, Vogelstein B, Kinzler KW (1998) Identification of c-MYC as a target of the APC pathway. Science 281, 1509–1512
178. Finch AJ, Soucek L, Junttila MR, Swigart LB, Evan GI (2009) Acute overexpression of Myc in intestinal epithelium recapitulates some but not all the changes elicited by Wnt/beta-catenin pathway activation. Mol Cell Biol 29:5306–5315
179. Muncan V, Sansom OJ, Tertoolen L, Phesse TJ, Begthel H, Sancho E, Cole AM, Gregorieff A, de Alboran IM, Clevers H et al (2006) Rapid loss of intestinal crypts upon conditional deletion of the Wnt/Tcf-4 target gene c-Myc. Mol Cell Biol 26:8418–8426
180. Sansom OJ, Meniel VS, Muncan V, Phesse TJ, Wilkins JA, Reed KR, Vass JK, Athineos D, Clevers H, Clarke AR (2007) Myc deletion rescues Apc deficiency in the small intestine. Nature 446:676–679

181. Shtutman M, Zhurinsky J, Simcha I, Albanese C, D'Amico M, Pestell R, Ben-Ze'ev A (1999) The cyclin D1 gene is a target of the beta-catenin/LEF-1 pathway. Proc Natl Acad Sci U S A 96:5522–5527
182. Tetsu O, McCormick F (1999) Beta-catenin regulates expression of cyclin D1 in colon carcinoma cells. Nature 398:422–426
183. Cole AM, Myant K, Reed KR, Ridgway RA, Athineos D, Van den Brink GR, Muncan V, Clevers H, Clarke AR, Sicinski P et al (2010) Cyclin D2-cyclin-dependent kinase 4/6 is required for efficient proliferation and tumorigenesis following Apc loss. Cancer Res 70:8149–8158
184. Hulit J, Wang C, Li Z, Albanese C, Rao M, Di Vizio D, Shah S, Byers SW, Mahmood R, Augenlicht LH et al (2004) Cyclin D1 genetic heterozygosity regulates colonic epithelial cell differentiation and tumor number in ApcMin mice. Mol Cell Biol 24:7598–7611
185. Kovall RA, Blacklow SC (2010) Mechanistic insights into Notch receptor signaling from structural and biochemical studies. Curr Top Dev Biol 92:31–71
186. Perdigoto CN, Bardin AJ (2013) Sending the right signal: Notch and stem cells. Biochim Biophys Acta 1830:2307–2322
187. Vooijs M, Liu Z, Kopan R (2011) Notch: architect, landscaper, and guardian of the intestine. Gastroenterology 141:448–459
188. Sander GR, Powell BC (2004) Expression of notch receptors and ligands in the adult gut. J Histochem Cytochem 52:509–516
189. Schroder N, Gossler A (2002) Expression of Notch pathway components in fetal and adult mouse small intestine. Gene Expr Patterns 2:247–250
190. Riccio O, van Gijn ME, Bezdek AC, Pellegrinet L, van Es JH, Zimber-Strobl U, Strobl LJ, Honjo T, Clevers H, Radtke F (2008) Loss of intestinal crypt progenitor cells owing to inactivation of both Notch1 and Notch2 is accompanied by derepression of CDK inhibitors p27Kip1 and p57Kip2. EMBO Rep 9:377–383
191. Benedito R, Duarte A (2005) Expression of Dll4 during mouse embryogenesis suggests multiple developmental roles. Gene Expr Patterns 5:750–755
192. Crosnier C, Vargesson N, Gschmeissner S, Ariza-McNaughton L, Morrison A, Lewis J (2005) Delta-Notch signalling controls commitment to a secretory fate in the zebrafish intestine. Development 132:1093–1104
193. Milano J, McKay J, Dagenais C, Foster-Brown L, Pognan F, Gadient R, Jacobs RT, Zacco A, Greenberg B, Ciaccio PJ (2004) Modulation of notch processing by gamma-secretase inhibitors causes intestinal goblet cell metaplasia and induction of genes known to specify gut secretory lineage differentiation. Toxicol Sci 82:341–358
194. van Es JH, van Gijn ME, Riccio O, van den Born M, Vooijs M, Begthel H, Cozijnsen M, Robine S, Winton DJ, Radtke F et al (2005b) Notch/gamma-secretase inhibition turns proliferative cells in intestinal crypts and adenomas into goblet cells. Nature 435:959–963
195. VanDussen KL, Carulli AJ, Keeley TM, Patel SR, Puthoff BJ, Magness ST, Tran IT, Maillard I, Siebel C, Kolterud A et al (2012) Notch signaling modulates proliferation and differentiation of intestinal crypt base columnar stem cells. Development 139:488–497
196. Wu Y, Cain-Hom C, Choy L, Hagenbeek TJ, de Leon GP, Chen Y, Finkle D, Venook R, Wu X, Ridgway J et al (2010) Therapeutic antibody targeting of individual Notch receptors. Nature 464:1052–1057
197. Vooijs M, Ong CT, Hadland B, Huppert S, Liu Z, Korving J, van den Born M, Stappenbeck T, Wu Y, Clevers H et al (2007) Mapping the consequence of Notch1 proteolysis in vivo with NIP-CRE. Development 134:535–544
198. Fre S, Huyghe M, Mourikis P, Robine S, Louvard D, Artavanis-Tsakonas S (2005) Notch signals control the fate of immature progenitor cells in the intestine. Nature 435:964–968
199. Stanger BZ, Datar R, Murtaugh LC, Melton DA (2005) Direct regulation of intestinal fate by Notch. Proc Natl Acad Sci U S A 102:12443–12448
200. Pellegrinet L, Rodilla V, Liu Z, Chen S, Koch U, Espinosa L, Kaestner KH, Kopan R, Lewis J, Radtke F (2011) Dll1- and dll4-mediated notch signaling are required for homeostasis of intestinal stem cells. Gastroenterology 140:1230–1240, e1231–e1237

201. Haines N, Irvine KD (2003) Glycosylation regulates Notch signalling. Nat Rev Mol Cell Biol 4:786–797
202. Shi S, Stanley P (2003) Protein O-fucosyltransferase 1 is an essential component of Notch signaling pathways. Proc Natl Acad Sci U S A 100:5234–5239
203. Waterhouse CC, Johnson S, Phillipson M, Zbytnuik L, Petri B, Kelly M, Lowe JB, Kubes P (2010) Secretory cell hyperplasia and defects in Notch activity in a mouse model of leukocyte adhesion deficiency type II. Gastroenterology 138:1079–1090, e1071–e1075
204. Guilmeau S, Flandez M, Bancroft L, Sellers RS, Tear B, Stanley P, Augenlicht LH (2008) Intestinal deletion of Pofut1 in the mouse inactivates notch signaling and causes enterocolitis. Gastroenterology 135:849–860, 860, e841–e846
205. Itoh M, Kim CH, Palardy G, Oda T, Jiang YJ, Maust D, Yeo SY, Lorick K, Wright GJ, Ariza-McNaughton L et al (2003) Mind bomb is a ubiquitin ligase that is essential for efficient activation of Notch signaling by Delta. Dev Cell 4:67–82
206. Koo BK, Lim HS, Song R, Yoon MJ, Yoon KJ, Moon JS, Kim YW, Kwon MC, Yoo KW, Kong MP et al (2005) Mind bomb 1 is essential for generating functional Notch ligands to activate Notch. Development 132:3459–3470
207. Koo BK, Lim HS, Chang HJ, Yoon MJ, Choi Y, Kong MP, Kim CH, Kim JM, Park JG, Kong YY (2009) Notch signaling promotes the generation of EphrinB1-positive intestinal epithelial cells. Gastroenterology 137:145–155, 155, e141–143
208. Yang Y, Zhu R, Bai J, Zhang X, Tian Y, Li X, Peng Z, He Y, Chen L, Ji Q et al (2011) Numb modulates intestinal epithelial cells toward goblet cell phenotype by inhibiting the Notch signaling pathway. Exp Cell Res 317:1640–1648
209. Rezza A, Skah S, Roche C, Nadjar J, Samarut J, Plateroti M (2010) The overexpression of the putative gut stem cell marker Musashi-1 induces tumorigenesis through Wnt and Notch activation. J Cell Sci 123:3256–3265
210. Sureban SM, May R, George RJ, Dieckgraefe BK, McLeod HL, Ramalingam S, Bishnupuri KS, Natarajan G, Anant S, Houchen CW (2008) Knockdown of RNA binding protein musashi-1 leads to tumor regression in vivo. Gastroenterology 134:1448–1458
211. Jensen J, Pedersen EE, Galante P, Hald J, Heller RS, Ishibashi M, Kageyama R, Guillemot F, Serup P, Madsen OD (2000) Control of endodermal endocrine development by Hes-1. Nat Genet 24:36–44
212. Ueo T, Imayoshi I, Kobayashi T, Ohtsuka T, Seno H, Nakase H, Chiba T, Kageyama R (2012) The role of Hes genes in intestinal development, homeostasis and tumor formation. Development 139:1071–1082
213. Yang Q, Bermingham NA, Finegold MJ, Zoghbi HY (2001) Requirement of Math1 for secretory cell lineage commitment in the mouse intestine. Science 294:2155–2158
214. Shroyer NF, Helmrath MA, Wang VY, Antalffy B, Henning SJ, Zoghbi HY (2007) Intestine-specific ablation of mouse atonal homolog 1 (Math1) reveals a role in cellular homeostasis. Gastroenterology 132:2478–2488
215. VanDussen KL, Samuelson LC (2010) Mouse atonal homolog 1 directs intestinal progenitors to secretory cell rather than absorptive cell fate. Dev Biol 346:215–223
216. Kazanjian A, Noah T, Brown D, Burkart J, Shroyer NF (2010) Atonal homolog 1 is required for growth and differentiation effects of notch/gamma-secretase inhibitors on normal and cancerous intestinal epithelial cells. Gastroenterology 139:918–928, 928, e911–e916
217. Kim TH, Shivdasani RA (2011) Genetic evidence that intestinal Notch functions vary regionally and operate through a common mechanism of Math1 repression. J Biol Chem 286:11427–11433
218. van Es JH, de Geest N, van de Born M, Clevers H, Hassan BA (2010) Intestinal stem cells lacking the Math1 tumour suppressor are refractory to Notch inhibitors. Nat Commun 1:18
219. Zheng X, Tsuchiya K, Okamoto R, Iwasaki M, Kano Y, Sakamoto N, Nakamura T, Watanabe M (2011) Suppression of hath1 gene expression directly regulated by hes1 via notch signaling is associated with goblet cell depletion in ulcerative colitis. Inflamm Bowel Dis 17:2251–2260
220. Zhang J, Li L (2005) BMP signaling and stem cell regulation. Dev Biol 284:1–11

221. Haramis AP, Begthel H, van den Born M, van Es J, Jonkheer S, Offerhaus GJ, Clevers H (2004) De novo crypt formation and juvenile polyposis on BMP inhibition in mouse intestine. Science 303:1684–1686
222. He XC, Zhang J, Tong WG, Tawfik O, Ross J, Scoville DH, Tian Q, Zeng X, He X, Wiedemann LM et al (2004) BMP signaling inhibits intestinal stem cell self-renewal through suppression of Wnt-beta-catenin signaling. Nat Genet 36:1117–1121
223. Latchford AR, Neale K, Phillips RK, Clark SK (2012) Juvenile polyposis syndrome: a study of genotype, phenotype, long-term outcome. Dis Colon Rectum 55:1038–1043
224. Batts LE, Polk DB, Dubois RN, Kulessa H (2006) Bmp signaling is required for intestinal growth and morphogenesis. Dev Dyn 235:1563–1570
225. Auclair BA, Benoit YD, Rivard N, Mishina Y, Perreault N (2007) Bone morphogenetic protein signaling is essential for terminal differentiation of the intestinal secretory cell lineage. Gastroenterology 133: 887–896
226. Kosinski C, Li VS, Chan AS, Zhang J, Ho C, Tsui WY, Chan TL, Mifflin RC, Powell DW, Yuen ST et al (2007) Gene expression patterns of human colon tops and basal crypts and BMP antagonists as intestinal stem cell niche factors. Proc Natl Acad Sci U S A 104:15418–15423
227. Hardwick JC, Van Den Brink GR, Bleuming SA, Ballester I, Van Den Brande JM, Keller JJ, Offerhaus GJ, Van Deventer SJ, Peppelenbosch MP (2004) Bone morphogenetic protein 2 is expressed by, and acts upon, mature epithelial cells in the colon. Gastroenterology 126:111–121
228. Salovaara R, Roth S, Loukola A, Launonen V, Sistonen P, Avizienyte E, Kristo P, Jarvinen H, Souchelnytskyi S, Sarlomo-Rikala M et al (2002) Frequent loss of SMAD4/DPC4 protein in colorectal cancers. Gut 51:56–59
229. Alberici P, Jagmohan-Changur S, De Pater E, Van Der Valk M, Smits R, Hohenstein P, Fodde R (2006) Smad4 haploinsufficiency in mouse models for intestinal cancer. Oncogene 25:1841–1851
230. Kitamura T, Kometani K, Hashida H, Matsunaga A, Miyoshi H, Hosogi H, Aoki M, Oshima M, Hattori M, Takabayashi A et al (2007) SMAD4-deficient intestinal tumors recruit CCR1+ myeloid cells that promote invasion. Nat Genet 39:467–475
231. Beck SE, Jung BH, Fiorino A, Gomez J, Rosario ED, Cabrera BL, Huang SC, Chow JY, Carethers JM (2006) Bone morphogenetic protein signaling and growth suppression in colon cancer. Am J Physiol Gastrointest Liver Physiol 291:G135–G145
232. Loh K, Chia JA, Greco S, Cozzi SJ, Buttenshaw RL, Bond CE, Simms LA, Pike T, Young JP, Jass JR et al (2008) Bone morphogenic protein 3 inactivation is an early and frequent event in colorectal cancer development. Genes Chromosomes Cancer 47:449–460
233. Deng H, Makizumi R, Ravikumar TS, Dong H, Yang W, Yang WL (2007) Bone morphogenetic protein-4 is overexpressed in colonic adenocarcinomas and promotes migration and invasion of HCT116 cells. Exp Cell Res 313:1033–1044
234. Motoyama K, Tanaka F, Kosaka Y, Mimori K, Uetake H, Inoue H, Sugihara K, Mori M (2008) Clinical significance of BMP7 in human colorectal cancer. Ann Surg Oncol 15:1530–1537
235. Merchant JL, Saqui-Salces M (2014) Inhibition of Hedgehog signaling in the gastrointestinal tract: targeting the cancer microenvironment. Cancer Treat Rev 40:12–21
236. van den Brink GR, Bleuming SA, Hardwick JC, Schepman BL, Offerhaus GJ, Keller JJ, Nielsen C, Gaffield W, van Deventer SJ, Roberts DJ et al (2004) Indian Hedgehog is an antagonist of Wnt signaling in colonic epithelial cell differentiation. Nat Genet 36:277–282
237. van Dop WA, Heijmans J, Buller NV, Snoek SA, Rosekrans SL, Wassenberg EA, van den Bergh Weerman MA, Lanske B, Clarke AR, Winton DJ et al (2010) Loss of Indian Hedgehog activates multiple aspects of a wound healing response in the mouse intestine. Gastroenterology 139:1665–1676, 1676, e1661–e1610
238. van Dop WA, Uhmann A, Wijgerde M, Sleddens-Linkels E, Heijmans J, Offerhaus GJ, van den Bergh Weerman MA, Boeckxstaens GE, Hommes DW, Hardwick JC et al (2009) Deple-

tion of the colonic epithelial precursor cell compartment upon conditional activation of the hedgehog pathway. Gastroenterology 136:2195–2203, e2191–e2197
239. Kosinski C, Stange DE, Xu C, Chan AS, Ho C, Yuen ST, Mifflin RC, Powell DW, Clevers H, Leung SY et al (2010) Indian hedgehog regulates intestinal stem cell fate through epithelial-mesenchymal interactions during development. Gastroenterology 139:893–903
240. Lemmon MA, Schlessinger J (2010) Cell signaling by receptor tyrosine kinases. Cell 141:1117–1134
241. Aliaga JC, Deschenes C, Beaulieu JF, Calvo EL, Rivard N (1999) Requirement of the MAP kinase cascade for cell cycle progression and differentiation of human intestinal cells. Am J Physiol 277:G631–G641
242. Suzuki A, Sekiya S, Gunshima E, Fujii S, Taniguchi H (2010) EGF signaling activates proliferation and blocks apoptosis of mouse and human intestinal stem/progenitor cells in long-term monolayer cell culture. Lab Invest 90:1425–1436
243. Konturek JW, Bielanski W, Konturek SJ, Bogdal J, Oleksy J (1989) Distribution and release of epidermal growth factor in man. Gut 30:1194–1200
244. Messa C, Russo F, Caruso MG, Di Leo A (1998) EGF, TGF-alpha, EGF-R in human colorectal adenocarcinoma. Acta Oncol 37:285–289
245. Biteau B, Jasper H (2011) EGF signaling regulates the proliferation of intestinal stem cells in Drosophila. Development 138:1045–1055
246. Houchen CW, George RJ, Sturmoski MA, Cohn SM (1999) FGF-2 enhances intestinal stem cell survival and its expression is induced after radiation injury. Am J Physiol 276:G249–G258
247. Khan WB, Shui C, Ning S, Knox SJ (1997) Enhancement of murine intestinal stem cell survival after irradiation by keratinocyte growth factor. Radiat Res 148:248–253
248. Arnaud-Dabernat S, Yadav D, Sarvetnick N (2008) FGFR3 contributes to intestinal crypt cell growth arrest. J Cell Physiol 216:261–268
249. Wang K, Degerny C, Xu M, Yang XJ (2009) YAP, TAZ, and Yorkie: a conserved family of signal-responsive transcriptional coregulators in animal development and human disease. Biochem Cell Biol 87:77–91
250. Yang, Z, Hata, Y (2013) What is the Hippo pathway? Is the Hippo pathway conserved in Caenorhabditis elegans? J Biochem 154:207–209
251. Lee JH, Kim TS, Yang TH, Koo BK, Oh SP, Lee KP, Oh HJ, Lee SH, Kong YY, Kim JM et al (2008) A crucial role of WW45 in developing epithelial tissues in the mouse. EMBO J 27:1231–1242
252. Barry ER, Morikawa T, Butler BL, Shrestha K, de la Rosa R, Yan KS, Fuchs CS, Magness ST, Smits R, Ogino S et al (2013) Restriction of intestinal stem cell expansion and the regenerative response by YAP. Nature 493:106–110
253. Cai J, Zhang N, Zheng Y, de Wilde RF, Maitra A, Pan D (2010) The Hippo signaling pathway restricts the oncogenic potential of an intestinal regeneration program. Genes Dev 24:2383–2388
254. Zhou D, Zhang Y, Wu H, Barry E, Yin Y, Lawrence E, Dawson D, Willis JE, Markowitz SD, Camargo FD et al (2011) Mst1 and Mst2 protein kinases restrain intestinal stem cell proliferation and colonic tumorigenesis by inhibition of Yes-associated protein (Yap) overabundance. Proc Natl Acad Sci U S A 108:E1312–E1320
255. Wehkamp J, Wang G, Kubler I, Nuding S, Gregorieff A, Schnabel A, Kays RJ, Fellermann K, Burk O, Schwab M et al (2007b) The Paneth cell alpha-defensin deficiency of ileal Crohn's disease is linked to Wnt/Tcf-4. J Immunol 179:3109–3118
256. Koslowski MJ, Kubler I, Chamaillard M, Schaeffeler E, Reinisch W, Wang G, Beisner J, Teml A, Peyrin-Biroulet L, Winter S et al (2009) Genetic variants of Wnt transcription factor TCF-4 (TCF7L2) putative promoter region are associated with small intestinal Crohn's disease. PLoS One 4:e4496
257. Wehkamp J, Salzman NH, Porter E, Nuding S, Weichenthal M, Petras RE, Shen B, Schaeffeler E, Schwab M, Linzmeier R et al (2005) Reduced Paneth cell alpha-defensins in ileal Crohn's disease. Proc Natl Acad Sci U S A 102:18129–18134

258. Wehkamp J, Schmid M, Stange EF (2007a) Defensins and other antimicrobial peptides in inflammatory bowel disease. Curr Opin Gastroenterol 23:370–378
259. Wehkamp J, Stange EF (2010) Paneth's disease. J Crohn's & Colitis 4:523–531
260. McCormick DA, Horton LW, Mee AS (1990) Mucin depletion in inflammatory bowel disease. J Clin Pathol 43:143–146
261. Strugala V, Dettmar PW, Pearson JP (2008) Thickness and continuity of the adherent colonic mucus barrier in active and quiescent ulcerative colitis and Crohn's disease. Int J Clin Pract 62:762–769
262. Gersemann M, Becker S, Kubler I, Koslowski M, Wang G, Herrlinger KR, Griger J, Fritz P, Fellermann K, Schwab M et al (2009) Differences in goblet cell differentiation between Crohn's disease and ulcerative colitis. Differentiation 77:84–94
263. Jemal A, Bray F, Center MM, Ferlay J, Ward E, Forman D (2011) Global cancer statistics. CA Cancer J Clin 61:69–90
264. Clarke MF, Dick JE, Dirks PB, Eaves CJ, Jamieson CH, Jones DL, Visvader J, Weissman IL, Wahl GM (2006) Cancer stem cells–perspectives on current status and future directions: AACR Workshop on cancer stem cells. Cancer Res 66:9339–9344
265. Todaro M, Francipane MG, Medema JP, Stassi G (2010) Colon cancer stem cells: promise of targeted therapy. Gastroenterology 138:2151–2162
266. Schepers AG, Snippert HJ, Stange DE, van den Born M, van Es JH, van de Wetering M, Clevers H (2012) Lineage tracing reveals Lgr5+ stem cell activity in mouse intestinal adenomas. Science 337:730–735
267. Schwitalla S, Fingerle AA, Cammareri P, Nebelsiek T, Goktuna SI, Ziegler PK, Canli O, Heijmans J, Huels DJ, Moreaux G et al (2013) Intestinal tumorigenesis initiated by dedifferentiation and acquisition of stem-cell-like properties. Cell 152:25–38
268. Kemper K, Prasetyanti PR, De Lau W, Rodermond H, Clevers H, Medema JP (2012) Monoclonal antibodies against Lgr5 identify human colorectal cancer stem cells. Stem Cells 30:2378–2386
269. O'Brien CA, Pollett A, Gallinger S, Dick JE (2007) A human colon cancer cell capable of initiating tumour growth in immunodeficient mice. Nature 445:106–110
270. Ricci-Vitiani L, Lombardi DG, Pilozzi E, Biffoni M, Todaro M, Peschle C, De Maria R (2007) Identification and expansion of human colon-cancer-initiating cells. Nature 445:111–115
271. Nakanishi Y, Seno H, Fukuoka A, Ueo T, Yamaga Y, Maruno T, Nakanishi N, Kanda K, Komekado H, Kawada M et al (2013) Dclk1 distinguishes between tumor and normal stem cells in the intestine. Nat Genet 45:98–103

The Role of PI3K Signaling Pathway in Intestinal Tumorigenesis

Jennifer W. Harris, Tianyan Gao and B. Mark Evers

Abstract The phosphoinositide 3-kinases (PI3K) are a family of signaling enzymes that regulate multiple key cellular functions including cell cycle progression, growth, migration, apoptosis, and vesicular trafficking. Components of this pathway are necessary for normal growth and development, but due to both genetic and metabolic changes, this pathway can become dysregulated as early as the fetal period. Continued alterations of the PI3K pathway can frequently lead to activation of cancer development and metastasis. As more information is discovered about the components of this pathway and their signaling mechanisms we have opportunities to exploit multiple sites of the cascade, and develop targeted inhibitors against these small molecules. A thorough understanding of this pathway and its signaling mechanisms is paramount to the development of new and improved treatments for intestinal cancers.

Keywords Phosphoinositide-3-kinase · Tumorigenesis · Protein kinase B · Small intestine · Colon · Kinase cascade · Tumor metabolism

Abbreviations

4E-BP	Eukaryotic translation initiation factor 4E-binding protein
5-FU	5-fluorouracil
BMP	Bone morphogenetic protein
CCK2R	Gastrin/cholecystokinin-2 receptor
c-FLIP	A master anti-apoptotic regulator
DNA-PK	DNA-dependent protein kinase

J. W. Harris (✉)
Department of Surgery, University of Kentucky,
Lexington, KY, USA
e-mail: jenn.whittington@uky.edu

T. Gao
Department of Molecular and Cellular Biochemistry, University of Kentucky,
Lexington, KY, USA

B. M. Evers
Department of Surgery, Markey Cancer Center, University of Kentucky, Lexington, KY, USA

V. W. Yang, A. B. Bialkowska (eds.), *Intestinal Tumorigenesis,*
DOI 10.1007/978-3-319-19986-3_4

EGF	Epidermal growth factor
EGFR	Epidermal growth factor receptors
EMT	Epithelial-mesenchymal transition
FasL	A type-II transmembrane protein of tumor necrosis factor
FGF	Fibroblast growth factor
FGFR	Fibroblast growth factor receptor
FOXO	Forkhead box subgroup O
GI	Gastrointestinal
GISTs	Gastrointestinal stromal tumors
GPCR	G-protein couple receptors
GSK3β	Glycogen synthase kinase-3 beta
GTP	Guanosine triphosphate
HER2	Human epidermal growth factor receptor 2
IGF-1	Insulin like growth factor 1
LY294002	2-(4-morpholinyl)-8-phenyl-4H-1-benzopyran-4-one
mTOR	Mammalian target of rapamycin
mTORC	mTOR complex (mTORC1 and mTORC2)
PDK1	3-phosphoinositide dependent protein kinase-1
PH	Pleckstrin homology
PHLPP	PH domain and leucine rich repeat protein phosphatases
PI3K	Phosphatidylinositide 3-kinase
PIK3CA	Gene encoding p110α
PIK3R1	Gene encoding p85α
PIP_2	Phosphatidylinositol 4,5-bisphosphate
PIP_3	Phosphatidylinositol (3, 4, 5) phosphate
PP2A	Protein phosphatase 2A
PTEN	Phosphatase and tensin homolog
Rheb	Ras homolog enriched in brain
RICTOR	Rapamycin insensitive component of mTOR
RTK	Receptor tyrosine kinase
S6K	S6 kinase
SCCE	Squamous cell carcinoma of the esophagus
SH2	Src-homology 2
TM	Transmembrane
TSC	Tuberous sclerosis complex
VEGF	Vascular endothelial growth factor
VEGFR	Vascular endothelial growth factor receptor

1 Introduction

The phosphoinositide 3-kinase (PI3K) signaling pathway regulates numerous processes including growth, metabolism, motility, and survival in the normal cell [1]. Multiple sites of this signaling pathway are prone to either activation of normal cellular pathways or inactivation of genes that usually act as tumor suppressors. These molecular changes lead to tumorigenesis and maintenance of a malignant phenotype and result in unregulated cell growth, tumor development, and metastasis [2, 3]. The PI3K signaling pathway has been strongly implicated in the development of cancers of the gastrointestinal (GI) tract. Examples of alterations of the molecular mechanisms and signaling dysfunction present in GI carcinogenesis include: gain of function mutations in growth receptors seen in esophageal, gastric, pancreatic, and colon cancers; somatic mutations of the *PIK3CA* gene seen predominantly in colorectal cancers; and inhibition of tumor suppressors seen in gastric cancer, hepatobiliary malignancies, small bowel adenocarcinoma, and colorectal cancers [4–7]. Multiple laboratories are focused on identifying novel ways to target the PI3K pathway components as anti-cancer therapeutics [8–11].

1.1 Overview of Pathway

The Class IA PI3Ks are the most widely implicated class of PI3Ks in cancer development. Class IA PI3Ks are usually activated by receptor tyrosine kinase (RTK) signaling, but can also be activated by G-protein couple receptors (GPCRs) if activated by the p110β enzymes located on the catalytic subunit of PI3K [12, 13]. This class of PI3Ks phosphorylate phosphatidylinositol 4,5-bisphosphate (PIP_2) on the plasma membrane to generate the second messenger phosphatidylinositol 3,4,5-trisphosphate (PIP_3) [14]. Phosphatase and tensin homolog (PTEN) is a protein that acts as a tumor suppressor by dephosphorylating PIP_3 and terminating PI3K signaling [7]. The class IA PI3K consists of a p85 regulatory subunit that is closely coupled to RTK signaling, and a p110 catalytic subunit that is involved with GPCR signaling. The Src-homology 2 (SH2) domain of the p85 subunit binds to phosphotyrosine residues on RTKs, which recruits the p85/p110 heterodimer to the plasma membrane where its substrate resides [15]. Pleckstrin homology (PH) domains play a role in recruiting proteins to the plasma membrane by enabling them to interact with other components of the signal transduction pathways after they are stimulated by the accumulation of PIP_3 residues [16]. AKT (also known as protein kinase B) and phosphoinositide-dependent protein kinase-1 (PDK1) directly bind to PIP_3 and are recruited to the plasma membrane by PH. The phosphorylation of AKT at threonine-308, and by PDK1 at the serine-473 motif by mammalian target of rapamycin complex 2 (mTORC2) results in the full activation of this protein kinase [17]. AKT then phosphorylates and coordinates the actions of multiple intracellular proteins that regulate cellular energy levels and metabolism, cell growth and proliferation, cell cycle, angiogenesis, and apoptosis [18–22] (Fig. 1).

1.2 Implications of Mutations in Intestinal Tumorigenesis

Neoplasia is caused by multiple genetic changes that take place over a period of decades and is driven by the progressive accumulation of genetic and epigenetic abnormalities that sequentially activate oncogenes or inactivate tumor suppressor

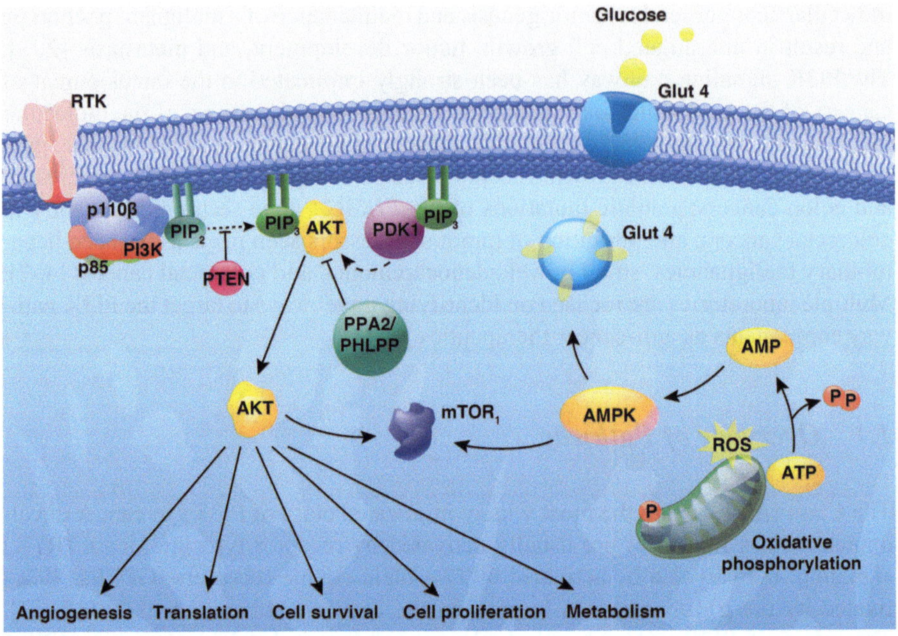

Fig. 1 Overview of PI3K pathway. Class IA PI3Ks are usually activated by RTK signaling, but can also be activated by GPCR. This class of PI3Ks phosphorylates PIP_2 on the plasma membrane to generate the second messenger PIP_3. These PI3Ks consist of a p85 regulatory subunit that is closely involved with RTK signaling, and a p110β catalytic subunit which is involved with GPCR signaling. PTEN is a protein encoded by a *PTEN* gene that acts as a tumor suppressor by dephosphorylating PIP_3 and terminating PI3K signaling. AKT and PDK1 directly bind to PIP_3 and are recruited to the plasma membrane. The phosphorylation of AKT at threonine-308 by PDK1 and at the serine-473 motif by mTORC2 results in the full activation of this protein kinase. AKT then phosphorylates mTORC1, which interacts with adenosine monophosphate kinase (AMPK) to impact cellular utilization of energy sources such as glucose. These interactions coordinate the multiple intracellular proteins that regulate cellular energy levels and metabolism, cell growth and proliferation, cell survival, angiogenesis, translation and apoptosis. Illustration provided by Matt Hazzard, Bio-Medical Illustrator, University of Kentucky.

genes [23, 24]. Dysregulation of the PI3K pathway offers molecular opportunities for tumorigenesis. Many studies have shown amplification of genomic regions coding for the *PIK3CA* gene in intestinal carcinogenesis [25–27].

Despite improved treatment strategies and multimodal therapy, the prevalence of GI cancers is increasing worldwide. In the U.S., the incidence of esophageal cancer has risen by 650 % in white males over the last 35 years [28, 29]. Overexpression of downstream effectors in the PI3K pathway has been linked to the poor prognosis in esophageal cancer. A recent study using tissue microarrays showed that mTOR, which is an important downstream effector of the PI3K/AKT signaling pathway, is overexpressed in patients with esophageal cancer with poor overall survival [30].

Increased PI3K signaling is related to both the development and resistance to therapeutic agents in gastric cancer. HER3 is a membrane-bound protein that can form heterodimers with other epidermal growth factor receptor (EGFR) family members with kinase activity. These HER3/EGFR heterodimers promote proliferation, tumor survival, and cell migration in gastric cancer mediated by the PI3K/AKT signaling pathway, and have increased expression in patients with decreased overall survival [31]. Cisplatin is an effective chemotherapy agent for gastric cancer, however, increased PI3K signaling independently increases resistance to this drug [32]. Although the incidence and mortality rates of gastric cancer have fallen dramatically over the last 50 years in many regions, the 5-year survival rate remains a dismal 20 % [33].

In industrialized countries, the incidence of pancreatic cancer ranks second after colorectal cancer among all GI malignancies. Surprisingly, the mortality rate for pancreatic cancer for males has decreased by 0.4 % over the past decade, but the mortality rate for females has increased by 4.4 % [34]. The loss of the tumor suppressor PTEN which acts upon PI3K and prevents overexpression of downstream effectors such as mTOR has been implicated in both cell progression and immunoresistance in pancreatic adenocarcinoma [35]. Interestingly *PIK3CA* gene mutations have been demonstrated in mucinous cystic neoplasms that are rare, potentially curable, mucin-producing neoplasms of the pancreas. A thorough understanding of *PIK3CA* status in these early lesions could be useful for understanding their progression to malignancy [36].

The prevalence of hepatocellular carcinoma (HCC) has been increasing in recent years and it is currently the third leading cause of cancer deaths worldwide, and the ninth leading cause of cancer death in the United States [37]. Both in vitro and in vivo studies of HCC have shown increased cell proliferation and tumor survival due to increased PI3K signaling. HCC cell lines displayed increased levels of p-IGF-IR, p-PI3K, p-Akt, Bcl-xL, and Bcl-2 while also inhibiting PTEN. Thus, HCC integrates multiple pathways for increased cell proliferation and tumor survival in this disease [38].

Although small bowel adenocarcinoma is relatively rare, its incidence has increased by almost 50 % from 1975 to 2000. This trend reflects increases for adenocarcinomas, malignant carcinoid tumors, and lymphomas in men, and for carcinoid tumors and lymphomas in women [39]. Overexpression of PI3K signaling, as well as PTEN inhibition, have also been linked to the development and progression of small bowel adenocarcinoma. Zhang et al. [40] evaluated 53 human small bowel adenocarcinoma samples, 11 non-cancer tissue samples and 11 normal small intestine tissues using quantum dots-based immunofluorescence histochemistry and traditional immunohistochemistry; PI3K and AKT expression were significantly higher in malignant tissue compared with the other tissue groups. Loss of PTEN has been implicated in colorectal cancer and a recent animal study suggests that intestinal epithelial-specific PTEN inactivation also results in tumor formation in the small intestine [41]. Colorectal cancers have been strongly linked to PI3K signaling pathway mutations and this pathway and its effectors have been extensively studied for cancer development and progression. Although rates for new colorectal cancer cases have been falling an average 2.9 % each year over the last 10 years and death rates have fallen an average 2.8 % each year over the same period, this disease still has significant morbidity and mortality and is the second most lethal type of cancer in men and women in the United States [42]. The onset of particularly aggressive colorectal cancers in an increasingly young patient population represents a worrisome trend and implicates a combination of environmental and genetic factors [43].

2 PI3K: Discovery, Treatment, and Role in Intestinal Tumorigenesis

2.1 Discovery and Determination of Classes and Isoforms

PI3K was identified in 1985 with the novel discovery of the existence of an oncoprotein-associated lipid kinase capable of phosphorylating the 3-hydroxyl group of the inositol ring of phosphatidylinositol [44]. Subsequent studies showed that the family of PI3K enzymes was an evolutionarily conserved group of enzymes involved in multiple cellular functions including cell growth, proliferation, motility, tumor survival, and intracellular transport [45, 46].

PI3K is further divided into three classes depending on the subunit structure, enzymatic regulation, and substrate selectivity [47]. PI3K Class IA, which is most frequently implicated in tumorigenesis, is composed of three catalytic subunits (p110α, p110β, p110δ) that form a heterodimer with one of five regulatory domains (p85α, p85β, p85γ, p50α, p55α). For example, the structure of PI3Kα is composed of the entire p110α subunit and the nSH2 and iSH2 domains of the regulatory p85 subunit [48].

2.2 The Role of PI3K, p110, and p85 in Intestinal Tumorigenesis

The *PIK3CA* gene encodes the p110α subunit, which is the major catalytic subunit of PI3K [49]. Mutations involving the *PIK3CA* and *PIK3R1* genes code for both of the PI3K subunits p110 and p85 respectively [50, 51]. Mutations of the *PIK3CA* gene as well as somatic mutations that affect the regulatory p85 subunit have been implicated in the development and progression of colon carcinogenesis [52, 53]. Tumors from patients with mutated *PIK3CA* genes tend to have worse overall outcomes compared to patients with wild type *PIK3CA* genes, and experience an increased colon cancer-specific mortality [54].

The p110 subunit is composed of five domains: an adaptor-binding domain, a Ras-binding domain, a C2 domain, a helical domain, and a kinase domain [55] (Fig. 2). In the inactivated state, p110α is bound to the regulatory subunit p85. The p85α regulatory subunit is the most frequently expressed form of the p85 subunits of Class IA PI3Kα, and generally has an *inhibitory effect* on the catalytic activity of p110. The two most important domains of the regulatory subunit are nSH2 and iSH2 [56]. The iSH2 subunit of p85 is responsible for binding to the C2 subunit of p110 [56]. The conformational loops formed by p110α allow for the successful interaction of p110α C2 domain with p85α iSH2 domain [57]. The iSH2 domain

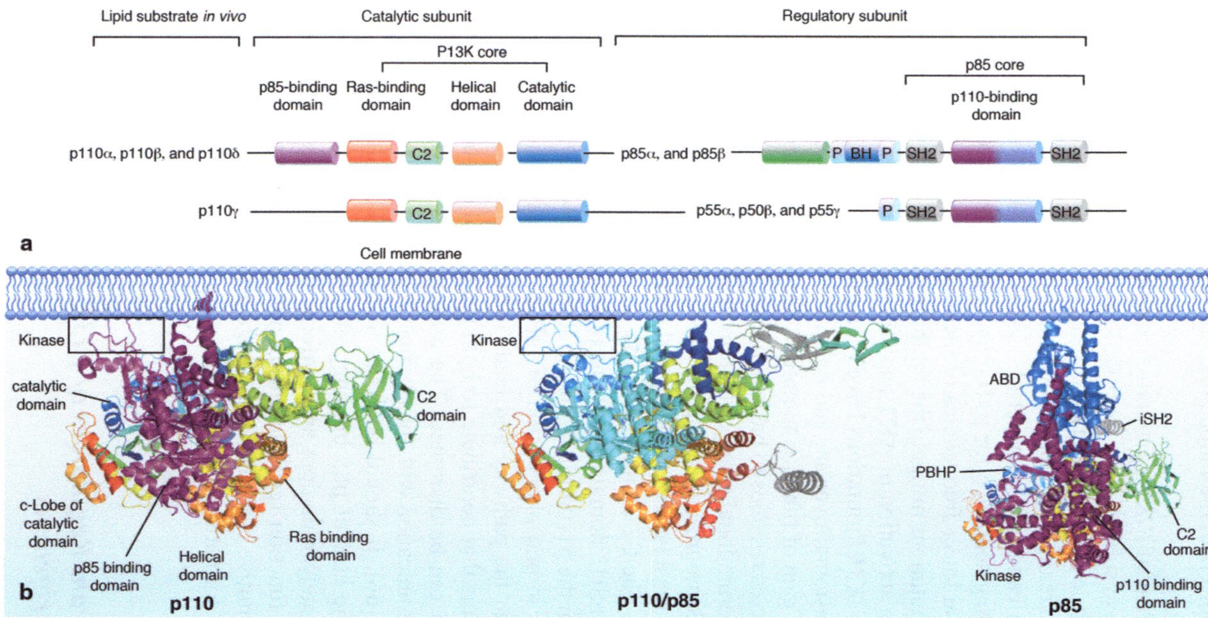

Fig. 2 Structural components of p110 and p85. The p110 catalytic subunit (*left*) is composed of five domains: an adaptor-binding domain (*ABD*), a Ras-binding domain (*RBD*), a C2 domain, a helical domain, and a kinase (catalytic) domain; whereas, the p85 regulatory subunit (*right*) is composed of three main domains: the iSH2 domain, nSH2 domain, and the p110 binding domain. The nSH2 domain has an inhibitory effect on the kinase activity of the molecule that is relieved when it binds to phosphorylated tyrosine residues involved in downstream signaling, which allow the p110α subunit to associate with the plasma membrane. The iSH2 domain has a structural coil that protects the larger subunit from protease degradation. The iSH2 binds between the C2 domain of p110α and in such a way that the iSH2, along with the C2 domain of p110α, provides a contact point to the membrane and adheres to the surface which is in line with positively charged lysine residues. In the inactivated state, p110α is bound to the regulatory subunit p85 as the iSH2 subunit of p85 is responsible for binding to the C2 subunit of p110 and preventing its catalytic interaction with the plasma membrane. The conformational loops formed by p110α allow for the successful interaction of p110α C2 domain with p85α iSH2 domain in order to form the p110α/niSH2 heterodimer. This heterodimer must be recruited to the plasma membrane in order to participate in PI3K signaling. The C2 domain of p110α makes contact with the plasma membrane once the nSH2 domain has released its inhibitory effect and it is bound to iSH2 to form the p110α/niSH2 heterodimer. Contact with the plasma membrane results in phosphorylation of PIP_2 to PIP_3, which is a very powerful second messenger in the pathway and leads to subsequent downstream signaling of mechanisms. Illustration provided by Matt Hazzard, Bio-Medical Illustrator, University of Kentucky.

has a structural coil that protects the larger subunit from protease degradation. The iSH2 binds between the C2 domain of p110α; this heterodimer then provides a contact point to the membrane and adheres to its surface [57, 58]. Somatic mutations in p85 promote an increase in signaling which contributes to cell survival, growth, proliferation and motility in cancers of the esophagus, stomach, colon, pancreas and liver [53].

The three most common areas of *PIK3CA* mutation are single nucleotide substitutions which lead to amino acid substitutions for E542K, E545K, and H1047R [5]. These areas of mutation result in two dysfunctional areas of the helical domain of p110α and one dysfunctional area in the catalytic domain, and occur in about 80% of mutated *PIK3CA* genes in cancer. German et al. [59] determined that gain of function mutations in the coding sequence for p110α show less enzymatic activity and may be responsible for the lower frequencies of these mutants found in colon cancer. The gain of function mutations mitigate the inhibitory action of the p85 subunit on p110α, thus leading to activation of AKT and stimulation of cell proliferation and tumor survival. The *PIK3R1* mutations have been most strongly implicated in endometrial cancer, glioblastoma, ovarian, and colon cancers [51, 60, 61]. The most prominent feature of the gain of function mutations seen in p110α and p85 are the changes of acidic amino acids residues to basic amino acid residues. These amino acid alterations cause conformational changes that prevent successful binding of p110α and p85, and thus block progression along the PI3K signaling pathway [62].

The p110/p85 heterodimer must be recruited to the plasma membrane in order to participate in PI3K signaling. The C2 domain of p110α normally provides contact with the membrane once the nSH2 domain (p85) has released its inhibitory effect and it is bound to iSH2 to form the p110α/niSH2 heterodimer [63, 64]. Contact with the plasma membrane results in phosphorylation of PIP_2 to PIP_3, which is a very powerful second messenger in the pathway and leads to subsequent downstream signaling [65]. The mechanisms that contribute to both the helical domain and N-terminal SH2 domain of p85 can be disrupted. These disruptions prevent p110α inhibition by p85, block communication with the plasma membrane and subsequent downstream PI3K signaling [66]. It should be noted that the helical domains of p110α are less affected by the lack of p85 binding, and mutations result in less enzymatic dysfunction. Ultimately, mutations and conformational changes that prohibit p110α/p85 heterodimer function result in an unregulated signaling cascade, and ultimately promote constitutive signaling and cell growth. This loss of control of signaling results in tumor formation and subsequent metastasis of GI cancers (Fig. 3) [52, 53, 67].

2.3 The Role of Upstream Regulators of PI3K Signaling in Intestinal Tumorigenesis

Activation of cell surface growth factor receptors occurs when growth factors, or receptor ligands, bind to the cell surface. The primary function of cell surface receptors is to recognize specific chemical signals from other substances and produce

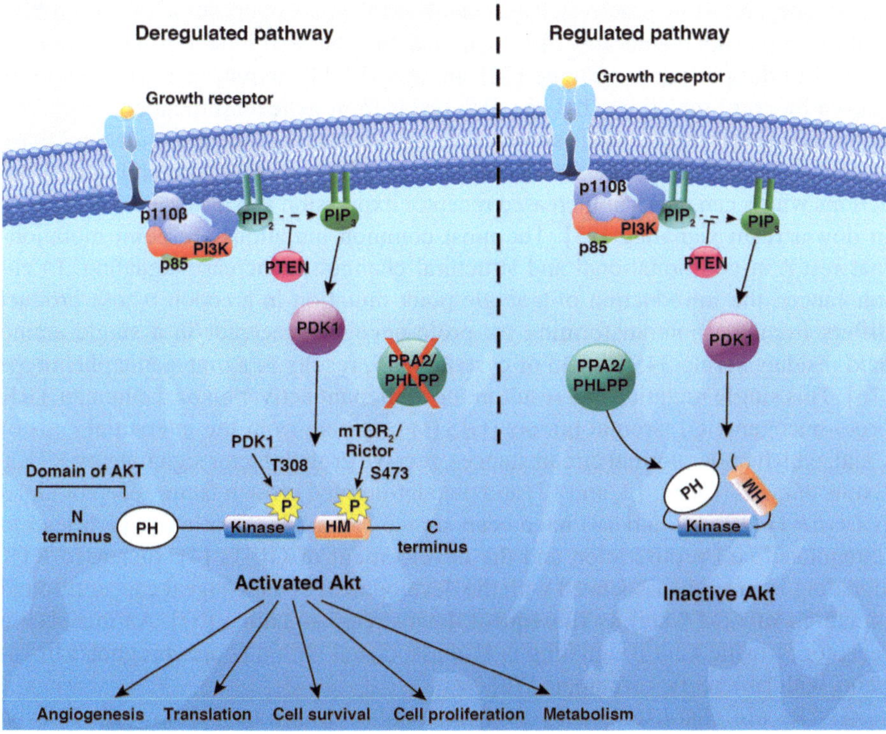

Fig. 3 Regulated and unregulated PI3K Signaling. AKT activation is the focal point for PI3K signaling. PTEN is a lipid phosphatase that dephosphorylates PIP_3 and interferes with AKT activation. AKT also becomes inactivated by phosphatases PP2A and PHLPP and returns to its inactive form in the cytosol (*right side*). When this complex array of mechanisms in the PI3K signaling pathway becomes unregulated (*left side*), uninhibited cell growth leads to tumorigenesis and metastasis of cellular variants with profound metabolic dysregulation. Illustration provided by Matt Hazzard, Bio-Medical Illustrator, University of Kentucky.

a biological response. The PI3K pathway is most commonly activated by the binding of hormones, cytokines, and extracellular ligands such as vascular endothelial growth factor (VEGF), fibroblast growth factor (FGF), and epidermal growth factor (EGF) which bind with their respective receptors, commonly a RTK, and initiate the signaling cascade [68]. The PI3K signaling pathway can also be activated by GPCRs [69]. Aberrant activation of cell surface receptors results in increased stimulation of downstream pathways that leads to increased oncogenicity and a malignant phenotype.

RTKs most commonly exist as single transmembrane subunit receptors, and have an extracellular binding domain that induces the formation of receptor dimers when the appropriate ligand accesses its extracellular N-terminal binding site. An intracellular C-terminal region, which serves as the catalytic portion of the receptor, autophosphorylates intracellular tyrosine residues [70, 71]. The autophosphorylation recruits the p110α/p85 heterodimer of the PIKα complex to the plasma membrane, and PI3K is activated as it catalyzes the phosphorylation of PIP_2 to PIP_3 within the plasma membrane. PIP_3 then mediates the recruitment of AKT from the cytosol to the plasma membrane [72]. Increased RTK activity is most commonly caused by point mutations, but can also result from gene amplification, or defects in the DNA replication/repair system [73–75]. Analysis of the mutations in kinase structure and activity suggests a disturbance of self-regulating autoinhibitory interactions which can result in increased receptor expression and a subsequent increase in downstream signaling [73]. The most common mutations are point mutations that result in conformational and structural changes to increase signaling. In colon cancer, the introduction of a single point mutation in a codon whose product differs from the nontransforming trk proto-oncogene receptor in a single amino acid residue, serine-345 instead of cysteine-345, results in a mutagenic phenotype [76]. This single mutation can result in the malignant activation of the human TRK proto-oncogene. GI stromal tumors (GISTs) originate from the interstitial cells of Cajal, which under normal circumstances, receive growth factor signaling by RTKs. Exons of *c-KIT* (9, 11, 13, and 17) and platelet-derived growth factor receptor alpha (*PDGFRA*) (12, 14, and 18) have been shown to harbor mutations which lead to unregulated cell proliferation and the development of GISTs [4]. Increased RTK signaling is strongly associated with the development of esophageal cancer through mutations of *EGFR* and its subsequent downstream signaling [77]. An increase in receptor tyrosine kinase signaling is also associated with a poorer prognosis in patients with pancreatic carcinoma [78].

GPCRs are characterized by an extracellular N-terminus, followed by seven transmembrane (7-TM) α-helices (TM-1 to TM-7) connected by three intracellular (IL-1 to IL-3) and three extracellular loops (EL-1 to EL-3), and finally an intracellular C-terminus [79]. The intracellular C-terminus contains serine and threonine residues which autophosphorylate and activate the G-protein subunits which can either positively or negatively regulate a variety of downstream effectors [80]. The GPCRs can also utilize G-protein independent mechanisms such as β-arrestin-mediated signaling via Src and mitogen-activated kinase pathways [81, 82]. In the event that the GPCR becomes mutated, it loses the ability to autoregulate, which leads to inactivation of inhibitory/activation of pathways that ultimately lead to increased and dysfunctional intestinal cell proliferation.

Similar to RTKs, GPCRs can also develop a tumorigenic phenotype due to point mutations. For example, the gastrin/cholecystokinin-2 receptor (CCK2R) is stimulated by gastrin and activates an intracellular signal transduction pathway in colon cancer cells. The valine-287 residue in third intracellular loop of CCK2R plays a pivotal role in CCK2R mediated intracellular signal transduction [83]. Aberrant activation of this signaling pathway leads to increased cell growth, proliferation, and metastasis [84]. GPR48 is an orphan receptor that also belongs to the GPCR family; overexpression of this receptor has been shown to exacerbate the development and progression of gastric and colorectal cancer. GPR48 increases nuclear β-catenin accumulation, T-cell factor 4 (TCF4) transcription activity, and expression of target genes *Cyclin D1* and *c-Myc* in colon cancer cells. Upregulation of GPR48 results in increased phosphorylation of glycogen synthase kinase 3β (GSK3β), AKT and the extracellular signal-regulated kinase 1/2 (ERK1/2) in colon cancer cells; inhibition of the PI3K/AKT and mitogen-activated protein kinase /ERK1/2 pathways was sufficient to abolish the effect of GPR48 on β-catenin/TCF signaling [85]. GPR120 is another multifunctional GPCR with the ability to promote both angiogenesis and tumor progression in colon cancer. It functions as a receptor for unsaturated long-chain free fatty acids and has an important role in regulating lipid and glucose metabolism. GPR120 promotes angiogenesis by inducing the expression and secretion of proangiogenic mediators such as VEGF, interleukin-8 and cyclooxygenase-2-derived prostaglandin E2 which activate the PI3K/AKT-NF-κB pathway. It has a dual function in that it can also enhance the motility of colon cancer cells by inducing epithelial-mesenchymal transition (EMT) through the Wnt/ß-catenin pathway [86]. Thus, GPCRs have multiple ways to exploit and activate the downstream signaling mechanisms of the PI3K pathway.

2.4 Downstream Activators of the PI3K Pathway

PDK1 is a serine/threonine kinase that consists of an N-terminal kinase domain and a C-terminal PH [87]. The intracellular C-terminal region serves as the catalytic portion of the receptor and is responsible for the enzymatic activity of the kinase cascade as it autophosphorylates intracellular tyrosine residues. Autophosphorylation leads to a conformational change in the catalytic subunit p110α with enhanced kinase activity as previously discussed. PDK1 mediates the cellular influence of growth factors and insulin by activating both RSK and S6K which are important factors in glucose metabolism and regulation, and is essential for activation of PKB/AKT [88]. Dysregulation of PDK1 is essential for initiating downstream signaling in intestinal carcinogenesis, and its mutation has been implicated in colorectal cancer development and progression [89].

AKT, which is activated by PDK1, exists in three isoforms, AKT1, AKT2, and AKT3, and is encoded by three separate genes. AKT1 is the predominant isoform in tissue and is most consistently associated with oncogenesis [90, 91]. For full activation, AKT requires binding at two amino acid sites, threonine 308 and serine-473; both sites can be phosphorylated by rapamycin insensitive component of mTOR (RICTOR, a component of mTORC2) and DNA-PK phosphorylates at the serine-473 moiety [92]. The AKT serine-473 motif activation by mTORC2 results in downstream mTORC1 activation, while under conditions of stress, DNA-PK stimulates AKT to elicit DNA damage dependent induction of $p21^{waf1}$, a cyclin-dependent kinase inhibitor, thus creating a prosurvival response [93]. This prosurvival mechanism has been noted in human pancreatic cancers, with nuclear accumulation of p53 correlated to adverse clinical features, and inversely correlated with the expression of inhibitor $p21^{waf1}$ [94]. Interestingly, more recent studies have shown $p21^{waf1}$ loss in colon cancer is associated with increased survival in patients $\geq$60-years-old, whereas it is associated with shorter survival in patients <60-years-old.

AKT is a prominent enzyme that is involved in multiple signaling pathways, many of which participate in cellular crosstalk [95–97]. For growth to be regulated, the pathways must have mechanisms in place to turn the molecular machinery off. In the PI3K/AKT/mTOR pathway, mTORC1 elicits a negative feedback regulation on the pathway by targeting insulin receptor substrate-1 inhibition through ribosomal S6K1 [98]. When this mechanism becomes dysregulated carcinogenesis prevails and previous authors have shown metabolic derangements which contribute to colon cancer, invasion and ultimately liver metastasis [99]. There is also cross talk between the TSC1/TSC2 and AKT pathways as AKT phosphorylates and inactivates TSC1/TSC2 and leads to formation of mTORC1 [97]. Complexity is increased with interaction of the proapoptotic transcription factor FOXO3a that regulates glycolysis. Foxo3a-deficient cells are associated with increased S6K1 phosphorylation, which has been linked to glycolysis regulation. Khatri et al. [100] showed that mTORC1-dependent glycolysis is increased in *Foxo3a* knockdown cells due to decreased expression of the TSC1 tumor suppressor that opposes mTORC1 activation. FOXO3a then binds to and transactivates the TSC1 promoter, indicating a key role for FOXO3a in regulating TSC1 expression [100]. The interaction of TSC1 and TSC2 with mTOR and other PI3K effectors plays a significant role in cell migration and the development of malignant phenotypes of the intestine [101]. Rheb is a small guanosine triphosphate (GTP) binding protein which functions to also activate mTORC1, but can be deactivated by the GTPase functions of the TSC1/TSC2 complex [102]. However, in the event that AKT deactivates mTOR and also stimulates its expression in the presence of Rheb, mTORC1 levels can increase since these mechanisms fail to regulate each other. Lacher et al. [103] demonstrated that Rheb regulates p27 in TSC cells by activating AMPK which in turn activates the TSC1-TSC2 tumor suppressor complex, thereby inhibiting Rheb and reducing mTORC1 activity and consequently suppressing protein synthesis in colon cancers. It is also important to note that RAS is an upstream regulator of the MAPK pathway that provides crosstalk with PI3K. RAS regulates RAF serine/threonine protein kinases which regulate MAPK signaling [104]. *RAS* is frequently mutated in intestinal cancers, most notably colorectal cancers, and leads to hyperactivation of the PI3K pathway via crosstalk with MAPK using TSC2 [102, 104] (see Figs. 3 and 4).

2.5 Tumor Suppressors Involved in the PI3K Pathway

Dysfunction of PTEN can occur with DNA methylation changes, transcriptional repression, and mRNA degradation and subsequent translational failure. This dysfunction leads to a plethora of cancer phenotypes [105–107]. Germline mutations of PTEN are associated with many neurological diseases, but with regards to intestinal tumorigenesis they exhibit variable penetrance in expression from benign intestinal hamartomas as in Cowden's disease to malignant GI carcinomas including colon, gastric, and pancreatic carcinomas [6, 108, 109–111]. PTEN mutations have been strongly associated with both the induction of EMT in colon cancer and progression to an aggressive and metastatic phenotype [112].

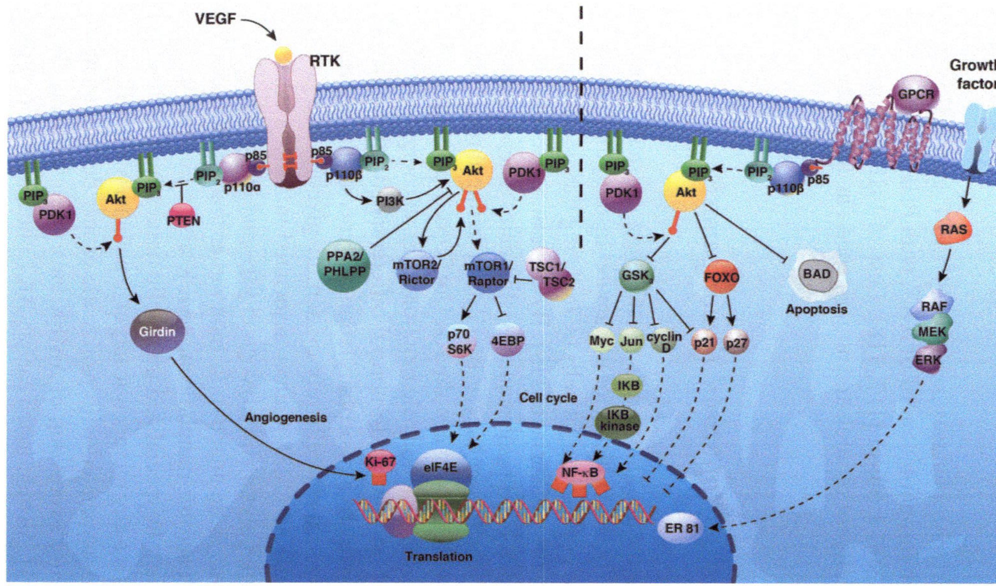

Fig. 4 PI3K/Akt pathway integrates with downstream signaling mechanisms through receptor tyrosine kinase (*RTK*) and G-protein couple receptors (*GPCR*). VEGF interacts with RTK to initiate this kinase cascade. Akt is subsequently activated by and phosphorylates the Girdin protein which induces vessel sprouting and branching during angiogenesis. One mechanism of cell growth is controlled by TSC1/TSC2 which are proteins encoded by tumor suppressor genes responsible for mutations in tuberous sclerosis. This complex normally suppresses the function of mTORC1 but when activated, Akt phosphorylates the TSC complex and releases its inhibitory effect, thus allowing mTORC1 to become activated. Once mTORC1 is activated, it promotes the phosphorylation of p70S6K and 4EBP, both of which act on transcription factor eIF4E. Phosphorylated Akt also ensures the progression of the cell cycle by preventing degradation of cyclin D1 through the phosphorylation and inhibition of GSK3β. In the event of Akt overexpression due to upstream mutations or loss of tumor suppressor genes, GSK3β can no longer exert control over the pathway because Akt will inhibit and allow cyclin D to continue with its cell cycle involvement. Both p21 and p53 act as tumor suppressors and can be inhibited by Akt. Inhibition of p53 activity allows the genetic machinery to continue without regulation. Akt can increase its expression and promote downstream activities by interfering with apoptotic mechanisms controlled by the phosphorylation and inhibition of BAD. Akt phosphorylates FOXO, a family of transcription factors, and promotes their interaction with 14-3-3 which results in their expulsion from the nucleus and their ability to continue to induce transcription factors. FOXO leaves the nucleus and pro-apoptotic genes such as *TRAIL*, *TRADD*, and *Fas* ligand cannot be transcribed. Akt can also inhibit apoptosis by upregulating anti-apoptotic transcription factors like NF-κB by phosphorylating and activating IκB kinase. This anti-apoptotic transcription factor is frequently upregulated and results in increased cell survival due to its anti-apoptotic mechanisms. Illustration provided by Matt Hazzard, Bio-Medical Illustration, University of Kentucky.

AKT signaling is induced by phosphorylation of its serine and threonine motifs and can be inactivated by protein phosphatase dephosphorylation. Protein phosphatase 2A (PP2A), which is encoded by the *PPTR-1* gene, is capable of dephosphorylating AKT at its activation loop, thus deactivating its downstream effects [113]. Loss of PP2A has been implicated in human colon cancers as this loss allows constitutive downstream signaling from AKT, and has been evaluated in gastric cancer [114, 115]. Heat shock protein-90 inhibits PP2A and permits AKT to function without inhibition [116]. Novel PP2A activators are currently in development as anti-oncogenic agents [114].

PH domain-containing protein phosphatases such as PHLPP1 and PHLPP2 specifically dephosphorylate the hydrophobic motif of AKT, with PHLPP2 preferentially dephosphorylating the AKT1 isoform [117, 118]. Loss of PHLPP expression has been shown to promote cell proliferation and tumorigenesis in colon cancer cells [117].

2.6 PI3K Signaling Effects on Intestinal Tumorigenesis

In order to maintain an oncogenic state, tumor cells must have a constant source of nutrients, preferentially in the form of glucose or glutamate. Insulin, through its respective RTK, increases glucose uptake into fat cells by initiating translocation of the glucose transporter GLUT4 to the cell surface [119]. By contrast, most cells use GLUT 1 as their glucose transporter. Insulin subsequently stimulates adiopogenesis by the AKT/mTORC1 pathway, which is enhanced by GLUT1 acting through the mTOR pathway. There is also crosstalk between the AKT/mTORC1 pathway and AMPK [120, 121]. AMPK plays a fundamental role functioning as an energy sensor, to regulate cell metabolism [122]. In contrast to normal differentiated cells which rely primarily on mitochondrial oxidative phosphorylation to generate the energy needed for cellular processes, most cancer cells rely on aerobic glycolysis, a phenomenon termed "the Warburg effect" [123]. Cancer-associated mutations which upregulate effectors inherent in glycolysis and glycogenolysis enable cancer cells to acquire and metabolize nutrients in an anabolic fashion rather than the efficient ATP production obtained through aerobic metabolism. AMPK also regulates lipid, cholesterol and glucose metabolism in specialized metabolic tissues such as liver, muscle, and adipose which makes it a key therapeutic target in patients with diabetes [124]. AMPK inhibition has been associated with pancreatic cancer development; factors which upregulate or increase expression of AMPK have been shown to be beneficial in colon cancer [95, 125]. As mentioned earlier, AKT phosphorylation inactivates GSK3β. This inhibition of GSK3β also promotes sterol regulatory element binding proteins that act as transcription factors to induce fatty acid synthesis. The dysfunction of fatty acid metabolism through effectors such as fatty acid synthase and malonic coenzyme A have been implicated in pancreatic cancer, esophageal cancer, and colorectal cancers [126–128]. Most sporadic cancers develop gradually as accumulating alterations in gene expression transform normal intestinal surfaces into a malignant phenotype. This process involves interplay between genetic and environmental factors that can be attributed to high fat diets and adipogenesis [128].

3 Intestinal Tumorigenesis: The Role of PI3K and Treatment Strategies

The PI3K signaling pathway has been implicated in intestinal tumorigenesis and the development of a malignant phenotype in essentially every organ of the GI tract. We will now describe the current knowledge regarding PI3K pathway derangement in specific organs of the Gi tract. Finally, potential treatment strategies will be discussed (Table 1).

Table 1 PI3K signaling pathway inhibitors discussed

Molecular target	Use in malignancy	Mechanism
EFGR		
Cituximab	Squamous cell esophagus, colon	EGFR inhibitor
Gefitinib	Squamous cell esophagus, gastric	EGFR inhibitor
Erlotinib	Pancreatic adenocarcinoma	EGFR inhibitor
Panitumumab	Colon cancer	EGFR inhibitor
VEGFR		
Bevacizumab	Squamous cell esophagus, esophageal adenocarcinoma, gastric, colon	VEGFR inhibitor
HER2/neu		
Trastuzumab	Squamous cell esophagus, esophageal adenocarcinoma, gastric cancer	HER2/neu inhibitor
PI3K		
LY294002	Gastric cancer, pancreatic adenocarcinoma, cholangiocarcinoma	PI3K inhibitor
Curcumin	Colon cancer	PI3K inhibitor
HS-159	Hepatocellular carcinoma	PI3K inhibitor
BKM120	Hepatocellular carcinoma	PI3K inhibitor
Rawq01	Squamous cell esophagus, colon cancer	Increases PTEN, decreases PI3K
HS-173	Pancreatic adenocarcinoma	PI3K inhibitor
Dual PI3K/mTOR		
NVP-BEZ235	Hepatocellular carcinoma, pancreatic adenocarcinoma, cholangiocarcinoma	Dual PI3K/mTOR
NVP-BGT226	Pancreatic adenocarcinoma	Dual PI3K/mTOR
Dual PI3K/AKT		
Arsenic trioxide	Gastric cancer	Dual PI3K/AKT
Ursolic acid	Pancreatic adenocarcinoma	PI3K/AKT/NF-κβ
mTOR		
RAD001	Hepatocellular carcinoma	mTOR inhibitor
Multikinase inhibitor		
Sorafenib	Hepatocellular carcinoma	Multikinase inhibitor
TRAIL		
Aldose reductase	Colon cancer	Increases TRAIL
miR-21		
Celastrol	Gastric cancer	miR-21 inhibits PI3K/AKT/NF-κβ

3.1 Esophageal Cancer

Esophageal cancer has a dismal prognosis with 5-year overall survival rates averaging approximately 15 % [129]. Squamous cell carcinoma of the esophagus (SCCE), the most prevalent histologic type worldwide, is commonly linked to smoking and alcohol abuse whereas adenocarcinoma is more closely associated with acid reflux disease and obesity [130].

Alterations in the PI3K/AKT signaling pathway have also been implicated in the occurrence and development of human intestinal cancers, including SCCE. SCCE abnormalities associated with PI3K signaling include a decreased expression of tumor suppressor gene *PTEN*, with subsequent increased PI3K and AKT signaling. These changes prevent apoptosis through inhibition of factors such as NF-κB, and increased EGFR expression [77, 131]. The most common genetic alterations identified in SCCE include EGFR (up to 70 %) and cyclin D1 oncogene overexpression (up to 60–70 %), inactivation of the p53 (50–60 %) and $p16^{INK4a}$ (40–50 %) tumor suppressor genes [132, 133]. Multiple studies, using different techniques to compare DNA copy number abnormalities, have been employed to elucidate the differences and potential targets between the two different types of esophageal cancer. Eleven of 17 (65 %) regions display higher frequencies of gain in SCCE, and some of these regions harbor known cancer-associated genes such as *SOX2, PIK3CA* [134]. Many molecular inhibitors targeting the PI3K pathway are thought to be less toxic that traditional chemotherapeutic agents and are being sought as potential therapeutic routes to target the PI3K signaling pathway. Apoptosis can selectively inhibit tumor cell survival using EGFR, MAPK and PI3K signaling, or by activating the cell surface death receptors Fas, TRAIL and TNF receptors [135]. Some of the most promising results from targeting the PI3K pathway in esophageal cancer are the use of EGFR antibodies (e.g., cituximab). Although the EFGR inhibitors alone have shown minimal efficacy against advanced esophageal cancers, there are promising results using them in conjunction with traditional chemotherapeutic agents [136, 137]. Another monoclonal antibody, trastuzumab, which acts against HER2/neu receptor, has been shown to have anti-tumor efficacy in human SCCE xenograft mouse models [138].

The mTOR pathway has been shown to be aberrantly activated in SCCE. Targeted therapies that interact with the mTOR pathway and its effectors have shown a therapeutic response in SCCE. Everolimus has been used as both a single agent and in combination with cisplatin to inhibit mTOR with success in SCCE. [139]. Another interesting application is the use of Organometallic arene Ru(II) complexes, such as Rawq01, as a substitute for platinum agents. These compounds have low toxicity to normal cells and high selectivity to tumor cells as they are selectively taken up by tumor cells, and have been studied in the treatment of SCCE. Weng et al. [140] showed that Rawq01 up-regulated the expression of PTEN through mir-21 inhibition and thereby inhibited the PI3K-AKT pathway.

3.2 *Adenocarcinoma of the Esophagus*

Esophageal adenocarcinoma has a poor prognosis, and its incidence has risen sharply in recent decades. Recent Western experiences report that the male:female ratio for patients undergoing esophagectomy for adenocarcinoma now exceeds 8:1. However, the identified risk factors for esophageal adenocarcinoma, including gastroesophageal reflux disease, Barrett's esophagus, obesity, alcohol, and tobacco consumption fail to explain the male predominance [141]. Bandla et al. [134] identified cancer loci, including gains of function in VEGFA, EGFR, CDK6, MET, KRAS, and ERBB2, and determined that MYB, TARP, GATA4, and GATA6 were more prominent in adenocarcinoma of the esophagus. VEGF, EGFR, and HER2/neu are all involved in the initial signaling of the PI3K signaling pathway and are implicated in the up-regulation and overexpression of effectors that promote tumorigenesis. Currently, there are not as many molecularly targeted treatment strategies for esophageal adenocarcinoma. Although HER2/neu is common to both types of esophageal cancer, mouse models show an increased resistance to trastuzumab due to the expression of a transcript variant of *DARPP* [142]. Treatment with EGFR inhibitors like gefitinib has demonstrated less success with esophageal adenocarcinoma compared to SCCE. Janmaat et al. [143] determined that gefitinib has a modest activity in second-line treatment, but the patient outcome was significantly better in female patients and in patients demonstrating high EGFR expression or SCC histology.

Bevacizumab is a monoclonal antibody directed against VEGFR that has shown the most promise for a targeted response to esophageal adenocarcinoma. A phase II trial which included distal esophageal and esophageal junctional adenocarcinomas showed a response rate of 65 % and median survival time of 12.3 mo, however an assessment of VEGFR expression was not made prior to treatment which suggests specifically targeting patients with increased VEGFR expression may provide an improved outcome to this select group [144].

3.3 *Gastric Cancer*

Gastric cancer is one of the most common causes of cancer deaths worldwide and has a poor 5 year survival rate, between 5–15 %, as many patients present with advanced disease [145]. Surgical resection to include either a subtotal or total gastrectomy is the preferred treatment if resection is possible, but the extent of appropriate lymph node basin resection and resulting outcomes is widely debated [146, 147]. Currently, the most common chemotherapeutic agents used for the treatment of resectable gastric cancer are relatively toxic and include 5-FU, cisplatin, doxorubicin, combination oxaliplatin, with leucovorin; etoposide, docetaxel and radiation treatment are reserved for stage IV disease and palliation [148–150].

Gastric cancers vary based on their histologic and anatomic presentation and can be classified into two types, intestinal or diffuse, based on the Lauren classification

[151]. Diffuse gastric cancers tend to occur in the proximal stomach while intestinal gastric cancers occur in both the proximal and distal stomach [152]. Interestingly, specific alterations in PI3K signaling have been described depending on tumor location. E-cadherin loss is more commonly noted in diffuse type gastric cancer. HER2 and EGFR are seen predominantly in proximal intestinal gastric cancer, and VEGF and VEGFR mutations are overexpressed in distal intestinal gastric cancer [153]. While most gastric cancers are sporadic, approximately 10% have a familial component; it is important to differentiate between these two as there is impact on clinical decisions [154]. Germline mutations seen in gastric cancer can include the p53, BRCA2, and E-cadherin genes [155]. Some very aggressive familial gastric cancers have a heterozygous nonsense mutation of exon 3 in the E-cadherin (*CDH1*) gene which produces a stop codon at position 95 and results in a truncated E-cadherin protein [156]. Due to the aggressiveness and high penetrance of this particular genetic mutation it is recommended that patients with this family history receive genetic counseling and at least consider prophylactic gastrectomy in young, asymptomatic carriers of the *CDH1* gene mutation [157, 158].

With regard to growth receptor inhibitors EGFR, VEGFR, and HER2/neu have all shown promise in the treatment of gastric cancer by decreasing ligand binding and thus inhibiting downstream signaling in the PI3K pathway. A phase II trial in metastatic gastric adenocarcinomas showed a response rate of 65% and median survival time of 12.3 months [144]. Trastuzumab is currently the only approved targeted agent for the subgroup of HER2+ advanced gastric cancer, based on results of the phase III Trastuzumab For Gastric Cancer trial [159]. Although an overall survival advantage was demonstrated, most patients did not respond to the combination of chemotherapy and trastuzumab in the first-line setting even with HER2+ disease. As with esophageal adenocarcinoma, the VEGFR inhibitor bevacizumab in addition to chemotherapy did not result in significant overall survival benefit in the phase III Avastin in Gastric Cancer trial [160]. Rojo et al. [161] showed that the EGFR inhibitor gefitinib reached the tumors at concentrations sufficient to inhibit EGFR activation in patients with advanced gastric carcinoma, but this did not translate into clinical benefit and intratumoral phosphorylation of MAPK and AKT was not significantly inhibited by the drug. However, the finding that decreases in pAKT correlated with enhanced apoptosis on a molecular level does warrant further investigation. A concerning effect of chemotherapeutic agents in use for gastric cancer is toxicity as well as resistance. Although cisplastin is currently one of the most commonly used agents for this malignancy, in vitro and in vivo mouse studies have shown that p21^{waf1}-activated kinase 4 expression can be elevated in cisplastin resistant gastric cancer via activation of MEK/ERK and PI3K /AKT pathways. These findings suggest the potential for initiating treatment with an inhibitor of the p21^{waf1}-activated kinase 4, such as the inhibitor PF-3758309 [32]. Other studies have shown that selective PI3K pathway inhibitors like 2-(4-morpholinyl)-8-phenyl-4H-1-benzopyran-4-one (LY294002) may potentiate chemotherapy agents like oxaliplatin by inducing activation of AKT and NF-κB, caspase-8. Bid and caspase-3, may increase expression of FasL, and inhibit c-FLIP$_S$ [162]. Other in vitro studies have shown that LY294002 induces synergistic cytotoxicity with 5-FU and can overcome intrinsic and acquired resistance by downregulation of activated p-AKT. Mitochondria-dependent apoptosis, demonstrated through the expression of p-AKT and p-NF-κB, was upregulated by 5-FU alone but sequential treatment of 5-FU and LY294002 decreased the expression of both p-AKT and p-NF-κB [163]. Other molecularly-target inhibitors of the PI3K pathway include agents such as arsenic trioxide that directly inactivates AKT and synergistically induces gastric cell apoptosis. When arsenic trioxide is combined with LY294002 and celastrol, a plant triterpenoid, there was an increase in gastric cell apoptosis due to miR-21 inhibition of the PI3K/AKT-dependent NF-κB pathway [164, 165].

3.4 *Small Bowel Adenocarcinoma*

Small bowel adenocarcinoma has a vague and insidious presentation and is relatively rare, but results in approximately 1170 deaths in the United States annually and has a 64.5% 5 year survival rate [166]. Its molecular tumorigenesis has common themes with other intestinal malignancies as it also utilizes PI3K signaling for progression. Investigation of novel molecular targets for this pathway could provide insight into better therapeutic agents and improve its prognosis. Zhang et al. [40] examined human specimens with quantum dots-based immunofluorescence histochemistry and traditional immunohistochemistry and showed increased PI3K and p-AKT expression were significantly higher in specimens with small bowel adenocarcinoma. These observations support the assumption that the PI3K/AKT pathway is activated in small bowel adenocarcinoma, and suggest that therapeutic targeting of the PI3K/AKT signaling pathway could provide novel and hopeful advances for this disease.

3.5 *Hepatobiliary/Pancreatic Malignancies*

Malignancies of the hepatobiliary tract include cancers of the liver, gallbladder, bile ducts, and pancreas, and are increasing in frequency worldwide. According to the National Cancer Institute, pancreatic cancer resulted in approximately 38,460 deaths in 2013 accounting for 6.6% of all cancer deaths, and had a 6.0% 5 year survival rate between 2003–2009, while liver and intrahepatic bile duct cancer resulted in 21,670 deaths, accounting for 3.7% of all cancer deaths, and had a 16.1% 5 year survival rate [166]. Although these cancers are not as prevalent as other intestinal malignancies, a typically vague presentation and advanced stage at diagnosis contribute to their insidious yet aggressive nature. For the purposes of this chapter we will limit our discussion to PI3K signaling of intestinal tumorigenesis in hepatocellular carcinoma, pancreatic adenocarcinoma, and cholangiocarcinoma.

Dysfunctional PI3K signaling has been implicated in hepatic cell dysplasia and malignancy. It is well-known that the liver plays a major role in glycolysis, glycogenolysis, and gluconeogenesis [123, 167]. 7-Ketocholesterol is elevated in patients with cancer and chronic liver disease. This elevation can cause up-regulation of an efflux pump through the activation of PI3K/mTOR pathway, activation of the downstream protein synthesis initiation factor 4E-BP1, and induce P-glycoprotein expression post-transcriptionally. Other studies show an increase in lactate production and fluorescence lifetime of protein-bound NADH was shortened suggesting a metabolic shift to glycolysis [168]. Thus far, sorafenib, an oral multikinase inhibitor that affects the PI3K signaling pathway, is the first agent that has demonstrated an improved overall survival benefit in advanced hepatocellular carcinoma, setting a new standard for first-line treatment [169, 170]. Multiple PI3K inhibitors such as N-(5-(3-(3 cyanophenyl)imidazo[1,2-a]pyridin-6-yl)pyridin-3-yl)benzenesulfonamide, the dual PI3K /mTOR inhibitor NVP-BEZ235, and inhibitors that combine pathways, such as the dual mTOR and PI3-kinase inhibitor BEZ235, the PI3K inhibitor BKM120, and mTOR inhibitor RAD001 show promise in pre-clinical studies for the treatment of hepatocellular carcinoma [171–173].

Pancreatic adenocarcinoma cancer susceptibility genes have been identified in recent years including *BRCA2*, *CDKN2A/p16*, *STK11*, *ERBB3*, and *EGFR* that interact with the PI3K pathway. It is of paramount importance to elucidate the family history of such malignancies and initiate appropriate genetic testing to identify patients who may respond to certain targeted molecular therapies [174–177]. Pancreatic adenocarcinoma has extremely high invasive potential. This trait is generated through cell dissociation mediated, in part, by MMP-7 with invasion of pancreatic cancer cells through disruption of tight junctions, and formation of a positive feedback loop with activation of the EGFR mediated MEK-ERK signaling pathway. The efficacy of EGFR inhibitors such as erlotinib has been evaluated [178, 179].

PI3K signaling pathway inhibitors such as combined treatment with BEZ235 and the pan-histone deacetylase inhibitor panobinostat increase the activity of the RAS-RAF-ERK1/2 and the PI3K-AKT-mTOR pathways in pancreatic adenocarcinoma [180]. NVP-BGT226 is another dual class PI3K /mTOR inhibitor that has entered Phase I/II clinical trials; this compound induces cell cycle arrest and regulates *Survivin* gene expression in human pancreatic cancer cell lines [181]. Another interesting and novel approach to target the PI3K pathway was utilized by Li et al. [182]. Ursolic acid can act as a novel and potent therapeutic agent via the inactivation of the PI3K/AKT/NF-κB pathway and activation of the c-Jun-terminal kinase pathway in gemcitabine-resistant pancreatic cancer and even as a promising candidate in other chemoresistant cancers [182].

Although fewer studies regarding cholangiocarcinoma and PI3K exist, in vitro studies have shown that NVP-BEZ235 upregulates PTEN expression and decreases the constitutive activation of PI3K/AKT pathway [183]. Other authors have also shown that the PI3K inhibitor LY294002 arrests of cell proliferation and enhances oxaliplatin-induced cytotoxicity through PI3K inhibition in cholangiocarcioma [184]. Further studies are required to identify additional targeted therapies for this malignancy.

3.6 Colorectal Cancers

Although colon cancer is more common than other intestinal cancers, its treatment has evolved significantly over the past decades. Increased surveillance, earlier surgical treatment and chemotherapy have improved disease free survival and overall survival rates significantly [166, 185]. When these patients experience recurrence or are initially diagnosed with advanced disease, the most common combination chemotherapy agents are 5-FU and leucovorin plus oxaliplatin, or FU and leucovorin plus irinotecan, which were studied in the Multicenter International Study of Oxaliplatin/5FU-LV in the Adjuvant Treatment of Colon Cancer and the National Surgical Adjuvant Breast and Bowel Project C-07 trial, respectively [186–189].

Genetic mutations and familial syndromes have been broadly studied in colon and rectal cancer and include defects in DNA mismatch repair genes, *RAS* gene family mutations, and *APC* gene polyposis associated malignancies [190–192]. Of the *PI3K* mutations, the *PIK3CA* mutation in either exon 9 or 20 is most often implicated in colon tumorigenesis [193]. Multiple PI3K pathway inhibitors have been studied in conjunction with chemotherapeutic agents and have shown improved survival in patients with cancers induced by PI3K pathway signaling. Multiple studies have shown benefit with bevacizumab plus irinotecan, fluorouracil, and leucovorin and also in combination with oxaliplatin, fluorouracil, and leucovorin [194, 195]. Cituximab, an EGFR inhibitor, has clinically significant activity when given alone or in combination with irinotecan in patients with irinotecan-refractory colorectal cancer, and an open phase III trial using the EGFR inhibitor panitumumab in chemotherapy-refractory metastatic colorectal cancer significantly improved progression free survival with manageable toxicity in patients with chemo-refractory colorectal cancer [196, 197].

Using in vivo mouse models Martinelli et al. [198] showed combined treatment with a selective MEK 1/2 inhibitor, pimasertib, and BEZ235 or with sorafenib caused significant tumor growth delays and increased survival in mice as compared to single agent treatment. These results suggest that dual blockade of MAPK and PI3K pathways could overcome intrinsic resistance to MEK inhibition. Rychahou et al. [199] sought to show that short interfering RNA (siRNA) directed to PI3K/AKT components can render colon cancer cells sensitive to treatment with tumor necrosis factor-related apoptosis-inducing ligand and activation of caspase-3 and caspase-8. This inactivation opens a window to explore yet another mechanism geared to inhibit this signaling pathway. Another interesting study showed that aldose potentiates TRAIL-induced apoptosis through downregulation of cell survival proteins and upregulation of death receptors; this process is mediated through activation of the AKT/FOXO3a pathway [200]. Other authors have used LY294002 in combination with curcumin to show a curcumin dose-dependent decrease in Bcl-2 mRNA expression in vitro, and suggest that this decrease may induce cytotoxic effects, partly by suppressing the PI3K/AKT signaling pathway [201]. Other authors suggest that stimulating weakened or inactivated tumor suppressor genes may assist in the restoration of a normal PI3K cascade. Liu et al. [117] found that the expression of PHLPP1 or PHLPP2 isoforms was lost or decreased in 78 and 86% of human colon cancer tissues. They also showed that stable overexpression of either PHLPP isoform in colon cancer cells decreased the rate of cell proliferation and sensitized the cells to growth inhibition induced by the PI3K inhibitor, LY294002, whereas knockdown of either PHLPP isoform by shRNA promoted cell proliferation. It was also demonstrated that the PHLPP-mediated growth inhibition in colon cancer cells was largely rescued by overexpression of a constitutively active AKT. Other authors have shown that β-TrCP-mediated ubiquitination and degradation of PHLPP1 are negatively regulated by Akt, and that inactivation of this tumor suppressor leads to increased colon cancer tumor motility and aggressiveness by increasing the expression of RAF1 [202]. This tumor suppressor, which has the opportunity to turn off the key of the PI3K signaling pathway, AKT, is an attractive option for new targeted therapies. Continued studies and trials targeting multiple areas of the PI3K signaling pathway provide interesting approaches to improve colon cancer treatment and tumor survival.

4 Conclusion

The PI3K family of proteins represents a complex and dynamic cell signaling pathway that is crucial for normal growth and development. Under normal conditions, cell cycle control is tightly regulated to suppress tumor formation and allow for intestinal cells to grow and function normally, participate in necessary metabolic functions, and endogenously address intestinal pathogens. Both endogenous and exogenous factors can induce irreversible alterations in the PI3K pathway resulting in an oncogenic phenotype and subsequent invasion and metastasis. Unfortunately, although extensive work has been performed to develop improved treatments for intestinal tumors, many GI malignancies continue to demonstrate dismal outcomes for both disease free survival and overall survival. A better understanding of the specific genetic mutations and alterations of this pathway in GI cancers will continue to provide for better, more directed and precise therapies targeting multiple points in this pathway.

References

1. Engelman JA, Luo J, Cantley LC (2006) The evolution of phosphatidylinositol 3-kinases as regulators of growth and metabolism. Nat Rev Genet 7(8):606–619. doi:10.1038/nrg1879
2. Edgar KA, Wallin JJ, Berry M, Lee LB, Prior WW, Sampath D, Friedman LS, Belvin M (2010) Isoform-specific phosphoinositide 3-kinase inhibitors exert distinct effects in solid tumors. Cancer Res 70(3):1164–1172. doi:10.1158/0008-5472.CAN-09-2525
3. Bader AG, Kang S, Zhao L, Vogt PK (2005) Oncogenic PI3K deregulates transcription and translation. Nat Rev Cancer 5(12):921–929. doi:10.1038/nrc1753

4. Soreide K, Sandvik OM, Soreide JA, Gudlaugsson E, Mangseth K, Haugland HK (2012) Tyrosine-kinase mutations in c-KIT and PDGFR-alpha genes of imatinib naive adult patients with gastrointestinal stromal tumours (GISTs) of the stomach and small intestine: relation to tumour-biological risk-profile and long-term outcome. Clin Transl Oncol 14(8):619–629. doi:10.1007/s12094-012-0851-x
5. Samuels Y, Velculescu VE (2004) Oncogenic mutations of PIK3CA in human cancers. Cell Cycle 3(10):1221–1224
6. Garcia-Carracedo D, Turk AT, Fine SA, Akhavan N, Tweel BC, Parsons R, Chabot JA, Allendorf JD, Genkinger JM, Remotti HE, Su GH (2013) Loss of PTEN expression is associated with poor prognosis in patients with intraductal papillary mucinous neoplasms of the pancreas. Clin Cancer Res 19(24):6830–6841. doi:10.1158/1078-0432.CCR-13-0624
7. Chen Z, Trotman LC, Shaffer D, Lin HK, Dotan ZA, Niki M, Koutcher JA, Scher HI, Ludwig T, Gerald W, Cordon-Cardo C, Pandolfi PP (2005) Crucial role of p53-dependent cellular senescence in suppression of Pten-deficient tumorigenesis. Nature 436(7051):725–730. doi:10.1038/nature03918
8. Fasolo A, Sessa C (2008) mTOR inhibitors in the treatment of cancer. Expert Opin Investig Drugs 17(11):1717–1734. doi:10.1517/13543784.17.11.1717
9. Stephens L, Williams R, Hawkins P (2005) Phosphoinositide 3-kinases as drug targets in cancer. Curr Opin Pharmacol 5(4):357–365. doi:10.1016/j.coph.2005.03.002
10. Koul D, Shen R, Bergh S, Sheng X, Shishodia S, Lafortune TA, Lu Y, de Groot JF, Mills GB, Yung WK (2006) Inhibition of Akt survival pathway by a small-molecule inhibitor in human glioblastoma. Mol Cancer Ther 5(3):637–644. doi:10.1158/1535-7163.MCT-05-0453
11. Yang L, Dan HC, Sun M, Liu Q, Sun XM, Feldman RI, Hamilton AD, Polokoff M, Nicosia SV, Herlyn M, Sebti SM, Cheng JQ (2004) Akt/protein kinase B signaling inhibitor-2, a selective small molecule inhibitor of Akt signaling with antitumor activity in cancer cells overexpressing Akt. Cancer Res 64(13):4394–4399. doi:10.1158/0008-5472.CAN-04-0343
12. Alessi DR, James SR, Downes CP, Holmes AB, Gaffney PR, Reese CB, Cohen P (1997) Characterization of a 3-phosphoinositide-dependent protein kinase which phosphorylates and activates protein kinase Balpha. Curr Biol 7(4):261–269
13. Deora AA, Win T, Vanhaesebroeck B, Lander HM (1998) A redox-triggered ras-effector interaction. Recruitment of phosphatidylinositol 3'-kinase to Ras by redox stress. J Biol Chem 273(45):29923–29928
14. Vogt PK, Kang S, Elsliger MA, Gymnopoulos M (2007) Cancer-specific mutations in phosphatidylinositol 3-kinase. Trends Biochem Sci 32(7):342–349. doi:10.1016/j.tibs.2007.05.005
15. Booker GW, Breeze AL, Downing AK, Panayotou G, Gout I, Waterfield MD, Campbell ID (1992) Structure of an SH2 domain of the p85 alpha subunit of phosphatidylinositol-3-OH kinase. Nature 358(6388):684–687. doi:10.1038/358684a0
16. Musacchio A, Gibson T, Rice P, Thompson J, Saraste M (1993) The PH domain: a common piece in the structural patchwork of signalling proteins. Trends Biochem Sci 18(9):343–348
17. Kennedy SG, Wagner AJ, Conzen SD, Jordan J, Bellacosa A, Tsichlis PN, Hay N (1997) The PI 3-kinase/Akt signaling pathway delivers an anti-apoptotic signal. Genes Dev 11(6): 701–713
18. Cho H, Mu J, Kim JK, Thorvaldsen JL, Chu Q, Crenshaw EB 3rd, Kaestner KH, Bartolomei MS, Shulman GI, Birnbaum MJ (2001) Insulin resistance and a diabetes mellitus-like syndrome in mice lacking the protein kinase Akt2 (PKB beta). Science 292(5522):1728–1731. doi:10.1126/science.292.5522.1728
19. Mayo LD, Donner DB (2001) A phosphatidylinositol 3-kinase/Akt pathway promotes translocation of Mdm2 from the cytoplasm to the nucleus. Proc Natl Acad Sci U S A 98(20):11598–11603. doi:10.1073/pnas.181181198
20. Goswami A, Ranganathan P, Rangnekar VM (2006) The phosphoinositide 3-kinase/Akt1/Par-4 axis: a cancer-selective therapeutic target. Cancer Res 66(6):2889–2892. doi:10.1158/0008-5472.CAN-05-4458
21. Kitamura T, Asai N, Enomoto A, Maeda K, Kato T, Ishida M, Jiang P, Watanabe T, Usukura J, Kondo T, Costantini F, Murohara T, Takahashi M (2008) Regulation of VEGF-mediated

angiogenesis by the Akt/PKB substrate Girdin. Nat Cell Biol 10(3):329–337. doi:10.1038/ncb1695
22. Maddika S, Panigrahi S, Wiechec E, Wesselborg S, Fischer U, Schulze-Osthoff K, Los M (2009) Unscheduled Akt-triggered activation of cyclin-dependent kinase 2 as a key effector mechanism of apoptin's anticancer toxicity. Mol Cell Biol 29(5):1235–1248. doi:10.1128/MCB.00668-08
23. Bishop JM (1991) Molecular themes in oncogenesis. Cell 64(2):235–248
24. Greenman C, Stephens P, Smith R, Dalgliesh GL, Hunter C, Bignell G, Davies H, Teague J, Butler A, Stevens C, Edkins S, O'Meara S, Vastrik I, Schmidt EE, Avis T, Barthorpe S, Bhamra G, Buck G, Choudhury B, Clements J, Cole J, Dicks E, Forbes S, Gray K, Halliday K, Harrison R, Hills K, Hinton J, Jenkinson A, Jones D, Menzies A, Mironenko T, Perry J, Raine K, Richardson D, Shepherd R, Small A, Tofts C, Varian J, Webb T, West S, Widaa S, Yates A, Cahill DP, Louis DN, Goldstraw P, Nicholson AG, Brasseur F, Looijenga L, Weber BL, Chiew YE, DeFazio A, Greaves MF, Green AR, Campbell P, Birney E, Easton DF, Chenevix-Trench G, Tan MH, Khoo SK, Teh BT, Yuen ST, Leung SY, Wooster R, Futreal PA, Stratton MR (2007) Patterns of somatic mutation in human cancer genomes. Nature 446(7132):153–158. doi:10.1038/nature05610
25. Bader AG, Kang S, Vogt PK (2006) Cancer-specific mutations in PIK3CA are oncogenic in vivo. Proc Natl Acad Sci U S A 103(5):1475–1479. doi:10.1073/pnas.0510857103
26. Day FL, Jorissen RN, Lipton L, Mouradov D, Sakthianandeswaren A, Christie M, Li S, Tsui C, Tie J, Desai J, Xu ZZ, Molloy P, Whitehall V, Leggett BA, Jones IT, McLaughlin S, Ward RL, Hawkins NJ, Ruszkiewicz AR, Moore J, Busam D, Zhao Q, Strausberg RL, Gibbs P, Sieber OM (2013) PIK3CA and PTEN gene and exon mutation-specific clinicopathologic and molecular associations in colorectal cancer. Clin Cancer Res 19(12):3285–3296. doi:10.1158/1078-0432.CCR-12-3614
27. Khan MW, Keshavarzian A, Gounaris E, Melson JE, Cheon EC, Blatner NR, Chen ZE, Tsai FN, Lee G, Ryu H, Barrett TA, Bentrem DJ, Beckhove P, Khazaie K (2013) PI3K/AKT signaling is essential for communication between tissue-infiltrating mast cells, macrophages, and epithelial cells in colitis-induced cancer. Clin Cancer Res 19(9):2342–2354. doi:10.1158/1078-0432.CCR-12-2623
28. Cook MB, Chow WH, Devesa SS (2009) Oesophageal cancer incidence in the United States by race, sex, and histologic type, 1977-2005. Br J Cancer 101(5):855–859. doi:10.1038/sj.bjc.6605246
29. Vizcaino AP, Moreno V, Lambert R, Parkin DM (2002) Time trends incidence of both major histologic types of esophageal carcinomas in selected countries, 1973–1995. Int J Cancer 99 (6):860–868. doi:10.1002/ijc.10427
30. Prins MJ, Verhage RJ, Ruurda JP, ten Kate FJ, van Hillegersberg R (2013) Over-expression of phosphorylated mammalian target of rapamycin is associated with poor survival in oesophageal adenocarcinoma: a tissue microarray study. J Clin Pathol 66(3):224–228. doi:10.1136/jclinpath-2012-201173
31. Wu X, Chen Y, Li G, Xia L, Gu R, Wen X, Ming X, Chen H (2014) Her3 is associated with poor survival of gastric adenocarcinoma: Her3 promotes proliferation, survival and migration of human gastric cancer mediated by PI3K/AKT signaling pathway. Med Oncol 31(4):903. doi:10.1007/s12032-014-0903-x
32. Fu X, Feng J, Zeng D, Ding Y, Yu C, Yang B (2014) PAK4 confers cisplatin resistance in gastric cancer cells via PI3K/Akt- and MEK/Erk-dependent pathways. Biosci Rep. doi:10.1042/bsr20130102
33. Nagini S (2012) Carcinoma of the stomach: a review of epidemiology, pathogenesis, molecular genetics and chemoprevention. World J Gastrointest Oncol 4(7):156–169. doi:10.4251/wjgo.v4.i7.156
34. Shaib YH, Davila JA, El-Serag HB (2006) The epidemiology of pancreatic cancer in the United States: changes below the surface. Aliment Pharmacol Ther 24(1):87–94. doi:10.1111/j.1365-2036.2006.02961.x

35. Zhang Y, Zhang J, Xu K, Xiao Z, Sun J, Xu J, Wang J, Tang Q (2013) PTEN/PI3K/mTOR/B7-H1 signaling pathway regulates cell progression and immuno-resistance in pancreatic cancer. Hepatogastroenterology 60(127):1766–1772
36. Garcia-Carracedo D, Chen ZM, Qiu W, Huang AS, Tang SM, Hruban RH, Su GH (2014) PIK3CA Mutations in Mucinous Cystic Neoplasms of the Pancreas. Pancreas 43(2):245–249. doi:10.1097/mpa.0000000000000034
37. Centers for Disease Control and Prevention. Hepatocellular carcinoma - United States, 2001–2006 (2010). MMWR Morb Mortal Wkly Rep 59(17):517–520
38. Hsu HH, Cheng LH, Ho TJ, Kuo WW, Lin YM, Chen MC, Lee NH, Tsai FJ, Tsai KH, Huang CY (2014) Apicidin-resistant HA22T hepatocellular carcinoma cells massively promote pro-survival capability via IGF-IR/PI3K/Akt signaling pathway activation. Tumour Biol 35(1):303–313. doi:10.1007/s13277-013-1041-3
39. Chow JS, Chen CC, Ahsan H, Neugut AI (1996) A population-based study of the incidence of malignant small bowel tumours: SEER, 1973-1990. Int J Epidemiol 25(4):722–728
40. Zhang Y, Yao X, Jiang C, Yue J, Guan J, Cheng H, Hajirashid M, Wang Y, Fan L (2013) Expression of PI3K, PTEN and Akt in small intestinal adenocarcinoma detected by quantum dots-based immunofluorescence technology. Cancer Biomark 13(4):299–305. doi:10.3233/cbm-130352
41. Byun DS, Ahmed N, Nasser S, Shin J, Al-Obaidi S, Goel S, Corner GA, Wilson AJ, Flanagan DJ, Williams DS, Augenlicht LH, Vincan E, Mariadason JM (2011) Intestinal epithelial-specific PTEN inactivation results in tumor formation. Am J Physiol Gastrointest Liver Physiol 301(5):G856–G864. doi:10.1152/ajpgi.00178.2011
42. Zhu C, Bassig BA, Zaridze D, Boyle P, Dai M, Li Q, Zheng T (2013) A birth cohort analysis of the incidence of ascending and descending colon cancer in the United States, 1973-2008. Cancer Cause Control 24(6):1147–1156. doi:10.1007/s10552-013-0193–1
43. Ahnen DJ, Wade SW, Jones WF, Sifri R, Mendoza Silveiras J, Greenamyer J, Guiffre S, Axilbund J, Spiegel A, You YN (2014) The increasing incidence of young-onset colorectal cancer: a call to action. Mayo Clin Proc 89(2):216–224. doi:10.1016/j.mayocp.2013.09.006
44. Whitman M, Kaplan DR, Schaffhausen B, Cantley L, Roberts TM (1985) Association of phosphatidylinositol kinase activity with polyoma middle-T competent for transformation. Nature 315(6016):239–242
45. Vanhaesebroeck B, Leevers SJ, Ahmadi K, Timms J, Katso R, Driscoll PC, Woscholski R, Parker PJ, Waterfield MD (2001) Synthesis and function of 3-phosphorylated inositol lipids. Annu Rev Biochem 70:535–602. doi:10.1146/annurev.biochem.70.1.535
46. Vanhaesebroeck B, Leevers SJ, Panayotou G, Waterfield MD (1997) Phosphoinositide 3-kinases: a conserved family of signal transducers. Trends Biochem Sci 22(7):267–272
47. Pirola L, Zvelebil MJ, Bulgarelli-Leva G, Van Obberghen E, Waterfield MD, Wymann MP (2001) Activation loop sequences confer substrate specificity to phosphoinositide 3-kinase alpha (PI3Kalpha). Functions of lipid kinase-deficient PI3Kalpha in signaling. J Biol Chem 276(24):21544–21554. doi:10.1074/jbc.M011330200
48. Sun M, Hillmann P, Hofmann BT, Hart JR, Vogt PK (2010) Cancer-derived mutations in the regulatory subunit p85alpha of phosphoinositide 3-kinase function through the catalytic subunit p110alpha. Proc Natl Acad Sci U S A 107(35):15547–15552. doi:10.1073/pnas.1009652107
49. Zhao L, Vogt PK (2008) Class I PI3K in oncogenic cellular transformation. Oncogene 27(41):5486–5496. doi:10.1038/onc.2008.244
50. Samuels Y, Ericson K (2006) Oncogenic PI3K and its role in cancer. Curr Opin Oncol 18(1):77–82
51. Philp AJ, Campbell IG, Leet C, Vincan E, Rockman SP, Whitehead RH, Thomas RJ, Phillips WA (2001) The phosphatidylinositol 3'-kinase p85alpha gene is an oncogene in human ovarian and colon tumors. Cancer Res 61(20):7426–7429
52. Oda K, Okada J, Timmerman L, Rodriguez-Viciana P, Stokoe D, Shoji K, Taketani Y, Kuramoto H, Knight ZA, Shokat KM, McCormick F (2008) PIK3CA cooperates with other phosphatidylinositol 3'-kinase pathway mutations to effect oncogenic transformation. Cancer Res 68(19):8127–8136. doi:10.1158/0008-5472.can-08-0755

53. Jaiswal BS, Janakiraman V, Kljavin NM, Chaudhuri S, Stern HM, Wang W, Kan Z, Dbouk HA, Peters BA, Waring P, Dela Vega T, Kenski DM, Bowman KK, Lorenzo M, Li H, Wu J, Modrusan Z, Stinson J, Eby M, Yue P, Kaminker JS, de Sauvage FJ, Backer JM, Seshagiri S (2009) Somatic mutations in p85alpha promote tumorigenesis through class IA PI3K activation. Cancer Cell 16(6):463–474. doi:10.1016/j.ccr.2009.10.016
54. Ogino S, Nosho K, Kirkner GJ, Shima K, Irahara N, Kure S, Chan AT, Engelman JA, Kraft P, Cantley LC, Giovannucci EL, Fuchs CS (2009) PIK3CA mutation is associated with poor prognosis among patients with curatively resected colon cancer. J Clin Oncol 27(9):1477–1484. doi:10.1200/JCO.2008.18.6544
55. Huang CH, Mandelker D, Schmidt-Kittler O, Samuels Y, Velculescu VE, Kinzler KW, Vogelstein B, Gabelli SB, Amzel LM (2007) The structure of a human p110alpha/p85alpha complex elucidates the effects of oncogenic PI3Kalpha mutations. Science 318(5857):1744–1748. doi:10.1126/science.1150799
56. Fu Z, Aronoff-Spencer E, Wu H, Gerfen GJ, Backer JM (2004) The iSH2 domain of PI 3-kinase is a rigid tether for p110 and not a conformational switch. Arch Biochem Biophys 432(2):244–251. doi:10.1016/j.abb.2004.09.032
57. Yu J, Wjasow C, Backer JM (1998) Regulation of the p85/p110alpha phosphatidylinositol 3'-kinase. Distinct roles for the n-terminal and c-terminal SH2 domains. J Biol Chem 273(46):30199–30203
58. Stintzing S, Lenz HJ (2013) A small cog in a big wheel: PIK3CA mutations in colorectal cancer. J Natl Cancer Inst 105(23):1775–1776. doi:10.1093/jnci/djt330
59. German S, Aslam HM, Saleem S, Raees A, Anum T, Alvi AA, Haseeb A (2013) Carcinogenesis of PIK3CA. Hered Cancer Clin Pract 11(1):5. doi:10.1186/1897-4287-11-5
60. Herrero-Gonzalez S, Di Cristofano A (2011) New routes to old places: PIK3R1 and PIK3R2 join PIK3CA and PTEN as endometrial cancer genes. Cancer Discov 1(2):106–107. doi:10.1158/2159-8290.CD-11-0116
61. Quayle SN, Lee JY, Cheung LW, Ding L, Wiedemeyer R, Dewan RW, Huang-Hobbs E, Zhuang L, Wilson RK, Ligon KL, Mills GB, Cantley LC, Chin L (2012) Somatic mutations of PIK3R1 promote gliomagenesis. PLoS One 7(11):e49466. doi:10.1371/journal.pone.0049466
62. Gymnopoulos M, Elsliger MA, Vogt PK (2007) Rare cancer-specific mutations in PIK3CA show gain of function. Proc Natl Acad Sci U S A 104(13):5569–5574. doi:10.1073/pnas.0701005104
63. Hofmann BT, Jucker M (2012) Activation of PI3K/Akt signaling by n-terminal SH2 domain mutants of the p85alpha regulatory subunit of PI3K is enhanced by deletion of its c-terminal SH2 domain. Cell Signal 24(10):1950–1954. doi:10.1016/j.cellsig.2012.06.009
64. Foukas LC, Daniele N, Ktori C, Anderson KE, Jensen J, Shepherd PR (2002) Direct effects of caffeine and theophylline on p110 delta and other phosphoinositide 3-kinases. Differential effects on lipid kinase and protein kinase activities. J Biol Chem 277(40):37124–37130. doi:10.1074/jbc.M202101200
65. Geering B, Cutillas PR, Nock G, Gharbi SI, Vanhaesebroeck B (2007) Class IA phosphoinositide 3-kinases are obligate p85-p110 heterodimers. Proc Natl Acad Sci U S A 104 (19):7809–7814. doi:10.1073/pnas.0700373104
66. Miled N, Yan Y, Hon WC, Perisic O, Zvelebil M, Inbar Y, Schneidman-Duhovny D, Wolfson HJ, Backer JM, Williams RL (2007) Mechanism of two classes of cancer mutations in the phosphoinositide 3-kinase catalytic subunit. Science 317(5835):239–242. doi:10.1126/science.1135394
67. Ogino S, Lochhead P, Giovannucci E, Meyerhardt JA, Fuchs CS, Chan AT (2013) Discovery of colorectal cancer PIK3CA mutation as potential predictive biomarker: power and promise of molecular pathological epidemiology. Oncogene. doi:10.1038/onc.2013.244
68. Yu JC, Heidaran MA, Pierce JH, Gutkind JS, Lombardi D, Ruggiero M, Aaronson SA (1991) Tyrosine mutations within the alpha platelet-derived growth factor receptor kinase insert domain abrogate receptor-associated phosphatidylinositol-3 kinase activity without affecting mitogenic or chemotactic signal transduction. Mol Cell Biol 11(7):3780–3785

69. O'Hayre M, Degese MS, Gutkind JS (2014) Novel insights into G protein and G protein-coupled receptor signaling in cancer. Curr Opin Cell Biol 27C:126–135. doi:10.1016/j.ceb.2014.01.005
70. Robinson DR, Wu YM, Lin SF (2000) The protein tyrosine kinase family of the human genome. Oncogene 19(49):5548–5557. doi:10.1038/sj.onc.1203957
71. Lemmon MA, Schlessinger J (2010) Cell signaling by receptor tyrosine kinases. Cell 141(7):1117–1134. doi:10.1016/j.cell.2010.06.011
72. Josse C, Bouznad N, Geurts P, Irrthum A, Huynh-Thu VA, Servais L, Hego A, Delvenne P, Bours V, Oury C (2014) Identification of a microRNA landscape targeting the PI3K/Akt signaling pathway in inflammation-induced colorectal carcinogenesis. Am J Physiol Gastrointest Liver Physiol 306(3):G229–G243. doi:10.1152/ajpgi.00484.2012
73. Yun CH, Mengwasser KE, Toms AV, Woo MS, Greulich H, Wong KK, Meyerson M, Eck MJ (2008) The T790M mutation in EGFR kinase causes drug resistance by increasing the affinity for ATP. Proc Natl Acad Sci U S A 105(6):2070–2075. doi:10.1073/pnas.0709662105
74. Zhang Z, Norris J, Schwartz C, Alexov E (2011) In silico and in vitro investigations of the mutability of disease-causing missense mutation sites in spermine synthase. PLoS One 6(5):e20373. doi:10.1371/journal.pone.0020373
75. Zhang Z, Wang L, Gao Y, Zhang J, Zhenirovskyy M, Alexov E (2012) Predicting folding free energy changes upon single point mutations. Bioinformatics 28(5):664–671. doi:10.1093/bioinformatics/bts005
76. Coulier F, Kumar R, Ernst M, Klein R, Martin-Zanca D, Barbacid M (1990) Human trk oncogenes activated by point mutation, in-frame deletion, and duplication of the tyrosine kinase domain. Mol Cell Biol 10(8):4202–4210
77. Okawa T, Michaylira CZ, Kalabis J, Stairs DB, Nakagawa H, Andl CD, Johnstone CN, Klein-Szanto AJ, El-Deiry WS, Cukierman E, Herlyn M, Rustgi AK (2007) The functional interplay between EGFR overexpression, hTERT activation, and p53 mutation in esophageal epithelial cells with activation of stromal fibroblasts induces tumor development, invasion, and differentiation. Genes Dev 21(21):2788–2803. doi:10.1101/gad.1544507
78. Koorstra JB, Karikari CA, Feldmann G, Bisht S, Rojas PL, Offerhaus GJ, Alvarez H, Maitra A (2009) The Axl receptor tyrosine kinase confers an adverse prognostic influence in pancreatic cancer and represents a new therapeutic target. Cancer Biol Ther 8(7):618–626
79. Locatelli-Hoops S, Yeliseev AA, Gawrisch K, Gorshkova I (2013) Surface plasmon resonance applied to G protein-coupled receptors. Biomed Spectrosc Imaging 2(3):155–181. doi:10.3233/BSI-130045
80. Neer EJ (1995) Heterotrimeric G proteins: organizers of transmembrane signals. Cell 80(2):249–257
81. Lefkowitz RJ, Rajagopal K, Whalen EJ (2006) New roles for beta-arrestins in cell signaling: not just for seven-transmembrane receptors. Mol Cell 24(5):643–652. doi:10.1016/j.molcel.2006.11.007
82. Connelly SF, Isley BA, Baker CH, Gallick GE, Summy JM (2010) Loss of tyrosine phosphatase-dependent inhibition promotes activation of tyrosine kinase c-Src in detached pancreatic cells. Mol Carcinog 49(12):1007–1021. doi:10.1002/mc.20684
83. Yu HG, Schafer H, Mergler S, Muerkoster S, Cramer T, Hocker M, Herzig KH, Schmidt WE, Schmitz F (2005) Valine-286 residue in the third intracellular loop of the cholecystokinin 2 receptor exerts a pivotal role in cholecystokinin 2 receptor mediated intracellular signal transduction in human colon cancer cells. Cell Signal 17(12):1505–1515. doi:10.1016/j.cellsig.2005.03.009
84. Yu W, Ma S, Wang L, Zuo B, Li M, Qiao Z, Pan X, Liu Y, Wang J (2013) Upregulation of GPR34 expression affects the progression and prognosis of human gastric adenocarcinoma by PI3K/PDK1/AKT pathway. Histol Histopathol 28(12):1629–1638
85. Wu Q, Wang H, Zhao X, Shi Y, Jin M, Wan B, Xu H, Cheng Y, Ge H, Zhang Y (2013) Identification of G-protein-coupled receptor 120 as a tumor-promoting receptor that induces angiogenesis and migration in human colorectal carcinoma. Oncogene 32(49):5541–5550. doi:10.1038/onc.2013.264

86. Wu J, Xie N, Xie K, Zeng J, Cheng L, Lei Y, Liu Y, Song L, Dong D, Chen Y, Zeng R, Nice EC, Huang C, Wei Y (2013) GPR48, a poor prognostic factor, promotes tumor metastasis and activates beta-catenin/TCF signaling in colorectal cancer. Carcinogenesis 34(12):2861–2869. doi:10.1093/carcin/bgt229
87. Hofler A, Nichols T, Grant S, Lingardo L, Esposito EA, Gridley S, Murphy ST, Kath JC, Cronin CN, Kraus M, Alton G, Xie Z, Sutton S, Gehring M, Ermolieff J (2011) Study of the PDK1/AKT signaling pathway using selective PDK1 inhibitors, HCS, and enhanced biochemical assays. Anal Biochem 414(2):179–186. doi:10.1016/j.ab.2011.03.013
88. Bayascas JR (2008) Dissecting the role of the 3-phosphoinositide-dependent protein kinase-1 (PDK1) signalling pathways. Cell Cycle 7(19):2978–2982
89. Slattery ML, Lundgreen A, Herrick JS, Wolff RK (2011) Genetic variation in RPS6KA1, RPS6KA2, RPS6KB1, RPS6KB2, and PDK1 and risk of colon or rectal cancer. Mutat Res 706(1–2):13–20. doi:10.1016/j.mrfmmm.2010.10.005
90. Meier R, Alessi DR, Cron P, Andjelkovic M, Hemmings BA (1997) Mitogenic activation, phosphorylation, and nuclear translocation of protein kinase Bbeta. J Biol Chem 272(48):30491–30497
91. Miao B, Skidan I, Yang J, Lugovskoy A, Reibarkh M, Long K, Brazell T, Durugkar KA, Maki J, Ramana CV, Schaffhausen B, Wagner G, Torchilin V, Yuan J, Degterev A (2010) Small molecule inhibition of phosphatidylinositol-3,4,5-triphosphate (PIP3) binding to pleckstrin homology domains. Proc Natl Acad Sci U S A 107(46):20126–20131. doi:10.1073/pnas.1004522107
92. Gan X, Wang J, Su B, Wu D (2011) Evidence for direct activation of mTORC2 kinase activity by phosphatidylinositol 3,4,5-trisphosphate. J Biol Chem 286(13):10998–11002. doi:10.1074/jbc.M110.195016
93. Bhaskar PT, Hay N (2007) The two TORCs and Akt. Dev Cell 12(4):487–502. doi:10.1016/j.devcel.2007.03.020
94. Harada N, Gansauge S, Gansauge F, Gause H, Shimoyama S, Imaizumi T, Mattfeld T, Schoenberg MH, Beger HG (1997) Nuclear accumulation of p53 correlates significantly with clinical features and inversely with the expression of the cyclin-dependent kinase inhibitor p21(WAF1/CIP1) in pancreatic cancer. Br J Cancer 76(3):299–305
95. Din FV, Valanciute A, Houde VP, Zibrova D, Green KA, Sakamoto K, Alessi DR, Dunlop MG (2012) Aspirin inhibits mTOR signaling, activates AMP-activated protein kinase, and induces autophagy in colorectal cancer cells. Gastroenterology 142(7):1504–1515 e1503. doi:10.1053/j.gastro.2012.02.050
96. Kumar A, Pandurangan AK, Lu F, Fyrst H, Zhang M, Byun HS, Bittman R, Saba JD (2012) Chemopreventive sphingadienes downregulate Wnt signaling via a PP2A/Akt/GSK3beta pathway in colon cancer. Carcinogenesis 33(9):1726–1735. doi:10.1093/carcin/bgs174
97. Slattery ML, Herrick JS, Lundgreen A, Fitzpatrick FA, Curtin K, Wolff RK (2010) Genetic variation in a metabolic signaling pathway and colon and rectal cancer risk: mTOR, PTEN, STK11, RPKAA1, PRKAG2, TSC1, TSC2, PI3K and Akt1. Carcinogenesis 31(9):1604–1611. doi:10.1093/carcin/bgq142
98. Destefano MA, Jacinto E (2013) Regulation of insulin receptor substrate-1 by mTORC2 (mammalian target of rapamycin complex 2). Biochem Soc Trans 41(4):896–901. doi:10.1042/BST20130018
99. Esposito DL, Aru F, Lattanzio R, Morgano A, Abbondanza M, Malekzadeh R, Bishehsari F, Valanzano R, Russo A, Piantelli M, Moschetta A, Lotti LV, Mariani-Costantini R (2012) The insulin receptor substrate 1 (IRS1) in intestinal epithelial differentiation and in colorectal cancer. PLoS One 7(4):e36190. doi:10.1371/journal.pone.0036190
100. Khatri S, Yepiskoposyan H, Gallo CA, Tandon P, Plas DR (2010) FOXO3a regulates glycolysis via transcriptional control of tumor suppressor TSC1. J Biol Chem 285(21):15960–15965. doi:10.1074/jbc.M110.121871
101. Larsen WJ, Sherman LS, Potter SS, Scott WJ (2001) Human embryology, 3rd edn. Churchill Livingstone, New York

102. Martin TD, Chen XW, Kaplan RE, Saltiel AR, Walker CL, Reiner DJ, Der CJ (2014) Ral and Rheb GTPase Activating Proteins Integrate mTOR and GTPase Signaling in Aging, Autophagy, and Tumor Cell Invasion. Mol Cell 53(2):209–220. doi:10.1016/j.molcel.2013.12.004
103. Lacher MD, Pincheira RJ, Castro AF (2011) Consequences of interrupted Rheb-to-AMPK feedback signaling in tuberous sclerosis complex and cancer. Small GTPases 2(4):211–216. doi:10.4161/sgtp.2.4.16703
104. Santarpia L, Lippman SM, El-Naggar AK (2012) Targeting the MAPK-RAS-RAF signaling pathway in cancer therapy. Expert Opin Ther Targets 16(1):103–119. doi:10.1517/14728222.2011.645805
105. Zheng J, Wu C, Lin Z, Guo Y, Shi L, Dong P, Lu Z, Gao S, Liao Y, Chen B, Yu F (2014) Curcumin up-regulates phosphatase and tensin homologue deleted on chromosome 10 through microRNA-mediated control of DNA methylation–a novel mechanism suppressing liver fibrosis. FEBS J 281(1):88–103. doi:10.1111/febs.12574
106. Guha M, Plescia J, Leav I, Li J, Languino LR, Altieri DC (2009) Endogenous tumor suppression mediated by PTEN involves survivin gene silencing. Cancer Res 69(12):4954–4958. doi:10.1158/0008-5472.CAN-09-0584
107. Vinciguerra M, Sgroi A, Veyrat-Durebex C, Rubbia-Brandt L, Buhler LH, Foti M (2009) Unsaturated fatty acids inhibit the expression of tumor suppressor phosphatase and tensin homolog (PTEN) via microRNA-21 up-regulation in hepatocytes. Hepatology 49(4):1176–1184. doi:10.1002/hep.22737
108. Chowdhury S, Ongchin M, Wan G, Sharratt E, Brattain MG, Rajput A (2013) Restoration of PTEN activity decreases metastases in an orthotopic model of colon cancer. J Surg Res 184(2):755–760. doi:10.1016/j.jss.2013.03.035
109. Ha M, Chung JW, Hahm KB, Kim YJ, Lee W, An J, Kim DK, Kim MG (2012) A case of Cowden syndrome diagnosed from multiple gastric polyposis. World J Gastroenterol 18(8):861–864. doi:10.3748/wjg.v18.i8.861
110. Pilarski R, Burt R, Kohlman W, Pho L, Shannon KM, Swisher E (2013) Cowden syndrome and the PTEN hamartoma tumor syndrome: systematic review and revised diagnostic criteria. J Natl Cancer Inst 105(21):1607–1616. doi:10.1093/jnci/djt277
111. Zhang LL, Liu J, Lei S, Zhang J, Zhou W, Yu HG (2014) PTEN inhibits the invasion and metastasis of gastric cancer via downregulation of FAK expression. Cell Signal 26(5):1011–1020. doi:10.1016/j.cellsig.2014.01.025
112. Bowen KA, Doan HQ, Zhou BP, Wang Q, Zhou Y, Rychahou PG, Evers BM (2009) PTEN loss induces epithelial–mesenchymal transition in human colon cancer cells. Anticancer Res 29(11):4439–4449
113. Wang JY, Chang CC, Chiang CC, Chen WM, Hung SC (2012) Silibinin suppresses the maintenance of colorectal cancer stem-like cells by inhibiting PP2A/AKT/mTOR pathways. J Cell Biochem 113(5):1733–1743. doi:10.1002/jcb.24043
114. Tsuchiya A, Kanno T, Shimizu T, Nakao S, Tanaka A, Tabata C, Nakano T, Nishizaki T (2014) A novel PP2A enhancer induces caspase-independent apoptosis of MKN28 gastric cancer cells with high MEK activity. Cancer Lett. doi:10.1016/j.canlet.2014.01.034
115. Chow JY, Barrett KE (2007) Role of protein phosphatase 2A in calcium-dependent chloride secretion by human colonic epithelial cells. Am J Physiol Cell Physiol 292(1):C452–C459. doi:10.1152/ajpcell.00034.2006
116. Polanowska-Grabowska R, Simon CG Jr, Falchetto R, Shabanowitz J, Hunt DF, Gear AR (1997) Platelet adhesion to collagen under flow causes dissociation of a phosphoprotein complex of heat-shock proteins and protein phosphatase 1. Blood 90(4):1516–1526
117. Liu J, Weiss HL, Rychahou P, Jackson LN, Evers BM, Gao T (2009) Loss of PHLPP expression in colon cancer: role in proliferation and tumorigenesis. Oncogene 28(7):994–1004. doi:10.1038/onc.2008.450
118. Brognard J, Sierecki E, Gao T, Newton AC (2007) PHLPP and a second isoform, PHLPP2, differentially attenuate the amplitude of Akt signaling by regulating distinct Akt isoforms. Mol Cell 25(6):917–931. doi:10.1016/j.molcel.2007.02.017

119. Call R, Grimsley M, Cadwallader L, Cialone L, Hill M, Hreish V, King ST, Riche DM (2010) Insulin–carcinogen or mitogen? Preclinical and clinical evidence from prostate, breast, pancreatic, and colorectal cancer research. Postgrad Med 122(3):158–165. doi:10.3810/pgm.2010.05.2153
120. Buller CL, Heilig CW, Brosius FC 3rd (2011) GLUT1 enhances mTOR activity independently of TSC2 and AMPK. Am J Physiol Renal Physiol 301(3):F588–596. doi:10.1152/ajprenal.00472.2010
121. Zhang HH, Huang J, Duvel K, Boback B, Wu S, Squillace RM, Wu CL, Manning BD (2009) Insulin stimulates adipogenesis through the Akt-TSC2-mTORC1 pathway. PLoS One 4(7):e6189. doi:10.1371/journal.pone.0006189
122. Hardie DG (2011) AMP-activated protein kinase: an energy sensor that regulates all aspects of cell function. Genes Dev 25(18):1895–1908. doi:10.1101/gad.17420111
123. Nijsten MW, van Dam GM (2009) Hypothesis: using the Warburg effect against cancer by reducing glucose and providing lactate. Med Hypotheses 73(1):48–51. doi:10.1016/j.mehy.2009.01.041
124. Shackelford DB, Shaw RJ (2009) The LKB1-AMPK pathway: metabolism and growth control in tumour suppression. Nat Rev Cancer 9(8):563–575. doi:10.1038/nrc2676
125. Duong HQ, Hwang JS, Kim HJ, Seong YS, Bae I (2012) BML-275, an AMPK inhibitor, induces DNA damage, G2/M arrest and apoptosis in human pancreatic cancer cells. Int J Oncol 41(6):2227–2236. doi:10.3892/ijo.2012.1672
126. Swierczynski J, Hebanowska A, Sledzinski T (2014) Role of abnormal lipid metabolism in development, progression, diagnosis and therapy of pancreatic cancer. World J Gastroenterol 20(9):2279–2303. doi:10.3748/wjg.v20.i9.2279
127. Hu S, Dong TS, Dalal SR, Wu F, Bissonnette M, Kwon JH, Chang EB (2011) The microbe-derived short chain fatty acid butyrate targets miRNA-dependent p21 gene expression in human colon cancer. PLoS One 6(1):e16221. doi:10.1371/journal.pone.0016221
128. Kubo A, Corley DA, Jensen CD, Kaur R (2010) Dietary factors and the risks of oesophageal adenocarcinoma and Barrett's oesophagus. Nutr Res Rev 23(2):230–246. doi:10.1017/S0954422410000132
129. Portale G, Hagen JA, Peters JH, Chan LS, DeMeester SR, Gandamihardja TA, DeMeester TR (2006) Modern 5-year survival of resectable esophageal adenocarcinoma: single institution experience with 263 patients. J Am Coll Surg 202(4):588–596; discussion 596–588. doi:10.1016/j.jamcollsurg.2005.12.022
130. Lindkvist B, Johansen D, Stocks T, Concin H, Bjorge T, Almquist M, Haggstrom C, Engeland A, Hallmans G, Nagel G, Jonsson H, Selmer R, Ulmer H, Tretli S, Stattin P, Manjer J (2014) Metabolic risk factors for esophageal squamous cell carcinoma and adenocarcinoma: a prospective study of 580 000 subjects within the Me-Can project. BMC Cancer 14(1):103. doi:10.1186/1471-2407-14-103
131. Li H, Gao Q, Guo L, Lu SH (2011) The PTEN/PI3K/Akt pathway regulates stem-like cells in primary esophageal carcinoma cells. Cancer Biol Ther 11(11):950–958
132. Metzger R, Schneider PM, Warnecke-Eberz U, Brabender J, Holscher AH (2004) Molecular biology of esophageal cancer. Onkologie 27(2):200–206. doi:10.1159/000076913
133. Sunpaweravong P, Sunpaweravong S, Puttawibul P, Mitarnun W, Zeng C, Baron AE, Franklin W, Said S, Varella-Garcia M (2005) Epidermal growth factor receptor and cyclin D1 are independently amplified and overexpressed in esophageal squamous cell carcinoma. J Cancer Res Clin Oncol 131(2):111–119. doi:10.1007/s00432-004-0610-7
134. Bandla S, Pennathur A, Luketich JD, Beer DG, Lin L, Bass AJ, Godfrey TE, Litle VR (2012) Comparative genomics of esophageal adenocarcinoma and squamous cell carcinoma. Ann Thorac Surg 93(4):1101–1106. doi:10.1016/j.athoracsur.2012.01.064
135. Stoner GD, Wang LS, Chen T (2007) Chemoprevention of esophageal squamous cell carcinoma. Toxicol Appl Pharmacol 224(3):337–349. doi:10.1016/j.taap.2007.01.030
136. Chan JA, Blaszkowsky LS, Enzinger PC, Ryan DP, Abrams TA, Zhu AX, Temel JS, Schrag D, Bhargava P, Meyerhardt JA, Wolpin BM, Fidias P, Zheng H, Florio S, Regan E, Fuchs

CS (2011) A multicenter phase II trial of single-agent cetuximab in advanced esophageal and gastric adenocarcinoma. Ann Oncol 22(6):1367–1373. doi:10.1093/annonc/mdq604
137. Tomblyn MB, Goldman BH, Thomas CR Jr., Benedetti JK, Lenz HJ, Mehta V, Beeker T, Gold PJ, Abbruzzese JL, Blanke CD, Committee SG (2012) Cetuximab plus cisplatin, irinotecan, and thoracic radiotherapy as definitive treatment for locally advanced, unresectable esophageal cancer: a phase-II study of the SWOG (S0414). J Thorac Oncol 7(5):906–912. doi:10.1097/JTO.0b013e31824c7bed
138. Wu X, Zhang J, Zhen R, Lv J, Zheng L, Su X, Zhu G, Gavine PR, Xu S, Lu S, Hou J, Liu Y, Xu C, Tan Y, Xie L, Yin X, He D, Ji Q, Hou Y, Ge D (2012) Trastuzumab anti-tumor efficacy in patient-derived esophageal squamous cell carcinoma xenograft (PDECX) mouse models. J Transl Med 10:180. doi:10.1186/1479-5876-10-180
139. Hirashima K, Baba Y, Watanabe M, Karashima RI, Sato N, Imamura Y, Nagai Y, Hayashi N, Iyama KI, Baba H (2012) Aberrant activation of the mTOR pathway and anti-tumour effect of everolimus on oesophageal squamous cell carcinoma. Br J Cancer 106(5):876–882. doi:10.1038/bjc.2012.36
140. Weng W, Wu Q, Yu Y, Mei W, Wang X (2013) A novel chemotherapeutic arene ruthenium(II) drug Rawq01 altered the effect of microRNA-21 on PTEN/AKT signaling pathway in esophageal cancer cells. Anticancer Res 33(12):5407–5414
141. Yang H, Sukocheva OA, Hussey DJ, Watson DI (2012) Estrogen, male dominance and esophageal adenocarcinoma: is there a link? World J Gastroenterol 18(5):393–400. doi:10.3748/wjg.v18.i5.393
142. Hong J, Katsha A, Lu P, Shyr Y, Belkhiri A, El-Rifai W (2012) Regulation of ERBB2 receptor by t-DARPP mediates trastuzumab resistance in human esophageal adenocarcinoma. Cancer Res 72(17):4504–4514. doi:10.1158/0008-5472.CAN-12-1119
143. Janmaat ML, Gallegos-Ruiz MI, Rodriguez JA, Meijer GA, Vervenne WL, Richel DJ, Van Groeningen C, Giaccone G (2006) Predictive factors for outcome in a phase II study of gefitinib in second-line treatment of advanced esophageal cancer patients. J Clin Oncol 24(10):1612–1619. doi:10.1200/jco.2005.03.4900
144. Shah MA, Ramanathan RK, Ilson DH, Levnor A, D'Adamo D, O'Reilly E, Tse A, Trocola R, Schwartz L, Capanu M, Schwartz GK, Kelsen DP (2006) Multicenter phase II study of irinotecan, cisplatin, and bevacizumab in patients with metastatic gastric or gastroesophageal junction adenocarcinoma. J Clin Oncol 24(33):5201–5206. doi:10.1200/jco.2006.08.0887
145. Huang CM, Wang HM, Zheng CH, Li P, Xie JW, Wang JB, Lin JX, Lu J (2013) Tumor size as a prognostic factor in patients with node-negative gastric cancer invading the muscularis propria and subserosa (pT2-3N0M0 stage). Hepatogastroenterology 60(124):699–703. doi:10.5754/hge12733
146. Diaz de Liano A, Yarnoz C, Artieda C, Aguilar R, Viana S, Artajona A, Ortiz H (2009) Results of R0 surgery with D2 lymphadenectomy for the treatment of localised gastric cancer. Clin Transl Oncol 11(3):178–182
147. Adachi Y, Mimori K, Mori M, Maehara Y, Sugimachi K (1997) Morbidity after D2 and D3 gastrectomy for node-positive gastric carcinoma. J Am Coll Surg 184(3):240–244
148. Chen S, Feng X, Li Y, Yuan X, Zhou Z, Chen Y (2012) Efficacy and safety of XELOX and FOLFOX6 adjuvant chemotherapy following radical total gastrectomy. Oncol Lett 3(4):781–786. doi:10.3892/ol.2012.577
149. Scartozzi M, Galizia E, Verdecchia L, Berardi R, Antognoli S, Chiorrini S, Cascinu S (2007) Chemotherapy for advanced gastric cancer: across the years for a standard of care. Expert Opin Pharmacother 8(6):797–808. doi:10.1517/14656566.8.6.797
150. Kulke MH, Wu B, Clark JW, Enzinger PC, Lynch TJ, Vincitore M, Michelini A, Fuchs CS (2006) A phase II study of doxorubicin, cisplatin, and 5-fluorouracil in patients with advanced adenocarcinoma of the stomach or esophagus. Cancer Invest 24(3):229–234. doi:10.1080/07357900600633924
151. Dicken BJ, Bigam DL, Cass C, Mackey JR, Joy AA, Hamilton SM (2005) Gastric adenocarcinoma: review and considerations for future directions. Ann Surg 241(1):27–39

152. Borch K, Jonsson B, Tarpila E, Franzen T, Berglund J, Kullman E, Franzen L (2000) Changing pattern of histological type, location, stage and outcome of surgical treatment of gastric carcinoma. Br J Surg 87(5):618–626. doi:10.1046/j.1365-2168.2000.01425.x
153. Wong H, Yau T (2013) Molecular targeted therapies in advanced gastric cancer: does tumor histology matter? Therap Adv Gastroenterol 6(1):15–31. doi:10.1177/1756283X12453636
154. Kang JM, Shin DW, Kwon YM, Park SM, Park MS, Park JH, Son KY, Cho BL (2011) Stomach cancer screening and preventive behaviors in relatives of gastric cancer patients. World J Gastroenterol 17(30):3518–3525. doi:10.3748/wjg.v17.i30.3518
155. Chun N, Ford JM (2012) Genetic testing by cancer site: stomach. Cancer J 18(4):355–363. doi:10.1097/PPO.0b013e31826246dc
156. Dussaulx-Garin L, Blayau M, Pagenault M, Le Berre-Heresbach N, Raoul JL, Campion JP, David V, Bretagne JF (2001) A new mutation of E-cadherin gene in familial gastric linitis plastica cancer with extra-digestive dissemination. Eur J Gastroenterol Hepatol 13(6):711–715
157. Huntsman DG, Carneiro F, Lewis FR, MacLeod PM, Hayashi A, Monaghan KG, Maung R, Seruca R, Jackson CE, Caldas C (2001) Early gastric cancer in young, asymptomatic carriers of germ-line E-cadherin mutations. N Engl J Med 344(25):1904–1909. doi:10.1056/NEJM200106213442504
158. Hebbard PC, Macmillan A, Huntsman D, Kaurah P, Carneiro F, Wen X, Kwan A, Boone D, Bursey F, Green J, Fernandez B, Fontaine D, Wirtzfeld DA (2009) Prophylactic total gastrectomy (PTG) for hereditary diffuse gastric cancer (HDGC): the Newfoundland experience with 23 patients. Ann Surg Oncol 16(7):1890–1895. doi:10.1245/s10434-009-0471-z
159. Bang YJ, Van Cutsem E, Feyereislova A, Chung HC, Shen L, Sawaki A, Lordick F, Ohtsu A, Omuro Y, Satoh T, Aprile G, Kulikov E, Hill J, Lehle M, Ruschoff J, Kang YK, To GATI (2010) Trastuzumab in combination with chemotherapy versus chemotherapy alone for treatment of HER2-positive advanced gastric or gastro-oesophageal junction cancer (ToGA): a phase 3, open-label, randomised controlled trial. Lancet 376(9742):687–697. doi:10.1016/S0140-6736(10)61121-X
160. Van Cutsem E, de Haas S, Kang YK, Ohtsu A, Tebbutt NC, Ming Xu J, Peng Yong W, Langer B, Delmar P, Scherer SJ, Shah MA (2012) Bevacizumab in combination with chemotherapy as first-line therapy in advanced gastric cancer: a biomarker evaluation from the AVAGAST randomized phase III trial. J Clin Oncol 30(17):2119–2127. doi:10.1200/JCO.2011.39.9824
161. Rojo F, Tabernero J, Albanell J, Van Cutsem E, Ohtsu A, Doi T, Koizumi W, Shirao K, Takiuchi H, Ramon y Cajal S, Baselga J (2006) Pharmacodynamic studies of gefitinib in tumor biopsy specimens from patients with advanced gastric carcinoma. J Clin Oncol 24(26):4309–4316. doi:10.1200/jco.2005.04.2424
162. Liu J, Fu XQ, Zhou W, Yu HG, Yu JP, Luo HS (2011) LY294002 potentiates the anti-cancer effect of oxaliplatin for gastric cancer via death receptor pathway. World J Gastroenterol 17(2):181–190. doi:10.3748/wjg.v17.i2.181
163. Shin JY, Kim JO, Lee SK, Chae HS, Kang JH (2010) LY294002 may overcome 5-FU resistance via down-regulation of activated p-AKT in Epstein-Barr virus-positive gastric cancer cells. BMC Cancer 10:425. doi:10.1186/1471-2407-10-425
164. Gao YH, Zhang HP, Yang SM, Yang Y, Ma YY, Zhang XY, Yang YM (2014) Inactivation of Akt by arsenic trioxide induces cell death via mitochondrial-mediated apoptotic signaling in SGC-7901 human gastric cancer cells. Oncol Rep. doi:10.3892/or.2014.2994
165. Sha M, Ye J, Zhang LX, Luan ZY, Chen YB, Huang JX (2014) Celastrol induces apoptosis of gastric cancer cells by miR-21 inhibiting PI3K/Akt-NF-kappaB signaling pathway. Pharmacology 93(1–2):39–46. doi:10.1159/000357683
166. Surveillance Epidemiology and End Results Program (2014) Cancer Stat Fact Sheets. GPO, Washington, D.C. http://seer.cancer.gov/statfacts/. Accessed 7 March 2014
167. Wittig R, Coy JF (2008) The role of glucose metabolism and glucose-associated signalling in cancer. Perspect Med Chem 1:64–82

168. Wang SF, Chou YC, Mazumder N, Kao FJ, Nagy LD, Guengerich FP, Huang C, Lee HC, Lai PS, Ueng YF (2013) 7-Ketocholesterol induces P-glycoprotein through PI3K/mTOR signaling in hepatoma cells. Biochem Pharmacol 86(4):548–560. doi:10.1016/j.bcp.2013.06.006
169. Meguro M, Mizuguchi T, Kawamoto M, Hirata K (2011) The molecular pathogenesis and clinical implications of hepatocellular carcinoma. Int J Hepatol 2011:818672. doi:10.4061/2011/818672
170. Dai XF, Ding J, Zhang RG, Ren JH, Ma CM, Wu G (2013) Radiosensitivity enhancement of human hepatocellular carcinoma cell line SMMC-7721 by sorafenib through the MEK/ERK signal pathway. Int J Radiat Biol 89(9):724–731. doi:10.3109/09553002.2013.791405
171. Chang Z, Shi G, Jin J, Guo H, Guo X, Luo F, Song Y, Jia X (2013) Dual PI3K/mTOR inhibitor NVP-BEZ235-induced apoptosis of hepatocellular carcinoma cell lines is enhanced by inhibitors of autophagy. Int J Mol Med 31(6):1449–1456. doi:10.3892/ijmm.2013.1351
172. Yun SM, Lee JH, Jung KH, Lee H, Lee S, Hong S, Hong SS (2013) Induction of apoptosis and suppression of angiogenesis of hepatocellular carcinoma by HS-159, a novel phosphatidylinositol 3-kinase inhibitor. Int J Oncol 43(1):201–209. doi:10.3892/ijo.2013.1912
173. Kirstein MM, Boukouris AE, Pothiraju D, Buitrago-Molina LE, Marhenke S, Schutt J, Orlik J, Kuhnel F, Hegermann J, Manns MP, Vogel A (2013) Activity of the mTOR inhibitor RAD001, the dual mTOR and PI3-kinase inhibitor BEZ235 and the PI3-kinase inhibitor BKM120 in hepatocellular carcinoma. Liver Int 33(5):780–793. doi:10.1111/liv.12126
174. Goggins M, Schutte M, Lu J, Moskaluk CA, Weinstein CL, Petersen GM, Yeo CJ, Jackson CE, Lynch HT, Hruban RH, Kern SE (1996) Germline BRCA2 gene mutations in patients with apparently sporadic pancreatic carcinomas. Cancer Res 56(23):5360–5364
175. Weiss GA, Rossi MR, Khushalani NI, Lo K, Gibbs JF, Bharthuar A, Cowell JK, Iyer R (2013) Evaluation of phosphatidylinositol-3-kinase catalytic subunit (PIK3CA) and epidermal growth factor receptor (EGFR) gene mutations in pancreaticobiliary adenocarcinoma. J Gastrointest Oncol 4(1):20–29. doi:10.3978/j.issn.2078–6891.2012.012
176. Permuth-Wey J, Egan KM (2009) Family history is a significant risk factor for pancreatic cancer: results from a systematic review and meta-analysis. Fam Cancer 8(2):109–117. doi:10.1007/s10689-008-9214–8
177. Liles JS, Arnoletti JP, Tzeng CW, Howard JH, Kossenkov AV, Kulesza P, Heslin MJ, Frolov A (2010) ErbB3 expression promotes tumorigenesis in pancreatic adenocarcinoma. Cancer Biol Ther 10(6):555–563
178. Kelley RK, Ko AH (2008) Erlotinib in the treatment of advanced pancreatic cancer. Biologics 2(1):83–95
179. Tan X, Egami H, Abe M, Nozawa F, Hirota M, Ogawa M (2005) Involvement of MMP-7 in invasion of pancreatic cancer cells through activation of the EGFR mediated MEK-ERK signal transduction pathway. J Clin Pathol 58(12):1242–1248. doi:10.1136/jcp.2004.025338
180. Venkannagari S, Fiskus W, Peth K, Atadja P, Hidalgo M, Maitra A, Bhalla KN (2012) Superior efficacy of co-treatment with dual PI3K/mTOR inhibitor NVP-BEZ235 and pan-histone deacetylase inhibitor against human pancreatic cancer. Oncotarget 3(11):1416–1427
181. Glienke W, Maute L, Wicht J, Bergmann L (2012) The dual PI3K/mTOR inhibitor NVP-BGT226 induces cell cycle arrest and regulates Survivin gene expression in human pancreatic cancer cell lines. Tumour Biol 33(3):757–765. doi:10.1007/s13277-011-0290-2
182. Li J, Liang X, Yang X (2012) Ursolic acid inhibits growth and induces apoptosis in gemcitabine-resistant human pancreatic cancer via the JNK and PI3K/Akt/NF-kappaB pathways. Oncol Rep 28(2):501–510. doi:10.3892/or.2012.1827
183. Yothaisong S, Dokduang H, Techasen A, Namwat N, Yongvanit P, Bhudhisawasdi V, Puapairoj A, Riggins GJ, Loilome W (2013) Increased activation of PI3K/AKT signaling pathway is associated with cholangiocarcinoma metastasis and PI3K/mTOR inhibition presents a possible therapeutic strategy. Tumour Biol 34(6):3637–3648. doi:10.1007/s13277-013-0945-2
184. Leelawat K, Narong S, Udomchaiprasertkul W, Leelawat S, Tungpradubkul S (2009) Inhibition of PI3K increases oxaliplatin sensitivity in cholangiocarcinoma cells. Cancer Cell Int 9:3. doi:10.1186/1475-2867-9-3

185. de Gramont A, Hubbard J, Shi Q, O'Connell MJ, Buyse M, Benedetti J, Bot B, O'Callaghan C, Yothers G, Goldberg RM, Blanke CD, Benson A, Deng Q, Alberts SR, Andre T, Wolmark N, Grothey A, Sargent D (2010) Association between disease-free survival and overall survival when survival is prolonged after recurrence in patients receiving cytotoxic adjuvant therapy for colon cancer: simulations based on the 20,800 patient ACCENT data set. J Clin Oncol 28(3):460–465. doi:10.1200/jco.2009.23.1407
186. Andre T, Boni C, Navarro M, Tabernero J, Hickish T, Topham C, Bonetti A, Clingan P, Bridgewater J, Rivera F, de Gramont A (2009) Improved overall survival with oxaliplatin, fluorouracil, and leucovorin as adjuvant treatment in stage II or III colon cancer in the MOSAIC trial. J Clin Oncol 27(19):3109–3116. doi:10.1200/jco.2008.20.6771
187. Andre T, Boni C, Mounedji-Boudiaf L, Navarro M, Tabernero J, Hickish T, Topham C, Zaninelli M, Clingan P, Bridgewater J, Tabah-Fisch I, de Gramont A (2004) Oxaliplatin, fluorouracil, and leucovorin as adjuvant treatment for colon cancer. N Engl J Med 350(23):2343–2351. doi:10.1056/NEJMoa032709
188. Kuebler JP, Wieand HS, O'Connell MJ, Smith RE, Colangelo LH, Yothers G, Petrelli NJ, Findlay MP, Seay TE, Atkins JN, Zapas JL, Goodwin JW, Fehrenbacher L, Ramanathan RK, Conley BA, Flynn PJ, Soori G, Colman LK, Levine EA, Lanier KS, Wolmark N (2007) Oxaliplatin combined with weekly bolus fluorouracil and leucovorin as surgical adjuvant chemotherapy for stage II and III colon cancer: results from NSABP C-07. J Clin Oncol 25(16):2198–2204. doi:10.1200/jco.2006.08.2974
189. Grothey A, Sargent D, Goldberg RM, Schmoll HJ (2004) Survival of patients with advanced colorectal cancer improves with the availability of fluorouracil-leucovorin, irinotecan, and oxaliplatin in the course of treatment. J Clin Oncol 22(7):1209–1214. doi:10.1200/jco.2004.11.037
190. Arrington AK, Heinrich EL, Lee W, Duldulao M, Patel S, Sanchez J, Garcia-Aguilar J, Kim J (2012) Prognostic and Predictive Roles of KRAS Mutation in Colorectal Cancer. Int J Mol Sci 13(10):12153–12168. doi:10.3390/ijms131012153
191. Kennedy RD, Potter DD, Moir CR, El-Youssef M (2014) The natural history of familial adenomatous polyposis syndrome: A 24year review of a single center experience in screening, diagnosis, and outcomes. J Pediatr Surg 49(1):82–86. doi:10.1016/j.jpedsurg.2013.09.033
192. Wielders EA, Hettinger J, Dekker R, Kets CM, Ligtenberg MJ, Mensenkamp AR, van den Ouweland AM, Prins J, Wagner A, Dinjens WN, Dubbink HJ, van Hest LP, Menko F, Hogervorst F, Verhoef S, Te Riele H (2014) Functional analysis of MSH2 unclassified variants found in suspected Lynch syndrome patients reveals pathogenicity due to attenuated mismatch repair. J Med Genet. doi:10.1136/jmedgenet-2013-101987
193. Ogino S, Liao X, Imamura Y, Yamauchi M, McCleary NJ, Ng K, Niedzwiecki D, Saltz LB, Mayer RJ, Whittom R, Hantel A, Benson AB 3rd, Mowat RB, Spiegelman D, Goldberg RM, Bertagnolli MM, Meyerhardt JA, Fuchs CS, Alliance for Clinical Trials in O (2013) Predictive and prognostic analysis of PIK3CA mutation in stage III colon cancer intergroup trial. J Natl Cancer Inst 105(23):1789–1798. doi:10.1093/jnci/djt298
194. Hurwitz H, Fehrenbacher L, Novotny W, Cartwright T, Hainsworth J, Heim W, Berlin J, Baron A, Griffing S, Holmgren E, Ferrara N, Fyfe G, Rogers B, Ross R, Kabbinavar F (2004) Bevacizumab plus irinotecan, fluorouracil, and leucovorin for metastatic colorectal cancer. N Engl J Med 350(23):2335–2342. doi:10.1056/NEJMoa032691
195. Giantonio BJ, Catalano PJ, Meropol NJ, O'Dwyer PJ, Mitchell EP, Alberts SR, Schwartz MA, Benson AB 3rd (2007) Bevacizumab in combination with oxaliplatin, fluorouracil, and leucovorin (FOLFOX4) for previously treated metastatic colorectal cancer: results from the Eastern Cooperative Oncology Group Study E3200. J Clin Oncol 25(12):1539–1544. doi:10.1200/jco.2006.09.6305
196. Van Cutsem E, Peeters M, Siena S, Humblet Y, Hendlisz A, Neyns B, Canon JL, Van Laethem JL, Maurel J, Richardson G, Wolf M, Amado RG (2007) Open-label phase III trial of panitumumab plus best supportive care compared with best supportive care alone in patients with chemotherapy-refractory metastatic colorectal cancer. J Clin Oncol 25(13):1658–1664. doi:10.1200/jco.2006.08.1620

197. Cunningham D, Humblet Y, Siena S, Khayat D, Bleiberg H, Santoro A, Bets D, Mueser M, Harstrick A, Verslype C, Chau I, Van Cutsem E (2004) Cetuximab monotherapy and cetuximab plus irinotecan in irinotecan-refractory metastatic colorectal cancer. N Engl J Med 351(4):337–345. doi:10.1056/NEJMoa033025
198. Martinelli E, Troiani T, D'Aiuto E, Morgillo F, Vitagliano D, Capasso A, Costantino S, Ciuffreda LP, Merolla F, Vecchione L, De Vriendt V, Tejpar S, Nappi A, Sforza V, Martini G, Berrino L, De Palma R, Ciardiello F (2013) Antitumor activity of pimasertib, a selective MEK 1/2 inhibitor, in combination with PI3K/mTOR inhibitors or with multi-targeted kinase inhibitors in pimasertib-resistant human lung and colorectal cancer cells. Int J Cancer 133(9):2089–2101. doi:10.1002/ijc.28236
199. Rychahou PG, Murillo CA, Evers BM (2005) Targeted RNA interference of PI3K pathway components sensitizes colon cancer cells to TNF-related apoptosis-inducing ligand (TRAIL). Surgery 138(2):391–397. doi:10.1016/j.surg.2005.05.012
200. Shoeb M, Ramana KV, Srivastava SK (2013) Aldose reductase inhibition enhances TRAIL-induced human colon cancer cell apoptosis through AKT/FOXO3a-dependent upregulation of death receptors. Free Radic Biol Med 63:280–290. doi:10.1016/j.freeradbiomed.2013.05.039
201. Jiang QG, Li TY, Liu DN, Zhang HT (2014) PI3K/Akt pathway involving into apoptosis and invasion in human colon cancer cells LoVo. Mol Biol Rep. doi:10.1007/s11033-014-3198-2
202. Li X, Stevens PD, Liu J, Yang H, Wang W, Wang C, Zeng Z, Schmidt MD, Yang M, Lee EY, Gao T (2014) PHLPP is a negative regulator of RAF1 that reduces colorectal cancer cell motility and prevents tumor progression in mice. Gastroenterology. doi:10.1053/j.gastro.2014.02.003

The Epigenetics in Intestinal Tumorigenesis

Leticia Moreira, Francesc Balaguer and Ajay Goel

Abstract Colorectal cancer develops as a result of stepwise accumulation of genetic and epigenetic alterations, which in turn, regulate the expression of various oncogenes and tumor suppressor genes. Epigenetic alterations include DNA methylation, histone modifications, chromatin remodeling, and noncoding RNAs. Data gathered in recent years clearly highlight that in contrast to genetic alterations, epigenetic alterations occur far more frequently, participate in virtually every step of colorectal cancer development, and are potentially reversible. More importantly, considering the ubiquitous nature of epigenetic alterations in human cancers, these molecular signatures have garnered significant attention for development as potential diagnostic, prognostic and predictive biomarkers in human colorectal cancer. This article reviews the current state of literature in the field, and focuses on the clinical relevance of various epigenetic alterations as potential biomarkers for the personalized treatment of patients suffering from this common malignancy worldwide.

Keywords Epigenetics · DNA methylation · CpG islands · Histone modifications · MicroRNAs · Non-coding RNAs · Colon cancer · Colorectal cancer · Biomarkers

A. Goel (✉)
Center for Gastrointestinal Cancer Research and Center for Epigenetics, Cancer Prevention and Cancer Genomics, Baylor Research Institute; Charles A Sammons Cancer Center, Baylor University Medical Center, 3500 Gaston Avenue, Suite H-250, Dallas, TX 75246, USA
e-mail: ajay.goel@baylorhealth.edu

L. Moreira · F. Balaguer
Department of Gastroenterology, Hospital Clínic, Centro de Investigación Biomédica en Red de Enfermedades Hepáticas y Digestivas (CIBERehd), Institut d'Investigacions Biomediques August Pi i Sunyer (IDIBAPS), University of Barcelona, Barcelona, Catalonia, Spain

V. W. Yang, A. B. Bialkowska (eds.), *Intestinal Tumorigenesis*,
DOI 10.1007/978-3-319-19986-3_5

1 Introduction

Colorectal cancer (CRC) is one of the most common cancers worldwide, and remains as the second most common malignancy in terms of its incidence and mortality in most developed countries [1]. Up to 75–80 % of CRCs are sporadic, while the remaining 20–25 % occurs in patients with some degree of family history, suggesting an inherited susceptibility [2]. The vast majority of sporadic CRC cases develop in a stepwise normal-adenoma-carcinoma sequence [3, 4]. This sequence is characterized by an accumulation of genetic and epigenetic alterations that cause disturbances in cellular differentiation, apoptosis, and promote tumor growth. These molecular events cause the initiation and establishment of advanced colorectal adenomas, which serve as precursor malignant lesions for CRC.

From a molecular standpoint, CRC acts as a complex and heterogeneous disease that ensues following accumulation of a series of genetic and epigenetic alterations in key tumor suppressor and oncogenes [5, 6, 7]. The term 'epigenetics' was introduced by Waddington [8], and referred to the study of the interactions between genes and the environment. Over the years, this definition has evolved significantly and is currently defined as the set of heritable changes in gene expression that are not a direct consequence of changes in the DNA sequence itself [9].

There is growing evidence that different types of human tumors with similar molecular characteristics, originate and behave quite similarly [10]. In view of this paradigm, the molecular classification of colorectal cancers has developed exponentially over the last decade, and currently one of the main objectives is to empirically predict the pathogenesis and biological behavior of each tumor [11], which will allow development of adequate diagnostic and therapeutic strategies for the management of this disease.

CRC remains one of the best studied malignancies and molecular characterization of these lesions has allowed classifying all CRCs into 3 broad categories: tumors with chromosomal instability (CIN), microsatellite instability (MSI), and with a methylator phenotype (CIMP, CpG island methylator phenotype). Tumors with CIN, MSI and CIMP represent 75–80 %, 15–20 % and 20–25 % of all CRCs, respectively. As illustrated in Fig. 1, while there is some overlap between sporadic MSI and CIMP CRCs, it is now well established that there is an inverse relationship between CIN and MSI/CIMP tumors [12, 13]. CIN is defined as gross chromosomal rearrangements, including widespread alterations in chromosomal copy numbers [14]. MSI is a direct manifestation of defects in DNA mismatch repair (MMR) system, which compromises the DNA proofreading ability of MMR proteins, and allows enhanced accumulation of frameshift mutations at the mono/di/tri-nucleotide repeat sequences, or "microsatellites" that can be found throughout the entire genome [15]. Finally, CIMP is defined as frequent and widespread hypermethylation of CpG dinucleotide sequences that are present within the promoter regions of > 50 % of tumor suppressor genes, also referred to as "CpG islands" [16, 17]. The importance of such a classification is that each subtype of CRC can now be identified with unique clinical, molecular and prognostic features.

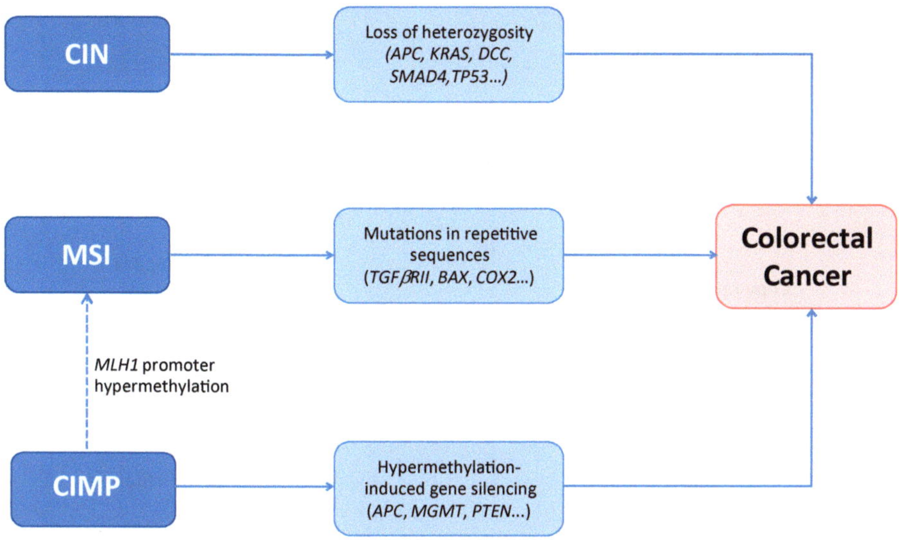

Fig. 1 The major mechanisms of genomic instability in colorectal cancer. In colorectal cancer, the two major mechanisms of genomic instability are chromosomal instability (CIN) and CpG Island Methylator Phenotype (CIMP). However, a subset of about 10–15 % of sporadic colorectal cancers demonstrates methylation of the *MLH1* gene, which results in acquisition of the microsatellite instability (MSI) phenotype

As mentioned above, epigenetic alterations are mitotically heritable modifications that alter gene expression without causing a change in the DNA sequence itself [18, 19]. Alterations in the human epigenome have now been virtually linked with most human diseases; but in particular, epigenetic alterations have been demonstrated to associate intimately with different types of cancer [9]. Epigenetic alterations in CRC that transform normal colonic epithelium into adenocarcinoma include aberrant DNA methylation, chromatin modifications, and non-coding RNAs as shown in Fig. 2 [7, 20]. Considering the frequency of epigenetic alterations in human cancer, particularly CRC, epigenetic aberrations are currently being pursued as promising candidates for the development of cancer biomarkers, and are perceived to be far superior to genetic alterations.

Abnormal gain of DNA methylation in normally unmethylated gene promoter CpG islands is associated with transcriptional repression and loss of function of the corresponding gene; and remains the most extensively studied epigenetic alteration in cancer [9, 21]. In addition to aberrant methylation of CpG islands residing within the gene promoters, key roles for DNA methylation have recently been observed in other regions, such as CpG island shores [22]. Besides DNA *hyper*-methylation, genome-wide loss of DNA methylation (or *hypo*-methylation) has long been recognized as a phenomenon that occurs in parallel to promoter hypermethylation in cancer cells [23, 24]. In addition to aberrant methylation, other types of post-translational modifications in histones have been described, including acetylation, methylation, phosphorylation, and ubiquitination [18, 25]. Finally, non-coding

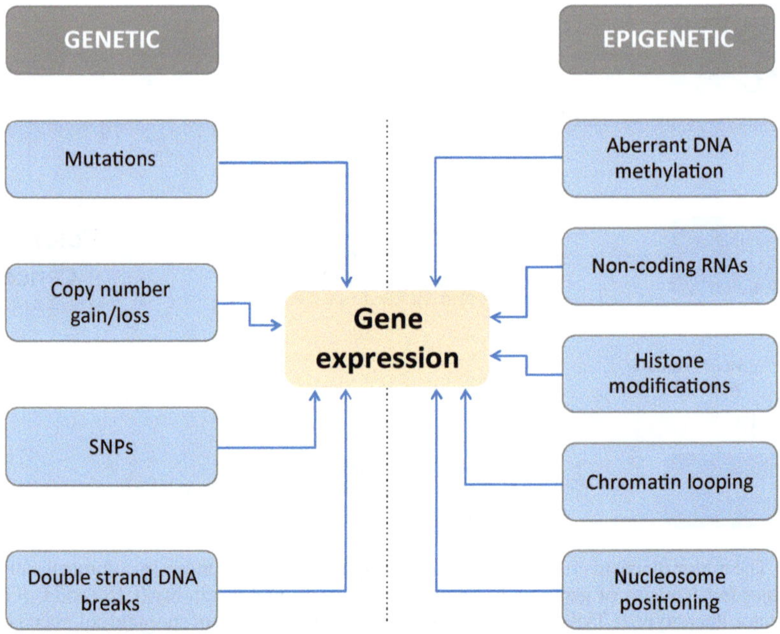

Fig. 2 Genetic and epigenetic alterations influencing gene expression in colorectal cancer. This figure illustrates the key genetic and epigenetic alterations that influence gene expression alterations in colorectal cancers. While most genetic alterations are static and irreversible, most epigenetic alterations are dynamic in nature and are potentially reversible. SNP, single nucleotide polymorphism

RNAs, including microRNAs also regulate gene expression and are part of the "epigenetic landscape", and play a central role in carcinogenesis [26].

The consequences of epigenetic changes include altered DNA transcription, aberrant activation of certain genes [27, 28], predisposition to genetic instability through dysregulated control of chromosomal replication [29–31], and transcriptional gene silencing – all of the reasons that are important in the cascade of initiation and progression of human cancer [18, 25, 32, 33]. Among these, the latter alteration is particularly relevant because it can simultaneously affect hundreds and thousands of genes, including tumor suppressors, genes involve in tissue remodelling, genes responsible for cell cycle control, and others that are involved in preventing abnormal activation of tumour development, anti-apoptosis and DNA mismatch repair genes. Taken together, it is becoming increasingly apparent that *epigenetic* mechanisms directly or indirectly influence a plethora of events linked to gene regulation and expression. More importantly, unlike genetic alterations, because epigenetic alterations are potentially 'reversible', these provide an attractive avenue for exploration as cancer diagnostics, prognostics, and may become in the future in one of the cornerstones of personalized medicine in human cancers, including colorectal cancer. This chapter integrates some of the recent data on epigenetic alterations in colorectal cancer pathogenesis and their potential as biomarkers and therapeutic targets in the future.

2 Epigenetic Alterations in Colorectal Carcinogenesis

2.1 DNA Methylation

Among various epigenetic alterations that regulate gene expression, methylation unequivocally remains the principle epigenomic mechanism in cancer, which controls overall hypomethylation or hypermethylation of various genes in different disease conditions, including cancer. Methylation is one of the best characterized epigenetic modifications and represents a covalent bond between a methyl group (CH_3) and the cytosine residues that occur in the context of CpG dinucleotides within gene promoter CpG islands, creating a 5-methyl-cytosine (m5C). This covalent bond between a methyl group and cytosines is catalyzed by a group of enzymes called DNA Methyl Transferases (DNMTs). CpG islands are sequences of at least 200 bases with more than 50 % of CpG content [34]. These CpG islands are present within ~50 % of human gene promoters, as well as within repetitive DNA sequences that include centromeric regions and other genomic repeat elements such as Alu and long interspersed nuclear element-1 or LINE-1 sequences [35]. In addition to methylation-inducing DNMTs, another family of proteins called methyl CpG-binding domain protein (MBD) play a follow-up role by binding to methylated CpGs and by blocking the transcriptional cascade [36]. Several elegant studies have already established the role of both DNMTs and MBD proteins in colorectal carcinogenesis [37–39].

2.1.1 DNA Hypomethylation

Although aberrant hypermethylation of genes/promoters have gained more attention in recent years, the first epigenetic change described in cancer was actually hypomethylation of genomic DNA [8, 40]. Global DNA hypomethylation, which means depletion of the overall 5-methyl cytosine content of colorectal tumors, is one of the first epigenetic changes recognized, and was linked to development of early stages of carcinogenesis.

DNA hypomethylation typically occurs at CpG dinucleotides in repetitive satellite sequences or peri-centromeric regions. Loss of methylation confers a risk for chromosome breakage that leads to enhanced genomic instability [41, 42]. The overall loss of methylation might have at least three consequences: the activation of transcription of oncogenes, activation of latent retrotransposons and chromosomal instability [43]. The level of global hypomethylation of the genome is related to the degree of malignancy, so this molecular event could also serve as a biological marker for prognostication of cancer patients [44].

Genome-wide DNA hypomethylation has been associated with the activation of proto-oncogenes (e.g. *S100A4, cMYC etc.)*, and loss of imprinting of genes (e.g. *IGF2A*), which results in the development of cancer following hypomethylation of a certain region of the gene [45, 46]. Global DNA hypomethylation can be measured by assessing the methylation status of LINE-1 repeat sequences [46–49]. LINE-1 methylation status is believed to be a robust surrogate marker of DNA-methylation level, due to its widespread repetition and distribution across the genome [13]. In normal cells, LINE-1 is usually methylated in order to prevent genomic instability and inadvertent activation of various proto-oncogenes and retrotransposon elements [50]. LINE-1 hypomethylation occurs early in the colorectal carcinogenesis [51, 52], and the degree of LINE-1 hypomethylation has been associated with CRC in younger patients with more aggressive form of the disease linked to poor outcomes [49, 53]. Also, a recent study has shown that CRC family history is associated with a higher risk of LINE-1 hypomethylation in CRC, suggesting a heritable predisposition to epigenetic alterations [44].

2.1.2 DNA Hypermethylation

As mentioned previously, one of the best studied epigenetic modification is aberrant DNA methylation, which typically occurs in the context of so-called CpG sites [32]. Although CpG dinucleotides are underrepresented in the human genome, certain regions of the genome are rich in CpG sites (called CpG islands) and usually occur within the promoter regions of the genes [9, 21, 23, 24]. In normal healthy cells, these CpG islands are usually unmethylated, favoring an active gene expression by allowing access for transcription factors to bind to various transcriptional elements located within gene promoters. However, this process is often compromised in a variety of diseases, including cancer, where aberrant DNA hypermethylation leads to transcriptional silencing of the corresponding gene (Fig. 3). In another situation, aberrant hypo- or demethylation of the repetitive DNA sequences within gene body leads to inadvertent activation of endoparasitic sequences that are evolutionarily silenced; and are essential for maintaining these methylation patterns for preserving chromosomal integrity. From a technological viewpoint, methylation analysis has undergone a revolution in the last decade, particularly since the implementation of microarray technology, and the emergence of next generation sequencing platforms [54, 55].

As it stands currently, methylated genes have been implicated during different steps of carcinogenesis. In cancer, regional hypermethylation of the CpG islands of several tumour suppressor genes was observed, leading to the inactivation of these genes. The first tumour suppressor gene shown to be silenced through DNA hypermethylation was the *RB* gene, the gene associated with retinoblastoma [31]. Since then, numerous other tumour suppressor genes have been associated with carcinogenesis. For instance, in some sporadic CRCs there is an epigenetic silencing of *STK11* (serine/threonine kinase 11), which regulates cell polarity and acts as a tumor suppressor gene. Besides this gene, a variety of tumour suppressor genes such as *MINT1, CDKN2A/p16, MGMT, p14ARF* and *HLTF* are aberrantly methylated in CRC [18, 56–58]. Also, it is known that some of these tumours are methylated at aberrant crypt foci, such as, *SFRP2*, *MGMT* and *SLC5A8*, implicating their involvement in early stages of CRC pathogenesis [59–61]. Some of the frequently methylated genes in colorectal cancer are listed in Table 1.

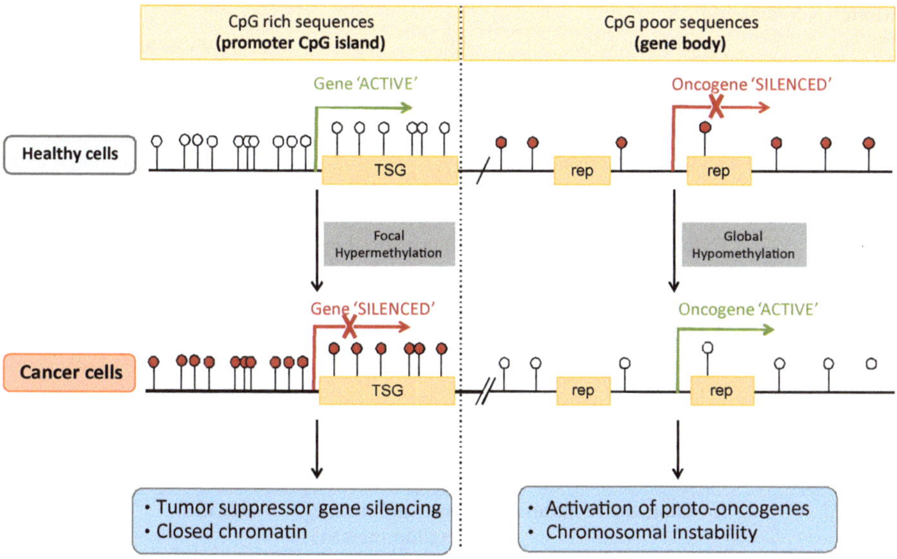

Fig. 3 DNA methylation patterns in normal and cancer cells. *Left panel*: Illustrates normal colonic epithelial cells that are generally unmethylated at the cytosine residues (*white circles*) within the CpG rich promoter CpG islands in the tumor suppressor genes (TSG), whereas these sequences frequently demonstrate "focal hypermethylation" of these sequences in the cancer cells (*red circles*)—a hallmark epigenetic phenomenon frequently observed in majority of cancer cells. Unmethylated CpG islands within the promoter region of genes, such as tumor suppressor genes (TSG), are correlated with an open chromatin structure (euchromatin), whereas methylated CpG islands are correlated with a condensed, closed chromatin structure (heterochromatin) and transcriptional silencing. *Right panel*: Repetitive sequences (depicted as "rep" boxes; i.e. LINE-1 and SINE-1/Alu sequences) are generally methylated at CpG poor sequences within the gene body of the normal cells. "Global hypomethylation" or loss of methylation in cancer cells leads to chromosomal instability and activation of endoparasitic elements and proto-oncogenes. Interestingly, an inverse association exists between local CpG island hypermethylation and global LINE-1 hypomethylation as colonic neoplasms progress

Table 1 Key genes that are aberrantly hypermethylated in colorectal cancer

Gene	Function in CRC	Protein	Chromosomal location	References
CDH1	Loss of cell adhesion. Invasion and metastasis	E-Cadherin	16q22.1	[62]
COX2	Tumor angiogenesis and metastasis. Inflammation	Prostaglandin-endoperoxide synthase 2	1q25.2-q25.3	[63]
HLTF	Transcription regulation. Chromatin remodelling factor	Helicase-like transcription factor	3q25.1-q26.1	[64]
SMAD4	Growth and differentiation	SMAD family member 4	18q21.1	[65]
TWIST	Transcription regulation. Growth and differentiation	Twist-related protein, also known as class A basic helix-loop-helix protein 38	7p21.2	[66]

Table 1 (continued)

Gene	Function in CRC	Protein	Chromosomal location	References
RAR-b	Growth and differentiation	Retinoic acid receptor	3p24.2	[67]
WIF1	Tumor suppressor gene. Aberrant activation of Wnt signal	Wnt inhibitory factor-1	12q14.2	[68]
SOCS1	Cytokine signaling	Suppressor of cytokine signaling 1	16p13.13	[69]
SEPT9	Cell cycle control	Septin 9	17q25.3	[70]
UNC5C	Axonal guidance	Netrin-1 receptor	4q21-q23	[71]
DKK1	Aberrant activation of Wnt signal	Dickkopf-1	10q11.2	[72]
CDH13	Regulator of adhesion	Cadherin 13	16q23.3	[73]
RUNX3	Transcription regulation. Apoptosis	Runt-related transcription factor 3	1p36	[74]
TIMP3	Invasion and metastasis	Tissue inhibitor of metalloproteinase 3	22q12.3	[75]
APC	Tumor suppressor gene. Wnt signaling pathway	Adenomatous polyposis coli	5q21-q22	[76]
MGMT	Involved in DNA repair system	O-6-methyl-guanine-DNA methyltransferase	10q26	[77]
MLH1	Involved in DNA repair system	MutL homolog 1 (hMLH1)	3p22.3	[7]
MSH2	Involved in DNA repair system	MutS homolog 2 (hMSH2)	2p21	[78]
RASSF1A	Tumor suppressor gene. An early event in CRC. Increased RAS/RAF/MAP kinase signaling	Sodium solute symporter family 5 member 8	3p21.3	[79]
VIM	Epithelial-mesenchymal transition	Vimentin	10p13	[80]
SFRP1/SFRP2	Aberrant activation of Wnt signaling	Secreted frizzled-related protein 1/protein 2	8p11.21/4q31.3	[81, 82]
ID4	Transcription factor	DNA binding protein	6p22.3	[83]
BNIP3	Apoptosis	Bcl-2/E1B 19-kDa interacting protein 3	8p21	[84]
CHFR	Cell cycle	E3 ubiquitin-protein ligase	12q24.33	[85]
DAPK	Early steps of tumor progression. Inhibition of apoptosis	Death-associated protein kinase	9q34.1	[86]
IGF2	Cell proliferation	Insulin-like growth factor 2	11p15.5	[87]
SLC5A8	Tumor suppressor gene	Sodium solute transporter family 5, member 8	12q23.1	[88]

Table 1 (continued)

Gene	Function in CRC	Protein	Chromosomal location	References
p14/ARF	Tumor suppressor gene. Increased cell proliferation. Decreased p53 stabilization	Protein 14. Cyclin-dependent kinase inhibitor 2A	9p21	[89]
IRF8	Transcription factor. Interferon signaling	Interferon regulatory factor 8	16q24.1	[90]
ALX4	Transcription regulation	Homeobox protein aristaless-like 4	11p11.2	[91]
GATA4/GATA5	Proliferation, migration, invasion. Transcription regulation	GATA binding protein	8p23.1-p22/20q13.33	[92]
CDKN2A/p16	Tumor suppressor gene. Cell growth and proliferation	Cyclin-dependent kinase inhibitor 2A	9p21.3	[93]
STK11	Tumor suppressor gene. Regulates cell polarity	Serine/threonine kinase 11	19p13.3	[94]
NEUROG1	Tumor suppressor gene	Neurogenin-1	5q23-q31	[95]

Among 10–15% of CRCs that exhibit microsatellite instability (MSI), approximately 70–80% show epigenetic silencing of *MLH1*, a gene that belongs to the DNA mismatch repair system. This epigenetic alteration has been associated with a unique subset of sporadic colorectal cancers, often referred to possess a methylator phenotype, which will be discussed further in subsequent section below.

2.1.3 Microsatellite Instability (MSI)

MSI refers to frameshift mutations that result in change of length (and ensuing instability) at short repetitive sequences or 'microsatellites' in the tumor DNA compared vis-à-vis normal mucosa. MSI occurs due to defects in the DNA repair system that consists of four key proteins (MLH1, MSH2, MSH6, PMS2), and somatic or germline mutations in these genes leads to compromised DNA MMR activity and increased acquisition of mutations at the microsatellites sequences. The process of carcinogenesis in these tumors is enhanced when one of the altered microsatellites is located within the coding region of a tumor suppressor gene, resulting in its inactivation, as is the case with the *BAX* or *TGFBRII* genes in colorectal cancer [96, 97].

MSI is one of the predominant mechanisms of genomic instability in Lynch syndrome patients, which represent ~2–3% of all CRCs. In these patients, MSI manifests through germline mutations in one of the DNA MMR genes, including *MLH1, MSH2, MSH6, and PMS2* [2, 7, 98]. However, in sporadic CRC, the most frequent cause of MSI is biallelic somatic hypermethylation associated transcriptional silencing of the *MLH1* gene. In contrast to Lynch syndrome patients who have a family history of CRC and generally younger, sporadic CRCs with MSI typically occurs in older patients with no family history of CRC [99]. MSI-positive sporadic CRCs represent about 12–15 % of all CRCs, and are usually associated with somatic mutations in the *BRAF* gene [100]. Tumors with MSI have specific pathological features including increased lymphocytic infiltration, mucinous component, signet ring cells and poor differentiation. Furthermore, MSI-positive CRCs are often associated with better outcome, are poorly metastatic, and are not responsive to 5-fluorouracil based chemotherapy [98, 101, 102]. The observation for somatic hypermethylation of the *MLH1* gene and a subsequent increased frequency of mutations at the growth regulatory genes with microsatellite sequences illustrates the interplay between genetic and epigenetic pathways. Although the exact timing of these two phenomena is not well established, both sporadic MSI and Lynch syndrome CRCs have a very similar phenotype.

Since loss of MLH1 expression can occur via germline mutations and its promoter hypermethylation, this presents a diagnostic challenge in identifying familial vs. no-familial patients with loss of MLH1 expression. To differentiate these two situations, the determination of somatic mutations in the *BRAF* gene (V600E) has recently proven to be quite useful, since the presence of somatic *BRAF* mutations has been shown to associate with *MLH1* hypermethylated tumors, which doesn't occur in the context of Lynch syndrome cancers. Similarly, it has been proposed the direct analysis of the methylation status of the *MLH1* gene promoter of by different techniques (MS-MLPA, MethyLight and bisulfite pyrosequencing) is an easy approach in differentiating between the two types of MSI tumors. Sporadic CRCs with such an epigenetic alteration are now being recognized to be associated with the so called, CpG island methylator phenotype (CIMP).

2.1.4 CpG Island Methylator Phenotype (CIMP)

In 1999, Toyota and colleagues [103] observed that certain colorectal tumors characteristically demonstrate a high frequency of hypermethylation in promoter regions of certain genes, that lead to be establishment of the concept that these tumors possess a unique epigenetic defect, known as methylator phenotype (CIMP, CpG island methylator phenotype). In the study by Toyota et al., tumors with CIMP did not associate with age, sex or developmental stage, but seemed to have a preferential location in the ascending colon or cecum (82 %). Later it was found that, up to 70 % of cases with CIMP also presented with MSI due to *MLH1* hypermethylation; it was also observed that MSI and CIMP tumors shared some clinical characteristics as right location, low grade differentiation, mucinous component, and high incidence in elderly women, better prognosis and often with *BRAF* and *KRAS* mutations [7, 13, 104–109].

Although CIMP was serendipitously discovered more than a decade ago, till date there isn't an established single panel of methylation markers that best defines CIMP tumors, particularly from a diagnostic and clinical standpoint [110, 111]. Until such a panel is identified, the most commonly used methylation markers were recently proposed by Weisenberger and colleagues that includes studying the methylation status of *CACNA1G, IGF2, NEUROG1, RUNX3,* and *SOCS1* [112]. This panel has replaced the original panel of CIMP markers that were proposed by Toyota and Issa, which included *p16, MLH1, MINT1, MINT2,* and *MINT31* [103].

The methylator phenotype has also been linked to the so-called serrated pathway of carcinogenesis, which is responsible for the development of up to 20–30% of CRCs. This pathway differs from the classic normal-adenoma-carcinoma pathways, both from a histologic viewpoint (the precursor lesion is the sessile serrated polyp instead of a conventional adenoma) and molecularly (usually MSI tumors with high frequency of somatic mutations in the *BRAF* gene and a low frequency of *TP53* mutations).

2.1.5 EPCAM Mutations

Recently, another epigenetic alteration has been discovered in another DNA repair gene, *MSH2*. It has been identified that germline mutations in the *EpCAM* gene (epithelial cell adhesion molecule) in families with Lynch syndrome can result in loss of expression of *MSH2* protein in patients that do not harbor germline mutations in this DNA MMR gene [78, 113]. Germline deletions affecting the region 3'- of the *EpCAM* gene located upstream of the 5'-region of *MSH2*, specifically, results in a fusion transcript between EpCAM-MSH2 that secondarily causes hypermethylation of the promoter region of *MSH2*, and subsequent silencing its expression [78, 114, 115]. As is the case in patients with the hypermethylation of the *MLH1* gene, those patients with secondary hypermethylation of *MSH2*, also have overlapping features of Lynch syndrome colorectal cancers.

2.1.6 Germline Epimutations or Constitutional Epimutation

A germline epimutation is an epigenetic aberration that occurs within normal somatic tissues and results in the silencing of the gene that is otherwise normally active, or conversely, it can lead to the reactivation of a gene that is normally silent. In the context of CRC germline hypermethylation of certain genes has been observed, and in particular, there are clear examples of constitutional epimutations involving DNA mismatch repair genes leading to a phenotype similar to families with Lynch syndrome.

The first constitutional epimutation described in CRC was discovered by identifying *MLH1* promoter hypermethylation in normal colonic mucosa of patients with CRC and resultant loss of MLH1 protein expression [116]. This hypermethylation occurs in the germline, and similar to what happens in Lynch syndrome, requires a "second hit" of the second wild-type allele. Therefore, patients with such epimutations are at a similar risk of developing cancer, as is the case with patients with a germline mutation [117]. Unlike germline mutations that are stably transmitted through the germline, constitutional *MLH1* epimutations are reversible between generations [116]. The incidence of constitutional *MLH1* epimutations remains to be accurately determined, but this defect appears to be rare.

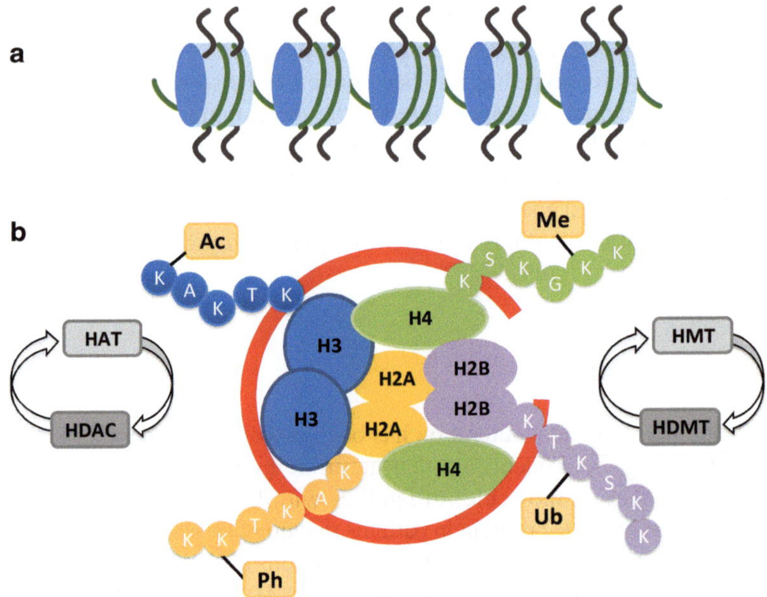

Fig. 4 Nucleosome and histone modifications in colorectal cancer. **a** Schematic representation of units of chromatin known as nucleosomes. Each nucleosome is comprised of 147 base pairs of DNA (in *green*) wrapped around the histone octamer, and the histone tails (in *black*) extending freely outside of the histone core (*blue circles*). Histone octamer is made of a heterotetramer of histones H3 and H4 and two heterodimers of H2A and H2B. **b** Most of the known histone modifications occur at the N-terminal tails of histones which are flexible and protrude outside the 'core' of the nucleosome. The modifications include acetylation (Ac), methylation (Me), phosphorylation (Ph) and ubiquitination (Ub). Globular domains of each core histone are represented as colored ovals. DNA is depicted as a red line. Aminoacids of each histone tails are represented as colored circles. Histone modifications occur preferentially at the lysine residues (K). Enzymes involved in the main histone modifications are depicted: acetylation (*HAT* histone acetyltransferases, *HDAC* histone deacetylases) and methylation (*HMT* histone methyltransferases, *HDMT* histone demethylases). K, lysine; T, threonine; A, alanine; S, serine; G, glycine

2.2 Histone Modifications

Among mammals, DNA is associated with proteins called histones, which constitutes a dynamic complex defined as a "chromatin complex". As illustrated in Fig. 4, the fundamental unit of chromatin, the nucleosome, consists of 146 base pairs of DNA wrapped around a histone octamer containing two H2A/H2B heterodimers and two H3 and H4 dimers. Histones are divided into two domains: a central folded region which interacts with the DNA, formed by a long helix-α, flanked by two short α-helices, and an N-terminal domain of 15–30 residues, called "histone tails". The N-terminal domain is flexible and goes outside the 'core' of the nucleosome. These "histone tails" undergo various post-translational modifications that affect chromatin structure and function [7].

In this complex, a myriad of molecular mechanisms are involved in chromatin conformation, which determine its condensation state and accordingly whether the DNA is accessible to proteins that control gene transcription. The patterns of histone modifications in colorectal carcinogenesis still remain unclear. Histones form the nucleosomes and undergo covalent modifications; and the effects of such post-transcriptional modifications depend on the type of amino acid, their position within the histone tail and the type of modification. Histones can regulate DNA sequences via post-translational modifications of their N-terminal tails. Histone modifications include acetylation, methylation, biotinylation, sumoylation, ubiquitination and phosphorylation, but in colorectal carcinogenesis, the histone (de) acetylation and (de)methylation are the two best known modifications [118, 119].

Covalent modifications (acetylation, methylation, phosphorylation and ubiquitinylation) of certain amino acids in these histones alter the chromatin structure, allowing the appearance of changes in the DNA, which finally impacts the expression of involved gene(s). Several enzymes regulate this process under what is known as 'histone code'. Among various enzymes, histone acetyltransferases (HAT) act by acetylating lysine residues in histones 3 and 4, and facilitate the access of transcription factors to bind to the promoter regions of chromatin, favoring gene expression [120]. In contrast, histone deacetylases (HDAC) cause repression of gene transcription by a complex mechanism in which these proteins interact with other enzymes termed histone methyltransferases, CpG islands binding proteins and phosphorylation enzymes. Therefore, the level of histone acetylation is determined by the activity of HAT and HDAC, and a controlled balance of both activities is essential for normal cell proliferation.

In contrast to the acetylation, histone methylation is usually catalyzed by particular enzymes to specific residues of histones. Histone methylation occurs at arginine, histidine, and more frequently at lysine residues. The enzymes involved are known as histone methyltransferases (HMTs) and demethylases (HDMTs) [121]. Histone methylation can have an activating or repressor effect on gene expression. H3K4, H3K36 and H3K79 are sites involved in transcriptional activation, while methylation of H3K9, H3K27 and H4K20 is associated with transcriptional repression [122, 123].

The histone modifications are regulated by transcriptional repressors, the polycomb group proteins, PRC1 and PRC2, which facilitate the epigenetic gene silencing process. In CRC, both proteins interact with DNMTs, suggesting that these proteins probably play a role in colorectal carcinogenesis. The role of some of histone modifications is well-established in CRC, and in this regard, the role of HDACs is best studied in CRC development. For instance, over-expression of some HDACs has been associated with CRC progression and poor survival. More specifically, over-expression of *HDAC2* is associated with hypoacetylation at H4K12 and H3K18 residues in colorectal cancer progression [124, 125]. Among other histone modifications in CRC, methylation of the lysine residues at positions 4, 14, 36 and 79 in histone 3, together with the phosphorylation of the residue 10 in serine of histone 3, have a positive effect on gene transcription. In contrast, methylation of residue 9 and 27 of histone 3 and residue 20 of histone 4 are associated with gene silencing, such as *p16, MLH1 and MGMT* [126]. More recently, it has been shown that expression patterns of some histone methylation markers (as H3K9m3 and H3K27me2) are associated with clinico-pathological features (e.g. TNM stage) in colorectal carcinogenesis [127, 128].

2.3 MicroRNAs

MicroRNAs (miRNAs) are noncoding RNA molecules of ~18–22 nucleotides that negatively regulate gene expression by inhibiting translation, or degradation of target messenger RNA (mRNA) encoding genes [129]. Although first described in 1993, it was not until 2001 when a better understanding of their role was recognized and their importance in various cellular processes was established [130,131]. Currently more than 1800 miRNAs have been identified in humans, and these constitute the best characterized family of non-coding RNAs. Recent studies suggest that miRNAs represent the most frequent process of post-transcriptional regulation, and it is estimated that about 60% of all protein coding mRNAs are regulated by various miRNAs [132, 133].

2.3.1 Biogenesis of miRNAs

In humans, miRNAs are produced as a result of multiple processing steps, during which first a larger transcript undergoes two maturation steps leading to synthesis of a functional miRNA (Fig. 5). Following transcription by the RNA polymerase (Pol II) in the nucleus, the primary transcript (pri-miR of around 1 kb in size) is processed by the enzyme complex Drosha (RNase III endonuclease) and DGCR8 protein to yield a smaller precursor (pre-miR, with about 70 nucleotides in length), which is exported from the nucleus by exportin 5 and the RNA-GTP protein. Once in the cytoplasm, another enzymatic process catalysed by DICER (RNase III enzyme) and protein TRBP, cleaves the two strands of pre-miR, forming a duplex molecule containing a string of a mature miRNA and another fragment known as miRNA* (derived from the complementary strand of pre-miR). Subsequently, the molecule miRNA:miRNA* integrates into the RISC complex (RNA-induced silencing complex), which is then activated. The miRNA* fragment is degraded, while the mature miRNA molecule guides the RISC complex to the messenger RNA. The latter complex inhibits protein expression in two ways, either by inhibiting translation or by the degradation of messenger RNA [134, 135]

2.3.2 miRNAs and Cancer

Role of miRNAs in Carcinogenesis

Since their initial discovery, miRNAs have been implicated in the regulation of almost all biological processes in mammals. Thus, the miRNAs are involved in

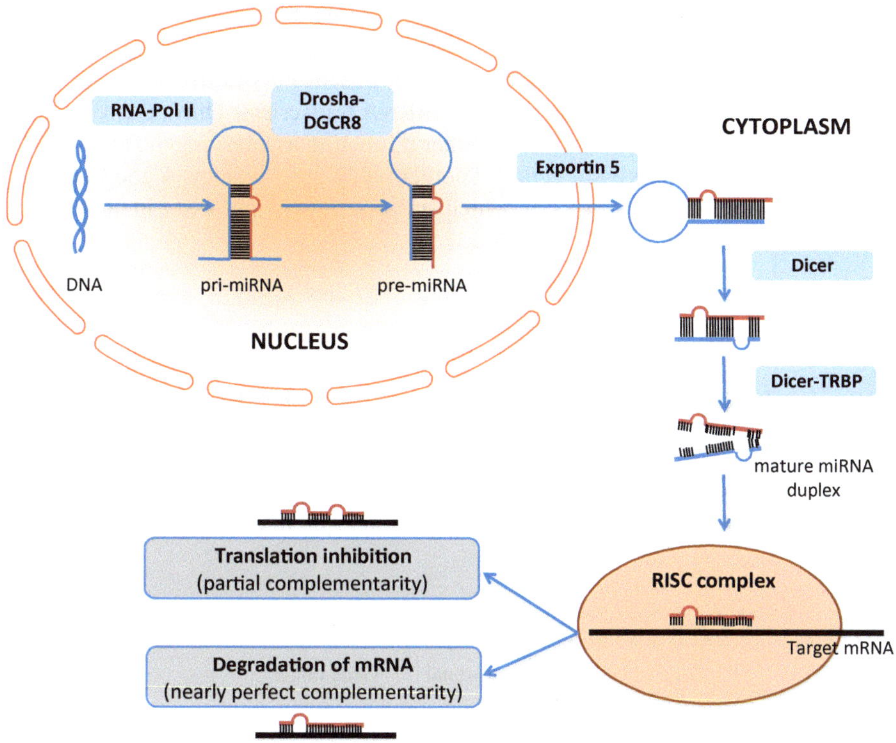

Fig. 5 miRNA biogenesis in colorectal cancer. RNA polymerase II transcribes miRNA genes, generating long primary transcripts (pri-miRNAs). Subsequently, the RNase III–type enzyme Drosha-DGCR8 processes pri-miRNAs yielding a hairpin precursors (pre-miRNA) consisting of approximately 70 nucleotides. In the nucleus, the pre-miRNA hairpins are exported to the cytoplasm by exportin-5 where these are further processed into mature 19–25 nucelotide miRNA duplex structures by the RNase III protein Dicer-TRBP. The duplex loads onto Argonaut ribonucleases in the RISC complex and separates. Mature miRNA strands (*red strand*) mediate small interfering RNA silencing by degrading the target mRNA or interfering with translation. The outcome of RISC formation varies with the degree of complementarity of the seed sequence of miRNA and 3′ untranslated regions of the target mRNA

many cellular processes related to carcinogenesis, such as proliferation, cell cycle, apoptosis, differentiation, migration and cell metabolism, so that its expression is altered in various types of tumours [26].

The first association between miRNAs and cancer was recognized in 2002 when the expression of two miRNAs (miR-16 and miR-15) was discovered to be altered in patients with chronic lymphocytic leukemia [136]. Since then, there have been many studies linking the pattern of expression of miRNAs with certain tumours [135]. Aberrant transcripts of miRNAs in cancer can be the result of activation of various oncogenic signalling cascades, genomic gains or losses, transcriptional repression, genetic mutations, secondary to inflammatory stimuli or epigenetic mechanisms [137, 138].

The study of miRNAs in cancer is quite interesting, but challenging at the same time due to multitude of reasons. First, within the same tumor, hundreds of miRNAs could be simultaneously altered. Second, because a specific miRNA is usually involved in the regulation of multiple mRNA transcripts. Third, the role of the miRNA may depend on the context, whether it is acting as an oncogene in a specific tumor and as tumor suppressor gene in another type of malignancy. A good example of such a process would be miR-29, which acts as a tumor suppressor in lung cancer and as an oncogene in breast cancer [133].

miRNAs and CRC

Michael and colleagues [139] were the first to observe that the expression patterns of miRNAs were altered in CRC, and identified that miR-143 and miR-145 expression was decreased in the tumor tissues and suggested their role as tumour suppressor genes. Since then, multiple studies have examined the role of miRNAs in CRC [7, 140–148], and some of the key miRNAs with a role in CRC have been listed in Table 2.

Table 2 Key miRNAs that are frequently altered in colorectal cancer

MicroRNA	Function in CRC	Chromosomal location	Gene target	References
let-7-a-1	Tumor suppressor gene	9q22.3	*KRAS, cMYC*	[149]
Cluster miR-17–92	Oncogene	13	*E2F1*	[150]
miR-21	Oncogene	17q23.2	*PTEN, PDCD4, RECK, TPM1, SPRY2, TIMP3*	[148, 151]
miR-29	Tumor suppressor gene	7q32	*DNMT3, DNMT3b, MCL1*	[152]
miR-30a	Tumor suppressor gene	6q13	*DTL*	[153]
miR-34a	Tumor suppressor gene	1p36.22	*CDK4, CDK6, E2F3, CyclinE2*	[154]
miR-95	Oncogene	4p16.1	*Nexin 1*	[155]
miR-101	Tumor suppressor gene	1p31.3	*COX2*	[156]
miR-135a/b	Oncogene	3p21, 12q23/ 1q32.1	*MSH2, APC*	[157, 158]
miR-137	Tumor suppressor gene	1p22	*LSD1, MITF, CDC42*	[144]
miR-143	Tumor suppressor gene	5q32–33	*DNMT3a, KRAS, ERK5, p53*	[159]
miR-145	Tumor suppressor gene	5q32–33	*IRS-1, cMYC, YES1, STAT1, OCT4, SOX, KLF4, p53*	[160]
miR-155	Oncogene	21q21.3	*MLH1, MSH2, MSH6*	[161]

Table 2 (continued)

MicroRNA	Function in CRC	Chromosomal location	Gene target	References
miR-221	Oncogene	Xp11.3	*Reck*	[162]
miR-342	Tumor suppressor gene	14q32.2	*DNMT1*	[163]
miR-365	Tumor suppressor gene	16p13.12	*Cyclin D1, Bcl-2*	[164]
miR-451	Tumor suppressor gene	17q11.2	*MIF*	[165]
miR-499	Oncogene	20q11.22	*PDCD4, FOXO4*	[166]
miR-675	Tumor suppressor gene	11p15.5	*pRB*	[167]

Role of miRNAs in the Initiation and in the Early Stages of CRC

It has been suggested that miRNAs are involved in the initiation of CRC since their expression pattern has been observed to be altered even in the premalignant adenomas, which server as precursor lesions of CRC [168]. It has been observed that miRNA expression in these colorectal adenomas is so distinct, that it allows discrimination between normal mucosa and colorectal cancer [168]. One example for such an observation is the expression pattern of miR-21, which increases in a stepwise fashion in early to late adenomas, followed by further increases in colorectal cancers; indicating its potential in tumour initiation and progression [169, 148]. Another finding that supports the involvement of miRNAs in the early stages of carcinogenesis is the epigenetic silencing of miR-137 in both the CRC and adenomas, but not in normal mucosa [144].

Role of miRNAs in Proliferation, Invasion and Metastasis in CRC

MiRNAs have been implicated also in the tumor progression and metastasis, by acting as oncogenes (*onco*miRs) or as tumor suppressors (*ts*miRs). Several such miRNAs associated with CRC progression have now been identified, which include: miR-365, which acts as a tumour suppressor by inhibiting cell cycle and inducing apoptosis [164]; miR-200c, involved in modulating the epithelial-mesenchymal transition and, therefore playing a key role in tumor metastasis [145]. Two recent studies have shown that miR-135b down-regulates *MTSS1* (metastasis suppressor-1) expression and contributes to CRC cell invasion and progression [158], as well as over-expression of miR-221 may regulate CRC migration and invasion by targeting *RECK* [162].

3 Clinical Applications of Epigenetic Biomarkers

Aberrant DNA methylation in cancer has been identified in virtually every type of tumor, including CRC. Analysis of methylation biomarkers in bodily fluids such as blood is a promising strategy that could radically change the non-invasive strategy for the early detection, stratification, prognosis and prediction of response to chemotherapy in cancer patients. Unlike other potential molecular biomarkers as mRNA and protein expression, DNA methylation patterns are inherently more stable, reproducible and consistent between individuals with colorectal neoplasia. Aberrant DNA methylation can also be readily detected in a variety of clinical specimens such as plasma, serum, sputum, feces, and formalin fixed paraffin embedded archival tissues—all the features making them attractive targets for development as potential biomarkers in CRC [170–172].

Likewise, in addition to their role in carcinogenesis, miRNAs are emerging as potential diagnostic, prognostic and therapeutic tools. In fact, the expression pattern of miRNA can also help classify various types and subtypes of cancer [173]. Just like DNA methylation signatures, miRNA expression has been established to be quite stable among a variety of biological and clinical specimens, their expression alterations occur in the earliest stages of carcinogenesis, and since a single miRNA can regulate the expression of hundreds of mRNAs- makes them as solid substrates for exploitation as cancer biomarkers [174, 175]. In this regard, it has been recently shown that it is possible to analyse miRNA expression in various biological fluids such as serum, plasma, urine, feces and colostrum, which opens the door for their use as non-invasive biomarkers for various diseases including cancer. Finally, several experimental studies in cell lines and other preclinical animal models have shown promising results regarding the role of miRNAs in cancer treatment, which raises the possibility for their sue as therapeutic targets through inhibition of oncogenic miRNAs or the reintroduction of tumor suppressor miRNAs [176–179].

3.1 Diagnosis

The disruption of normal pattern of methylation is an important epigenetic mechanism that is linked to carcinogenesis, tumor progression and early detection of CRC [80, 180, 181]. In recent years, there has been an increasing interest in exploring the feasibility and clinical utility for developing assays for specific methylated genes as potential biomarkers for the diagnosis of CRC [182–184]. Several studies have attempted to determine the usefulness for the detection of aberrantly methylated genes in blood and fecal specimens as potential biomarkers in CRC. For instance, methylation of *Vimentin* gene is frequently methylated in stool DNA of patients with CRC but rarely in faeces from healthy individuals, and this test is now clinically available [80]. A recent meta-analysis included 20 studies of methylated genes in stool as potential biomarkers in CRC, finding a model of DNA testing for multiple markers with a sensitivity of 75 % [185]. In addition, a recent multitarget stool DNA test including analysis of aberrant *NDRG4* and *BMP3* methylation has shown a high sensitivity for CRC and advanced adenoma detection [186].

Another example is aberrant methylation of the *SFRP2* gene, an inhibitor of Wnt signaling, in aberrant crypt foci suggesting that it is an early event in colorectal carcinogenesis [187, 188]. Likewise, aberrant methylation of *SFRP2* and *RASSF2A* genes in fecal DNA was able to accurately diagnose 80 % of colorectal cancers, and the methylation of these genes allowed discrimination between colorectal and gastric cancers [184]. A clear example is the detection of methylated *SEPT9* in a blood test and this is already commercially available, showing a sensitivity up to 90 % for CRC diagnosis [189]. Other genes, such as *ALX4* and *TMEFF2* are also frequently methylated in early stages of colorectal cancer [190, 191]. A recent study showed that hypermethylation of transcriptional factors *GATA4* and *GATA5* are potential faecal biomarkers [92], as well as *RUNX3, SDC2* and *NEUROG1* in serum, for CRC diagnosis [5].

Besides DNA methylation, recent studies have evaluated the role of miRNAs as diagnostic biomarkers in CRC. The first study that detected circulating miRNAs differentially expressed in the plasma of CRC patients and controls was published in 2009 [192], and in 2010 another study detected altered expression of miRNAs in stool of patients with colorectal adenomas and cancers [193]. Since then, several studies have evaluated the role of miRNAs as potential biomarkers as non-invasive diagnostic tools for CRC screening either alone or in combination with the fecal occult blood tests [194]. It has been suggested that miR-137 acts as a tumour suppressor in the colon, and its silencing in early stage CRCs suggests that methylation of miR-137 could be a marker of early tumorigenesis [144]. More recently, it was shown that serum levels of miR-21 alone were highly accurate in identifying advanced adenomas and cancers, and that using miR-21 alone, or in a panel of miRNAs may serve as a clinically viable non-invasive test in the foreseeable future [148].

Beyond of the role of miRNAs in the diagnosis of CRC, new data have shown promise for their ability to differentiate between tumor subtypes. Few studies have specifically analysed miRNA signatures in different CRC subtypes based on the presence or absence of MSI [141, 195–198]. One of the miRNAs that have been associated with the MSI phenotype in CRC is miR-155, since it has been observed that its overexpression has an inverse correlation with the protein expression of MLH1, MSH2, and MSH6, suggesting that this miRNA is involved in altering the DNA repair system in the CRC [199]. Recently, another study discovered specific miRNA expression profiles that can successfully discriminate between Lynch syndrome, sporadic MSI, and sporadic MSS colorectal cancers [198].

3.2 Prognosis and Prediction of Treatment Response

The clinical outcome and survival of cancer patients largely depends on the early diagnosis of malignant lesions. Recently, epigenetic changes have been considered as one of the most promising tools in improving patient survival. An example of the role of epigenetic alterations as biomarkers of prognosis is the epigenetic status of LINE-1 that is prognostic in CRC, with a potential use in determining clinical outcome [200]. Promoter hypermethylation of several other genes also have been associated with poor prognosis, such as *CDKN2A* or *p16* [201, 202], as well as *p14, RASSF1A* and *APC* [76]. In contrast, hypermethylation of *MGMT* and *MLH1* have been linked to a better survival in CRC patients [203].

With respect to treatment response, several studies have demonstrated associations between 5-fluorouracil resistance (standard of care therapy in CRC) and epigenetic alterations. For example, methylation of *DAPK* or *BNIP3* has been related with a lower response to 5-FU treatment [84, 204]. On similar lines, it was recently shown that *TFAP2E* hypermethylation is associated with clinical non-responsiveness to regular chemotherapy (5-FU, irinotecan or oxaliplatin) in colorectal cancer [205], although these data need to be validated in future studies that utilize more uniform patient populations and treatments.

At present, several studies have identified differential expression patterns of miRNAs as prognostic biomarkers and as predictors of response to treatment in CRC. The best example is expression of miR-21, which has been associated as an independent marker of poor prognosis [137, 140, 161, 206, 207]. Another example is miR-148a as an independent prognostic and predictive biomarker of advanced CRC, and low levels of miR-148a associated with worse therapeutic response and poor overall survival [146]. Recently, a study has shown that low expression level of miR-30b closely related to poor differentiation, advanced TNM stage and poor prognosis of CRC, suggesting it could be also a potential prognostic marker or therapeutic target for CRC [208].

4 Epigenetic Therapy

One of the most attractive attributes of epigenetic alterations is that these are potentially *reversible*, making them compelling targets for consideration as therapeutic targets. DNA methylation and histone modifications are attractive for the development of new therapeutics targets, since the treatment with drugs that can inhibit cytosine methylation and histone deacetylation may facilitate re-establishment of gene transcription [209]. Blocking DNA methyltransferases have been the most studied mechanism till date, and the use of various DNMT inhibitors as potential cancer therapeutics is being explored in various phase I and II clinical trials in patients with haematological malignancies, with some success [108, 210]. At present, two DNMTs inhibitors (azacytidine and decitabine) have been approved as drugs for epigenetic cancer therapy, although these drugs have not found their use in the treatment of solid cancers, including CRC [211]. One of the major limitations of DNMT inhibitors relates to the non-specificity and toxicity, and a result, these can accidentally result in the activation of various oncogenes and other endoparasitic elements in cancer patients. Another strategy that is of interest includes the combination treatment with DNMT and HDAC inhibitors to enhance the demethylating effect. Thus a study combining treatment with 5-azacytidine and valproic acid showed a significant reduction in DNA methylation and increased histone deacetylation in some solid tumors, including CRC [212].

Based on the involvement of miRNAs in different processes of carcinogenesis and their role as oncogenes or tumour suppressor genes, these molecules are very attractive therapeutic targets in the CRC, which can be accomplished through their inhibition (through antisense oligonucleotides) or restitution of their expression. In this regard, studies have been performed experimentally in animal models where the reintroduction of miR-145 and miR-33a has been associated with significant antitumor effects [213].

Molecular heterogeneity of CRC is partially the cause of the variability of treatment response and prognosis between subtypes of tumours, and epigenetics may help to define and understand these subgroups better. The main goal of epigenetic therapy is to identify specific targets that could be regulated but with minimal side effects and lower toxicity. Although it may seem like an insurmountable task currently, considering the sophistication and finesse of technological and intellectual tools available to us, the future for epigenetic signatures to serve as diagnostic, prognostic, predictive and therapeutic tools seems very bright. It is just a matter of time before we can solve some additional pieces of this puzzle and develop non-invasive epigenetic assays for the early detection of cancers, and establish safe and effective epigenetic therapies that will open new doors to the personalized medicine of colorectal cancer.

Disclosures The authors have no conflicts of interest to disclose.

Grant Funding The present work was supported by grants R01 CA72851 and CA181572 from the National Cancer Institute, National Institutes of Health, a pilot grant from the Charles A Sammons Cancer Center, and funds from the Baylor Research Institute. Francesc Balaguer is supported by a grant from the Instituto de Salud Carlos III (PI13/00719). “Cofinanciado por el Fondo Europeo de Desarrollo Regional (FEDER)”. European Union.

Conflict of Interests The authors have no conflict of interests to disclose.

References

1. Jemal A, Siegel R, Ward E, Hao Y, Xu J, Thun MJ (2009) Cancer statistics, 2009. CA Cancer J Clin 59:225–249
2. Boland CR, Shike M (2010) Report from the Jerusalem workshop on Lynch syndrome-hereditary nonpolyposis colorectal cancer. Gastroenterology 138:2197, e1–7
3. Fearon ER, Vogelstein B (1990) A genetic model for colorectal tumorigenesis. Cell 61:759–767
4. Jasperson KW, Tuohy TM, Neklason DW, Burt RW (2010) Hereditary and familial colon cancer. Gastroenterology 138:2044–2058
5. Coppede F, Lopomo A, Spisni R, Migliore L (2014) Genetic and epigenetic biomarkers for diagnosis, prognosis and treatment of colorectal cancer. World J Gastroenterol 20:943–956
6. Ewing I, Hurley JJ, Josephides E, Millar A (2014) The molecular genetics of colorectal cancer. Frontline Gastroenterol 5:26–30
7. Goel A, Boland CR (2012) Epigenetics of colorectal cancer. Gastroenterology 143:1442–1460, e1

8. Holliday R (2006) Epigenetics: a historical overview. Epigenetics 1:76–80
9. Esteller M (2008) Epigenetics in cancer. N Engl J Med 358:1148–1159
10. Simons CC, Hughes LA, Smits KM, Khalid-de Bakker CA, de Bruine AP, Carvalho B, Meijer GA, Schouten LJ, van den Brandt PA, Weijenberg MP, van Engeland M (2013) A novel classification of colorectal tumors based on microsatellite instability, the CpG island methylator phenotype and chromosomal instability: implications for prognosis. Ann Oncol 24:2048–2056
11. Markowitz SD, Bertagnolli MM (2009) Molecular origins of cancer: Molecular basis of colorectal cancer. N Engl J Med 361:2449–2460
12. Goel A, Nagasaka T, Arnold CN, Inoue T, Hamilton C, Niedzwiecki D, Compton C, Mayer RJ, Goldberg R, Bertagnolli MM, Boland CR (2007) The CpG island methylator phenotype and chromosomal instability are inversely correlated in sporadic colorectal cancer. Gastroenterology 132:127–138
13. Ogino S, Goel A (2008) Molecular classification and correlates in colorectal cancer. J Mol Diagn 10:13–27
14. Pritchard CC, Grady WM (2011) Colorectal cancer molecular biology moves into clinical practice. Gut 60:116–129
15. Boland CR, Thibodeau SN, Hamilton SR, Sidransky D, Eshleman JR, Burt RW, Meltzer SJ, Rodriguez-Bigas MA, Fodde R, Ranzani GN, Srivastava S (1998) A National Cancer Institute workshop on microsatellite instability for cancer detection and familial predisposition: development of international criteria for the determination of microsatellite instability in colorectal cancer. Cancer Res 58:5248–5257
16. Donehower LA, Creighton CJ, Schultz N, Shinbrot E, Chang K, Gunaratne PH, Muzny D, Sander C, Hamilton SR, Gibbs RA, Wheeler D (2013) MLH1-silenced and non-silenced subgroups of hypermutated colorectal carcinomas have distinct mutational landscapes. J Pathol 229:99–110
17. Issa JP (2008) Colon cancer: it's CIN or CIMP. Clin Cancer Res 14:5939–5940
18. Baylin SB (2012) The cancer epigenome: its origins, contributions to tumorigenesis, and translational implications. Proc Am Thorac Soc 9:64–65
19. Baylin SB, Jones PA (2011) A decade of exploring the cancer epigenome – biological and translational implications. Nat Rev Cancer 11:726–734
20. van Engeland M, Derks S, Smits KM, Meijer GA, Herman JG (2011) Colorectal cancer epigenetics: complex simplicity. J Clin Oncol 29:1382–1391
21. Esteller M (2007) Cancer epigenomics: DNA methylomes and histone-modification maps. Nat Rev Genet 8:286–298
22. Irizarry RA, Ladd-Acosta C, Wen B, Wu Z, Montano C, Onyango P, Cui H, Gabo K, Rongione M, Webster M, Ji H, Potash JB, Sabunciyan S, Feinberg AP (2009) The human colon cancer methylome shows similar hypo- and hypermethylation at conserved tissue-specific CpG island shores. Nat Genet 41:178–186
23. Feinberg AP (2004) The epigenetics of cancer etiology. Semin Cancer Biol 14:427–432
24. Feinberg AP, Tycko B (2004) The history of cancer epigenetics. Nat Rev Cancer 4:143–153
25. Jones PA, Baylin SB (2002) The fundamental role of epigenetic events in cancer. Nat Rev Genet 3:415–428
26. Calin GA, Croce CM (2006) MicroRNA signatures in human cancers. Nat Rev Cancer 6:857–866
27. Bestor TH (1998) Gene silencing. Methylation meets acetylation. Nature 393:311–312
28. Bird A (2002) DNA methylation patterns and epigenetic memory. Genes Dev 16:6–21
29. Okano M, Bell DW, Haber DA, Li E (1999) DNA methyltransferases Dnmt3a and Dnmt3b are essential for de novo methylation and mammalian development. Cell 99:247–257
30. Tuck-Muller CM, Narayan A, Tsien F, Smeets DF, Sawyer J, Fiala ES, Sohn OS, Ehrlich M (2000) DNA hypomethylation and unusual chromosome instability in cell lines from ICF syndrome patients. Cytogenet Cell Genet 89:121–128
31. Xu GL, Bestor TH, Bourc'his D, Hsieh CL, Tommerup N, Bugge M, Hulten M, Qu X, Russo JJ, Viegas-Pequignot E (1999) Chromosome instability and immunodeficiency syndrome caused by mutations in a DNA methyltransferase gene. Nature 402:187–191

32. Herman JG, Baylin SB (2003) Gene silencing in cancer in association with promoter hypermethylation. N Engl J Med 349:2042–2054
33. Jones PA, Laird PW (1999) Cancer epigenetics comes of age. Nat Genet 21:163–167
34. Gardiner-Garden M, Frommer M (1987) CpG islands in vertebrate genomes. J Mol Biol 196:261–282
35. Chen T, Li E (2006) Establishment and maintenance of DNA methylation patterns in mammals. Curr Top Microbiol Immunol 301:179–201
36. Parry L, Clarke AR (2011) The roles of the methyl-CpG binding proteins in cancer. Genes Cancer 2:618–630
37. Howard JH, Frolov A, Tzeng CW, Stewart A, Midzak A, Majmundar A, Godwin A, Heslin M, Bellacosa A, Arnoletti JP (2009) Epigenetic downregulation of the DNA repair gene MED1/MBD4 in colorectal and ovarian cancer. Cancer Biol Ther 8:94–100
38. Bader S, Walker M, Hendrich B, Bird A, Bird C, Hooper M, Wyllie A (1999) Somatic frameshift mutations in the MBD4 gene of sporadic colon cancers with mismatch repair deficiency. Oncogene 18:8044–8047
39. Bader S, Walker M, Mcqueen HA, Sellar R, Oei E, Wopereis S, Zhu Y, Peter A, Bird AP, Harrison DJ (2003) MBD1, MBD2 and CGBP genes at chromosome 18q21 are infrequently mutated in human colon and lung cancers. Oncogene 22:3506–3510
40. Holliday R, Jeggo PA (1985) Mechanisms for changing gene expression and their possible relationship to carcinogenesis. Cancer Surv 4:557–581
41. Ji W, Hernandez R, Zhang XY, Qu GZ, Frady A, Varela M, Ehrlich M (1997) DNA demethylation and pericentromeric rearrangements of chromosome 1. Mutat Res 379:33–41
42. Qu GZ, Grundy PE, Narayan A, Ehrlich M (1999) Frequent hypomethylation in Wilms tumors of pericentromeric DNA in chromosomes 1 and 16. Cancer Genet Cytogenet 109:34–39
43. Silver A, Sengupta N, Propper D, Wilson P, Hagemann T, Patel A, Parker A, Ghosh A, Feakins R, Dorudi S, Suraweera N (2012) A distinct DNA methylation profile associated with microsatellite and chromosomal stable sporadic colorectal cancers. Int J Cancer 130:1082–1092
44. Ogino S, Nishihara R, Lochhead P, Imamura Y, Kuchiba A, Morikawa T, Yamauchi M, Liao X, Qian ZR, Sun R, Sato K, Kirkner GJ, Wang M, Spiegelman D, Meyerhardt JA, Schernhammer ES, Chan AT, Giovannucci E, Fuchs CS (2013) Prospective study of family history and colorectal cancer risk by tumor LINE-1 methylation level. J Natl Cancer Inst 105:130–140
45. Cui H, Onyango P, Brandenburg S, Wu Y, Hsieh CL, Feinberg AP (2002) Loss of imprinting in colorectal cancer linked to hypomethylation of H19 and IGF2. Cancer Res 62:6442–6446
46. Hur K, Cejas P, Feliu J, Moreno-Rubio J, Burgos E, Boland CR, Goel A (2014) Hypomethylation of long interspersed nuclear element-1 (LINE-1) leads to activation of proto-oncogenes in human colorectal cancer metastasis. Gut 63:635–646
47. Speek M (2001) Antisense promoter of human L1 retrotransposon drives transcription of adjacent cellular genes. Mol Cell Biol 21:1973–1985
48. Baba Y, Huttenhower C, Nosho K, Tanaka N, Shima K, Hazra A, Schernhammer ES, Hunter DJ, Giovannucci EL, Fuchs CS, Ogino S (2010a) Epigenomic diversity of colorectal cancer indicated by LINE-1 methylation in a database of 869 tumors. Mol Cancer 9:125
49. Antelo M, Balaguer F, Shia J, Shen Y, Hur K, Moreira L, Cuatrecasas M, Bujanda L, Giraldez MD, Takahashi M, Cabanne A, Barugel ME, Arnold M, Roca EL, Andreu M, Castellvi-Bel S, Llor X, Jover R, Castells A, Boland CR, Goel A (2012) A high degree of LINE-1 hypomethylation is a unique feature of early-onset colorectal cancer. PLoS ONE 7:e45357
50. Kazazian HH Jr (2004) Mobile elements: drivers of genome evolution. Science 303:1626–1632
51. Pavicic W, Joensuu EI, Nieminen T, Peltomaki P (2012) LINE-1 hypomethylation in familial and sporadic cancer. J Mol Med 90:827–835
52. Sunami E, de Maat M, Vu A, Turner RR, Hoon DS (2011) LINE-1 hypomethylation during primary colon cancer progression. PLoS ONE 6:e18884
53. Rhee YY, Kim MJ, Bae JM, Koh JM, Cho NY, Juhnn YS, Kim D, Kang GH (2012) Clinical outcomes of patients with microsatellite-unstable colorectal carcinomas depend on L1 methylation level. Ann Surg Oncol 19:3441–3448

54. Hinoue T, Weisenberger DJ, Lange CP, Shen H, Byun HM, van den Berg D, Malik S, Pan F, Noushmehr H, van Dijk CM, Tollenaar RA, Laird PW (2012) Genome-scale analysis of aberrant DNA methylation in colorectal cancer. Genome Res 22:271–282
55. Laird PW (2010) Principles and challenges of genomewide DNA methylation analysis. Nat Rev Genet 11:191–203
56. Kim JC, Choi JS, Roh SA, Cho DH, Kim TW, Kim YS (2010a). Promoter methylation of specific genes is associated with the phenotype and progression of colorectal adenocarcinomas. Ann Surg Oncol 17:1767–1776
57. Kim MS, Lee J, Sidransky D (2010b) DNA methylation markers in colorectal cancer. Cancer Metastasis Rev 29:181–206
58. Jacinto FV, Ballestar E, Ropero S, Esteller M (2007) Discovery of epigenetically silenced genes by methylated DNA immunoprecipitation in colon cancer cells. Cancer Res 67:11481–11486
59. Li H, Myeroff L, Smiraglia D, Romero MF, Pretlow TP, Kasturi L, Lutterbaugh J, Rerko RM, Casey G, Issa JP, Willis J, Willson JK, Plass C, Markowitz SD (2003) SLC5A8, a sodium transporter, is a tumor suppressor gene silenced by methylation in human colon aberrant crypt foci and cancers. Proc Natl Acad Sci U S A 100:8412–8417
60. Moinova HR, Chen WD, Shen L, Smiraglia D, Olechnowicz J, Ravi L, Kasturi L, Myeroff L, Plass C, Parsons R, Minna J, Willson JK, Green SB, Issa JP, Markowitz SD (2002) HLTF gene silencing in human colon cancer. Proc Natl Acad Sci U S A 99:4562–4567
61. Chan AO, Broaddus RR, Houlihan PS, Issa JP, Hamilton SR, Rashid A (2002) CpG island methylation in aberrant crypt foci of the colorectum. Am J Pathol 160:1823–1830
62. Li YX, Lu Y, Li CY, Yuan P, Lin SS (2014) Role of CDH1 promoter methylation in colorectal carcinogenesis: a meta-analysis. DNA Cell Biol 33(7):455–462
63. Elzagheid A, Emaetig F, Alkikhia L, Buhmeida A, Syrjanen K, EL-Faitori O, Latto M, Collan Y, Pyrhonen S (2013) High cyclooxygenase-2 expression is associated with advanced stages in colorectal cancer. Anticancer Res 33:3137–3143
64. Yamamoto H, Min Y, Itoh F, Imsumran A, Horiuchi S, Yoshida M, Iku S, Fukushima H, Imai K (2002) Differential involvement of the hypermethylator phenotype in hereditary and sporadic colorectal cancers with high-frequency microsatellite instability. Genes Chromosomes Cancer 33:322–325
65. Papageorgis P, Cheng K, Ozturk S, Gong Y, Lambert AW, Abdolmaleky HM, Zhou JR, Thiagalingam S (2011) Smad4 inactivation promotes malignancy and drug resistance of colon cancer. Cancer Res 71:998–1008
66. Ruppenthal RD, Nicolini C, Filho AF, Meurer R, Damin AP, Rohe A, Alexandre CO, Damin DC (2011) TWIST1 promoter methylation in primary colorectal carcinoma. Pathol Oncol Res 17:867–872
67. Youssef EM, Estecio MR, Issa JP (2004) Methylation and regulation of expression of different retinoic acid receptor beta isoforms in human colon cancer. Cancer Biol Ther 3:82–86
68. Taniguchi H, Yamamoto H, Hirata T, Miyamoto N, Oki M, Nosho K, Adachi Y, Endo T, Imai K, Shinomura Y (2005) Frequent epigenetic inactivation of Wnt inhibitory factor-1 in human gastrointestinal cancers. Oncogene 24:7946–7952
69. David M, Naudin C, Letourneur M, Polrot M, Renoir JM, Lazar V, Dessen P, Roche S, Bertoglio J, Pierre J (2014) Suppressor of cytokine signaling 1 modulates invasion and metastatic potential of colorectal cancer cells. Mol Oncol 8(5):942–955
70. Church TR, Wandell M, Lofton-Day C, Mongin SJ, Burger M, Payne SR, Castanos-Velez E, Blumenstein BA, Rosch T, Osborn N, Snover D, Day RW, Ransohoff DF (2014) Prospective evaluation of methylated SEPT9 in plasma for detection of asymptomatic colorectal cancer. Gut 63:317–325
71. Shin SK, Nagasaka T, Jung BH, Matsubara N, Kim WH, Carethers JM, Boland CR, Goel A (2007) Epigenetic and genetic alterations in Netrin-1 receptors UNC5C and DCC in human colon cancer. Gastroenterology 133:1849–1857
72. Huang Z, Li S, Song W, Li X, Li Q, Zhang Z, Han Y, Zhang X, Miao S, Du R, Wang L (2013) Lysine-specific demethylase 1 (LSD1/KDM1A) contributes to colorectal tumorigenesis via activation of the Wnt/beta-catenin pathway by down-regulating Dickkopf-1 (DKK1) [corrected]. PloS ONE 8:e70077

73. Wang Z, Yuan X, Jiao N, Zhu H, Zhang Y, Tong J (2012) CDH13 and FLBN3 gene methylation are associated with poor prognosis in colorectal cancer. Pathol Oncol Res 18:263–70
74. Peng Z, Tang H, Wang X, Zhou C, Fan J, Wang L, Jia Z, Li Q, Le X, Wei D, Xie K (2008) Inhibition of the growth and metastasis of human colon cancer by restoration of RUNX3 expression in cancer cells. Int J Oncol 33:979–984
75. Lin H, Zhang Y, Wang H, Xu D, Meng X, Shao Y, Lin C, Ye Y, Qian H, Wang S (2012) Tissue inhibitor of metalloproteinases-3 transfer suppresses malignant behaviors of colorectal cancer cells. Cancer Gene Ther 19:845–851
76. Nilsson TK, Lof-Ohlin ZM, Sun XF (2013) DNA methylation of the p14arf, RASSF1A and APC1A genes as an independent prognostic factor in colorectal cancer patients. Int J Oncol 42:127–133
77. Nagasaka T, Goel A, Notohara K, Takahata T, Sasamoto H, Uchida T, Nishida N, Tanaka N, Boland CR, Matsubara N (2008) Methylation pattern of the O6-methylguanine-DNA methyltransferase gene in colon during progressive colorectal tumorigenesis. Int J Cancer 122:2429–2436
78. Ligtenberg MJ, Kuiper RP, Geurts van Kessel A, Hoogerbrugge N (2013) EPCAM deletion carriers constitute a unique subgroup of Lynch syndrome patients. Fam Cancer 12:169–174
79. Hesson LB, Wilson R, Morton D, Adams C, Walker M, Maher ER, Latif F (2005) CpG island promoter hypermethylation of a novel Ras-effector gene RASSF2A is an early event in colon carcinogenesis and correlates inversely with K-ras mutations. Oncogene 24:3987–3994
80. Chen WD, Han ZJ, Skoletsky J, Olson J, Sah J, Myeroff L, Platzer P, Lu S, Dawson D, Willis J, Pretlow TP, Lutterbaugh J, Kasturi L, Willson JK, Rao JS, Shuber A, Markowitz SD (2005) Detection in fecal DNA of colon cancer-specific methylation of the nonexpressed vimentin gene. J Natl Cancer Inst 97:1124–1132
81. Caldwell GM, Jones C, Gensberg K, Jan S, Hardy RG, Byrd P, Chughtai S, Wallis Y, Matthews GM, Morton DG (2004) The Wnt antagonist sFRP1 in colorectal tumorigenesis. Cancer Res 64:883–888
82. Takeda M, Nagasaka T, DONG-Sheng S, Nishie H, Oka T, Yamada E, Mori Y, Shigeyasu K, Morikawa T, Mizobuchi S, Fujiwara T (2011) Expansion of CpG methylation in the SFRP2 promoter region during colorectal tumorigenesis. Acta Med Okayama 65:169–177
83. Umetani N, Takeuchi H, Fujimoto A, Shinozaki M, Bilchik AJ, Hoon DS (2004) Epigenetic inactivation of ID4 in colorectal carcinomas correlates with poor differentiation and unfavorable prognosis. Clin Cancer Res 10:7475–7483
84. Deng Q, Huang CM, Chen N, Li L, Wang XD, Zhang W, Bi F, Tang QL, Li ZP, Wang W (2012) Chemotherapy and radiotherapy downregulate the activity and expression of DNA methyltransferase and enhance Bcl-2/E1B-19-kDa interacting protein-3-induced apoptosis in human colorectal cancer cells. Chemotherapy 58:445–453
85. Pelosof L, Yerram SR, Ahuja N, Delmas A, Danilova L, Herman JG, Azad NS (2014) CHFR silencing or microsatellite instability is associated with increased antitumor activity of docetaxel or gemcitabine in colorectal cancer. Int J Cancer 134:596–605
86. Mittag F, Kuester D, Vieth M, Peters B, Stolte B, Roessner A, Schneider-Stock R (2006) DAPK promotor methylation is an early event in colorectal carcinogenesis. Cancer Lett 240:69–75
87. Baba Y, Nosho K, Shima K, Huttenhower C, Tanaka N, Hazra A, Giovannucci EL, Fuchs CS, Ogino S (2010b) Hypomethylation of the IGF2 DMR in colorectal tumors, detected by bisulfite pyrosequencing, is associated with poor prognosis. Gastroenterology 139:1855–1864
88. Miyauchi S, Gopal E, Fei YJ, Ganapathy V (2004) Functional identification of SLC5A8, a tumor suppressor down-regulated in colon cancer, as a Na(+)-coupled transporter for short-chain fatty acids. J Biol Chem 279:13293–13296
89. Martin V, Jorgensen HF, Chaubert AS, Berger J, Barr H, Shaw P, Bird A, Chaubert P (2008) MBD2-mediated transcriptional repression of the p14ARF tumor suppressor gene in human colon cancer cells. Pathobiology 75:281–287
90. Slattery ML, Lundgreen A, Bondurant KL, Wolff RK (2011) Interferon-signaling pathway: associations with colon and rectal cancer risk and subsequent survival. Carcinogenesis 32:1660–1667

91. Ebert MP, Model F, Mooney S, Hale K, Lograsso J, Tonnes-Priddy L, Hoffmann J, Csepregi A, Rocken C, Molnar B, Schulz HU, Malfertheiner P, Lofton-Day C (2006) Aristaless-like homeobox-4 gene methylation is a potential marker for colorectal adenocarcinomas. Gastroenterology 131:1418–1430
92. Hellebrekers DM, Lentjes MH, van den Bosch SM, Melotte V, Wouters KA, Daenen KL, Smits KM, Akiyama Y, Yuasa Y, Sanduleanu S, Khalid-de Bakker CA, Jonkers D, Weijenberg MP, Louwagie J, van Criekinge W, Carvalho B, Meijer GA, Baylin SB, Herman JG, de Bruine AP, van Engeland M (2009) GATA4 and GATA5 are potential tumor suppressors and biomarkers in colorectal cancer. Clin Cancer Res 15:3990–3997
93. Psofaki V, Kalogera C, Tzambouras N, Stephanou D, Tsianos E, Seferiadis K, Kolios G (2010) Promoter methylation status of hMLH1, Mgmt, and CDKN2A/p16 in colorectal adenomas. World J Gastroenterol 16:3553–3560
94. Shackelford DB, Shaw RJ (2009) The LKB1-AMPK pathway: metabolism and growth control in tumour suppression. Nat Rev Cancer 9:563–575
95. Herbst A, Rahmig K, Stieber P, Philipp A, Jung A, Ofner A, Crispin A, Neumann J, Lamerz R, Kolligs FT (2011) Methylation of NEUROG1 in serum is a sensitive marker for the detection of early colorectal cancer. Am J Gastroenterol 106:1110–1118
96. Rampino N, Yamamoto H, Ionov Y, Li Y, Sawai H, Reed JC, Perucho M (1997) Somatic frameshift mutations in the BAX gene in colon cancers of the microsatellite mutator phenotype. Science 275:967–969
97. Markowitz S, Wang J, Myeroff L, Parsons R, Sun L, Lutterbaugh J, Fan RS, Zborowska E, Kinzler KW, Vogelstein B et al. (1995) Inactivation of the type II TGF-beta receptor in colon cancer cells with microsatellite instability. Science 268:1336–1338
98. Lynch HT, Boland CR, Gong G, Shaw TG, Lynch PM, Fodde R, Lynch JF, de la Chapelle A (2006) Phenotypic and genotypic heterogeneity in the Lynch syndrome: diagnostic, surveillance and management implications. Eur J Hum Genet 14:390–402
99. Cunningham JM, Christensen ER, Tester DJ, Kim CY, Roche PC, Burgart LJ, Thibodeau SN (1998) Hypermethylation of the hMLH1 promoter in colon cancer with microsatellite instability. Cancer Res 58:3455–3460
100. Kane MF, Loda M, Gaida GM, Lipman J, Mishra R, Goldman H, Jessup JM, Kolodner R (1997) Methylation of the hMLH1 promoter correlates with lack of expression of hMLH1 in sporadic colon tumors and mismatch repair-defective human tumor cell lines. Cancer Res 57:808–811
101. Goel A, Xicola RM, Nguyen TP, Doyle BJ, Sohn VR, Bandipalliam P, Rozek LS, Reyes J, Cordero C, Balaguer F, Castells A, Jover R, Andreu M, Syngal S, Boland CR, Llor X (2010) Aberrant DNA methylation in hereditary nonpolyposis colorectal cancer without mismatch repair deficiency. Gastroenterology 138:1854–1862
102. Jover R, Nguyen TP, Perez-Carbonell L, Zapater P, Paya A, Alenda C, Rojas E, Cubiella J, Balaguer F, Morillas JD, Clofent J, Bujanda L, Rene JM, Bessa X, Xicola RM, Nicolas-Perez D, Castells A, Andreu M, Llor X, Boland CR, Goel A (2011) 5-Fluorouracil adjuvant chemotherapy does not increase survival in patients with CpG island methylator phenotype colorectal cancer. Gastroenterology 140:1174–1181
103. Toyota M, Ahuja N, OHE-Toyota M, Herman JG, Baylin SB, Issa JP (1999) CpG island methylator phenotype in colorectal cancer. Proc Natl Acad Sci U S A 96:8681–8686
104. Ang PW, Loh M, Liem N, Lim PL, Grieu F, Vaithilingam A, Platell C, Yong WP, Iacopetta B, Soong R (2010) Comprehensive profiling of DNA methylation in colorectal cancer reveals subgroups with distinct clinicopathological and molecular features. BMC Cancer 10:227
105. Ogino S, Cantor M, Kawasaki T, Brahmandam M, Kirkner GJ, Weisenberger DJ, Campan M, Laird PW, Loda M, Fuchs CS (2006) CpG island methylator phenotype (CIMP) of colorectal cancer is best characterised by quantitative DNA methylation analysis and prospective cohort studies. Gut 55:1000–1006
106. Ogino S, Nosho K, Kirkner GJ, Kawasaki T, Meyerhardt JA, Loda M, Giovannucci EL, Fuchs CS (2009) CpG island methylator phenotype, microsatellite instability, BRAF mutation and clinical outcome in colon cancer. Gut 58:90–96

107. Bae JM, Kim JH, Cho NY, Kim TY, Kang GH (2013) Prognostic implication of the CpG island methylator phenotype in colorectal cancers depends on tumour location. Br J Cancer 109:1004–1012
108. Gronbaek K, Hother C, Jones PA (2007) Epigenetic changes in cancer. APMIS 115:1039–1059
109. Lao VV, Grady WM (2011) Epigenetics and colorectal cancer. Nat Rev Gastroenterol Hepatol 8:686–700
110. Berg M, Hagland HR, Soreide K (2014) Comparison of CpG island methylator phenotype (CIMP) frequency in colon cancer using different probe- and gene-specific scoring alternatives on recommended multi-gene panels. PLoS ONE 9:e86657
111. Ashktorab H, Rahi H, Wansley D, Varma S, Shokrani B, Lee E, Daremipouran M, Laiyemo A, Goel A, Carethers JM, Brim H (2013) Toward a comprehensive and systematic methylome signature in colorectal cancers. Epigenetics 8:807–815
112. Weisenberger DJ, Siegmund KD, Campan M, Young J, Long TI, Faasse MA, Kang GH, Widschwendter M, Weener D, Buchanan D, Koh H, Simms L, Barker M, Leggett B, Levine J, Kim M, French AJ, Thibodeau SN, Jass J, Haile R, Laird PW (2006) CpG island methylator phenotype underlies sporadic microsatellite instability and is tightly associated with BRAF mutation in colorectal cancer. Nat Genet 38:787–793
113. Hitchins MP (2013) The role of epigenetics in Lynch syndrome. Fam Cancer 12:189–205
114. Rumilla K, Schowalter KV, Lindor NM, Thomas BC, Mensink KA, Gallinger S, Holter S, Newcomb PA, Potter JD, Jenkins MA, Hopper JL, Long TI, Weisenberger DJ, Haile RW, Casey G, Laird PW, LE Marchand L, Thibodeau SN (2011) Frequency of deletions of EPCAM (TACSTD1) in MSH2-associated lynch syndrome cases. J Mol Diagn 13:93–99
115. Kovacs ME, Papp J, Szentirmay Z, Otto S, Olah E (2009) Deletions removing the last exon of TACSTD1 constitute a distinct class of mutations predisposing to Lynch syndrome. Hum Mutat 30:197–203
116. Hitchins MP, Wong JJ, Suthers G, Suter CM, Martin DI, Hawkins NJ, Ward RL (2007) Inheritance of a cancer-associated MLH1 germ-line epimutation. N Engl J Med 356:697–705
117. Peltomaki P (2014) Epigenetic mechanisms in the pathogenesis of Lynch syndrome. Clin Genet 85(5):403–412
118. Nakazawa T, Kondo T, Ma D, Niu D, Mochizuki K, Kawasaki T, Yamane T, Iino H, Fujii H, Katoh R (2012) Global histone modification of histone H3 in colorectal cancer and its precursor lesions. Hum Pathol 43:834–842
119. Binder H, Steiner L, Przybilla J, Rohlf T, Prohaska S, Galle J (2013) Transcriptional regulation by histone modifications: towards a theory of chromatin re-organization during stem cell differentiation. Phys Biol 10:026006
120. Ramsahoye BH, Biniszkiewicz D, Lyko F, Clark V, Bird AP, Jaenisch R (2000) Non-CpG methylation is prevalent in embryonic stem cells and may be mediated by DNA methyltransferase 3a. Proc Natl Acad Sci U S A 97:5237–5242
121. Dawson MA, Kouzarides T (2012) Cancer epigenetics: from mechanism to therapy. Cell 150:12–27
122. Vakoc CR, Sachdeva MM, Wang H, Blobel GA (2006) Profile of histone lysine methylation across transcribed mammalian chromatin. Mol Cell Biol 26:9185–9195
123. Kouzarides T (2007). Chromatin modifications and their function. Cell 128:693–705
124. Ashktorab H, Belgrave K, Hosseinkhah F, Brim H, Nouraie M, Takkikto M, Hewitt S, Lee EL, Dashwood RH, Smoot D (2009) Global histone H4 acetylation and HDAC2 expression in colon adenoma and carcinoma. Dig Dis Sci 54:2109–2117
125. Ishihama K, Yamakawa M, Semba S, Takeda H, Kawata S, Kimura S, Kimura W (2007) Expression of HDAC1 and CBP/p300 in human colorectal carcinomas. J Clin Pathol 60:1205–1210
126. Torroja C, Gorfinkiel N, Guerrero I (2005) Mechanisms of Hedgehog gradient formation and interpretation. J Neurobiol 64:334–356
127. Tamagawa H, Oshima T, Numata M, Yamamoto N, Shiozawa M, Morinaga S, Nakamura Y, Yoshihara M, Sakuma Y, Kameda Y, Akaike M, Yukawa N, Rino Y, Masuda M, Miyagi Y (2013) Global histone modification of H3K27 correlates with the outcomes in patients with metachronous liver metastasis of colorectal cancer. Eur J Surg Oncol 39:655–661

128. Yokoyama Y, Hieda M, Nishioka Y, Matsumoto A, Higashi S, Kimura H, Yamamoto H, Mori M, Matsuura S, Matsuura N (2013) Cancer-associated upregulation of histone H3 lysine 9 trimethylation promotes cell motility in vitro and drives tumor formation in vivo. Cancer Sci 104:889–895
129. Ambros V (2004) The functions of animal microRNAs. Nature 431:350–355
130. Lagos-Quintana M, Rauhut R, Lendeckel W, Tuschl T (2001) Identification of novel genes coding for small expressed RNAs. Science 294:853–858
131. Lee RC, Feinbaum RL, Ambros V (1993) The *C. elegans* heterochronic gene lin-4 encodes small RNAs with antisense complementarity to lin-14. Cell 75:843–854
132. Bartel DP (2009) MicroRNAs: target recognition and regulatory functions. Cell 136:215–233
133. Jansson MD, Lund AH (2012) MicroRNA and cancer. Mol Oncol 6:590–610
134. Esquela-Kerscher A, Slack FJ (2006) Oncomirs – microRNAs with a role in cancer. Nat Rev Cancer 6:259–269
135. Lynam-Lennon N, Maher SG, Reynolds JV (2009) The roles of microRNA in cancer and apoptosis. Biol Rev Camb Philos Soc 84:55–71
136. Calin GA, Dumitru CD, Shimizu M, Bichi R, Zupo S, Noch E, Aldler H, Rattan S, Keating M, Rai K, Rassenti L, Kipps T, Negrini M, Bullrich F, Croce CM (2002) Frequent deletions and down-regulation of micro- RNA genes miR15 and miR16 at 13q14 in chronic lymphocytic leukemia. Proc Natl Acad Sci U S A 99:15524–15529
137. Schetter AJ, Okayama H, Harris CC (2012) The role of microRNAs in colorectal cancer. Cancer J 18:244–252
138. Yamamoto H, Adachi Y, Taniguchi H, Kunimoto H, Nosho K, Suzuki H, Shinomura Y (2012) Interrelationship between microsatellite instability and microRNA in gastrointestinal cancer. World J Gastroenterol 18:2745–2755
139. Michael MZ, Sm OC, van Holst Pellekaan NG, Young GP, James RJ (2003) Reduced accumulation of specific microRNAs in colorectal neoplasia. Mol Cancer Res 1:882–891
140. Schetter AJ, Leung SY, Sohn JJ, Zanetti KA, Bowman ED, Yanaihara N, Yuen ST, Chan TL, Kwong DL, Au GK, Liu CG, Calin GA, Croce CM, Harris CC (2008) MicroRNA expression profiles associated with prognosis and therapeutic outcome in colon adenocarcinoma. JAMA 299:425–436
141. Schepeler T, Reinert JT, Ostenfeld MS, Christensen LL, Silahtaroglu AN, Dyrskjot L, Wiuf C, Sorensen FJ, Kruhoffer M, Laurberg S, Kauppinen S, Orntoft TF, Andersen CL (2008) Diagnostic and prognostic microRNAs in stage II colon cancer. Cancer Res 68:6416–6424
142. Bandres E, Cubedo E, Agirre X, Malumbres R, Zarate R, Ramirez N, Abajo A, Navarro A, Moreno I, Monzo M, Garcia-Foncillas J (2006) Identification by real-time PCR of 13 mature microRNAs differentially expressed in colorectal cancer and non-tumoral tissues. Mol Cancer 5:29
143. Cummins JM, He Y, Leary RJ, Pagliarini R, Diaz LA Jr, Sjoblom T, Barad O, Bentwich Z, Szafranska AE, Labourier E, Raymond CK, Roberts BS, Juhl H, Kinzler KW, Vogelstein B, Velculescu VE (2006) The colorectal microRNAome. Proc Natl Acad Sci U S A 103:3687–3692
144. Balaguer F, Link A, Lozano JJ, Cuatrecasas M, Nagasaka T, Boland CR, Goel A (2010) Epigenetic silencing of miR-137 is an early event in colorectal carcinogenesis. Cancer Res 70:6609–6618
145. Hur K, Toiyama Y, Takahashi M, Balaguer F, Nagasaka T, Koike J, Hemmi H, Koi M, Boland CR, Goel A (2013) MicroRNA-200c modulates epithelial-to-mesenchymal transition (EMT) in human colorectal cancer metastasis. Gut 62:1315–1326
146. Takahashi M, Cuatrecasas M, Balaguer F, Hur K, Toiyama Y, Castells A, Boland CR, Goel A (2012) The clinical significance of MiR-148a as a predictive biomarker in patients with advanced colorectal cancer. PLoS ONE 7:e46684
147. Toiyama Y, Hur K, Tanaka K, Inoue Y, Kusunoki M, Boland CR, Goel A (2014) Serum miR-200c Is a novel prognostic and metastasis-predictive biomarker in patients with colorectal cancer. Ann Surg 259:735–743

148. Toiyama Y, Takahashi M, Hur K, Nagasaka T, Tanaka K, Inoue Y, Kusunoki M, Boland CR, Goel A (2013) Serum miR-21 as a diagnostic and prognostic biomarker in colorectal cancer. J Natl Cancer Inst 105:849–859
149. Akao Y, Nakagawa Y, Naoe T (2006) let-7 microRNA functions as a potential growth suppressor in human colon cancer cells. Biol Pharm Bull 29:903–906
150. Diosdado B, van de Wiel MA, Terhaar Sive Droste JS, Mongera S, Postma C, Meijerink WJ, Carvalho B, Meijer GA (2009) MiR-17–92 cluster is associated with 13q gain and c-myc expression during colorectal adenoma to adenocarcinoma progression. Br J Cancer 101:707–714
151. Asangani IA, Rasheed SA, Nikolova DA, Leupold JH, Colburn NH, Post S, Allgayer H (2008) MicroRNA-21 (miR-21) post-transcriptionally downregulates tumor suppressor Pdcd4 and stimulates invasion, intravasation and metastasis in colorectal cancer. Oncogene 27:2128–2136
152. Ding Q, Chang CJ, Xie X, Xia W, Yang JY, Wang SC, Wang Y, Xia J, Chen L, Cai C, Li H, Yen CJ, Kuo HP, Lee DF, Lang J, Huo L, Cheng X, Chen YJ, Li CW, Jeng LB, Hsu JL, Li LY, Tan A, Curley SA, Ellis LM, Dubois RN, Hung MC (2011) APOBEC3G promotes liver metastasis in an orthotopic mouse model of colorectal cancer and predicts human hepatic metastasis. J Clin Invest 121:4526–4536
153. Baraniskin A, Birkenkamp-Demtroder K, Maghnouj A, Zollner H, Munding J, Klein-Scory S, Reinacher-Schick A, Schwarte-Waldhoff I, Schmiegel W, Hahn SA (2012) MiR-30a-5p suppresses tumor growth in colon carcinoma by targeting DTL. Carcinogenesis 33:732–739
154. Wu J, Wu G, Lv L, Ren YF, Zhang XJ, Xue YF, Li G, Lu X, Sun Z, Tang KF (2012) MicroRNA-34a inhibits migration and invasion of colon cancer cells via targeting to Fra-1. Carcinogenesis 33:519–528
155. Huang Z, Huang S, Wang Q, Liang L, Ni S, Wang L, Sheng W, He X, Du X (2011) MicroRNA-95 promotes cell proliferation and targets sorting Nexin 1 in human colorectal carcinoma. Cancer Res 71:2582–2589
156. Strillacci A, Valerii MC, Sansone P, Caggiano C, Sgromo A, Vittori L, Fiorentino M, Poggioli G, Rizzello F, Campieri M, Spisni E (2013) Loss of miR-101 expression promotes Wnt/beta-catenin signalling pathway activation and malignancy in colon cancer cells. J Pathol 229:379–389
157. Nagel R, le Sage C, Diosdado B, van der Waal M, Oude Vrielink JA, Bolijn A, Meijer GA, Agami R (2008) Regulation of the adenomatous polyposis coli gene by the miR-135 family in colorectal cancer. Cancer Res 68:5795–5802
158. Wu W, Wang Z, Yang P, Yang J, Liang J, Chen Y, Wang H, Wei G, Ye S, Zhou Y (2014) MicroRNA-135b regulates metastasis suppressor 1 expression and promotes migration and invasion in colorectal cancer. Mol Cell Biochem 388:249–259
159. Akao Y, Nakagawa Y, Hirata I, Iio A, Itoh T, Kojima K, Nakashima R, Kitade Y, Naoe T (2010) Role of anti-oncomirs miR-143 and -145 in human colorectal tumors. Cancer Gene Ther 17:398–408
160. Arndt GM, Dossey L, Cullen LM, Lai A, Druker R, Eisbacher M, Zhang C, Tran N, Fan H, Retzlaff K, Bittner A, Raponi M (2009) Characterization of global microRNA expression reveals oncogenic potential of miR-145 in metastatic colorectal cancer. BMC Cancer 9:374
161. Shibuya H, Iinuma H, Shimada R, Horiuchi A, Watanabe T (2010) Clinicopathological and prognostic value of microRNA-21 and microRNA-155 in colorectal cancer. Oncology 79:313–320
162. Qin J, Luo M (2014) MicroRNA-221 promotes colorectal cancer cell invasion and metastasis by targeting RECK. FEBS Lett 588:99–104
163. Wang H, Wu J, Meng X, Ying X, Zuo Y, Liu R, Pan Z, Kang T, Huang W (2011) MicroRNA-342 inhibits colorectal cancer cell proliferation and invasion by directly targeting DNA methyltransferase 1. Carcinogenesis 32:1033–1042
164. Nie J, Liu L, Zheng W, Chen L, Wu X, Xu Y, Du X, Han W (2012) microRNA-365, down-regulated in colon cancer, inhibits cell cycle progression and promotes apoptosis of colon cancer cells by probably targeting Cyclin D1 and Bcl-2. Carcinogenesis 33:220–225

165. Bitarte N, Bandres E, Boni V, Zarate R, Rodriguez J, Gonzalez-Huarriz M, Lopez I, Javier Sola J, Alonso MM, Fortes P, Garcia-Foncillas J (2011) MicroRNA-451 is involved in the self-renewal, tumorigenicity, and chemoresistance of colorectal cancer stem cells. Stem Cells 29:1661–1671
166. Liu X, Zhang Z, Sun L, Chai N, Tang S, Jin J, Hu H, Nie Y, Wang X, Wu K, Jin H, Fan D (2011) MicroRNA-499-5p promotes cellular invasion and tumor metastasis in colorectal cancer by targeting FOXO4 and PDCD4. Carcinogenesis 32:1798–1805
167. Tsang WP, Ng EK, Ng SS, Jin H, Yu J, Sung JJ, Kwok TT (2010) Oncofetal H19-derived miR-675 regulates tumor suppressor RB in human colorectal cancer. Carcinogenesis 31:350–358
168. Oberg AL, French AJ, Sarver AL, Subramanian S, Morlan BW, Riska SM, Borralho PM, Cunningham JM, Boardman LA, Wang L, Smyrk TC, Asmann Y, Steer CJ, Thibodeau SN (2011) miRNA expression in colon polyps provides evidence for a multihit model of colon cancer. PLoS ONE 6:e20465
169. Slaby O, Svoboda M, Fabian P, Smerdova T, Knoflickova D, Bednarikova M, Nenutil R, Vyzula R (2007) Altered expression of miR-21, miR-31, miR-143 and miR-145 is related to clinicopathologic features of colorectal cancer. Oncology 72:397–402
170. Dote H, Toyooka S, Tsukuda K, Yano M, Ouchida M, Doihara H, Suzuki M, Chen H, Hsieh JT, Gazdar AF, Shimizu N (2004) Aberrant promoter methylation in human DAB2 interactive protein (hDAB2IP) gene in breast cancer. Clin Cancer Res 10:2082–2089
171. Lofton-Day C, Model F, Devos T, Tetzner R, Distler J, Schuster M, Song X, Lesche R, Liebenberg V, Ebert M, Molnar B, Grutzmann R, Pilarsky C, Sledziewski A (2008) DNA methylation biomarkers for blood-based colorectal cancer screening. Clin Chem 54:414–423
172. Yano M, Toyooka S, Tsukuda K, Dote H, Ouchida M, Hanabata T, Aoe M, Date H, Gazdar AF, Shimizu N (2005) Aberrant promoter methylation of human DAB2 interactive protein (hDAB2IP) gene in lung cancers. Int J Cancer 113:59–66
173. Volinia S, Calin GA, Liu CG, Ambs S, Cimmino A, Petrocca F, Visone R, Iorio M, Roldo C, Ferracin M, Prueitt RL, Yanaihara N, Lanza G, Scarpa A, Vecchione A, Negrini M, Harris CC, Croce CM (2006) A microRNA expression signature of human solid tumors defines cancer gene targets. Proc Natl Acad Sci U S A 103:2257–2261
174. Xi Y, Nakajima G, Gavin E, Morris CG, Kudo K, Hayashi K, Ju J (2007) Systematic analysis of microRNA expression of RNA extracted from fresh frozen and formalin-fixed paraffin-embedded samples. RNA 13:1668–1674
175. Zhang X, Chen J, Radcliffe T, Lebrun DP, Tron VA, Feilotter H (2008) An array-based analysis of microRNA expression comparing matched frozen and formalin-fixed paraffin-embedded human tissue samples. J Mol Diagn 10:513–519
176. Elmen J, Lindow M, Schutz S, Lawrence M, Petri A, Obad S, Lindholm M, Hedtjarn M, Hansen HF, Berger U, Gullans S, Kearney P, Sarnow P, Straarup EM, Kauppinen S (2008) LNA-mediated microRNA silencing in non-human primates. Nature 452:896–899
177. Ma L, Teruya-Feldstein J, Weinberg RA (2007) Tumour invasion and metastasis initiated by microRNA-10b in breast cancer. Nature 449:682–688
178. Johnson SM, Grosshans H, Shingara J, Byrom M, Jarvis R, Cheng A, Labourier E, Reinert KL, Brown D, Slack FJ (2005) RAS is regulated by the let-7 microRNA family. Cell 120:635–647
179. Kitade Y, Akao Y (2010) MicroRNAs and their therapeutic potential for human diseases: microRNAs, miR-143 and -145, function as anti-oncomirs and the application of chemically modified miR-143 as an anti-cancer drug. J Pharmacol Sci 114:276–280
180. Samowitz WS, Albertsen H, Herrick J, Levin TR, Sweeney C, Murtaugh MA, Wolff RK, Slattery ML (2005) Evaluation of a large, population-based sample supports a CpG island methylator phenotype in colon cancer. Gastroenterology 129:837–845
181. Gonzalo V, Lozano JJ, Munoz J, Balaguer F, Pellise M, Rodriguez de Miguel C, Andreu M, Jover R, Llor X, Giraldez MD, Ocana T, Serradesanferm A, Alonso-Espinaco V, Jimeno M, Cuatrecasas M, Sendino O, Castellvi-Bel S, Castells A (2010) Aberrant gene promoter methylation associated with sporadic multiple colorectal cancer. PLoS ONE 5:e8777

182. Gyparaki MT, Basdra EK, Papavassiliou AG (2013) DNA methylation biomarkers as diagnostic and prognostic tools in colorectal cancer. J Mol Med 91:1249–1256
183. Nagasaka T, Goel A, Matsubara N, Tanaka N (2006) Detection of fecal DNA methylation for colorectal neoplasia: does it lead to an optimal screening test? Acta Med Okayama 60:249–256
184. Nagasaka T, Tanaka N, Cullings HM, Sun DS, Sasamoto H, Uchida T, Koi M, Nishida N, Naomoto Y, Boland CR, Matsubara N, Goel A (2009) Analysis of fecal DNA methylation to detect gastrointestinal neoplasia. J Natl Cancer Inst 101:1244–1258
185. Yang H, Xia BQ, Jiang B, Wang G, Yang YP, Chen H, Li BS, Xu AG, Huang YB, Wang XY (2013) Diagnostic value of stool DNA testing for multiple markers of colorectal cancer and advanced adenoma: a meta-analysis. Can J Gastroenterol 27:467–475
186. Imperiale TF, Ransohoff DF, Itzkowitz SH, Levin TR, Lavin P, Lidgard GP, Ahlquist DA, Berger BM (2014) Multitarget stool DNA testing for colorectal-cancer screening. N Engl J Med 370:1287–1297
187. Muller HM, Oberwalder M, Fiegl H, Morandell M, Goebel G, Zitt M, Muhlthaler M, Ofner D, Margreiter R, Widschwendter M (2004) Methylation changes in faecal DNA: a marker for colorectal cancer screening? Lancet 363:1283–1285
188. Oberwalder M, Zitt M, Wontner C, Fiegl H, Goebel G, Kohle O, Muhlmann G, Ofner D, Margreiter R, Muller HM (2008) SFRP2 methylation in fecal DNA–a marker for colorectal polyps. Int J Colorectal Dis 23:15–19
189. Lange CP, Laird PW (2013) Clinical applications of DNA methylation biomarkers in colorectal cancer. Epigenomics 5:105–108
190. Mulero-Navarro S, Esteller M (2008) Epigenetic biomarkers for human cancer: the time is now. Crit Rev Oncol Hematol 68:1–11
191. Lujambio A, Esteller M (2009) How epigenetics can explain human metastasis: a new role for microRNAs. Cell Cycle 8:377–382
192. Ng EK, Chong WW, Jin H, Lam EK, Shin VY, Yu J, Poon TC, Ng SS, Sung JJ (2009) Differential expression of microRNAs in plasma of patients with colorectal cancer: a potential marker for colorectal cancer screening. Gut 58:1375–1381
193. Link A, Balaguer F, Shen Y, Nagasaka T, Lozano JJ, Boland CR, Goel A (2010) Fecal MicroRNAs as novel biomarkers for colon cancer screening. Cancer Epidemiol Biomarkers Prev 19:1766–1774
194. Giraldez MD, Lozano JJ, Ramirez G, Hijona E, Bujanda L, Castells A, Gironella M (2013) Circulating microRNAs as biomarkers of colorectal cancer: results from a genome-wide profiling and validation study. Clin Gastroenterol Hepatol 11:681–688, e3
195. Earle JS, Luthra R, Romans A, Abraham R, Ensor J, Yao H, Hamilton SR (2010) Association of microRNA expression with microsatellite instability status in colorectal adenocarcinoma. J Mol Diagn 12:433–440
196. Lanza G, Ferracin M, Gafa R, Veronese A, Spizzo R, Pichiorri F, Liu CG, Calin GA, Croce CM, Negrini M (2007) mRNA/microRNA gene expression profile in microsatellite unstable colorectal cancer. Mol Cancer 6:54
197. Sarver AL, French AJ, Borralho PM, Thayanithy V, Oberg AL, Silverstein KA, Morlan BW, Riska SM, Boardman LA, Cunningham JM, Subramanian S, Wang L, Smyrk TC, Rodrigues CM, Thibodeau SN, Steer CJ (2009) Human colon cancer profiles show differential microRNA expression depending on mismatch repair status and are characteristic of undifferentiated proliferative states. BMC Cancer 9:401
198. Balaguer F, Moreira L, Lozano JJ, Link A, Ramirez G, Shen Y, Cuatrecasas M, Arnold M, Meltzer SJ, Syngal S, Stoffel E, Jover R, Llor X, Castells A, Boland CR, Gironella M, Goel A (2011) Colorectal cancers with microsatellite instability display unique miRNA profiles. Clin Cancer Res 17:6239–6249
199. Valeri N, Gasparini P, Fabbri M, Braconi C, Veronese A, Lovat F, Adair B, Vannini I, Fanini F, Bottoni A, Costinean S, Sandhu SK, Nuovo GJ, Alder H, Gafa R, Calore F, Ferracin M, Lanza G, Volinia S, Negrini M, Mcilhatton MA, Amadori D, Fishel R, Croce CM (2010) Modulation of mismatch repair and genomic stability by miR-155. Proc Natl Acad Sci U S A 107:6982–6987

200. Benard A, van de Velde CJ, Lessard L, Putter H, Takeshima L, Kuppen PJ, Hoon DS (2013) Epigenetic status of LINE-1 predicts clinical outcome in early-stage rectal cancer. Br J Cancer 109:3073–3083
201. Xing X, Cai W, Shi H, Wang Y, Li M, Jiao J, Chen M (2013a) The prognostic value of CDKN2A hypermethylation in colorectal cancer: a meta-analysis. Br J Cancer 108:2542–2548
202. Xing XB, Cai WB, Luo L, Liu LS, Shi HJ, Chen MH (2013b) The prognostic value of hypermethylation in cancer: a meta-analysis. PLoS ONE 8:e66587
203. Jensen LH, Rasmussen AA, Byriel L, Kuramochi H, Cruger DG, Lindebjerg J, Danenberg PV, Jakobsen A, Danenberg K (2013) Regulation of MLH1 mRNA and protein expression by promoter methylation in primary colorectal cancer: a descriptive and prognostic cancer marker study. Cell Oncol 36:411–419
204. Sugita H, Iida S, Inokuchi M, Kato K, Ishiguro M, Ishikawa T, Takagi Y, Enjoji M, Yamada H, Uetake H, Kojima K, Sugihara K (2011) Methylation of BNIP3 and DAPK indicates lower response to chemotherapy and poor prognosis in gastric cancer. Oncol Rep 25:513–518
205. Ebert MP, Tanzer M, Balluff B, Burgermeister E, Kretzschmar AK, Hughes DJ, Tetzner R, Lofton-Day C, Rosenberg R, Reinacher-Schick AC, Schulmann K, Tannapfel A, Hofheinz R, Rocken C, Keller G, Langer R, Specht K, Porschen R, Stohlmacher-Williams J, Schuster T, Strobel P, Schmid RM (2012) TFAP2E-DKK4 and chemoresistance in colorectal cancer. N Engl J Med 366:44–53
206. Kulda V, Pesta M, Topolcan O, Liska V, Treska V, Sutnar A, Rupert K, Ludvikova M, Babuska V, Holubec L Jr, Cerny R (2010) Relevance of miR-21 and miR-143 expression in tissue samples of colorectal carcinoma and its liver metastases. Cancer Genet Cytogenet 200:154–160
207. Nielsen BS, Jorgensen S, Fog JU, Sokilde R, Christensen IJ, Hansen U, Brunner N, Baker A, Moller S, Nielsen HJ (2011) High levels of microRNA-21 in the stroma of colorectal cancers predict short disease-free survival in stage II colon cancer patients. Clin Exp Metastasis 28:27–38
208. Liao WT, Ye YP, Zhang NJ, Li TT, Wang SY, Cui YM, Qi L, Wu P, Jiao HL, Xie YJ, Zhang C, Wang JX, Ding YQ (2014) MicroRNA-30b functions as a tumour suppressor in human colorectal cancer by targeting Kras, PIK3CD and BCL2. J Pathol 232:415–427
209. Nagaraju GP, EL-Rayes BF (2013) SPARC and DNA methylation: possible diagnostic and therapeutic implications in gastrointestinal cancers. Cancer Lett 328:10–17
210. Foulks JM, Parnell KM, Nix RN, Chau S, Swierczek K, Saunders M, Wright K, Hendrickson TF, Ho KK, Mccullar MV, Kanner SB (2012) Epigenetic drug discovery: targeting DNA methyltransferases. J Biomol Screen 17:2–17
211. Wodarz D, Boland CR, Goel A, Komarova NL (2013) Methylation kinetics and CpG-island methylator phenotype status in colorectal cancer cell lines. Biol Direct 8:14
212. Braiteh F, Soriano AO, Garcia-Manero G, Hong D, Johnson MM, Silva lde P, Yang H, Alexander S, Wolff J, Kurzrock R (2008) Phase I study of epigenetic modulation with 5-azacytidine and valproic acid in patients with advanced cancers. Clin Cancer Res 14:6296–6301
213. Ibrahim AF, Weirauch U, Thomas M, Grunweller A, Hartmann RK, Aigner A (2011) MicroRNA replacement therapy for miR-145 and miR-33a is efficacious in a model of colon carcinoma. Cancer Res 71:5214–5224

MicroRNA Involvement in Intestinal Tumorigenesis

Naohiro Nishida and George A. Calin

Abstract Intestinal tumorigenesis is characterized by accumulation of genetic and epigenetic alterations that lead to enhanced malignant potential of epithelial cells. Recent evidence suggests that 18–22 nt small non-coding RNAs, microRNAs (miRNAs) play a crucial role in epigenetic regulation during carcinogenesis. Regulating hundreds of targets simultaneously, miRNAs can affect multiple signaling pathways and serve as a "hub" in gene regulatory network. A large part of oncogenes and tumor suppressors are controlled by miRNAs, and conversely, these genes regulate miRNA expression by transcriptional control and by interfering with the miRNA biogenesis process. In addition, somatic mutations and single nucleotide polymorphisms in miRNA genes and in target messenger RNAs attenuate miRNA-mediated gene suppression efficiency. These complicated gene–miRNA interactions create aberrant gene expression profiles, which lead to disruption of cellular homeostasis and eventually to carcinogenesis. In this chapter, we summarize a variety of dysregulation mechanisms of miRNAs and present an overview of miRNA involvement in intestinal tumorigenesis, mainly in colorectal cancer.

Keywords MicroRNA · Intestinal tumorigenesis · Epigenetic regulation · miRNA biogenesis · Carcinogenesis · Single nucleotide polymorphisms (SNPs) · Colorectal cancer

1 Introduction

Cancer initiation and development can be explained by two basic concepts—genetic alterations and epigenetic modification. MicroRNAs (miRNAs) can be classified as one of the epigenetic regulators, as well as DNA methylation and histone modification factors, due to their posttranscriptional target gene regulatory mechanism.

G. A. Calin (✉) · N. Nishida
Department of Experimental Therapeutics and The Center for RNA Interference and Non-Coding RNAs, The University of Texas MD Anderson Cancer Center, South Campus Research Bldg 3, 1881 East Road, Houston, TX 77030, USA
e-mail: gcalin@mdanderson.org

V. W. Yang, A. B. Bialkowska (eds.), *Intestinal Tumorigenesis*,
DOI 10.1007/978-3-319-19986-3_6

miRNAs are a class of 18–22 nucleotide (nt) non-coding RNAs that predominantly serve as translational repressors by binding to complementary sequences in the 3′ untranslated region (UTR) of their target messenger RNAs (mRNAs) [1]. The target recognition of miRNAs depends on base pairing between the seed sequence of the miRNA (nucleotides 2–8 from the 5′ end) and the complementary sequence of the target genes [1]. Since only 5–7 nucleotides are essential for target recognition, the number of candidate target is very large. Although suppression of each target is modest [2], alterations of miRNAs can affect the expression of numerous target genes, leading to significant phenotypic changes.

The expressions of miRNAs are regulated by transcriptional factors in the similar manner as that of protein coding genes. miRNAs genes are broadly distributed in intergenic and intragenic regions of genome, and about half of mammalian miRNAs reside within host genes in the same strand, and two thirds of them share the same promoter and are coordinately transcribed with them [3]. miRNAs co-expressed with host genes are likely to be evolutionally conserved and, importantly, miRNAs and their host genes are considered to be functionally related and to work in a synergistic or antagonistic manner [4, 5]. Some miRNAs in introns are also shown to control host gene expressions through regulation of transcriptional factors [6], indicating the importance of the interaction between intronic miRNAs and its host genes.

The important thing is that miRNA expression is precisely controlled by a processing step as well as by transcriptional regulation. In the canonical pathway, primary miRNA transcripts (pri-miRs) are transcribed from intergenic regions or introns of protein-coding genes by RNA polymerase II (Pol II) and partially by RNA polymerase III (Pol III). Ribonuclease III (RNase III)-type endonuclease Drosha with its partner double-stranded RNA-binding domain protein DGCR8 (DiGeorge critical region 8) cleaves pri-miRNA and generates 70-nt stem loop transcripts called precursor miRNAs (pre-miRNAs) [7, 8]. The Drosha complex also contains several auxiliary factors including p68 (DDX5) and p72 (DDX17), which interact with a variety of cell signaling pathways [8]. After active transport from the nucleus to the cytoplasm by Exportin 5 (XPO5), Dicer, another RNase III–type endonuclease, cleaves pre-miRNAs and generates a 22-nt mature form of miRNAs. One of the double-stranded miRNAs is selectively incorporated into the RNA-induced silencing complex (RISC), which contains the Argonaute (Ago) family protein as the core component. miRNAs recognize their target mRNA in a sequence-dependent manner and suppress target expression within the RISC [7].

miRNAs are involved in a variety of biological processes including initiation and progression of human malignancies [9]. Reliable evidence that miRNAs are crucially involved in cancer initiation comes from the fact that miRNA genes are frequently located at fragile sites or in cancer-associated genomic regions including loss of heterozygosity, genomic-amplified regions and common breakpoint regions [10]. Furthermore, miRNA structural clusters, which are coordinately transcribed from close genomic regions, are likely to be located in the proximity of genes involved in essential cellular functions, such as immune systems, signal transduction and development [11]. Therefore, chromosomal events in close proximity to miRNA genes have substantial biological effects, leading to possible carcinogenesis.

miRNAs exhibit a wide variety of species and expression patterns of miRNAs differ between individual tissues, as well as between cancer and normal tissues [12, 13], making miRNAs attractive biomarkers as well as therapeutic targets. In colorectal cancer, growing evidence suggests that miRNAs play a crucial role in every step of cancer development. In this chapter, we summarize a variety of dysregulation mechanisms of miRNAs and their involvement in the initiation and progression of colorectal cancer.

2 miRNA Dysregulation in Colorectal Cancer

miRNA expression is regulated by a variety of interacting molecules and biological pathways. In human cancers, aberrant gene expressions lead to miRNA dysregulation, and subsequently, miRNAs themselves work as oncogenes or tumor suppressors by regulating hundreds of targets, leading to coordinated regulation of downstream targets by genes and miRNAs. Also, a positive-and-negative-feedback loop exists in the miRNA-gene regulatory network, which enhances or maintains malignant potential. Below, we discuss various miRNA dysregulation mechanisms in colorectal cancer.

2.1 Transcriptional Regulation by Oncogenes and Tumor Suppressors

In human cancers, oncogenic or tumor-suppressive transcriptional factors directly regulate specific miRNA expression and coordinately work with them. For example, the well-characterized oncogenic miRNA cluster, miR-17-92a, is transcriptionally activated by direct binding of c-MYC to its promoter. Subsequently, overexpressed miR-17-92a suppresses a large number of targets, including cell cycle regulator p21 and proapoptotic BIM, leading to enhanced malignant potential [14]. Likewise, miR-143 and miR-145, which are significantly downregulated in colorectal cancer, are transcriptionally regulated by oncogenic KRAS in a Ras responsive element binding protein (RREB1)-dependent manner. In turn, miR-143 or miR-145 targets multiple components of KRAS signaling, forming a negative-feedback loop [15]. In another example, p53 transcriptionally activate tumor-suppressive miRNAs such as miR-34, miR-15a/16-1, and miR-145, which also exhibit antitumorigenic effects by inhibiting oncogenic MYC, cell cycle–dependent kinesis CDK4 and CDK6, and antiapoptotic factors such as SIRT1 and BCL-2 [16] (Table 1).

Table 1 Overexpressed and downregulated miRNAs in colorectal cancer and their mechanisms of dysregulation

miRNA	Upreulgation/ downregulation	Crucial targets	Transcriptional regualtion	Epigenetic regulation	Alterations of biogenesis	Ref
let-7	Down	KRAS, c-MYC	–	–	LIN28	[17, 18]
miR-15a	Down	BCL-2, Cyclin E	p53	–	p53	[19, 20]
miR-16	Down	BCL-2	p53	–	p53	[19]
miR-34a/34b/34c	Down	CCNE2, CDK4, E2F3,MYC. BCL-2, survivin, SNAIL, SIRT1	p53	DNA methylation	p53	[16]
miR-124a	Down	CDK6	–	DNA methylation	–	[21]
miR-126	Down	CXCR4, VEGF	–	DNA methylation	–	[22, 23]
miR-137	Down	CDC42, LSD1	–	DNA methylation	–	[24, 25]
miR-143	Down	KRAS, ERK5	p53	DNA methylation	p53	[16, 26]
miR-145	Down	KRAS, MYC, OCT4, SOX, KLF4, p53	p53	DNA methylation	p53	[27, 16, 26]
miR-192	Down	MDM2	p53	–	–	[28]
miR-200,miR-429	Down	ZEB1, ZEB2	p53	DNA methylation	–	[29]
miR-215	Down	MDM2	p53	–	–	[28]
miR-342	Down	DNMT1	–	DNA methylation	–	[30, 31]
miR-497	Down	IGF1-R	–	–	–	[32]
miR-10b	Up	HOXD10, BCL2L11	TWIST	–	–	[33, 34]
miR-17-92a cluster	Up	MYC, E2F1, PTEN, p21, BIM	MYC, E2F1	–	–	[14]
miR-21	Up	PTEN, PDCD4, SPRY, TIMP3	–	–	–	[35, 36]
miR-29a	Up	DNMT3A, DNMT3B	–	–	–	[37]
miR-106b-25 cluster	Up	E2F1, p21, BIM	MYC, E2F2	–	–	[14]
miR-125b	Up	p53, Mcl-1, Bcl-2, IL-6R	–	–	–	[38]
miR-135	Up	APC	–	–	–	[39]
miR-181	up	HOXA11, PBX3	KRAS	–	–	[40–42]
miR-210	Up	E2F3, GPD1L	KRAS, HIF1A	–	–	[42–45]
miR-221/222	Up	P27, P57	KRAS	–	–	[42, 46]
miR-223	Up	IGF1R, FOXO1	–	–	–	[47, 48]

2.2 Epigenetic Dysregulation

DNA methylation and histone modification constitute an important regulatory mechanism for miRNA regulation. In human malignancies, miRNA promoters are known to be highly methylated throughout whole genome [49]. Down regulation of putative tumor-suppressive miRNAs miR-34b and miR-34c is predominantly caused by hypermethylation of their CpG island, which is observed in 90% of primary colorectal cancer samples [27, 50]. In another example, miR-342, which is down regulated in colorectal cancer, and its host gene EVL undergo methylation in 86% of colorectal adenocarcinomas. The reconstitution of miR-342 induces apoptosis in colorectal cancer cells, suggesting that miR-342 works as a tumor suppressor [30]. Balaguer et al. [24] reported that the miR-137 CpG island is frequently hypermethylated in colorectal cancer samples, which leads to derepression of several target oncogenes [24]. Importantly, a large part of adenomas exhibit miR-137 and miR-342 hypermethylation, indicating that these miRNAs are involved in colorectal carcinogenesis in its early stage [30]. Likewise, the miR-124a CpG island is densely methylated in cancer cells, and restoration of miR-124a inhibits the expression of cyclin-dependent kinase CDK6, which leads to suppression of phosphorylation of the tumor suppressor Rb protein [21]. It has also been demonstrated that the miR-9 family CpG island is frequently methylated in various types of human malignancies, including colorectal cancer, and that miR-9-1 expression is associated with lymph node metastasis in these cancers [51]. Interestingly, epigenetic regulators, such as histone deacetylases (HDACs), DNA methyltransferases (DNMTs) and EZH2, can be direct targets of miRNAs, forming a complex regulatory network. For example, miR-342 has been shown to directly inhibit the expression of DNMT1 in colorectal cancer cells [31]. Given that miR-342 expression is itself partially regulated by promoter methylation, a double negative-feedback loop exists to maintain DNA methylation status. EZH2, a histone modification enzyme in polycomb repressive complex 2 (PRC2), is also regulated by miRNAs such as miR-101, miR-26a, and miR-124 [52], although these pathways have not been validated in colorectal cancer. These miRNAs, termed epi-miRNA, possibly change histone and DNA methylation status throughout whole genome, representing a new layer of gene regulation [53]. Taking into account that post-transcriptional regulation of miRNAs is itself an epigenetic process, miRNAs are considered to be an important component of a global epigenetic network (Fig. 1).

2.3 Genetic Variations in miRNA Genes and miRNA Binding Sites

Genetic variations, including single nucleotide polymorphisms (SNPs) and germline mutations in miRNA genes and their precursors, target sites, and genes encoding components of processing machinery, can affect phenotypic variation and disease susceptibility [54]. In consistent with the high conservation of miRNA loci, single

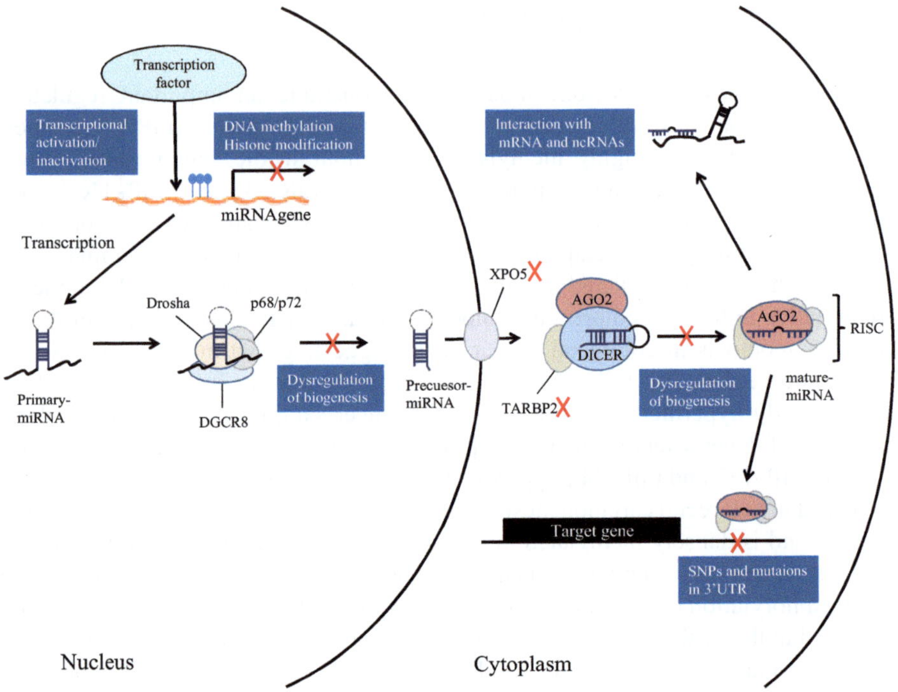

Fig. 1 Mechanisms of miRNA biogenesis, post-transcriptional gene regulation, and their alterations in cancer. Primary miRNA (pri-miRs) are originally transcribed from genome, and subsequently cleaved by endonuclease Drosha complex, which consists with Drosha, DGCR8 and auxiliary factors including p68 and p72, to generate 70-nt stem loop transcripts called precursor miRNAs (pre-miRNAs). Exportin 5 (XPO5) actively transports Pre-miR from the nucleus to the cytoplasm, where Dicer, another endonuclease, cleaves pre-miRNAs to create 22-nt mature form of miRNAs. Alterations of pri-miR transcription by oncogenes, tumor suppressors and epigenetic factors, lead to aberrant miRNA gene expressions. miRNA processing steps by Drosha and Dicer also interrupted by a variety of interacting factors. Inactivating mutations of *XPO5* and *TARBP2* are shown to result in the reduced miRNA processing efficiency. After loading to RNA-induced silencing complex (RISC), miRNAs inhibit their target mRNA expression by destabilization and translational repression. SNPs and mutations of miRNA genes and/or miRNA binding sites on target genes can lead to the attenuation of miRNA mediated gene suppression. Other non-coding RNAs and mRNAs, which have complementary sequence to miRNAs, can bind to them, and keep miRNAs from original target, thereby disturb the normal miRNA-mRNA interaction

nucleotide polymorphism (SNP) density in miRNA genes and miRNA binding sites is relatively lower [54]. However, genetic changes in these regions lead to critical cause of carcinogenesis, as in the case of protein-coding genes. The following four types of genetic alterations are associated with miRNA expression: SNPs in primary transcript of miRNA (pri-miR) and precursor miRNA (pre-miR), which can affect the processing efficiency, SNPs in miRNA genes (miR-SNPs), SNPs in miRNA binding sites on target genes, SNPs in genes that are involved in miRNA processing [54, 55].

One well-understood example of SNPs in the miRNA recognition element is LCS6 of the *KRAS* 3′ UTR, which is located on the let-7 binding element. The risk allele of this SNP, which enables *KRAS* to avoid being regulated by let-7, is associated with shorter survival in patients with advanced colorectal cancer. Of interest, this SNP has also been shown to affect drug resistance to the epidermal growth factor receptor (EGFR) inhibitor cetuximab in KRAS wild-type metastatic cases [56]. The rs712 polymorphism is another example of a polymorphism of the let-7-*KRAS* binding site, which is linked to the risk of advanced colorectal cancer [57]. Landi et al. investigated SNPs in the 3′ UTRs of 104 genes that are highly responsible for colorectal cancer and found that rs17281995 in the CD86 gene and rs1051690 in the INSR gene could affect the binding of miRNAs to their 3′ UTR region and were associated with colorectal cancer risk [58]. SNPs in miRNA processing factors, including DROSHA, DICER, RAN, GEMIN3, and GEMIN4, may also be associated with cancer risk [54]. However, these SNPs have not been investigated in large independent cohorts and needs to be further validated [54]. Recently identified alternative processing versions of miRNA, termed isomiR, represent a novel class of miRNA genetic variation [59]. Although the association between isomiR and human cancers needs to be further explored, several reports have demonstrated distinct distribution of isomiRs in cancer versus normal tissue, indicating its potential involvement in cancer pathogenesis [60, 61].

2.4 Alterations in miRNA Biogenesis

miRNA expression is regulated by its biogenesis process as well as by transcriptional control. Multistep cleavage by two types of endonucleases, Drosha and Dicer, is a key processing step in miRNA processing. Seminal work by Kumar et al. revealed that suppression of Drosha, Dicer, and cofactor for Drosha, DGCR8, promote cellular transformation and tumorigenesis [62], indicating that these processing factors generally exhibit a tumor-suppressive function. Also, truncating mutations in TARBP2, a cofactor for Dicer, are frequently found in sporadic and hereditary colorectal cancer with microsatellite instability (MSI) [63]. Impaired TARBP2 expression is shown to cause the destabilization of the DICER protein, which leads to blocking of the miRNA-generating process [63]. Likewise, XPO5, which plays a crucial role in nucleocytoplasmic transport of pre-miRNAs, possesses loss-of-function mutations in 22 % of colorectal cancer cases [64]. Recent evidence also showed that genetic ablation of *DICER1* confers tumor stemness properties, as well as an epithelial-to-mesenchymal transition (EMT) phenotype, to colorectal cancer cells [65]. Despite this tumor-suppressive behavior of miRNA processing factors in in vitro and animal model studies, Dicer expression at both the mRNA and protein level has been associated with advanced clinical stages and poor prognosis in colorectal cancer patients [66–68]. This finding requires further consideration that Dicer function is also affected by the activity of interacting molecules including AGO2 [69] and TARBP2 [70].

In addition to the alterations of miRNA processing factors, a wide variety of molecules are known to interfere with the miRNA biogenesis cascade. For example, tumor suppressor protein p53 interacts with binding to p68, a Drosha co-factor, and facilitates the processing of specific primary miRNAs, such as pri-miR-143, pro-miR-145, and pri-miR-16. These miRNAs have been shown to function as tumor suppressors and are downregulated in colorectal cancer [71], suggesting that p53 exerts its tumor-suppressive effect by regulating miRNA processing machinery. In another well-known example, EGFR was shown to interact with the RISC complex and inhibit miRNA maturation under hypoxic conditions. In hypoxia, EGFR directly binds to AGO2 protein and promotes AGO2-Y393 phosphorylation, which reduces the binding of Dicer to AGO2 and inhibits miRNA processing [69]. As observed in these examples, important signaling pathways can cross-talk with the miRNA-generating cascade and regulate miRNA expression in a transcription-independent manner.

Although the mechanism by which the subsets of specific miRNAs are recognized is not fully understood, there are evidence that terminal loop of the precoursor miRNAs are crucial for the selection of specific miRNAs. RNA binding proteins, such as hnRNP A1 [72], KSRP [73], and Lin-28 [74], have been shown to recognize the terminal loop of precursor miRNAs and to promote or impair miRNA biogenesis [8]. Lin-28, which is overexpressed in colorectal cancer, inhibits the processing of tumor-suppressive let-7 miRNA in two different mechanisms. Lin-28A recognizes the specific terminal loop sequences of precursor let-7 and recruits a uridylyl transferase to inhibit the processing by Dicer [75]. Lin-28B, however, directly binds to primary let-7 transcripts and prevents them from translocating into the cytoplasm, which blocks the maturation process [75].

2.5 Interactions with Other Non-coding RNAs (ncRNAs) and Protein-Coding Genes

Recent evidence suggests that highly abundant ncRNAs are transcribed from the whole genome [76], and some of them have been shown to interact with miRNAs. Seminal work from Croce group demonstrated that ncRNAs transcribed from genomic ultraconserved regions (UCRs) are regulated by miRNAs through direct binding in a sequence-dependent manner [77]. Identification of miRNA-ncRNA interaction leads to the concept that ncRNA serves as a miRNA 'sponge' and interrupts access to their targets for mRNA decay. For example, one of the recently identified special class of long ncRNAs, named *circular RNA* (circRNA), which contains more than 70 selectively conserved miRNA-7 target sites, directly binds miR-7 in a sequence-dependent manner, leading to deregulation of miR-7 targets [78, 79]. In another example, linc-RoR, which is highly expressed in pluripotent stem cells, shares miRNA-response elements with pluripotency-associated genes including Oct4, Sox2, and Nanog [80]. As a result, linc-RoR prevents these transcriptional factors from miRNA-mediated suppression [80]. These ncRNAs, termed *competing endogenous RNAs* (ceRNAs), can control the distribution of miRNAs on these targets, thereby exerting a substantial effect on protein output [81]. Tay et al. revealed that genes sharing the same miRNA binding site in their 3′ UTR absorb regulating miRNAs competitively and act as ceRNAs toward each other [82]. They validated their hypothesis by using the phosphatase and tensin homolog (PTEN), a negative regulator of PI3K/AKT signaling, and revealed that genes that share the same miRNA binding site with PTEN can affect protein expression and the tumor-suppressive function of PTEN. Importantly, these genes display concordant expression patterns with PTEN and copy number loss in colorectal cancer samples [82]. Although these mechanisms need to be further validated in a larger number clinical samples, miRNA-ceRNAs interactions could provide another layer of miRNA regulation.

3 Alterations of miRNAs in Signal Transduction

miRNAs target multiple components of essential signaling pathways, such as Wnt, EGFR, TGFβ, and p53 signaling, and serve as a crucial regulator of these pathways. Below we discuss the various roles of miRNAs in these signal transductions that are crucial for each step of colorectal carcinogenesis.

3.1 Canonical Wnt Signaling

Wnt signaling is known to be involved in a variety of biological process, including cell migration and proliferation and stem cell maintenance; Wnt signaling also plays a central role in the initiation of human malignancies [83]. High-throughput sequencing revealed that more than 90 % of colorectal cancer cases exhibit dysregulation of the Wnt signaling pathway [84]. APC, which regulates nuclear translocation of β-catenin, is a key regulator of Wnt signaling. Almost 80 % of MSI tumors and 50 % of non-MSI tumors possess mutations in the *APC* gene [84], and inactivation of APC and subsequent overstimulation of Wnt signaling is an early event in colorectal cancer carcinogenesis [85]. Multiple components of Wnt signaling, such as AXIN2, WNT, LRP6 and β-catenin are regulated by p53 induced miR-34a [16, 86]. This suggests p53 and Wnt signaling are interacting through miRNA regulation. β-catenin is also known to be suppressed by miR-200, which plays a pivotal role in EMT regulation [87]. Crucial tumor suppressor gene in Wnt signaling, *APC* are shown to be regulated by miR-135a/135b, regardless of the mutational status of APC [39]. Consistent with the early inactivation of APC in the carcinogenesis model, relatively high expression of miR-135b was observed in adenoma as well as in carcinoma, suggesting that posttranscriptional modification by miRNAs as well as genetic alteration is important to regulating APC expression. In another example, well-characterized tumor-suppressive miR-145 regulates c-MYC, which is a crucial Wnt signaling target, and inhibits the proliferation of colorectal cancer cells. miR-145 has also been shown to regulate SOX9 expression in various types of tissues

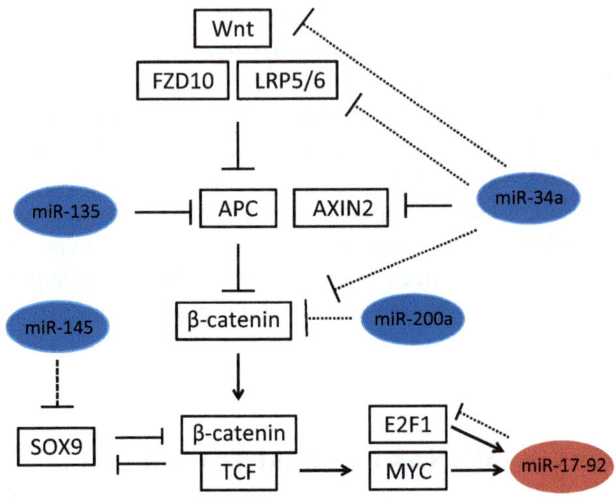

Fig. 2 Involvement of miRNAs in the dysregulation of canonical Wnt signaling. *Dotted lines* indicate pathways, which are not confirmed in colorectal cancer

including head and neck cancer [88], gliomas [89], and melanoma [90]. SOX9 is repressed by Wnt signaling, and conversely, facilitates β-catenin degradation [84].

Wnt signaling also promotes oncogenic miR-17-92a cluster expression through activation of transcriptional factors, such as MYC and E2F1 [14]. The miR-17-92a cluster is one of the most characterized oncogenic miRNAs, and its overexpression is reported in a variety of human malignancies including acute lymphocytic leukemia, chronic lymphocytic leukemia, lung, ovarian, gastric, liver, and colorectal cancers [91]. The miR-17-92a cluster exerts its oncogenic ability by inhibiting crucial tumor suppressors, such as cyclin-dependent kinase inhibitor CDKN1A p21, pro-apoptotic BCL2L11 (BIM), and PTEN [14, 27] (Fig. 2).

3.2 EGFR Signaling

EGFR signaling is involved in cell proliferation, migration, and survival, and plays a pivotal role in every step of cancer initiation and development. Both of the two major downstream signaling pathways RAS–RAF–mitogen-activated protein kinase (MAPK) and phosphatidylinositide 3-kinase (PI3K)–AKT cascades [92] have been shown to be significantly affected by miRNA expression. For example, tumor-suppressive miR-143 and the miR-145 cluster work as negative regulators of the MAPK pathway by inhibiting KRAS and BRAF, which leads to the suppression of proliferation and migration in colorectal cancer cells [26, 93]. KRAS is also directly targeted by tumor-suppressive let-7. This is supported by the fact that SNPs in let-7 and the KRAS binding site have been associated with increased risk of colorectal cancer [57]. KRAS has also been shown to upregulate expression of oncogenic miRNAs such as miR-210, miR-221/222, and miR-181a in colorectal cancer cells [42].

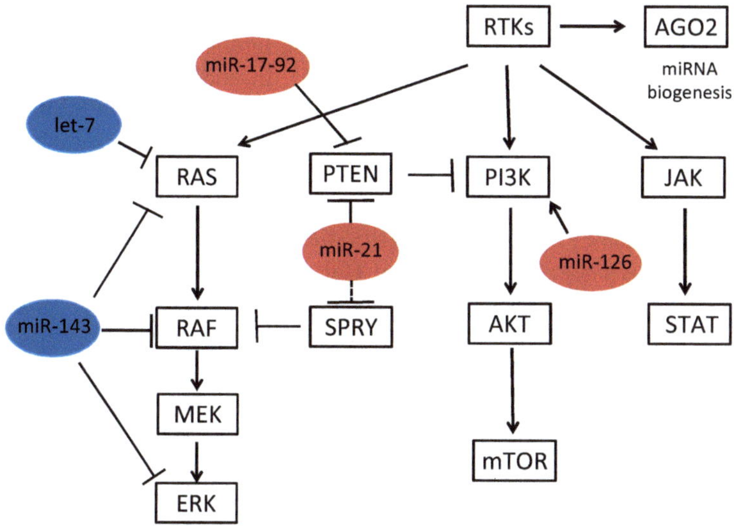

Fig. 3 Involvement of miRNAs in the dysregulation of EGFR signaling. *Dotted lines* indicate pathways, which are not confirmed in colorectal cancer

PTEN, a crucial negative regulator of the PI3K pathway, is directly targeted by oncogenic miR-21 and miR-17-92a cluster. miR-21 also works as a stimulator of the MAPK pathway through regulation of sprouty homologue 1 (SPRY1) [35], leading to strong activation of EGFR signal transduction. Consistent with the alterations of the EGFR signaling pathway in human cancer, miR-21 is upregulated in almost every kind of human malignancy, including colorectal cancer. Adenoma tissues exhibit higher expression of miR-21 compared with normal tissues [36], suggesting early involvement of miR-21 in carcinogenesis. miR-21 expression is associated with disease progression and shorter survival, as well as poor therapeutic response in colorectal cancer [36]. The PI3K pathway is also mediated by miR-126 expression, which regulates p85β, a stabilizing factor for PI3K, and therefore inhibits the proliferation of cancer cells [94] (Fig. 3).

3.3 *TGFβ Signaling*

TGFβ plays a crucial role in physiological and pathological processes including embryonic development, cell differentiation, cell survival, organ homeostasis, and initiation and metastasis of human malignancies. Frequent inactivation of TGFβ signaling in a variety of cancers supports the notion that TGFβ signaling generally works as a tumor suppressor in carcinogenesis. However, this signaling is also known to enhance the invasive ability of cancer cells by promoting epithelial-mesenchymal transition (EMT) and contributes to cancer metastasis. This indicates two distinct roles of TGFβ signaling in cancer development. The oncogenic miRNA cluster, miR-17-92a, affects TGFβ signaling via regulation of multiple components of this pathway. miR-17/20a directly inhibits transforming growth factor beta receptor II (TGFBR2), and miR-18a targets crucial signal transducer of TGFβ signaling, SMAD2 and SMAD4 [95]. In addition, miR-17-92a and its sequence homolog, miR-106b-25, coordinately target cell cycle inhibitor p21 and apoptotic BIM, which are important downstream targets of TGFβ signaling [14]. In colorectal cancer, miRNA-130a/301a/454 family has been shown to target SMAD4. Furthermore, these miRNAs were highly expressed and inversely correlated with SMAD4 expression in colorectal cancer tissues [95]. Oncogenic miR-21 regulates SMAD7, which is a negative regulator of TGF-β signaling, and thereby works as an activator of this signaling cascade [96].

The miR-200 family is a crucial regulator of TGFβ-induced EMT. miR-200 inhibits a downstream target of TGFβ signaling, ZEB1/ZEB2 [29], which results in derepression of epithelial marker CDH1 and interrupts the EMT program. In situ hybridization analysis of colorectal cancer tissues revealed reduced expression of miR-200c in invasive front of primary tumors, and overall miR-200c expression was downregulated in metastatic colorectal cancers compared with nonmetastatic cases [97]. Downregulation of the miR-200 family in cancer tissues is partially explained by DNA hypermethylation. Therefore, EMT phenotypes in tumor progression are possibly controlled by dynamic epigenetic regulation [98]. Interestingly, miR-200 family members miR-200a and miR-200b, and the related miR-429, are transcriptionally controlled by their target, ZEB1/ZEB2 [99], forming a double-negative-feedback loop, which could allow mutually exclusive expression patterns of epithelial and mesenchymal phenotypes [35] (Fig. 4).

3.4 p53 Signaling

p53, a central regulator of the cell cycle and cell survival, can transcriptionally modulate a subset of miRNA expressions. In 2007, several groups reported that miR-34a is a direct transcriptional target of p53 [16]. When DNA is damaged, p53 induces miR-34a, which causes cell cycle arrest by regulating crucial cell cycle regulators including CCNE2, CDK4, E2F3, and MYC [100]. miR-34a also induces apoptosis by targeting BCL-2, survivin, SNAIL, and SIRT1 [16]. Interestingly, one of the miR-34a targets, NAD-dependent deacetylase SIRT1, inhibits p53 protein expression by deacetylating, suggesting the existence of a positive-feedback loop between miR-34a and p53 [101]. p53 also binds to promoters of other tumor-suppressive miRNAs including miR-16, miR-200, miR-192, miR-215, and miR-145, all of which are reported to be downregulated in colorectal cancer. Importantly, miRNA-192 and miR-215 target MDM2, a crucial regulator of p53 protein expression [102]. P53 itself is regulated by miRNA. miR-125b directly binds to the 3′UTR of p53 mRNA and inhibit its expression. Overexpression of miR-125b was shown to be associated with poor prognosis in colorectal cancer cases [103].

p53 is also involved in EMT and stem cell properties by regulating miRNAs. miR-200c, which is transcriptionally induced by p53, has been shown to work as a negative regulator of the EMT phenotype and stem cell properties by inhibiting

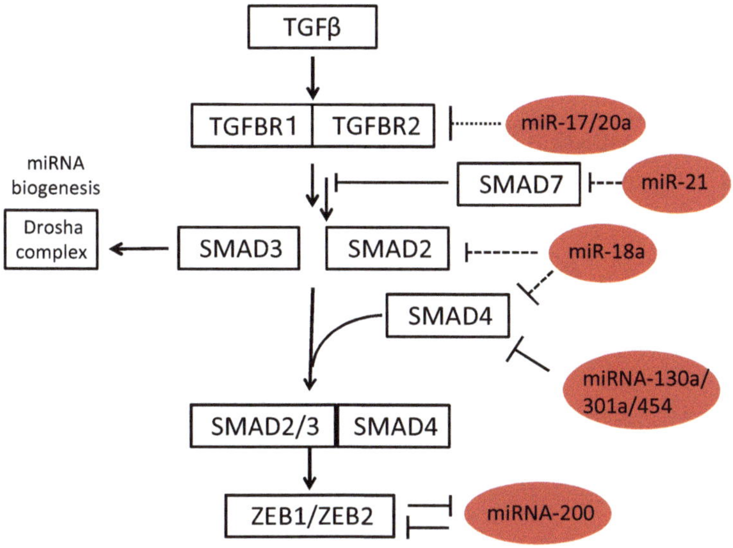

Fig. 4 Involvement of miRNAs in the dysregulation of TGFβ signaling. *Dotted lines* indicate pathways, which are not confirmed in colorectal cancer

ZEB1/ZEB2, pluripotency-related Krüppel-like factor 4 (KLF4), and the polycomb repressor BMI1 [104]. Mutated p53 competes with wild-type p53 in the regulation of EMT and stem cell markers [104]. Siemens et al. revealed that p53 downregulates the EMT-inducing transcription factor SNAIL via induction of the miR-34a/b/c genes in colorectal cancer cells [105]. Conversely, SNAIL inhibits miR-34a/b/c expression, indicating a double-negative feedback loop stabilizes EMT/MET status [105].

p53 expression also affects miRNA biogenesis. As mentioned in the previous section, p53 interacts with the Drosha processing complex by interacting with DEAD-box RNA helicase p68 and facilitates the processing of tumor-suppressive primary miRNAs to precursor miRNAs [71] (Fig. 5).

4 Conclusions and Perspectives

In this chapter, we have summarized complex epigenetic dysregulation by miRNAs as a critical mechanism underlying establishment of cancer. Alterations of protein-coding gene and miRNA expressions are like two wheels of a cart to lead carcinogenesis. Their interaction plays an important role in initiation, maintaining and enhancing the malignant potential of cancer cells. Importantly, the number of miRNA gene is relatively fewer compared with protein coding genes, and miRNAs regulate hundreds of targets simultaneously, suggesting they are crucial therapeutic targets in cancer treatment. Delivery methods of small oligonucleotide to each organ are being developed, and miRNA-targeted therapy has been increasingly

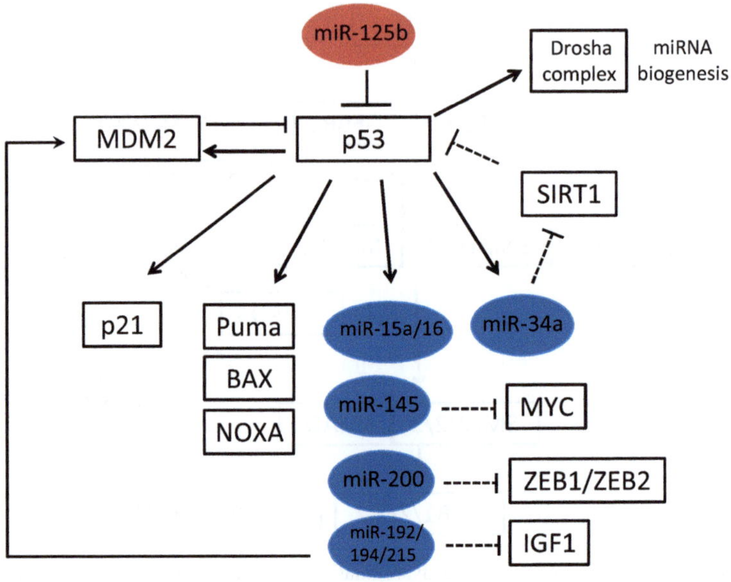

Fig. 5 Involvement of miRNAs in the dysregulation of p53 signaling. *Dotted lines* indicate pathways, which are not confirmed in colorectal cancer

recognized as a promising therapeutic strategy [106]. In addition, recent evidence suggest miRNAs are unexpectedly stable in a range of specimen types including formalin-fixed paraffin-embedded tissues, plasma/serum and other body fluids, and serves as a useful biomarkers in cancer diagnosis [107]. Conventional protein biomarkers in colorectal caner, such as CEA and CA19-9, exhibit high sensitivity for various types of cancers, but low tissue specificity for each cancer type. Given that miRNA profiles are extremely unique according to tissue types, and also differs between cancer and normal [12, 13], miRNAs could be potential biomarkers, which can detect each cancer types with high accuracy. Further understanding of complex miRNA-gene regulatory network can lead to effective clinical application of miRNAs as useful biomarkers and therapeutic targets.

References

1. Bartel DP (2009) MicroRNAs: target recognition and regulatory functions. Cell 136(2):215–233. doi:10.1016/j.cell.2009.01.002
2. Baek D, Villén J, Shin C, Camargo FD, Gygi SP, Bartel DP (2008) The impact of microRNAs on protein output. Nature 455(7209):64–71. doi:10.1038/nature07242
3. Ozsolak F, Poling LL, Wang Z, Liu H, Liu XS, Roeder RG et al (2008) Chromatin structure analyses identify miRNA promoters. Genes Dev 22(22):3172–3183. doi:10.1101/gad.1706508
4. Godnic I, Zorc M, Jevsinek Skok D, Calin GA, Horvat S, Dovc P et al (2013) Genome-wide and species-wide in silico screening for intragenic microRNAs in human, mouse and chicken. PLoS ONE 8(6):e65165. doi:10.1371/journal.pone.0065165

5. Lutter D, Marr C, Krumsiek J, Lang EW, Theis FJ (2010) Intronic microRNAs support their host genes by mediating synergistic and antagonistic regulatory effects. BMC Genomics 11:224. doi:10.1186/1471-2164-11-224
6. Petrocca F, Visone R, Onelli MR, Shah MH, Nicoloso MS, de Martino I et al (2008) E2F1-regulated microRNAs impair TGFbeta-dependent cell-cycle arrest and apoptosis in gastric cancer. Cancer Cell 13(3):272–286. doi:10.1016/j.ccr.2008.02.013
7. Krol J, Loedige I, Filipowicz W (2010) The widespread regulation of microRNA biogenesis, function and decay. Nat Rev Genetics 11(9):597–610. doi:10.1038/nrg2843
8. Suzuki HI, Miyazono K (2011) Emerging complexity of microRNA generation cascades. J Biochem 149(1):15–25. doi:10.1093/jb/mvq113
9. Spizzo R, Nicoloso MS, Croce CM, Calin GA (2009) SnapShot: microRNAs in cancer. Cell 137(3):586–586, e1. doi:10.1016/j.cell.2009.04.040
10. Calin GA, Sevignani C, Dumitru CD, Hyslop T, Noch E, Yendamuri S et al (2004) Human microRNA genes are frequently located at fragile sites and genomic regions involved in cancers. Proc Natl Acad Sci U S A 101(9):2999–3004. doi:10.1073/pnas.0307323101
11. Mathelier A, Carbone A (2013) Large scale chromosomal mapping of human microRNA structural clusters. Nucl Acids Res 41(8):4392–4408. doi:10.1093/nar/gkt112
12. Liang Y, Ridzon D, Wong L, Chen C (2007) Characterization of microRNA expression profiles in normal human tissues. BMC Genomics 8:166. doi:10.1186/1471-2164-8-166
13. Lu J, Getz G, Miska EA, Alvarez-Saavedra E, Lamb J, Peck D et al (2005) MicroRNA expression profiles classify human cancers. Nature 435(7043):834–838. doi:10.1038/nature03702
14. Petrocca F, Vecchione A, Croce CM (2008) Emerging role of miR-106b-25/miR-17-92 clusters in the control of transforming growth factor beta signaling. Cancer Res 68(20):8191–8194. doi:10.1158/0008-5472.CAN-08-1768
15. Kent OA, Fox-Talbot K, Halushka MK (2013) RREB1 repressed miR-143/145 modulates KRAS signaling through downregulation of multiple targets Oncogene 32(20):2576–2585. doi:10.1038/onc.2012.266
16. Hermeking H (2012) MicroRNAs in the p53 network: micromanagement of tumour suppression. Nat Rev Cancer 12(9):613–626. doi:10.1038/nrc3318
17. Chang T-C, Zeitels LR, Hwang H-W, Chivukula RR, Wentzel EA, Dews M et al (2009) Lin-28B transactivation is necessary for Myc-mediated let-7 repression and proliferation. Proc Natl Acad Sci U S A 106(9):3384–3389. doi:10.1073/pnas.0808300106
18. Johnson SM, Grosshans H, Shingara J, Byrom M, Jarvis R, Cheng A et al (2005) RAS is regulated by the let-7 microRNA family. Cell 120(5):635–647. doi:10.1016/j.cell.2005.01.014
19. Calin GA, Ferracin M, Cimmino A, Di Leva G, Shimizu M, Wojcik SE et al (2005) A MicroRNA signature associated with prognosis and progression in chronic lymphocytic leukemia. New Eng J Med 353(17):1793–1801. doi:10.1056/NEJMoa050995
20. Ofir M, Hacohen D, Ginsberg D (2011) MiR-15 and miR-16 are direct transcriptional targets of E2F1 that limit E2F-induced proliferation by targeting cyclin E. Mol Cancer Res 9(4):440–447. doi:10.1158/1541-7786.MCR-10-0344
21. Lujambio A, Ropero S, Ballestar E, Fraga MF, Cerrato C, Setién F et al (2007) Genetic unmasking of an epigenetically silenced microRNA in human cancer cells. Cancer Res 67(4):1424–1429. doi:10.1158/0008-5472.CAN-06-4218
22. Liu Y, Zhou Y, Feng X, An P, Quan X, Wang H et al (2014) MicroRNA-126 functions as a tumor suppressor in colorectal cancer cells by targeting CXCR4 via the AKT and ERK1/2 signaling pathways. Int J Oncol 44(1):203–210. doi:10.3892/ijo.2013.2168
23. Zhang Y, Wang X, Xu B, Wang B, Wang Z, Liang Y et al (2013) Epigenetic silencing of miR-126 contributes to tumor invasion and angiogenesis in colorectal cancer. Oncol Rep 30(4):1976–1984. doi:10.3892/or.2013.2633
24. Balaguer F, Link A, Lozano JJ, Cuatrecasas M, Nagasaka T, Boland CR, Goel A (2010) Epigenetic silencing of miR-137 is an early event in colorectal carcinogenesis. Cancer Res 70(16):6609–6618. doi:10.1158/0008-5472.CAN-10-0622
25. Liu M, Lang N, Qiu M, Xu F, Li Q, Tang Q et al (2011) miR-137 targets Cdc42 expression, induces cell cycle G1 arrest and inhibits invasion in colorectal cancer cells. Int J Cancer 128(6):1269–1279. doi:10.1002/ijc.25452

26. Pagliuca A, Valvo C, Fabrizi E, di Martino S, Biffoni M, Runci D et al (2013) Analysis of the combined action of miR-143 and miR-145 on oncogenic pathways in colorectal cancer cells reveals a coordinate program of gene repression. Oncogene 32(40):4806–4813. doi:10.1038/onc.2012.495
27. Goel A, Boland CR (2012) Epigenetics of colorectal cancer. Gastroenterology 143(6):1442–1460, e1. doi:10.1053/j.gastro.2012.09.032
28. Pichiorri F, Suh S-S, Rocci A, De Luca L, Taccioli C, Santhanam R et al (2010) Downregulation of p53-inducible microRNAs 192, 194, and 215 impairs the p53/MDM2 autoregulatory loop in multiple myeloma development. Cancer Cell 18(4):367–381
29. Gregory PA, Bert AG, Paterson EL, Barry SC, Tsykin A, Farshid G et al (2008) The miR-200 family and miR-205 regulate epithelial to mesenchymal transition by targeting ZEB1 and SIP1. Nat Cell Biol 10(5):593–601. doi:10.1038/ncb1722
30. Grady WM, Parkin RK, Mitchell PS, Lee JH, Kim Y-H, Tsuchiya KD et al (2008) Epigenetic silencing of the intronic microRNA hsa-miR-342 and its host gene EVL in colorectal cancer. Oncogene 27(27):3880–3888. doi:10.1038/onc.2008.10
31. Wang H, Wu J, Meng X, Ying X, Zuo Y, Liu R et al (2011) MicroRNA-342 inhibits colorectal cancer cell proliferation and invasion by directly targeting DNA methyltransferase 1. Carcinogenesis 32(7):1033–1042. doi:10.1093/carcin/bgr081
32. Guo ST, Jiang CC, Wang GP, Li YP, Wang CY, Guo XY et al (2013) MicroRNA-497 targets insulin-like growth factor 1 receptor and has a tumour suppressive role in human colorectal cancer. Oncogene 32(15):1910–1920. doi:10.1038/onc.2012.214
33. Ma L, Teruya-Feldstein J, Weinberg RA (2007) Tumour invasion and metastasis initiated by microRNA-10b in breast cancer. Nature 449(7163):682–688. doi:10.1038/nature06174
34. Nishida N, Yamashita S, Mimori K, Sudo T, Tanaka F, Shibata K et al (2012) MicroRNA-10b is a prognostic indicator in colorectal cancer and confers resistance to the chemotherapeutic agent 5-fluorouracil in colorectal cancer cells. Ann Surg Oncol. doi:10.1245/s10434-012-2246-1
35. Inui M, Martello G, Piccolo S (2010) MicroRNA control of signal transduction. Nat Rev Mol Cell Biol 11(4):252–263. doi:10.1038/nrm2868
36. Schetter AJ, Leung SY, Sohn JJ, Zanetti KA, Bowman ED, Yanaihara N et al (2008) MicroRNA expression profiles associated with prognosis and therapeutic outcome in colon adenocarcinoma. JAMA 299(4):425–436. doi:10.1001/jama.299.4.425
37. Fabbri M, Garzon R, Cimmino A, Liu Z, Zanesi N, Callegari E et al (2007) MicroRNA-29 family reverts aberrant methylation in lung cancer by targeting DNA methyltransferases 3A and 3B. Proc Natl Acad Sci U S A 104(40):15805–15810. doi:10.1073/pnas.0707628104
38. Zhou M, Liu Z, Zhao Y, Ding Y, Liu H, Xi Y et al (2010) MicroRNA-125b confers the resistance of breast cancer cells to paclitaxel through suppression of pro-apoptotic Bcl-2 antagonist killer 1 (Bak1) expression. J Biol Chem 285(28):21496–21507. doi:10.1074/jbc.M109.083337
39. Nagel R, le Sage C, Diosdado B, van der Waal M, Oude Vrielink, JAF, Bolijn A et al (2008) Regulation of the adenomatous polyposis coli gene by the miR-135 family in colorectal cancer. Cancer Res 68(14):5795–5802. doi:10.1158/0008-5472.CAN-08-0951
40. Li Z, Huang H, Li Y, Jiang X, Chen P, Arnovitz S et al (2012) Up-regulation of a HOXA-PBX3 homeobox-gene signature following down-regulation of miR-181 is associated with adverse prognosis in patients with cytogenetically abnormal AML. Blood 119(10):2314–2324. doi:10.1182/blood-2011-10-386235
41. Naguibneva I, Ameyar-Zazoua M, Polesskaya A, Ait-Si-Ali S, Groisman R, Souidi M et al (2006) The microRNA miR-181 targets the homeobox protein Hox-A11 during mammalian myoblast differentiation. Nat Cell Biol 8(3):278–284. doi:10.1038/ncb1373
42. Ota T, Doi K, Fujimoto T, Tanaka Y, Ogawa M, Matsuzaki H et al (2012) KRAS up-regulates the expression of miR-181a, miR-200c and miR-210 in a three-dimensional-specific manner in DLD-1 colorectal cancer cells. Anticancer Res 32(6):2271–2275.
43. Giannakakis A, Sandaltzopoulos R, Greshock J, Liang S, Huang J, Hasegawa K et al (2008) miR-210 links hypoxia with cell cycle regulation and is deleted in human epithelial ovarian cancer. Cancer Biol Ther 7(2):255–264

44. Huang X, Ding L, Bennewith KL, Tong RT, Welford SM, Ang KK et al (2009) Hypoxia-inducible mir-210 regulates normoxic gene expression involved in tumor initiation. Mol Cell 35(6):856–867. doi:10.1016/j.molcel.2009.09.006
45. Kelly TJ, Souza AL, Clish CB, Puigserver P (2011) A hypoxia-induced positive feedback loop promotes hypoxia-inducible factor 1alpha stability through miR-210 suppression of glycerol-3-phosphate dehydrogenase 1-like. Mol Cell Biol 31(13):2696–2706. doi:10.1128/MCB.01242-10
46. Medina R, Zaidi SK, Liu C-G, Stein JL, van Wijnen AJ, Croce CM, Stein GS (2008) MicroRNAs 221 and 222 bypass quiescence and compromise cell survival. Cancer Res 68(8):2773–2780. doi:10.1158/0008-5472.CAN-07-6754
47. Jia CY, Li HH, Zhu XC, Dong YW, Fu D, Zhao QL et al (2011) MiR-223 suppresses cell proliferation by targeting IGF-1R. PLoS ONE 6(11):e27008. doi:10.1371/journal.pone.0027008
48. Wu L, Li H, Jia CY, Cheng W, Yu M, Peng M et al (2012) MicroRNA-223 regulates FOXO1 expression and cell proliferation. FEBS Lett 586(7):1038–1043. doi:10.1016/j.febslet.2012.02.050
49. Suzuki H, Takatsuka S, Akashi H, Yamamoto E, Nojima M, Maruyama R et al (2011) Genome-wide profiling of chromatin signatures reveals epigenetic regulation of microRNA genes in colorectal cancer. Cancer Res 71(17):5646–5658. doi:10.1158/0008-5472.CAN-11-1076
50. Toyota M, Suzuki H, Sasaki Y, Maruyama R, Imai K, Shinomura Y, Tokino T (2008) Epigenetic silencing of microRNA-34b/c and B-cell translocation gene 4 is associated with CpG island methylation in colorectal cancer. Cancer Res 68(11):4123–4132. doi:10.1158/0008-5472.CAN-08-0325
51. Bandres E, Agirre X, Bitarte N, Ramirez N, Zarate R, Roman-Gomez J et al (2009) Epigenetic regulation of microRNA expression in colorectal cancer. Int J Cancer 125(11):2737–2743. doi:10.1002/ijc.24638
52. Suzuki H, Maruyama R, Yamamoto E, Kai M (2013) Epigenetic alteration and microRNA dysregulation in cancer. Front Genet 4(December):258. doi:10.3389/fgene.2013.00258
53. Fabbri M, Calin GA (2010) Epigenetics and miRNAs in human cancer. Adv Genet 70:87–99 (1st ed, Elsevier Inc.). doi:10.1016/B978-0-12-380866-0.60004-6
54. Ryan BM, Robles AI, Harris CC (2010) Genetic variation in microRNA networks: the implications for cancer research. Nat Rev Cancer 10(6):389–402. doi:10.1038/nrc2867
55. Yang R, Schlehe B, Hemminki K, Sutter C, Bugert P, Wappenschmidt B et al (2010) A genetic variant in the pre-miR-27a oncogene is associated with a reduced familial breast cancer risk. Breast Cancer Res Treat 121(3):693–702. doi:10.1007/s10549-009-0633-5
56. Zhang W, Winder T, Ning Y, Pohl A, Yang D, Kahn M et al (2011) A let-7 microRNA-binding site polymorphism in 3†²-untranslated region of KRAS gene predicts response in wild-type KRAS patients with metastatic colorectal cancer treated with cetuximab monotherapy. Annal Oncol 22(1):104–109. doi:10.1093/annonc/mdq315
57. Pan X-M, Sun R-F, Li Z-H, Guo X-M, Zhang Z, Qin H-J et al (2014) A let-7 KRAS rs712 polymorphism increases colorectal cancer risk. Tumour Biol 35(1):831–835. doi:10.1007/s13277-013-1114-3
58. Landi D, Gemignani F, Naccarati A, Pardini B, Vodicka P, Vodickova L et al (2008) Polymorphisms within micro-RNA-binding sites and risk of sporadic colorectal cancer. Carcinogenesis 29(3):579–584. doi:10.1093/carcin/bgm304
59. Lee LW, Zhang S, Etheridge A, Ma L, Martin D, Galas D, Wang K (2010) Complexity of the microRNA repertoire revealed by next-generation sequencing. RNA 16(11):2170–2180. doi:10.1261/rna.2225110
60. Chang H-T, Li S-C, Ho M-R, Pan H-W, Ger L-P, Hu L-Y et al (2012) Comprehensive analysis of microRNAs in breast cancer. BMC Genomics 13(Suppl 7):S18. doi:10.1186/1471-2164-13-S7-S18
61. Li S-C, Liao Y-L, Ho M-R, Tsai K-W, Lai C-H, Lin W (2012) miRNA arm selection and isomiR distribution in gastric cancer. BMC Genomics 13(Suppl 1):S13. doi:10.1186/1471-2164-13-S1-S13

62. Kumar MS, Lu J, Mercer KL, Golub TR, Jacks T (2007) Impaired microRNA processing enhances cellular transformation and tumorigenesis. Nat Genet 39(5):673–677. doi:10.1038/ng2003
63. Melo SA, Ropero S, Moutinho C, Aaltonen LA, Yamamoto H, Calin GA et al (2009) A TARBP2 mutation in human cancer impairs microRNA processing and DICER1 function. Nat Genet 41(3):365–370. doi:10.1038/ng317
64. Melo SA, Moutinho C, Ropero S, Calin GA, Rossi S, Spizzo R et al (2010) A genetic defect in exportin-5 traps precursor microRNAs in the nucleus of cancer cells. Cancer Cell 18(4):303–315. doi:10.1016/j.ccr.2010.09.007
65. Iliou MS, da Silva-Diz V, Carmona FJ, Ramalho-Carvalho J, Heyn H, Villanueva A et al (2013) Impaired DICER1 function promotes stemness and metastasis in colon cancer. Oncogene (August):1–13. doi:10.1038/onc.2013.398
66. Faber C, Horst D, Hlubek F, Kirchner T (2011) Overexpression of Dicer predicts poor survival in colorectal cancer. Eur J Cancer 47(9):1414–1419. doi:10.1016/j.ejca.2011.01.006
67. Papachristou DJ, Korpetinou A, Giannopoulou E, Antonacopoulou AG, Papadaki H, Grivas P et al (2011) Expression of the ribonucleases drosha, dicer, and ago2 in colorectal carcinomas virchows archiv. Int J Pathol 459(4):431–440. doi:10.1007/s00428-011-1119-5
68. Stratmann J, Wang C-J, Gnosa S, Wallin A, Hinselwood D, Sun X-F, Zhang H (2011) Dicer and miRNA in relation to clinicopathological variables in colorectal cancer patients. BMC Cancer 11(1):345. doi:10.1186/1471-2407-11-345
69. Shen J, Xia W, Khotskaya YB, Huo L, Nakanishi K, Lim S-O et al (2013) EGFR modulates microRNA maturation in response to hypoxia through phosphorylation of AGO2. Nature 497(7449):383–387. doi:10.1038/nature12080
70. Paroo Z, Ye X, Chen S, Liu Q (2009) Phosphorylation of the human microRNA-generating complex mediates MAPK/Erk signaling. Cell 139(1):112–122. doi:10.1016/j.cell.2009.06.044
71. Suzuki HI, Yamagata K, Sugimoto K, Iwamoto T, Kato S, Miyazono K (2009) Modulation of microRNA processing by p53. Nature 460(7254):529–533. doi:10.1038/nature08199
72. Michlewski G, Guil S, Semple CA, Cáceres JF (2008) Posttranscriptional regulation of miRNAs harboring conserved terminal loops. Mol Cell 32(3):383–393. doi:10.1016/j.molcel.2008.10.013
73. Trabucchi M, Briata P, Garcia-Mayoral M, Haase AD, Ramos A, Gherzi R, Rosenfeld MG (2009) The RNA-binding protein ksrp promotes the biogenesis of a subset of miRNAs. Nature 459(7249):1010–1014. doi:10.1038/nature08025.The
74. Viswanathan SR, Daley GQ, Gregory RI (2008) Selective blockade of microRNA processing by Lin28. Science 320(5872):97–100. doi:10.1126/science.1154040
75. Piskounova E, Polytarchou C, Thornton JE, Hagan JP, Lapierre J, Pothoulakis C et al (2012) Oncogenic Lin28A and Lin28B inhibit let-7 microRNA biogenesis by distinct mechanisms. Cell 147(5):730–748. doi:10.1016/j.cell.2011.10.039. Oncogenic
76. Rinn JL, Chang HY (2012) Genome regulation by long noncoding RNAs. Annu Rev Biochem 81:145–166. doi:10.1146/annurev-biochem-051410-092902
77. Calin GA, Liu C, Ferracin M, Hyslop T, Spizzo R, Sevignani C et al (2007) Ultraconserved regions encoding ncRNAs are altered in human leukemias and carcinomas. Cancer Cell 12(3):215–229. doi:10.1016/j.ccr.2007.07.027
78. Hansen TB, Jensen TI, Clausen BH, Bramsen JB, Finsen B, Damgaard CK, Kjems J (2013) Natural RNA circles function as efficient microRNA sponges. Nature 495(7441):384–388. doi:10.1038/nature11993
79. Memczak S, Jens M, Elefsinioti A, Torti F, Krueger J, Rybak A et al (2013) Circular RNAs are a large class of animal RNAs with regulatory potency. Nature 495(7441):333–338. doi:10.1038/nature11928
80. Wang Y, Xu Z, Jiang J, Xu C, Kang J, Xiao L et al (2013) Endogenous miRNA sponge lincRNA-RoR regulates Oct4, Nanog, and Sox2 in human embryonic stem cell self-renewal. Dev Cell 25(1):69–80. doi:10.1016/j.devcel.2013.03.002
81. De Giorgio A, Krell J, Harding V, Stebbing J, Castellano L (2013) Emerging roles of ceRNAs in cancer: insights from the regulation of PTEN. Mol Cell Biol. doi:10.1128/MCB.00683-13

82. Tay Y, Kats L, Salmena L, Weiss D, Tan SM, Ala U et al (2011) Coding-independent regulation of the tumor suppressor PTEN by competing endogenous mRNAs. Cell 147(2):344–357. doi:10.1016/j.cell.2011.09.029
83. Klaus A, Birchmeier W (2008) Wnt signalling and its impact on development and cancer Nat Rev Cancer 8(5):387–398. doi:10.1038/nrc2389
84. Muzny DM, Bainbridge MN, Chang K, Dinh HH, Drummond JA, Fowler G et al (2012) Comprehensive molecular characterization of human colon and rectal cancer. Nature 487(7407):330–337. doi:10.1038/nature11252
85. Fearon ER, Vogelstein B (1990) A genetic model for colorectal tumorigenesis. Cell 61(5):759–767
86. Kim NH, Cha YH, Kang SE, Lee Y, Lee I, Cha SY et al (2013) p53 regulates nuclear GSK-3 levels through miR-34-mediated Axin2 suppression in colorectal cancer cells. Cell Cycle 12(10):1578–1587. doi:10.4161/cc.24739
87. Mongroo PS, Rustgi AK (2010) The role of the miR-200 family in epithelial-mesenchymal transition. Cancer Biol Ther 10(3):219–222
88. Yu C-C, Tsai L-L, Wang M-L, Yu C-H, Lo W-L, Chang Y-C et al (2013) miR145 targets the SOX9/ADAM17 axis to inhibit tumor-initiating cells and IL-6-mediated paracrine effects in head and neck cancer. Cancer Res 73(11):3425–3440. doi:10.1158/0008-5472.CAN-12-3840
89. Rani SB, Rathod SS, Karthik S, Kaur N, Muzumdar D, Shiras AS (2013) MiR-145 functions as a tumor-suppressive RNA by targeting Sox9 and adducin 3 in human glioma cells. Neuro Oncol 15(10):1302–1316. doi:10.1093/neuonc/not090
90. Dynoodt P, Speeckaert R, De Wever O, Chevolet I, Brochez L, Lambert J, Van Gele M (2013) miR-145 overexpression suppresses the migration and invasion of metastatic melanoma cells. Int J Oncol 42(4):1443–1451. doi:10.3892/ijo.2013.1823
91. Mogilyansky E, Rigoutsos I (2013) The miR-17/92 cluster: a comprehensive update on its genomics, genetics, functions and increasingly important and numerous roles in health and disease. Cell Death Different 20(12):1603–1614. doi:10.1038/cdd.2013.125
92. Pao W, Chmielecki J (2010) Rational, biologically based treatment of EGFR-mutant non-small-cell lung cancer. Nat Rev Cancer 10(11):760–774. doi:10.1038/nrc2947
93. Mlcochova J, Faltejskova P (2013) MicroRNAs targeting EGFR signalling pathway in colorectal cancer. J Cancer Res Clin Oncol 1615–1624. doi:10.1007/s00432-013-1470-9
94. Guo C, Sah JF, Beard L, Willson JKV, Markowitz SD, Guda K (2008) The Noncoding RNA, miR-126, suppresses the growth of neoplastic cells by targeting phosphatidylinositol 3-kinase signaling and is frequently lost in colon cancers. Genes Chromosome Cancer 946(April):939–946. doi:10.1002/gcc
95. Mestdagh P, Boström A-K, Impens F, Fredlund E, Van Peer G, De Antonellis P et al (2010) The miR-17-92 microRNA cluster regulates multiple components of the TGF-β pathway in neuroblastoma. Mol Cell 40(5):762–773. doi:10.1016/j.molcel.2010.11.038
96. Li Q, Zhang D, Wang Y, Sun P, Hou X, Larner J et al (2013) MiR-21/Smad 7 signaling determines TGF-β1-induced CAF formation. Sci Rep 3:2038. doi:10.1038/srep02038
97. Hur K, Toiyama Y, Takahashi M, Balaguer F, Nagasaka T, Koike J et al (2013) MicroRNA-200c modulates epithelial-to-mesenchymal transition (EMT) in human colorectal cancer metastasis. Gut 62(9):1315–1326. doi:10.1136/gutjnl-2011-301846
98. Davalos V, Moutinho C, Villanueva A, Boque R, Silva P, Carneiro F, Esteller M (2012) Dynamic epigenetic regulation of the microRNA-200 family mediates epithelial and mesenchymal transitions in human tumorigenesis. Oncogene 31(16):2062–2074. doi:10.1038/onc.2011.383
99. Bracken CP, Gregory PA, Kolesnikoff N, Bert AG, Wang J, Shannon MF, Goodall GJ (2008) A double-negative feedback loop between ZEB1-SIP1 and the microRNA-200 family regulates epithelial-mesenchymal transition. Cancer Res 68(19):7846–7854. doi:10.1158/0008-5472.CAN-08-1942
100. He L, He X, Lim LP, de Stanchina E, Xuan Z, Liang Y et al (2007) A microRNA component of the p53 tumour suppressor network. Nature 447(7148):1130–1134. doi:10.1038/nature05939

101. Yamakuchi M, Ferlito M, Lowenstein CJ (2008) miR-34a repression of SIRT1 regulates apoptosis. Proc Natl Acad Sci U S A 105(36):13421–13426. doi:10.1073/pnas.0801613105
102. Braun CJ, Zhang X, Savelyeva I, Wolff S, Moll UM, Schepeler T et al. (2008) p53-Responsive micrornas 192 and 215 are capable of inducing cell cycle arrest. Cancer Res 68(24):10094–10104. doi:10.1158/0008-5472.CAN-08-1569
103. Nishida N, Yokobori T, Mimori K, Sudo T, Tanaka F, Shibata K et al (2011) MicroRNA miR-125b is a prognostic marker in human colorectal cancer. Int J Oncol 38(5):1437–1443. doi:10.3892/ijo.2011.969
104. Chang C-J, Chao C-H, Xia W, Yang J-Y, Xiong Y, Li C-W et al (2011) p53 regulates epithelial-mesenchymal transition and stem cell properties through modulating miRNAs. Nat Cell Biol 13(3):317–323. doi:10.1038/ncb2173
105. Siemens H, Jackstadt R, Hünten S, Kaller M, Menssen A, Götz U, Hermeking H (2011) miR-34 and SNAIL form a double-negative feedback loop to regulate epithelial-mesenchymal transitions. Cell Cycle 10(24):4256–4271. doi:10.4161/cc.10.24.18552
106. Garzon R, Marcucci G, Croce CM (2010) Targeting microRNAs in cancer: rationale, strategies and challenges. Nat Rev Drug Discov 9(10):775–789. doi:10.1038/nrd3179
107. Schwarzenbach H, Nishida N, Calin GA, Pantel K (2014) Clinical relevance of circulating cell-free microRNAs in cancer. Nat Rev Clin Oncol. doi:10.1038/nrclinonc.2014.5

Microbiome Implications in Intestinal Tumorigenesis

David Posocco, Oxana Dmitrieva and Sergei I. Grivennikov

Abstract Host inflammatory responses are essential for tumor growth and progression. Multiple mechanisms by which inflammation affects cancer have recently come to light and questions of how inflammatory reactions are induced in the tumors are currently under scrutiny. There is increasing understanding that various infectious agents as well as commensal components of host microbiome are important regulators of normal and tumor-associated inflammatory responses. Here we review recent findings on quantitative and qualitative composition of bacteria, its properties and its ability to activate various arms of the host immune system to promote or suppress inflammation and tumorigenesis. We also summarize recent discoveries on molecular mechanisms underlying this newly emerging connection between microbiota, inflammation and cancer. This review primarily describes colitis-associated and spontaneous colon cancers, but the general mechanisms described here are likely to apply to other cancers, which arise at epithelial surfaces colonized with bacteria, such as skin or gastrointestinal cancers; or to those cancers in which potential exposure to either microbial-derived and other inflammation-inducing stimuli is possible, such as liver and lung cancers.

Keywords Cancer · Inflammation · Bacteria · Cytokines · Microbiome · Signaling pathways · Immune cells · Dysbiosis

1 Introduction

Chronic inflammation has been linked to a variety of autoimmune diseases and auto-inflammatory disorders. The presence of immune and inflammatory cells (leukocytes) within tumors was first observed in the nineteenth century by Rudolf Virchow and provided the first indication of a now much more widely accepted

---

S. I. Grivennikov (✉) · D. Posocco · O. Dmitrieva
Cancer Prevention and Control Program, Fox Chase Cancer Center, 333 Cottman Avenue, Philadelphia, PA 19111, USA
e-mail: sergey.grivennikov@fccc.edu

V. W. Yang, A. B. Bialkowska (eds.), *Intestinal Tumorigenesis*,
DOI 10.1007/978-3-319-19986-3_7

phenomenon [1, 2]. An inflammatory microenvironment is an essential component of all tumors, including some in which a direct causal relationship with inflammation has not yet been proven [2–5]. Hence, in some types of cancer (probably in 20–25 % of them) chronic inflammation precedes tumor development and therefore is a common suspect in disease initiation [6, 7]. In other tumors, inflammation can be induced during tumor growth and progression, or during anti-tumor therapy, and thus may be dispensable for tumor initiation but quite essential for growth, metastasis and recurrence [8].

If we assume that our immune system has been evolutionary selected and shaped to protect us against cancer just as it does from pathogens, then this seemingly 'strange' tumor-promoting function of the immune system may appear puzzling. However, it is important to realize that the process of inflammation exists to protect the body by isolating the damaged area, attracting immune cells to the site, eliminating possible invading pathogens and, in later stages, promoting the wound healing of affected tissues. As a diversion of the process, in some cases, the body can sustain a long-term inflammatory state in response to a lingering, low-grade infection (or even commensal bacteria) like in cases of inflammatory bowel diseases (IBD), such as ulcerative colitis (UC) and Crohn's disease (CD) [9, 10]. These inflammatory conditions may later on predispose an individual to colon cancer development (to its variation called colitis-associated cancer (CAC)) [11], similar to how *Helicobacter pylori*-induced gastritis predisposes one to stomach cancer, and how chronic viral, obesity-related or alcohol-induced hepatitis promotes the development of hepatocellular carcinoma (HCC) the most common form of liver cancer [12–14]. By now, the field of cancer predisposition by inflammation has been extensively reviewed and many examples for a particular organ site and cancer type exist in the literature [2–5].

2 Inflammation and Cancer: What Is the Mechanistic Connection?

Most, if not all, solid tumors have infiltrates of immune and inflammatory cells and exhibit enhanced production of a variety of pro-inflammatory cytokines and chemokines [2, 15, 16]. The hallmarks of cancer are important features required for tumor growth, progression and metastasis [17]. Interestingly, the six cornerstone hallmarks of cancer proposed in 2000 by Hanahan and Weinberg are almost entirely cancer cell autonomous characteristics, including resistance of cancer cells to cell death and tumor suppression evasion, increased proliferation capacity and ability to sustain telomere length, as well as the induction of invasion and metastasis and de novo angiogenesis [17]. However, we now can appreciate that inflammatory reactions in the tumor are now added as a distinct hallmark of cancer [18]. Moreover, an increasing number of reports suggests that inflammation can act to initiate, mimic or enhance each and every hallmark of cancer, by a variety

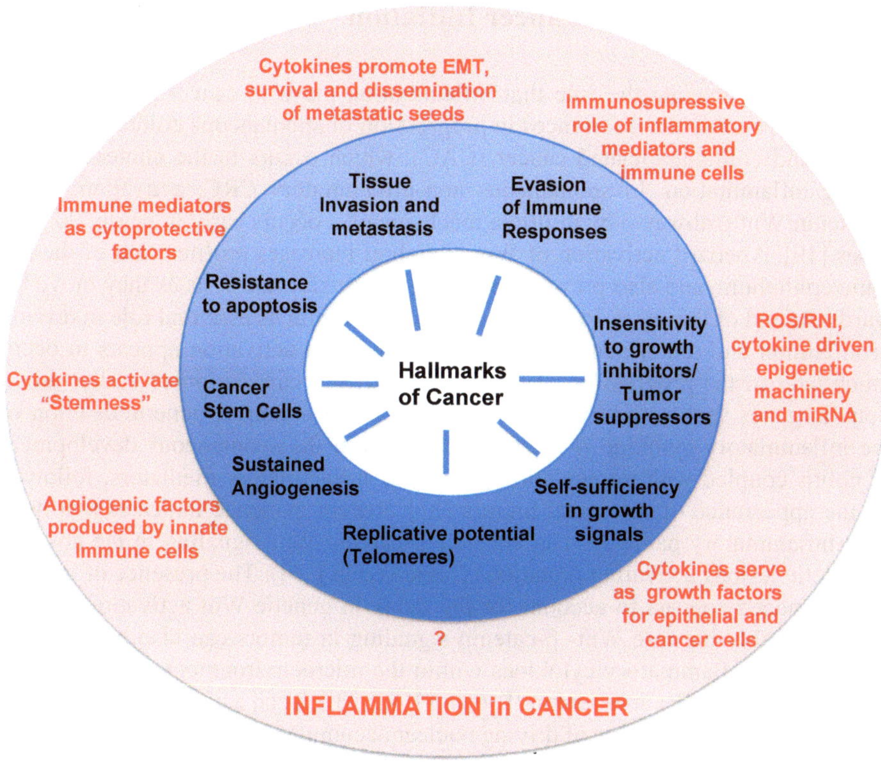

Fig. 1 This figure illustrates that virtually all proposed Hallmarks of Cancer [17, 18] can be modulated or significantly enhanced by cancer cell extrinsic mechanisms attributed to inflammation. Examples of how inflammation may affect virtually every defined Hallmark of Cancer (in *blue*) are presented (in *pink*). The only example where the role of inflammation is not yet established is potential ability to control telomere length

of mechanisms (Fig. 1). For instance, cytokines can serve as growth factors and protect cancer cells from apoptosis, thereby enabling cancer cells with resistance to cell death and uncontrolled proliferation. Chemokines and cytokines produced by intra-tumoral immune cells can activate angiogenesis. Reactive oxygen (ROS) and nitrogen species (RNI) produced by immune cells can further trigger mutagenesis and the ability of cytokine signaling to activate epigenetic and miRNA-based pathways of gene silencing is essential for oncogenic mutations and epigenetic loss of tumor suppressors. Importantly, various factors produced by immune cells, in addition to the cells themselves, aid invasion, metastatic spread and survival of cancer cells during the process of metastasis (by forming cell-to-cell contacts) (further reviewed in [8, 19]).

3 Inflammation and Cancer Initiation

In order to appreciate the role that inflammation plays in cancer initiation, it is helpful to examine the differences in progression of spontaneous colorectal cancer (CRC) and colitis-associated cancer (CAC), which occurs in the context of preceding inflammation. In spontaneous non-inflammatory CRC, activation of the β-catenin/Wnt pathway—by multiple mechanisms—occurs early in nearly 90% of cases [19]. Aberrant activation of Wnt signaling increases proliferation of the colonic epithelium, and also prevents shedding of transformed cells as they move towards the end of the crypt. Rather paradoxically, despite its essential role in driving proliferation and tumorigenesis, mutation-driven Wnt activation appears to occur much later in the progression of CAC—only after accumulation of earlier mutations in genes such as *p53* and *KRAS* [19, 20]. For example, genetic deletion of the inflammatory cytokine *Il-10* in mice results in the spontaneous development of colitis coupled with upregulation of various inflammatory mediators, followed by the appearance of colorectal tumors [21]. Recent evidence indicates that multiple inflammatory pathways can enhance β-catenin/Wnt signaling in the absence of genetic lesions activating mutations (reviewed in [19]). The presence of inflammation indeed appears to account for the delay in genetic Wnt activation seen in CAC, probably because Wnt- β-catenin signaling in tumors can also be activated by abundant inflammatory cytokines within the microenvironment, such as tumor necrosis factor (TNF), interleukin IL-6 and IL-1, which can activate pathways such as NF-κB and AKT, capable of driving nuclear accumulation of β-catenin (reviewed in [19]). Thus, the presence of inflammation diminishes the early requirement for Wnt activating mutations in spontaneous CRC by enhancing extrinsically driven β-catenin signaling. Through the activation of compensatory pathways capable of driving β-catenin accumulation, inflammation may increase susceptibility to colorectal cancer by providing an alternative mechanism, without requirement for genetic alterations, by which tumor initiation can commence. While the aforementioned mechanism may be essential, the ability of immune and inflammatory cells to trigger mutagenesis in epithelial and cancer cells by virtue of production of ROS and RNI, should not be discounted [19, 22–24]. Furthermore, inflammatory cytokines produced in the microenvironment, have been shown to activate DNA and histone-modifying epigenetic programs, as well as transcription of different species of microRNA's, which can contribute to knocking down the expression of essential tumor-suppressors, leading to subsequent tumor initiation [19, 25–27].

4 Injury, Cell Death and Excessive Regeneration may underlie Inflammation Driven Tumor Promotion and Growth

Local inflammation, in response to tissue damage and injury, is a vital mechanism required for repair and regeneration of the damaged area [28]. However, inflammation in response to chronic tissue damage has many parallels that may implicate it in both tumor promotion and initiation [1, 28]. Shortly after injury, immune cells migrate to the affected area via chemokine gradients and promote healing by releasing cytokines such as TNF, IL-1, and IL-6. These cytokines support the healing and regenerative processes by preventing further cell death, activating stem cells, and promoting epithelial proliferation [1, 19]. Importantly, tumor-associated immune cells often also express this cytokine profile; and, not coincidentally, tumors were named as "wounds that never heal" [28]. The mechanism by which cell death induces cancer proceeds as follows: although the majority of cells constituting the bulk of the tissue die in the process, the death of some pre-malignant cells is prevented by ensuing inflammation and their proliferation is further induced as the tissue regenerative process commences.

While tumor necrosis as a result of hypoxic conditions at a tumor's core may be perceived as an "injury" that can drive tumor-elicited inflammation, recurring tissue injury and regeneration may play an important role early in the initial stages of tumor development. In the context of repeated injury, infiltrating immune cells trigger the proliferation of intestinal stem cells in order to commence regeneration of damaged tissue. However, when this process becomes chronic, and exposure to both internal and external carcinogens occurs concurrently, mutated and eventually transformed cells survive and contribute to the newly regenerated tissue—providing a mechanism by which chronic wound healing can contribute to tumor initiation [1, 19].

Necrotic cell death at a tumor's core can promote inflammation by a number of routes, and while some likely remain to be discovered, one important contributor is the release of danger associated molecular patterns (DAMPs) or "danger signals" by necrotic cells [29]. DAMP's are endogenous proteins, such as HMGB1 and the S100A8 and 9 proteins, which upon release are recognized by and subsequently activate immune cells such as neutrophils, macrophages and dendritic cells [1, 30–33]. DAMPs also influence epithelial and cancer cells directly, in addition to their role as mediators of the immune response [1]. The release of DAMPs by necrotic cells thus aids in shaping the pro-inflammatory tumor microenvironment, and provides yet another mechanism able to act in parallel with pro-inflammatory cytokines, mediating immune networks in such a way as to exacerbate intestinal tumorigenesis [1, 19, 34].

5 Microbial Influence on Inflammation and Cancer

The majority of all cancers (>90%) are caused by somatic mutations and are strongly influenced by a variety of environmental factors [35]. Among these factors, the influence of bacteria and other microorganisms appears to be especially important for cancers that arise at epithelial surfaces. Several lines of evidence—based on a range of findings, from epidemiological studies of patients (20–25% of cancer preceded by chronic inflammation and infection) to molecular studies of genetically modified mice—have led to a general acceptance that inflammation (often microbial driven) and cancer are linked [2, 5, 7, 8, 36–38]. As discussed above, there are many potential distinct triggers of chronic inflammation that increase the

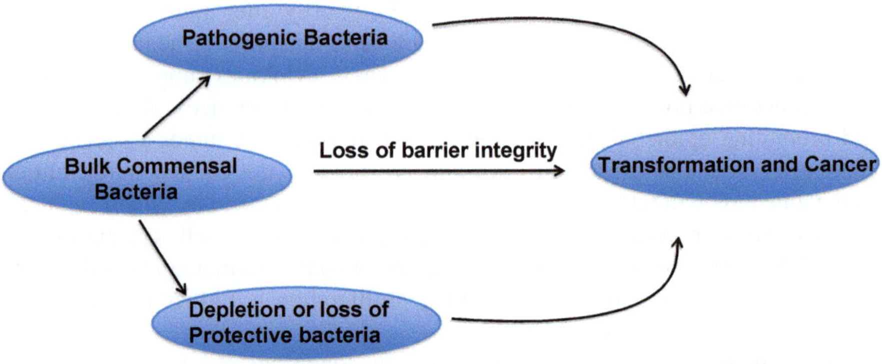

Fig. 2 This figure illustrates the multifaceted role of specific and "bulk" components of microbiome in CRC tumorigenesis. First, distinct pathogenic or conditionally-pathogenic bacteria can directly drive Transformation and Cancer. Potentially, the loss of yet to be defined "cancer protective" members of community may result in the same effect. Due to the tumor-specific loss of protective epithelial barriers, "bulk" commensal microbiota can drive inflammation and tumorigenesis in CRC. In addition, "bulk" commensal microbiota may form and/or occupy microecological niches for pathogenic or protective species, thereby indirectly regulating their abundance and effect on tumorigenesis

risk of developing cancer. Such triggers include microbial infections, autoimmune diseases (also often distantly controlled by microbial-derived stimuli) and inflammatory conditions of unknown origin (for example, prostatitis is associated with prostate cancer).

Host-microbe interactions occur continuously and are vital for maintaining a wide array of indispensable biological functions [39–42]. Colonization with commensal microbes, although mutually beneficial to both species at the time, presented the host with a difficult evolutionary problem: how to tolerate commensal and often beneficial microbes while maintaining the ability to identify and eliminate potentially harmful pathogens? Elaborate mechanisms have evolved towards this end; however, when this delicate balance is tipped and the commensal flora is altered, it appears that disease often follows. Colonization with commensal microbiota, and the species-specific make-up of the colonizing populations, is especially important in the digestive tract, where the aggregate number of bacterial genes and cells outnumbers all human cells and genes by at least a factor of 10 [43]. Whether an altered microbiome leads to disease or is a product of the disease itself is often not clear and likely depends on the context and the specific disease in question (Fig. 2). Recent studies provide substantial evidence that both the microbiome at large as well as specific, pathogenic microbial strains likely play a key role in colon cancer, mediating its progression in both direct and indirect ways [33, 34, 44–46]. In most instances, however, the microbiota exerts its deleterious (or protective) effect by one of three distinct mechanisms: by direct interaction with colonic epithelial cell (IECs), by production of immune and epithelial-modulatory metabolites, or by stimulation of unwanted, often pro-inflammatory immune responses (Fig. 3).

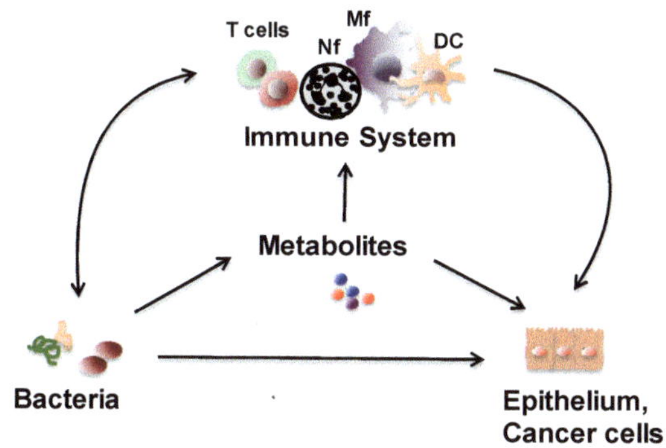

Fig. 3 This figure illustrates the highly interactive cross talk that occurs between the host and the colonic microbial community. Three mechanisms of microbial influence are clearly defined: (1) Ability of the microbiome to interact and influence directly IECs, (2) The production of metabolites, which in turn influences both the immune system and IECs, and (3) Modulation of, and interaction with, the immune system by bacteria (Mf-macrophages, Nf-neutrophils, DC-dendritic cells)

6 Colon Cancer: Bacteria Mediate Tumorigenesis Through Direct Interaction with Intestinal Epithelial Cells (IECs)

Although the microbiome is now accepted as an important factor in exacerbating and, perhaps, initiating colonic tumorigenesis, the intricate mechanistic details of this contribution remain unresolved. It is still unclear whether chronic infection with specific microbes is sufficient to cause cancer, or if perhaps a single microbe (or a collection of microbial species with specific distinct properties) only set in motion a series of alterations in the microbiome at large—all of which contribute to eventual tumorigenesis (Fig. 2). While multiple mechanisms exist where the microbiome can influence tumorigenesis, cases in which bacteria have a direct effect on IECs provide insight into the selective advantage afforded to these microbes in the context of cancer.

One example of a microbe that exerts its influence, at least in part, by acting directly on IECs is provided by the example of enterotoxigenic *Bacteroides fragilis* (ETBF). *B. fragilis* is a common commensal microbe, present in an estimated 30–70 % of individuals [11, 47]. Although present in a large proportion of the population *B. fragilis* comprises a small minority (only 1–2 %) of the bacterial community at large [47]. *B. fragilis* is distinguished by its ability to secrete a 20 kDa zinc-dependent metalloprotease toxin, known as the *B. fragilis* toxin (BFT) [47]. It is this toxin, and its ability to cleave E-cadherin on the surface of host epithelial cells, that makes ETBF a formidable microbe in contributing to intestinal tumorigenesis.

E-cadherin is a structural protein that aids in the formation of tight-junctions between colonic epithelial cells, thus its cleavage results in increased intestinal permeability [47]. As E-cadherin cleavage disrupts tight-junctions, it also alters β-catenin/

Wnt signaling; consequently stimulating proliferation and migration of intestinal epithelial cells (IEC) while simultaneously increasing barrier permeability [48]. The β-catenin/Wnt pathway is up-regulated in nearly all colon cancers; therefore BFT's ability to drive intestinal tumorigenesis is two-fold: it is capable of up-regulating Wnt signaling, at the same time also increases the translocation of bacterial products leading to subsequent immune cell infiltration and activation of unwanted inflammatory programs.

The importance of E-cadherin in mediating the pro-tumorigenic effect of certain microbes is further illustrated by *Fusobacterium nucleatum* (Fn). Fn is a commensal anaerobe that predominately occupies the oral cavity; however, it opportunistically disseminates throughout the body via its ability to attach to and invade epithelial cells [49]. This capacity, to adhere to and invade certain cell types, is made possible via adhesion FadA, a virulence factor that, in its mature form, binds E-cadherin and mediates the invasion of Fn into specific cell types [50]. The binding of mature FadA to E-cadherin has been shown to increase the internalization of its active, phosphorylated form and thus to play a role in modulating β-catenin/Wnt signaling providing a possible mechanism by which Fn drives intestinal tumorigenesis [50].

It remains unclear whether the FadA/E-cadherin interaction is solely responsible for the pro-tumorigenic properties of Fn, or whether invasion of IECs with Fn plays a separate, distinct role in addition to the modulation of Wnt signaling. Several studies have shown that Fn is vastly enriched in colonic adenomas relative to surrounding normal tissue, and, additionally, that FadA expression is up-regulated anywhere from 10–100 fold in these same samples [49, 50]. While it appears that, theoretically, Fn possesses the ability to initiate tumorigenesis directly via modulating Wnt signaling, its presence in colon tumors may be probabilistic potentially only occurring once adenomas have already developed by alternative mechanisms.

Under "normal" conditions, Fn only periodically takes up successful residence in other anatomical locations outside the oral cavity. In order for Fn to colonize the intestinal tract, it must first be swallowed and successfully transverse the acidic (pH 1) environment, a tough task without reliable frequency. Moreover, once Fn does transverse the stomach and intestines, it is normally barred from actually attaching to and invading IECs, as thick layers of intestinal mucous presumably block access of mature FadA to E-cadherin. Fn can therefore only realistically bind IECs in locations where the intestinal mucosa has been degraded primarily in regions where inflammation and tumorigenesis have already commenced. Thus, it is possible that Fn may preferentially bind to pre-cancerous lesions and act only to exacerbate tumorigenesis, as prior initiation serves as a prerequisite for its successful attachment and invasion.

As illustrated in the above example, the ability of preceding inflammation to alter the commensal flora by giving specific, often pathogenic strains of bacteria a local, selective advantage can often play a profound role in driving intestinal tumorigenesis. Altered colonization with the commensal adhesive strain of *Escherichia coli* (named NC101) provides yet another example of this phenomenon. Adherent-invasive *E. coli,* which fall within the Enterobacteriaceae family, have been shown to be upregulated to a significant extent in *Il10*$^{-/-}$murine models of intestinal inflammation [51]. In addition, when the carcinogen azoxymethane (AOM) is administered to germ-free (GF) *Il10*$^{-/-}$ mice mono-associated with *E. coli* NC101, the mice develop invasive adenocarcinomas. This result, however, is not seen in the same model mono-associated with the commensal *Enterococcus faecalis* OG1RF—even though both commensals are able to induce chronic inflammation in *Il10*$^{-/-}$ mice, which potentially precedes tumor development [33, 51, 52].

This indicates that *E. coli* NC101 must possess a distinct, carcinogenic function not possessed by *E. faecalis*—as chronic inflammation, on its own, is rarely capable of inducing carcinogenesis. Importantly *E. coli* NC101 has a 54 kb polyketide synthases (pks) pathogenicity island, which encodes machinery capable of producing the genotoxin—Colibactin. This toxin has been shown to induce double-stranded DNA-breaks in mammalian cells, as well as cell-cycle arrest and often, cell death [51]. It is likely that the ability to produce this genotoxic element provides *E. coli* NC101 with its pro-carcinogenic capacity when compared to other commensal pathogens. While inflammation itself does not appear to have a pro-carcinogenic capacity, its ability to shape the microbiome and to select for the expansion of strains with the ability to damage genomic DNA certainly may.

In addition to the specific strains discussed above, gram-negative bacteria, in general, possess the ability to influence IECs through production of lipopolysaccharide (LPS). IECs also express pattern recognition receptor (PRR)s to recognize and respond to intestinal microbes through chemokines secretion and cell homeostasis regulation. Experiments on mice that have IEC-specific knockout of signals downstream of PRRs demonstrated that IECs have a fundamental role in intestinal immune homeostasis and responses to pathogens [53], as well as in the pathogenesis of CAC [54]. Recent studies have shown that Toll-like receptor 4 (TLR4)—the receptor for gram-negative LPS—is overexpressed in nearly a third of sporadic colorectal cancer cases [55]. In addition to correlative human studies, direct causality has been shown using transgenic Villin-TLR4 transgenic mice, which express *TLR4* specifically in the intestinal epithelium. In a AOM-induced model of CRC, these mice exhibited increased epithelial proliferation, and an increase in tumor load when compared to wild-type (WT) mice [55]. Moreover, activation of β-catenin and Wnt signaling was upregulated indicating that signaling through TLR4 by LPS can induce tumorigenesis through the canonical Wnt pathway [55].

The examples presented above clearly illustrate that certain bacteria possess the unique ability to influence tumorigenesis by direct interaction with IECs. However, the overall contribution of this mechanism to the aggregate influence of the microbiome at large is difficult to ascertain. Attempting to account for the contribution of separate mechanisms may be futile, as both direct and indirect effects likely act synergistically and in parallel. Nonetheless, it is important to examine and to understand fully the distinct pathways by which the microbiome can modulate tumorigenesis.

7 Colon Cancer: Bacteria Act on Immune Cells and Cause Chronic Inflammation, Thereby Promoting Cancer

While we have discussed the direct effect of bacteria on epithelial and cancer cells, also an alternative and formidable possibility is that bacteria can modulate the immune system to promote inflammation capable of driving the development of cancer (Fig. 3). The high density of microbes in the intestinal lumen leads to continuous communication between host cells and microbes. These interactions can be mutually beneficial or can have adverse effects and contribute to intestinal inflammation. The intestinal immune system must, on the one hand, defend against pathogens and entry of excessive intestinal microbes and, on the other hand, allow tolerance to resident intestinal microbes [56]. The most frequently found immune cells within the tumor microenvironment are tumor-associated macrophages (TAMs), and, in more general terms, tumor-associated myeloid cells, including various subsets of neutrophils, macrophages, monocytes, and dendritic cells. Among the cellular components of adaptive immunity, the most prominent are various subsets of T cells. TAMs primarily promote tumor growth and may be obligatory for angiogenesis, invasion, and metastasis [57, 58]. High TAM content also generally correlates with poor prognosis [59]. Enterocytes (IECs) can produce enzymes, such as intestinal alkaline phosphatase and acyloxyacyl hydrolase that modify LPS to reduce its inflammatory effects [60, 61]. Paneth cells produce antimicrobial proteins such as defensins and lectins [62] and goblet cells produce mucins, trefoil peptides and resistin-like molecule β (RELMβ) [63–66]. Altogether, these mechanisms would decrease the ability of sentinel myeloid cells to recognize bacteria. Microbe sensing is required for the optimal regulation of antimicrobial proteins, which in turn, are mediated through PRR pathways, as demonstrated in studies of MyD88- and Nod2-deficient mice [67, 68]. Likewise, MyD88-dependent recognition of commensal microbiota by tumor myeloid cells regulates the production of pro-tumorigenic cytokines IL-23, IL-17 and IL-6 [69]. The interaction of bacteria and bacterial-derived products with mucosal immune cells sets in motion a series of both chronic and acute inflammatory responses. These responses likely play a vital role in exacerbating intestinal tumorigenesis and, as such, a selection of representative examples is explored below.

8 Microbiota-Bacteria Driven Promotion of Pro-inflammatory, Pro-tumorigenic Pathways

Recently, two bacterial pathogens have been identified that promote colitis-associated colorectal cancer. Enterotoxigenic *Bacteroides fragilis* induces colitis and colonic tumors in $Apc^{Min/+}$ mice by triggering a T helper type 17 (Th17) inflammatory response [48], while adherent-invasive *Escherichia coli* strain NC101 promotes colitis-associated colorectal cancer in monocolonized, azoxymethane-injected *Il10*$^{-/-}$ mice [51]. These two models allow comparing effect on epithelial and immune cells. However, the majority of human colorectal cancers do not arise in the setting of inflammatory bowel disease. The gut microbiota may be a driver in colorectal cancers that are not associated with colitis. Metagenomic analyses using whole-genome sequencing [50], transcriptome sequencing [70], or bacterial 16S ribosomal RNA (rRNA) gene DNA sequencing [71, 72] have shown enrichment of *Fusobacterium* species in colorectal cancers relative to adjacent normal tissue. *Fusobacterium nucleatum* may be able to persist and slowly replicate in hypoxic tumor microenvironment. Adhesive molecules that contribute to invasiveness in Fn can promote bacterial aggregation and biofilm formation that also enhance oxygen tolerance [73]. Alternatively, it is selected by the absence of mucus layer and requirements for strong adhesive properties in colon adenomas and carcinomas. In response to the presence of Fusobacteria attached to IEC's, several immune cell type (macrophages, DC, neutrophils, monocytes) may be recruited, and that would make the tumor microenvironment more tumor permissive over time by directly promoting tumor cell proliferation or immune cell infiltration. It was shown that an increased pro-inflammatory gene expression signature is associated with the accumulation and abundance of *Fusobacterium* in CRC tumors [49, 50].

However, an important and, as of yet, unanswered question remains: namely, whether bacterial-induced inflammation is sufficient to cause CRC progression? As discussed previously, although *E. coli* NC101 and *E. faecalis* are both able to induce identical levels of inflammation in experimental models, only the genotoxic *E. coli* strain exacerbates intestinal tumorigenesis [51]. These results indicate that although inflammation may be a necessary mediator of tumorigenesis, it is likely not sufficient to exacerbate tumor growth without accompanying contributions from either genotoxic microbiota or carcinogenic environmental conditions. Thus, while an altered microbiota or specific expansion in pro-carcinogenic strains appears to exacerbate tumorigenesis in the context of inflammation, inflammation appears equally dependent on the microbiota for its pro-tumorigenic capacity.

While certain microbes may drive tumorigenesis through unique adhesive properties or genotoxic capabilities, gram-negative bacteria, in general, may influence tumorigenesis through their production of LPS. LPS is only one example of a bacterial-derived product capable of activating the host immune system. It is very likely that other bacterial products, particularly those to be sensed by TLR or NOD PRRs, also contribute to the induction of inflammatory responses and tumorigenesis. In fact, recent studies have implicated innate PRRs as vital contributors to the initiation of pro-tumorigenic inflammatory programs. When bone marrow from *Myd88*$^{-/-}$ or *Tlr2,4,9*$^{-/-}$ mice is transplanted into lethally irradiated WT mice—so that innate myeloid cells in recipients lack the requisite components to signal through TLRs—colorectal tumor growth, as well as intratumoral Il-6 and Il-17 expression, are reduced when compared to TLR-competent WT controls [69]. While the above case appears to demonstrate that signaling through innate PRRs, such as TLRs, contributes detrimentally to intestinal tumorigenesis, similar ablation some of intracellular Nod-like receptors [NLRs] (and thus, inflammasome assembly) appears to have the opposite effect—actually exacerbating tumorigenesis.

Inflammasomes are intracellular multi-protein complexes comprised of specific NLRs, pro-caspase-1, and often an adaptor protein (ASC) necessary for recognition of pro-caspase-1 by NLRs [74]. Signaling through inflammasomes can be activated through a diverse range of cell stress and microbial danger signals; activation leads to the subsequent cleavage of pro-caspase-1 and to the maturation, by proteolytic cleavage, of the pro-inflammatory cytokines IL-1β and IL-18 [74]. Somewhat paradoxically, due to its involvement in inducing inflammatory responses, ablating the ability to assemble inflammasomes in mice appears to exacerbate tumorigenesis.

In an AOM-DSS model of CAC, *Asc*$^{-/-}$ mice, which lack the ability to assemble most inflammasomes, exhibit a significantly greater degree of tumorigenesis than do WT mice [74]. Interestingly, this increased susceptibility to tumorigenesis can be transferred by co-housing WT mice with *Asc*$^{-/-}$ mice—implicating the microflora as potentially responsible for the exacerbation of tumor growth seen in *Asc*$^{-/-}$ mice alone. The phenotype observed in *Asc*$^{-/-}$ mice, as well as in co-housed WT [*Asc*$^{-/-}$] mice, was abrogated upon treatment with broad-spectrum antibiotics—lending more credence to the idea that the altered microflora may be the causative agent responsible for inducing increased tumorigenesis.

In this same model, experiments with *Il-1β* and *Il-18* knockouts indicate that IL-18, but not IL-1β, is involved in the induction of the aberrant microflora, as WT mice co-housed with *Il18*$^{-/-}$ mice exhibited a similar degree of tumorigenesis as seen in WT [*Asc*$^{-/-}$] mice [74]. Ultimately, signaling through the chemokine Ccl5 with subsequent up-regulation in Il-6 production was found to mediate the aberrant phenotype observed in inflammasome-deficient (*Asc*$^{-/-}$) mice [74]. It thus appears that inflammasome-deficient mice lack the ability to maintain normal biosis in the intestine, and that altered microbiota is ultimately the causative agent responsible for exacerbating tumorigenesis (Fig. 2). Importantly, while opposite to the ameliorated phenotype observed in the triple TLR-knockout mice, these results serve to illustrate that the crosstalk between innate immune cells and the intestinal microbiota is a complex, context-dependent phenomenon with profound implications for intestinal tumorigenesis.

Distinct microbial species can preferentially accumulate in tumors and drive local inflammation and CRC tumorigenesis, the question is whether many other commensal microbes are capable of doing the same. Depletion of commensal microflora with broad-spectrum antibiotics results in reduction of CRC tumorigenesis in different mouse models [69, 75, 76] when mice are housed in specific pathogen free facility. Arguably, many bacteria, which can closely attach to the tumor area, would be able to shed off microbial products (such as LPS) and activate local inflammatory response in CRC tumors because these tumors lack adequate protective mucus coverage and display defective tight junctions [69]. This drives pro-inflammatory and pro-tumorigenic cytokine responses, which are initially regulated by IL-23, and subsequently, executed by the pro-inflammatory, IL-23-dependent cytokines IL-17, IL-22, IL-6, and IL-11 [69, 77–79].

Accumulating evidence clearly implicates bacterial-induced inflammation as a central factor in driving intestinal tumorigenesis; however, the diverse range of mechanisms through which inflammation regulates tumorigenesis are only recently coming to light. In addition to the IL-23 and IL-6 pathways, it also appears that inflammation may influence cancer through up-regulation of local nitric oxide (NO) concentrations. Studies have shown that *Rag2*$^{-/-}$ mice infected with *Helicobacter hepaticus* serve as a useful model for inflammation-driven colorectal cancer, and were thus chosen to study the effect of NO production on cancer progression [80]. *Rag2*$^{-/-}$ injected with *H. hepaticus* exhibited a time-dependent increase in nitrate production when compared to uninfected mice, which correlated well with the time-course exacerbation of both IBD and cancer symptoms. Importantly, this increase in nitrate production was not observed in infected WT mice, indicating that injection with *H. hepaticus* alone is not sufficient to drive increased nitrate production or IBD progression [80]. Local increases in NO concentration were found to result from the up-regulation of inducible NO synthase (iNOS), primarily by infiltrating Gr1^{+} neutrophils—as evidenced by the fact that depletion of Gr1^{+} cells with a neutralizing antibody served to decrease iNOS expression as well as the exacerbation of intestinal tumorigenesis seen in infected controls. This phenotype could be similarly abrogated by way of systemic treatment with the iNOS inhibitor NMA—indicating that local NO production is infected a causative agent in producing the tumorigenic phenotype observed in indeed *Rag2*$^{-/-}$ mice [80]. Increased expression of Tnfα and Il-6 also correlated with increased iNOS expression, and treatment with a Tnfα neutralizing antibody served to decrease the exacerbation of tumorigenesis seen in untreated controls. It thus appears that bacterial infection may increase TNFα expression, leading to subsequent up-regulation of iNOS expression and ultimately to a pro-tumorigenic inflammatory microenvironment [80]. This data indicates that bacterial-induced modulation of host immune programs is complex and that multiple avenues exist by which inflammation can ultimately influence and exacerbate intestinal tumorigenesis.

9 Direct vs Indirect Role of Bacteria: Role of Bacterial Metabolites in Tumorigenesis

The co-evolution of the host-microbe interaction had a number of positive results for both species: the ability to extend the metabolic capacity of humans and thus the range of viable food sources being key among these. While commensal bacteria aid in human (animal) metabolism in terms of nutrient extraction and digestion, it also becomes increasingly clear that these bacterial-derived metabolites play additional roles in maintaining host immune homeostasis (Fig. 3) [81]. For instance, it has been shown that short-chain fatty acids (SCFAs), products of the bacterial breakdown of dietary fiber (one of the components of a "healthy ration"), have the capacity to influence host gene expression by a number of different routes and modulate host inflammatory responses [82]. While it appears that multiple SCFAs possess biological relevance, butyrate has been recently implicated immunoregulation and its anti-inflammatory and anti-tumorigenic properties were uncovered in mouse models of colitis, CAC [81]. Multiple studies have illustrated butyrate's ability to act as a histone-deactylase (HDAC) inhibitor, interestingly both in epithelial/cancer cells and in various subsets of immune cells in health and disease, thereby affecting expression levels of thousands of genes [81, 82]. A recent study demonstrated that increased concentrations of butyrate correlate with an increase in H3 acetylation pattern of the *Foxp3* gene locus—an important transcription factor that acts as a master regulator governing T-regulatory (Treg) cell differentiation [82]. By inhibiting deacetylases, butyrate increases the acetylation pattern at the *Foxp3* locus and thus also increases *Foxp3* expression, resulting in an increase in the proportion of fully differentiated Treg cells and potentially pacing down inflammatory responses. Butyrate is able to induce the differentiation of naïve T cells into Treg's both in vitro and in vivo. When homeostasis is altered, and Treg differentiation is diminished, the severity of colon cancer—measured both in tumor load and number—increases rather drastically [81, 82]. While butyrate appears to exert its effect, at least in part, by way of its histone deacetylase activity—it is also able to signal directly through G-protein coupled receptor—Gpr109 [82]. Both mechanisms ultimately lead to an increase in Foxp3 expression and Treg differentiation; this phenotype can be abrogated either by altering fiber content of the diet or by directly knocking out the receptor Gpr109 [81, 82], with concurrent increase in production of pro-tumorigenic cytokine Il-17a [81, 82]. The loss of Treg cell differentiation diminishes immune system's ability to suppress tumor-promoting inflammatory programs in the colon, as evidenced by the increase in Il-17 production seen in models lacking butyrate production or Gpr109 [81, 82].

Whereas SCFAs, and butyrate in particular, are the best researched bacterial metabolites implicated in colorectal cancer, aggregate metabolome analysis has revealed many additional metabolites that are altered in adenoma tissue when compared to normal tissue. Among these metabolites, prostaglandin E2 (PGE2) was most significantly increased in adenoma cases [11]. PGE2 is synthesized from the precursor arachidonic acid by cyclooxygenase enzymes (COX1 and COX2) [83]. PGE2 is capable of signaling through phosphoinositide 3-kinase as well as AKT, both of which have been shown to promote tumor growth by activating proliferation, migration, abnormal apoptosis, and angiogenesis [11]. In addition to PGE2, the antioxidant related metabolites, 5-oxoproline and diketogulonic acid, varied significantly between adenoma and normal tissue [11]. Thus, it appears that alterations in microbial metabolites—whether by modulating immune function, by direct influence on IECs or by creating an environment conducive to DNA damage—play a key role in maintaining homeostasis and preventing the bacterial-driven exacerbation of intestinal tumorigenesis.

10 The Role of Dysbiosis in Colorectal Cancer

Inflammation, as a result of both endogenous and environmental triggers, can drive dysbiosis. While the expansion of specific microbes with pro-carcinogenic capabilities undoubtedly contributes to carcinogenesis, other factors inherent to dysbiosis itself likely contribute as well (Fig. 2). The selective advantage afforded to opportunistic, pro-inflammatory pathogens such as *Fusobacterium*, allows for the displacement of essential, commensal microbes involved in maintaining intestinal homeostasis [49]. This shift in microbiome composition is seen in correlative studies, where populations of "protective" bifidobacteria and the butyrate-producing strains *F. prausnitzii* and *Roseburia*, are depleted at the expense of expansion in pro-carcinogenic populations [84]. The pro-carcinogenic effect of these specific pathogenic microbes is thus exacerbated by the simultaneous loss of homeostasis-promoting commensals (Fig. 2). Much of these mechanisms are not yet clearly demonstrated in experimental system, but are highly suspected.

Comparative analyses of the microbiome from cancerous tissue and matched normal tissue (as well as stool [luminal] microbiome of normal controls and cancer bearing patients and experimental animals) have revealed an overall loss of microbial diversity in diseased tissue, which is characterized by a distinct pro-inflammatory profile [83]. As previously mentioned, it is difficult to ascertain causation concerning dysbiosis; specifically, whether the loss in microbial diversity and the expansion of certain pro-carcinogenic microbes is the cause or merely the product of intestinal inflammation and tumorigenesis is unknown. One hypothesis concerning this dilemma, known as the "bacterial driver-passenger model", defines two distinct stages in the progression of dysbiosis [83]. Initially, pro-carcinogenic microbes termed 'drivers' contribute to epithelial breakdown, the accumulation of DNA damage, and ultimately to the initiation of colorectal cancer. Subsequently, as a result of initiation, the local microenvironment is altered in such a way as to provide competitive advantages to certain microbes, termed "passengers", which expand in population and contribute to the continued shaping of a pro-tumorigenic microenvironment [83]. Thus, according to this logic, by preventing dysbiosis and by re-establishing homeostasis once alterations have already occurred, one may be able to decrease the severity and prevalence of intestinal tumorigenesis.

One non-invasive method for shaping the microbiome currently being investigated is the use of both pre- and probiotics. Probiotics are often preparations containing viable microbes that exert a beneficial effect on the host, while prebiotics—often comprised of non-digestible carbohydrates such as fructo-oligosaccharides—are meant to selectively promote the growth and expansion of beneficial commensals [49]. The use of probiotics may exert its effect by selective exclusion of pro-carcinogenic or pathogenic microbes. Studies have shown that administration of products such as fermented milk, containing *Lactobacillus* among other beneficial microbes, aids in reducing inflammation in mouse-models of colitis and can increase the abundance of butyrate-producing commensals in the stool, which again would produce anti-inflammatory and anti-carcinogenic butyrate [49].

The exact mechanism by which probiotics selectively exclude pro-carcinogenic microbes is unclear, but likely has to do with competition for nutrients and for the specific increased capacity to adhere to colonic epithelial cells or their neoplastic and malignant counterparts. In vitro studies have demonstrated that *L. acidophilus L casei and L. planetarium* all exhibit extremely high adhesiveness to the colonic epithelium, and are thus able to outcompete and reduce adhesion of the pathogenic

Salmonella typhimurium [49]. It is also possible that probiotics may exert a beneficial effect by modulating the activity of certain bacterial enzymes. Bacterial enzymes such as β-glucuronidase and nitroreductase have the ability to modify dietary metabolites into often mutagenic and thus carcinogenic compounds. By outcompeting commensals with high enzyme activity (i.e. *Clostridium spp., B. fragilis, and E.coli*), probiotics can aid in reducing the production of carcinogenic metabolites in colon [34]. Although still in its infancy, the tailored use of specific pre- and probiotic may hold significant promise in the future prevention, and possibly pretreatment, of intestinal tumorigenesis.

11 Conclusions

The recent accumulation of evidence clearly indicates that the microbiome is inextricably linked to multiple stages of intestinal tumorigenesis. Although at this point in time no specific microbial composition is mechanistically linked in a causal manner to the development of colon cancer, certain important trends have emerged. A decrease in microbial diversity, with the concurrent expansion in certain pro-carcinogenic species such as genotoxic *E. coli,* appears to reliably lead to an exacerbation of colon tumorigenesis. Although global dysbiosis is clearly present during CRC development and even likely involved in the etiology of colon cancer, the exact mechanisms behind this observation remain to be elucidated. Most importantly, it remains unclear whether the expansion in specific pro-carcinogenic strains is sufficient to initiate and drive tumorigenesis, or whether the aggregate influence of a globally-altered microflora is necessary in addition to the outgrowth (or depletion) of a few select strains. However, given all the data concerning the microbial influence on tumorigenesis, one factor in particular, inflammation, seems central in practically all microbial-linked mechanisms driving tumorigenesis [85, 86].

In rather simplistic terms, roughly three pathways exist by which the microbiota can exert influence over intestinal tumorigenesis; specifically, the microbiota can interact with and directly influence IECs, bacterial-derived metabolites can modulate inflammatory programs and, in some cases, act directly as mutagenic compounds, and the microbiota can influence and modulate crosstalk with the mucosal immune system—modulating immune responses so as to exacerbate tumorigenesis. Although a range of distinct mechanisms exist by which the microbiota can influence colon cancer, it appears that inflammation may be the central element on which all of these disparate mechanisms ultimately depend.

The role that inflammation plays in mediating the effect of the microbiota on colon tumorigenesis appears complex and context dependent. Preceding inflammation may provide selective advantages to certain pro-carcinogenic strains and ultimately drive global dysbiosis. In other cases, often once dysbiosis has already been established, the altered microbiota is capable of influencing host immune programs in such as way as to upregulate established inflammatory responses. Subsequently, a positive feedback loop may commence, wherein inflammation selects for an altered microbiota, which itself is then capable of inducing further inflammation—providing an environment even more hospitable to selection of an altered pro-tumorigenic microflora.

Whether dysbiosis is the cause or the result of preceding inflammation may be an irrelevant question, as both are likely able to influence and to exacerbate one another. However, recognizing the roles that both dysbiosis and inflammation play in exacerbating intestinal tumorigenesis once established, may provide promising future avenues for potential interventions. Positive preliminary data have already been seen with the use of tailored pre- and pro-biotic regimens as well as with dietary modulation.

Acknowledgments Supported by the FCCC-Temple University Nodal grant and by the Pew Scholar in Biomedical Sciences Program to S.G.

References

1. Kuraishy A, Karin M, Grivennikov SI (2011) Tumor promotion via injury- and death-induced inflammation. Immunity 35:467–477
2. Mantovani A, Allavena P, Sica A, Balkwill F (2008) Cancer-related inflammation. Nature 454:436–444
3. Coussens LM, Zitvogel L, Palucka AK (2013) Neutralizing tumor-promoting chronic inflammation: a magic bullet? Science 339:286–291
4. Hanahan D, Coussens LM (2012) Accessories to the crime: functions of cells recruited to the tumor microenvironment. Cancer Cell 21:309–322
5. Trinchieri G (2012) Cancer and inflammation: an old intuition with rapidly evolving new concepts. Annu Rev Immunol 30:677–706
6. Grivennikov SI, Karin M (2010) Inflammation and oncogenesis: a vicious connection. Curr Opin Genet Dev 20:65–71
7. Mantovani A (2010) Molecular pathways linking inflammation and cancer. Curr Mol Med 10:369–373
8. Grivennikov SI, Greten FR, Karin M (2010) Immunity, inflammation, and cancer. Cell 140:883–899
9. Strober W (2013) Impact of the gut microbiome on mucosal inflammation. Trends Immunol 34:423–430
10. Strober W, Fuss I, Mannon P (2007) The fundamental basis of inflammatory bowel disease. J Clin Invest 117:514–521
11. Sears CL, Garrett WS (2014) Microbes, microbiota, and colon cancer. Cell Host Microbe 15:317–328
12. Eaden JA, Mayberry JF, British Society for G, Association of Coloproctology for Great B, and Ireland (2002) Guidelines for screening and surveillance of asymptomatic colorectal cancer in patients with inflammatory bowel disease. Gut 51(Suppl 5):V10–V12
13. El-Serag HB (2012) Epidemiology of viral hepatitis and hepatocellular carcinoma. Gastroenterology 142:1264–1273e1261
14. Fox JG, Wang TC (2007) Inflammation, atrophy, and gastric cancer. J Clin Invest 117:60–69
15. Mantovani A, Savino B, Locati M, Zammataro L, Allavena P, Bonecchi R (2010) The chemokine system in cancer biology and therapy. Cytokine Growth Factor Rev 21:27–39
16. Taniguchi K, Karin M (2014) IL-6 and related cytokines as the critical lynchpins between inflammation and cancer. Semin Immunol 26:54–74
17. Hanahan D, Weinberg RA (2000) The hallmarks of cancer. Cell 100:57–70

18. Hanahan D, Weinberg RA (2011) Hallmarks of cancer: the next generation. Cell 144:646–674
19. Grivennikov SI (2013) Inflammation and colorectal cancer: colitis-associated neoplasia. Semin Immunopathol 35:229–244
20. Fearon ER, Vogelstein B (1990) A genetic model for colorectal tumorigenesis. Cell 61:759–767
21. Berg DJ, Davidson N, Kuhn R, Muller W, Menon S, Holland G, Thompson-Snipes L, Leach MW, Rennick D (1996) Enterocolitis and colon cancer in interleukin-10-deficient mice are associated with aberrant cytokine production and CD4(+) TH1-like responses. J Clin Investig 98:1010–1020
22. Hussain SP, Hofseth LJ, Harris CC (2003) Radical causes of cancer. Nat Rev Cancer 3:276–285
23. Meira LB, Bugni JM, Green SL, Lee CW, Pang B, Borenshtein D, Rickman BH, Rogers AB, Moroski-Erkul CA, McFaline JL et al (2008) DNA damage induced by chronic inflammation contributes to colon carcinogenesis in mice. J Clin Invest 118:2516–2525
24. Shaked H, Hofseth LJ, Chumanevich A, Chumanevich AA, Wang J, Wang Y, Taniguchi K, Guma M, Shenouda S, Clevers H et al (2012) Chronic epithelial NF-kappaB activation accelerates APC loss and intestinal tumor initiation through iNOS up-regulation. Proc Natl Acad Sci U S A 109:14007–14012
25. Blandino G, Fazi F, Donzelli S, Kedmi M, Sas-Chen A, Muti P, Strano S, Yarden Y (2014) Tumor suppressor microRNAs: a novel non-coding alliance against cancer. FEBS Lett 588(16):2639–2652
26. Gomez GG, Wykosky J, Zanca C, Furnari FB, Cavenee WK (2013) Therapeutic resistance in cancer: microRNA regulation of EGFR signaling networks. Cancer Biol Med 10:192–205
27. Sun X, Jiao X, Pestell TG, Fan C, Qin S, Mirabelli E, Ren H, Pestell RG 2014 MicroRNAs and cancer stem cells: the sword and the shield. Oncogene 33(42):4967–77
28. Dvorak HF (1986) Tumors: wounds that do not heal. Similarities between tumor stroma generation and wound healing. N Engl J Med 315:1650–1659
29. Srikrishna G, Freeze HH (2009) Endogenous damage-associated molecular pattern molecules at the crossroads of inflammation and cancer. Neoplasia 11:615–628
30. Kang R, Zhang Q, Zeh HJ 3rd, Lotze MT, Tang D (2013) HMGB1 in cancer: good, bad, or both? Clin Cancer Res 19:4046–4057
31. Matzinger P, Kamala T (2011) Tissue-based class control: the other side of tolerance. Nat Rev Immunol 11:221–230
32. Ostrand-Rosenberg S, Sinha P, Chornoguz O, Ecker C (2012) Regulating the suppressors: apoptosis and inflammation govern the survival of tumor-induced myeloid-derived suppressor cells (MDSC). Cancer Immunol Immunother 61:1319–1325
33. Schwabe RF, Jobin C (2013) The microbiome and cancer. Nat Rev Cancer 13:800–812
34. Francescone R, Hou V, Grivennikov S (2014). Microbiome, Inflammation and Cancer. Cancer J 20(3):181–189
35. Weinberg RA (2007) The biology of cancer. Garland Science, NY
36. Balkwill F, Mantovani A (2001) Inflammation and cancer: back to Virchow? Lancet 357:539–545
37. Ben-Neriah Y, Karin M (2011) Inflammation meets cancer, with NF-kappaB as the matchmaker. Nat Immunol 12:715–723
38. Mantovani A (2009) Cancer: inflaming metastasis. Nature 457:36–37
39. Human Microbiome Project C (2012). A framework for human microbiome research. Nature 486:215–221
40. Koren O, Goodrich JK, Cullender TC, Spor A, Laitinen K, Backhed HK, Gonzalez A, Werner JJ, Angenent LT, Knight R et al (2012) Host remodeling of the gut microbiome and metabolic changes during pregnancy. Cell 150:470–480
41. Lee SM, Donaldson GP, Mikulski Z, Boyajian S, Ley K, Mazmanian SK (2013) Bacterial colonization factors control specificity and stability of the gut microbiota. Nature 501:426–429

42. Slack E, Hapfelmeier S, Stecher B, Velykoredko Y, Stoel M, Lawson MA, Geuking MB, Beutler B, Tedder TF, Hardt WD et al (2009) Innate and adaptive immunity cooperate flexibly to maintain host-microbiota mutualism. Science 325:617–620
43. Maynard CL, Elson CO, Hatton RD, Weaver CT (2012) Reciprocal interactions of the intestinal microbiota and immune system. Nature 489:231–241
44. Arthur JC, Jobin C (2011) The struggle within: microbial influences on colorectal cancer. Inflamm Bowel Dis 17:396–409
45. Candela M, Turroni S, Biagi E, Carbonero F, Rampelli S, Fiorentini C, Brigidi P (2014) Inflammation and colorectal cancer, when microbiota-host mutualism breaks. World J Gastroenterol 20:908–922
46. Plottel CS, Blaser MJ (2011) Microbiome and malignancy. Cell Host Microbe 10:324–335
47. Sears CL, Pardoll DM (2011) Perspective: alpha-bugs, their microbial partners, and the link to colon cancer. J Infect Dis 203:306–311
48. Wu S, Rhee KJ, Albesiano E, Rabizadeh S, Wu X, Yen HR, Huso DL, Brancati FL, Wick E, McAllister F et al (2009) A human colonic commensal promotes colon tumorigenesis via activation of T helper type 17 T cell responses. Nat Med 15:1016–1022
49. Rubinstein MR, Wang X, Liu W, Hao Y, Cai G, Han YW (2013) *Fusobacterium nucleatum* promotes colorectal carcinogenesis by modulating E-Cadherin/beta-catenin signaling via its FadA adhesin. Cell Host Microbe 14:195–206
50. Kostic AD, Chun E, Robertson L, Glickman JN, Gallini CA, Michaud M, Clancy TE, Chung DC, Lochhead P, Hold GL et al (2013) *Fusobacterium nucleatum* potentiates intestinal tumorigenesis and modulates the tumor-immune microenvironment. Cell Host Microbe 14:207–215
51. Arthur JC, Perez-Chanona E, Muhlbauer M, Tomkovich S, Uronis JM, Fan TJ, Campbell BJ, Abujamel T, Dogan B, Rogers AB et al (2012) Intestinal inflammation targets cancer-inducing activity of the microbiota. Science 338:120–123
52. Cougnoux A, Dalmasso G, Martinez R, Buc E, Delmas J, Gibold L, Sauvanet P, Darcha C, Dechelotte P, Bonnet M et al (2014) Bacterial genotoxin colibactin promotes colon tumour growth by inducing a senescence-associated secretory phenotype. Gut 63(12):1932–42
53. Nenci A, Becker C, Wullaert A, Gareus R, van Loo G, Danese S, Huth M, Nikolaev A, Neufert C, Madison B et al (2007) Epithelial NEMO links innate immunity to chronic intestinal inflammation. Nature 446:557–561
54. Greten FR, Eckmann L, Greten TF, Park JM, Li ZW, Egan LJ, Kagnoff MF, Karin M (2004) IKKb links inflammation and tumorigenesis in a mouse model of colitis-associated cancer. Cell 118:285–296
55. Santaolalla R, Sussman DA, Ruiz JR, Davies JM, Pastorini C, Espana CL, Sotolongo J, Burlingame O, Bejarano PA, Philip S et al (2013) TLR4 activates the beta-catenin pathway to cause intestinal neoplasia. PLoS ONE 8:e63298
56. Abraham C, Medzhitov R (2012) Interactions between the host innate immune system and microbes in inflammatory bowel disease. Gastroenterology 140:1729–1737
57. Condeelis J, Pollard JW (2006) Macrophages: obligate partners for tumor cell migration, invasion, and metastasis. Cell 124:263–266
58. Hagemann T, Lawrence T (2009) Investigating macrophage and malignant cell interactions in vitro. Methods Mol Biol 512:325–332
59. Murdoch C, Muthana M, Coffelt SB, Lewis CE (2008) The role of myeloid cells in the promotion of tumour angiogenesis. Nat Rev Cancer 8:618–631
60. Bates JM, Akerlund J, Mittge E, Guillemin K (2007) Intestinal alkaline phosphatase detoxifies lipopolysaccharide and prevents inflammation in zebrafish in response to the gut microbiota. Cell Host Microbe 2:371–382
61. Munford RS, Varley AW (2006) Shield as signal: lipopolysaccharides and the evolution of immunity to gram-negative bacteria. PLoS Pathog 2:e67
62. Dessein R, Gironella M, Vignal C, Peyrin-Biroulet L, Sokol H, Secher T, Lacas-Gervais S, Gratadoux JJ, Lafont F, Dagorn JC et al (2009) Toll-like receptor 2 is critical for induction of Reg3 beta expression and intestinal clearance of *Yersinia pseudotuberculosis*. Gut 58:771–776

63. Furuta GT, Turner JR, Taylor CT, Hershberg RM, Comerford K, Narravula S, Podolsky DK, Colgan SP (2001) Hypoxia-inducible factor 1-dependent induction of intestinal trefoil factor protects barrier function during hypoxia. J Exp Med 193:1027–1034
64. McVay LD, Keilbaugh SA, Wong TM, Kierstein S, Shin ME, Lehrke M, Lefterova MI, Shifflett DE, Barnes SL, Cominelli F et al (2006) Absence of bacterially induced RELMbeta reduces injury in the dextran sodium sulfate model of colitis. J Clin Invest 116:2914–2923
65. Taupin D, Podolsky DK (2003) Trefoil factors: initiators of mucosal healing. Nat Rev Mol Cell Biol 4:721–732
66. Van der Sluis M, De Koning BA, De Bruijn AC, Velcich A, Meijerink JP, Van Goudoever JB, Buller HA, Dekker J, Van Seuningen I, Renes IB et al (2006) Muc2-deficient mice spontaneously develop colitis, indicating that MUC2 is critical for colonic protection. Gastroenterology 131:117–129
67. Kobayashi KS, Chamaillard M, Ogura Y, Henegariu O, Inohara N, Nunez G, Flavell RA (2005) Nod2-dependent regulation of innate and adaptive immunity in the intestinal tract. Science 307:731–734
68. Vaishnava S, Behrendt CL, Ismail AS, Eckmann L, Hooper LV (2008) Paneth cells directly sense gut commensals and maintain homeostasis at the intestinal host-microbial interface. Proc Natl Acad Sci U S A 105:20858–20863
69. Grivennikov SI, Wang K, Mucida D, Stewart CA., Schnabl B, Jauch D, Taniguchi K, Yu GY, Osterreicher CH, Hung KE et al (2012) Adenoma-linked barrier defects and microbial products drive IL-23/IL-17-mediated tumour growth. Nature 491:254–258
70. Castellarin M, Warren RL, Freeman JD, Dreolini L, Krzywinski M, Strauss J, Barnes R, Watson P, Allen-Vercoe E, Moore RA et al (2012) *Fusobacterium nucleatum* infection is prevalent in human colorectal carcinoma. Genome Res 22:299–306
71. Kostic AD, Gevers D, Pedamallu CS, Michaud M, Duke F, Earl AM, Ojesina AI, Jung J, Bass AJ, Tabernero J et al (2012) Genomic analysis identifies association of *Fusobacterium* with colorectal carcinoma. Genome Res 22:292–298
72. Marchesi JR, Dutilh BE, Hall N, Peters WH, Roelofs R, Boleij A, Tjalsma H (2011) Towards the human colorectal cancer microbiome. Plos One 6:e20447
73. Gursoy UK, Pollanen M, Kononen E, Uitto VJ (2010) Biofilm formation enhances the oxygen tolerance and invasiveness of *Fusobacterium nucleatum* in an oral mucosa culture model. J Periodontol 81:1084–1091
74. Hu B, Elinav E, Huber S, Strowig T, Hao L, Hafemann A, Jin C, Wunderlich C, Wunderlich T, Eisenbarth SC et al (2013) Microbiota-induced activation of epithelial IL-6 signaling links inflammasome-driven inflammation with transmissible cancer. Proc Natl Acad Sci U S A 110:9862–9867
75. Bongers G, Pacer ME, Geraldino TH, Chen L, He Z, Hashimoto D, Furtado GC, Ochando J, Kelley KA, Clemente JC et al (2014) Interplay of host microbiota, genetic perturbations, and inflammation promotes local development of intestinal neoplasms in mice. J Exp Med 211:457–472
76. Song X, Gao H, Lin Y, Yao Y, Zhu S, Wang J, Liu Y, Yao X, Meng G, Shen N et al (2014) Alterations in the microbiota drive interleukin-17C production from intestinal epithelial cells to promote tumorigenesis. Immunity 40:140–152
77. Huber S, Gagliani N, Zenewicz LA., Huber FJ., Bosurgi L, Hu B, Hedl M, Zhang W, O'Connor W Jr, Murphy AJ et al (2012) IL-22BP is regulated by the inflammasome and modulates tumorigenesis in the intestine. Nature 491:259–263
78. Kirchberger S, Royston DJ, Boulard O, Thornton E, Franchini F, Szabady RL., Harrison O, Powrie F (2013) Innate lymphoid cells sustain colon cancer through production of interleukin-22 in a mouse model. J Exp Med 210:917–931
79. Putoczki TL, Thiem S, Loving A, Busuttil RA, Wilson NJ, Ziegler PK, Nguyen PM, Preaudet A, Farid R, Edwards KM et al (2013) Interleukin-11 is the dominant IL-6 family cytokine during gastrointestinal tumorigenesis and can be targeted therapeutically. Cancer Cell 24:257–271

80. Erdman SE, Rao VP, Poutahidis T, Rogers AB, Taylor CL, Jackson EA, Ge Z, Lee CW, Schauer DB, Wogan GN et al (2009) Nitric oxide and TNF-alpha trigger colonic inflammation and carcinogenesis in *Helicobacter hepaticus*-infected, Rag2-deficient mice. Proc Natl Acad Sci U S A 106:1027–1032
81. Yoshimoto S, Loo TM, Atarashi K, Kanda H, Sato S, Oyadomari S, Iwakura Y, Oshima K, Morita H, Hattori M et al (2013) Obesity-induced gut microbial metabolite promotes liver cancer through senescence secretome. Nature 499:97–101
82. Singh N, Gurav A, Sivaprakasam S, Brady E, Padia R, Shi H, Thangaraju M, Prasad PD, Manicassamy S, Munn DH et al (2014) Activation of gpr109a, receptor for niacin and the commensal metabolite butyrate, suppresses colonic inflammation and carcinogenesis. Immunity 40:128–139
83. Nugent JL, McCoy AN, Addamo CJ, Jia W, Sandler RS, Keku TO (2014) Altered tissue metabolites correlate with microbial dysbiosis in colorectal adenomas. J Proteome Res 13:1921–1929
84. Chong ES (2014) A potential role of probiotics in colorectal cancer prevention: review of possible mechanisms of action. World J Microbiol Biotechnol 30:351–374
85. Schwitalla S, Fingerle AA, Cammareri P, Nebelsiek T, Goktuna SI, Ziegler PK, Canli O, Heijmans J, Huels DJ, Moreaux G et al (2013a) Intestinal tumorigenesis initiated by dedifferentiation and acquisition of stem-cell-like properties. Cell 152:25–38
86. Schwitalla S, Ziegler PK, Horst D, Becker V, Kerle I, Begus-Nahrmann Y, Lechel A, Rudolph KL, Langer R, Slotta-Huspenina J et al (2013b) Loss of p53 in enterocytes generates an inflammatory microenvironment enabling invasion and lymph node metastasis of carcinogen-induced colorectal tumors. Cancer Cell 23:93–106

Inflammation and Colorectal Cancer

Rishu Dheer, Julie M. Davies and Maria T. Abreu

Abstract Chronic intestinal Inflammation occurs in response to environmental factors, infection and genetics; and plays a critical role in initiation, promotion, progression and metastasis of colon cancer. Colitis associated colon cancer (CAC) is a classic example of multifactorial, multi-step colorectal cancer associated with inflammatory bowel diseases. In recent years, the generation of animal models of CAC and recognition of the importance of the gut microbiota, altered immune system, and other environmental factors in CAC, has expanded the basic understanding of inflammation associated colon cancer. In this chapter, we discuss the cellular alterations and mechanisms by which inflammation contributes towards the development of colon cancer using CAC as a model system. We have also explored some of the promising strategies for preventing progression of inflammation to colon cancer. The emerging role of dietary factors, obesity and gut microbiota in colon cancer is also reviewed.

Keywords Colitis associated colon cancer · Microbiota · Innate immunity · Molecular pathways · Inflammatory bowel disease · Obesity · Diet · Chemoprevention · Cytokines

1 Introduction

Colorectal cancer (CRC) is the third most prevalent form of cancer in humans. It is also one of the most common causes of cancer-related mortality in the world [1]. CRC is a multi-factorial disease and only between 7.4–35 % of the CRC cases are

M. T. Abreu (✉) · R. Dheer · J. M. Davies
Department of Medicine, Division of Gastroenterology, University of Miami Miller School of Medicine, Miami, FL, USA
e-mail: Mabreu1@med.miami.edu

R. Dheer
e-mail: rdheer@med.miami.edu

J. M. Davies
e-mail: jdavies@med.miami.edu

V. W. Yang, A. B. Bialkowska (eds.), *Intestinal Tumorigenesis*,
DOI 10.1007/978-3-319-19986-3_8

due to specific genetic causes and the rest are mainly associated with environmental factors leading secondarily to acquired genetic mutations [2]. The risk factors for CRC include genetic predisposition, diet, obesity, intestinal microbes and pathogens and chronic intestinal inflammation. Numerous epidemiologic, molecular and animal based studies support the link between chronic inflammation and cancer.

Colitis-associated cancer (CAC) is an exemplary case of inflammation based cancer mainly found in patients with inflammatory bowel diseases that includes Crohn's disease and ulcerative colitis [3]. In the US alone, between 1 and 1.5 million people suffer from IBD and of these, approximately 18 % UC and 8.3 % Crohn's disease patients are at risk of developing CAC within 30 years of the disease onset [4–6]. The risk of developing cancer correlates with severity, extent and duration of the active inflammatory disease [7, 8]. Based on recent meta-analysis of population based cohorts, the standardized incidence ratio of CAC is equal for UC and CD colitis patients with a value of 1.7. The cumulative risks of developing CAC are 1 %, 3 % and 7 % after 10, 20 and more than 20 years of IBD, respectively. However, extensive colitis raises this cumulative risk to 2 %, 12 % and 21 % respectively [8]. Interestingly, other forms of colitis such as microscopic or lymphocytic/collagenous colitis, do not appear to increase the risk of CAC leading to the conclusion that not all types of inflammation increase CRC risk [9]. A strong link between inflammation and colon cancer is also evident in animal models. In the AOM-DSS murine model of colon cancer, a single dose of genotoxic agent azoxymethane (AOM) coupled with chronic colitis results in multiple tumors in a shorter duration whereas multiple doses of AOM without chronic inflammation takes a longer time for tumorigenesis suggesting that inflammation promotes tumorigenesis [10].

Understanding the factors and mechanisms by which inflammation contributes towards colon cancer is the key towards developing treatment and possibly preventive strategies against CAC. In this chapter, we describe the central role of inflammation in colon cancer development using CAC as a model system. We also focused on how the interaction of diet, obesity, microbiota and even genetics can affect the influence of inflammation on colon cancer.

2 CAC and CRC: Similarities and Differences

Development of CRC and CAC is a slow process where procession from normal epithelium to invasive carcinoma may take decades. Over the years, numerous similarities and dissimilarities have been described in the clinical and pathological features of CAC and other forms of CRC especially sporadic colorectal cancer, which occurs in people with no family history of the disease. While both CRC and CAC share aberrant crypt foci, polyps, adenoma and carcinoma stages of cancer development, dysplastic lesions in CRC are generally focal adenomatous polyp whereas in CAC these can be multi-focal, flat or polyploid without being a clear pedunculated polyp [3, 11]. Moreover, both CRC and CAC are multi-stage processes wherein

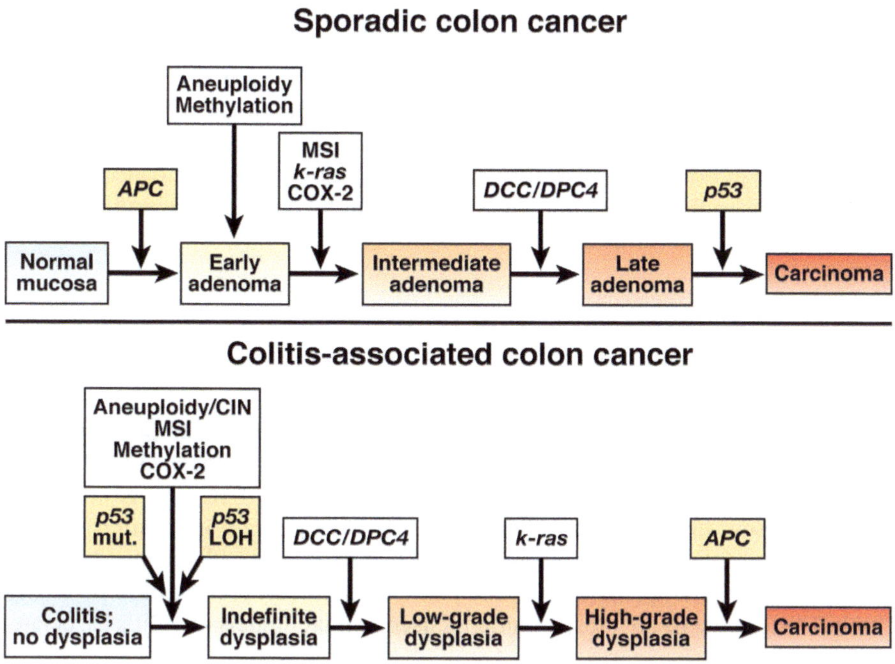

Fig. 1 Common genetic and chromosomal alterations in sporadic and inflammation driven colon cancer. (Reprinted from Ref. [11], copyright 2011, with permission from Elsevier)

the progression from adenoma to carcinoma in other types of CRC is replaced by inflammation and injury-dysplasia-carcinoma type of sequence in CAC. Molecular genetics and sequencing of colorectal tumors have depicted CRC to be a sequence of events involving successive accumulation of genetic and epigenetic abnormalities wherein specific events are associated with particular stages of cancer [12–14]. Figure 1 compares the sequence of occurrence of chromosomal instability (CIN), microsatellite instability (MSI) and other common genetic mutations in CRC and CAC. Both CRC and CAC involve mutations in common genes and signaling pathways such as Wnt/β-catenin, p53, KRAS and TGF-β [12, 14]. However the number of mutations and the cancer stage at which these mutations appear differ between the types of cancer; for example, *APC* mutations are one of the earliest events in CRC and are functional drivers of CRC. However *APC* mutations are probably passengers in CAC since they are less frequent, generally appear much later in cancer progression and perhaps do not have any functional role in CAC [11, 13]. On the other hand, there are commonalities between CAC and CRC. Most importantly, CRC tumors may also exhibit inflammatory cell infiltration and increased release of inflammatory cytokines [15, 3, 16]. In fact, the density of tumor in-filtrating T-cell subset $CD45RO^+$ is a prognostic biomarker in colon cancer patients [17]. Like CAC, a great majority of CRC tumors exhibit constitutive activation of transcription factors such as NF-κB and STAT3 that are essential components of inflammatory pathways. In CRC, inflammation is probably a consequence of tumorigenesis whereas in CAC inflammation it is the driver of tumorigenesis. Finally, environmental factors such as the intestinal microbiota and diet influence the development and progression of both CRC and CAC [18–21]. Thus it is important to analyze colon cancer in the context of inflammation especially when the knowledge gained from this classical inflammation based cancer can be disseminated and applied to other more common forms of cancer.

3 Animal Models of CAC

A number of colitis based animal models have been developed to understand the pathogenesis of inflammation and cancer. The usefulness of these animal models also lies in the ability to determine the novel markers for different stages of cancer and in the possibility of identifying targets for therapeutics and testing their efficacy. Most of these models are either based on treatment with chemical carcinogens or genetic alterations of immune pathways and cell proliferation pathways. Table 1 lists some of the most commonly used animal models of colitis associated cancer.

Table 1 Chemical and genetic models of CAC

Model	Characteristics	References
Chemical models		
DSS	Causes direct epithelial injury. Induces dose and duration dependent colonic inflammation, dysplasia and cancer	[26]
Carrageenan	Degraded sulfated polysaccharide from sea weed leads to duration dependent ulcerative lesions and tumors in rats	[282]
AOM	DNA alkylating agent used usually with DSS to generate tumors in short period of time	[10]
DMH	Being a precursor of AOM its mode of action is similar to AOM.	[283]
Genetic models		
$Il\text{-}10^{-/-}$	Develop spontaneous colitis and carcinoma in the colon	[284]
$G\alpha i2^{-/-}$	Loss of G-protein subunit results in UC like colitis and carcinoma in the colon	[285]
$Rag2^{-/-}$+*H. hepaticus*	Develop colitis and colon carcinoma in the absence of functional lymphocytes	[35]
$Gpx1/Gpx2^{-/-}$	The glutathione peroxidase deficient mice develop ileocolitis, adenocarcinoma and signet ring cell carcinoma only in the presence of host bacteria	[286]
$TCR\beta/p53^{-/-}$	Develop dysplasia and adenocarcinoma in ileocecum and cecum only in the presence of conventional bacteria	[287]
$IL\text{-}2/\beta 2M^{-/-}$	Spontaneously develop UC-like carcinoma in proximal colon	[288]
$Tbet/Rag2^{-/-}$	UC gradually progresses to dysplasia and carcinoma	[59]

3.1 Chemical Models

Induction of colitis and associated neoplasia using chemicals offers advantages of ease, reproducibility and often being able to generate CAC in short duration. A variety of chemical compounds that either disrupt the epithelial barrier (e.g. DSS, carrageenan) or cause DNA mutations (e.g. DMH, AOM) are used to experimentally reproduce intestinal inflammation and cancer [22–25]. The most commonly used mouse model of CAC is the AOM-DSS model. In this model a single intra-peritoneal injection of the DNA alkylating agent azoxymethane (AOM) is followed by oral exposure to dextran sodium sulfate (DSS) for epithelial injury. The dysplasia in this model correlates with the degree of inflammation. Increased levels of Cox-2, PGE_2 and nuclear translocation of β-catenin are observed in the AOM-DSS model [26–28]. The tumorigenic potential of this model is dose dependent and strain specific with mice of A/J or BALB/c genetic background being more susceptible to tumors than C57Bl/6J mice [29, 30].

3.2 Genetic models

A number of single and double gene knockout mice that develop spontaneous colitis and colon cancer have been described in recent years (Table 1) [31, 23, 22]. In some cases genetic models are used in conjunction with chemical carcinogens to mimic human colitis-associated neoplasia mechanisms e.g. *Apc*$^{Min/+}$ mice which in itself is a genetic model of sporadic colorectal cancer and the inflammation is not as obvious as in typical CAC models. The *Apc*$^{Min/+}$ mice are heterozygous for mutation in the *Apc* gene and develop numerous tumors in the small intestine [32]. However in the presence of DSS these mice develop colonic tumors and mimic CAC [33]. Other frequently used genetic models are those involved in innate immunity (e.g. *Tlr4*$^{-/-}$, *Myd88*$^{-/-}$), signal transduction (e.g. *Stat3*$^{-/-}$), stress response (e.g. *Xbp1*$^{-/-}$), cytokines based (e.g. *Il-6*$^{-/-}$) and immune regulation (e.g. *IKKβ*$^{-/-}$) and are used in combination with a chemical model (AOM-DSS) to understand their role in colon cancer etiology [31, 23, 22]. Some of these genetic models are developed to understand the role of inflammatory pathways in specific cell types of the intestine. For e.g. Villin-TLR4 mice were developed to understand the role of overexpression of *Tlr4* in the epithelial cells of the intestine [27]. These mice show increased susceptibility towards colitis and tumorigenesis. Besides this, genetic models are frequently used to determine the role of other factors such as microbes and diet on inflammation and cancer. Infection of *Apc*$^{Min/+}$ mice with human commensal enterotoxigenic *Bacteroides fragilis* (ETBF) triggers inflammatory colitis and tumors in distal colon and has been used to explain the pathogenic pathways involved in microbe-mediated inflammation and cancer [34].

3.3 Adoptive Transfer Models

These models are based on selective transfer of specific cell types into immunodeficient mice such as *Rag2*$^{-/-}$ to determine the role of transferred cell types in inflammation and cancer [35, 36]. Until now these models have been mainly used to elucidate the role of T-cells in CAC and have helped in identifying both pathogenic and protective T cell subpopulations [37]. Adoptive transfer of CD4^{+} CD45RB^{lo} CD25^{+} regulatory T-cells (Tregs) from uninfected WT mice to *H. hepaticus* infected *Rag2*$^{-/-}$ mice significantly reduced *H. hepaticus* induced inflammation and cancer in the latter but not with T effector cells [35, 38]. Educated Tregs that were obtained from infected WT mice were more effective towards protection against inflammation associated colon cancer than the Tregs obtained from hygienic uninfected WT mice. These results demonstrated the protective role of Tregs in inflammation and suggested that adaptive immune cells retain their anti-inflammatory phenotype after infection.

Although the majority of animal based models discussed above do not exactly reproduce all the clinical features of typical human CAC disease, each one of them has played a role in the current understanding of inflammation and its progression towards cancer. However the need of animal models that can recapitulate invasion and metastasis of human colon cancer still remains.

4 The Microenvironment and Immune System in CAC

The inflamed mucosal microenvironment as observed in IBD patients activates transcription factors (e.g. NF-κB, STAT3) and the release of mediators such as cytokines (e.g. TNFα, IL-6) and chemokines (CCL2, CXCL8) into the microenvironment. Within an inflamed region, these mediators may tip the immune system towards either anti-tumorigenic or pro-tumorigenic by regulating cell growth, survival, differentiation and apoptosis. A typical CAC tumor is a complex tissue composed of transformed epithelial cells, normal stromal cells and infiltrated innate and adaptive immune cells. The interaction of tumor cells with macrophages, B and T cells, NK cells, fibroblasts, mast cells and others creates a microenvironment conducive for cancer progression. Various studies have identified both positive and negative regulatory mediators that act in the same direction in IBD and CAC. At the same time there are other mediators that work in the opposite direction in IBD and CAC wherein they exacerbate one disease but are protective for the other [39]. Here we briefly describe cells and factors that modulate the balance of pro and anti-inflammatory immune responses in CAC and contrast this with sporadic colon cancer.

4.1 *Innate Immune Mediators*

4.1.1 Macrophages

Macrophages function in intestinal innate immunity as professional phagocytic cells which reside in the mucosa close to the epithelial layer. Their primary purpose is to phagocytize components of the microbiota that make it across the epithelial layer providing a second layer of defense against the environment. Intestinal macrophages do not transit to the draining lymph nodes to present antigen to T cells but perform their effector functions in the tissue [40]. Intestinal macrophages have limited responsiveness to pathogen recognition receptors (PRRs) and during homeostasis respond to microbial patterns by inducing tolerance [41, 42].

Human studies have determined that macrophages in IBD are a major source of TNFα, iNOS and IL-6 [43, 44]. As such, they are important cellular sources of reactive oxygen species (ROS) and reactive nitrogen species (RNS) which are implicated in DNA damage. During inflammation in murine models $Ly6C^{hi}$ macrophage precursors flood the intestine and are capable of inducing fulminant inflammation. Loss of these cells prevents the development of acute and chronic colitis [45, 46].

Studies examining the importance of tumor-associated macrophages (TAMs) in the progression and growth of tumors as diverse as breast, prostate ovarian and cervical [47], have overwhelmingly associated TAM infiltration with a poor prognosis and worse outcome. TAMs are alternatively activated (M2), and prevent the generation of protective immune responses. This is postulated to be due to the immunomodulatory action of the tumor itself, most often through production of TGFβ. TAMs also induce the production of vascular endothelial growth factor (VEGF) and matrix metalloproteinases (MMPs) which drive tissue remodeling and promote angiogenesis [47]. The colon however, seems to play by different rules. Numerous studies have correlated the presence of CD68+ macrophages at the tumor invasive front in sporadic CRC with better prognosis, and reduced risk ratios. Five year survival in patients that have high numbers of CD68+ cells at the invasive front in "hotspots" is significantly increased compared to patients with lower numbers of CD68+ cells [48, 49]. These hotspots were also correlated to fewer liver metastases [48]. The number of CD68+ cells at the invasive front was found to be an independent prognostic indicator by several studies [49–53]. High numbers of activated macrophages (CD14+ CD80+ CD86+) are correlated to better 5 year survival in patients with lymph node metastases [54]. The colon tumor microenvironment is a mixture of both M1 (inflammatory—iNOS+) and M2 (regulatory—CD163+) macrophages. In a recent study, high numbers of M1 and M2 macrophages were correlated to each other meaning that both were present at high numbers in the mixed tumor environment. High numbers of macrophages, either M1 or M2, were positively correlated with disease free survival. In examining the importance of either subset the ratio of either subset did not impact survival [55]. However, another study did find that a low ratio of M2 macrophages (described as Clever-1/Stabilin-1+) to the total macrophage pool (total CD68+ cells) was correlated to a low frequency of distant recurrences, providing some indication that the type of macrophage present in the microenvironment was important in patient outcome [56].

A possible mechanism for the protective effect of macrophages in colon cancer was investigated by examining TGFβ in the tumors which were surrounded by CD68+ cells. This study found that low numbers of infiltrating macrophages were associated with high levels of tumor TGFβ, SMAD4 and TGFβRII. The authors postulated that tumor production of TGFβ attenuated the recruitment and activation of macrophages allowing tumor growth [51]. A postulated mechanism of macrophages control of tumors in colon tumors is through induction of apoptosis. Macrophages along the invasive front were found to express FasL and correlated to increased numbers of dead or dying tumor cells [57].

Thus, macrophages are intimately involved in all aspects of inflammation and tumorigenesis in both human and mouse CAC, and efforts to control their function and prevent inflammation will be important not only for inducing remission in IBD, but also for the future potential of preventing tumor development.

4.1.2 Dendritic cells

Dendritic cells are important bridges between the innate and adaptive immune systems. Dendritic cells capture antigen in the mucosa and traffic to the draining lymph nodes (or Peyer's patches in the small intestine or isolated lymphoid follicles in the colon) to present antigen to T cells which shape the subsequent T cell and B cell antigen specific reactions. In the normal intestine DCs induce regulatory T cell responses to commensal and food antigens maintaining homeostasis in the intestine. Similar to macrophages, human colon cancer patients with high numbers of dendritic cells in the tumor stroma have better overall survival. Low numbers of dendritic cells predict a poorer outcome and demonstrate the importance of dendritic cells in coordinating protective immune responses to the tumor [58]. A model of the importance of dendritic cells in colitis and CAC is a spontaneous colitis model in which the transcription factor *Tbet (Tbx21)* is knocked out in *Rag*$^{-/-}$ mice (lacking T cells and B cells) (TRUC mice) [59]. Loss of dendritic cells in these mice completely abrogates inflammation. The investigators found that Tbet was a negative regulator of Tnfα [60] and in these mice loss of Tbet in CD103-CD11b+dendritic cells leads to high levels of Tnfα and worse colitis and CAC. Despite a requirement for the microbiota in the development of colitis in these mice, loss of MyD88 signaling in TRUC mice did not lessen colitis or CAC development highlighting the multitude of pathogen recognition pathways that may lead to Tnfα production. Dendritic cell Tnfα production in this model was a potent inducer of Il-17a from innate lymphoid cells which was found to be the effector cell type as antibodies against CD90 (marker of innate lymphoid cells) lead to reduced colitis inflammation scores [61]. Thus loss of regulation of TNFα production from dendritic cells in the intestine is an important factor in the promotion of IL-17A responses and colitis and CAC.

4.1.3 Neutrophils

Neutrophils are the most abundant leukocyte in the blood and represent between 50–70% of circulating leukocytes. They are generally short-lived and are professional phagocytes with limited functional plasticity, and as such have generally been disregarded as important mediators of tumor growth. However, renewed interest in their importance in tumor growth has been based on their prognostic value in many cancers. Patients with high numbers of tumor-infiltrating neutrophils (tumor-associated neutrophils, TANs) have a poorer outcome and decreased survival [62]. Neutrophils may contribute to cancer initiation and progression by a number of pathways which they normally use to great protective effect during acute infection. First, neutrophils produce large quantities of ROS and RNS, effectors that are used to kill phagocytized bacteria. Some of these mediators are also released to control extracellular bacteria. However, ROS and in particular H_2O_2, which is cell permeant, causes DNA damage by inducing alterations to bases and by impeding DNA damage repair enzymes [63, 62]. This has the potential to contribute to tumor initiation. Second, neutrophils produce other effector proteins including the serine protease neutrophil elastase (NE) and matrix metalloproteinase- 8 (MMP-8) and MMP-9. Neutrophil elastase has been proposed to induce cell proliferation at low doses and cell death at higher concentrations [64, 63]. MMP-9 which is also produced by other cell types enhances growth and metastases by promoting angiogenesis [65]. Finally, neutrophils are potent secretors of inflammatory cytokines including Il-1β, TNFα and IL-12. They also secrete IL-8 and several other CXC chemokines in a feed-forward loop resulting in recruitment of more neutrophils. In humans with CRC, a high intra-tumoral CD66b+neutrophil count was correlated with poorer outcomes [66], and in mouse AOM-DSS models of CAC, neutrophil depletion using an antibody to Ly6G lead to decreased tumor growth, even when given after tumor induction [67]. Despite their relative short life span and limited functional repertoire, neutrophils have the potential to participate in tumor initiation through ROS production and inducing DNA damage, through promotion by producing high levels of cytokines and through metastases by promoting angiogenesis. Future studies would do well not to dismiss neutrophils as unimportant in the immune response to tumors.

4.1.4 Mast cells

Mast cells are found in human colon adenomas and polyps. Mast cells infiltrate polyps and are a significant contributor to the pool of TNFα in the tumor environment. Anti-TNFα antibodies administered to $Apc^{\Delta 468}$ mice, which spontaneously develop polyps, results in decreased mast cell accumulation [68] in the polyps and fewer polyps. Mast cell function and lymphocyte function are closely intertwined. Mast cells capture antigens and can be potent antigen presenting cells to T cells. Recent evidence has also suggested that Treg cells can limit the accumulation of mast cells in the tumor environment resulting in decreased polyp size. Transfer of WT Tregs to *Apc*$^{\Delta468}$ mice reduces tumor growth. Natural Tregs produced in *Apc*$^{\Delta468}$ mice are defective in production of Il-10 and controlling inflammation and instead produce Il-17a and may contribute to defective resolution of local inflammation, thus all Tregs do not function the same and merely enumerating Treg numbers, rather than function, will not advance our understanding or our ability to target Tregs as a strategy to treat CAC [69].

4.1.5 Myeloid-Derived Suppressor Cells

MDSCs are a heterogeneous population of myeloid progenitor cells with the capacity to differentiate to dendritic cells and macrophages under appropriate signaling. However, during chronic inflammation and cancer progenitor cells are arrested in their differentiation and remain immature myeloid cells with potent immunosuppressive properties. MDSCs are observed in chronic inflammatory conditions including arthritis [70], autoimmune encephalomyelitis [71] and are important mediators of transplant tolerance [72] and healing of spinal cord injuries [73]. MDSCs have been extensively studied in tumor models. The main mechanism of suppressive activity by MDSCs is the production of NO and ROS to inhibit T and NK cell anti-tumor activity [74]. S100A8/9 (also known as calprotectin) is a protein heterodimer that has been described as essential for MDSC accumulation and function. *S100A9*$^{-/-}$ mice do not accumulate MDSCs and have much lower tumor burdens following tumor transplantation [75]. The receptors for S100A8/9 is RAGE and TLR4 [76] and stimulation of MDSCs with LPS and IFN-γ increases their suppressive function [77]. *S100A9*$^{-/-}$, *Tlr4*$^{-/-}$, and *MyD88*$^{-/-}$ mice all have reduced MDSC accumulation [78], fewer tumors, and their MDSCs have reduced suppressive effects on T cells [79]. MDSCs also promote tumor growth by producing VEGF, bFGF, TGFβ and MMP-9 [80], factors which enhance angiogenesis.

There has been little study of MDSCs in human IBD. A single study by Haile et al. has demonstrated increased frequency of MDSCs in the blood in both CD and UC patients [81]. In several diverse mouse colitis models, increased MDSCs were correlated with lower inflammation scores and their transfer reduced colitis induction [81–83]. Several reports have found MDSCs in the dysplastic regions of mice given AOM-DSS, and inhibition of tumor growth in *S100A9*$^{-/-}$ mice demonstrating the importance of this cell type to colitis-associated tumor growth [84, 85]. TNFα has recently been identified as an important driver of MDSCs generation in the mouse [86]. It was shown that TNFα blocked the differentiation of monocytes to effector macrophages and dendritic cells and instead drove MDSCs generation in response to chronic inflammation. Mice given etanercept during the course of chronic inflammation had fewer MDSCs and more robust T and NK cell responses indicating a lack of inflammatory resolution [87]. While this study was not using a chronic colitis inflammation model it will be important to remember that the use of anti-TNF therapy in UC may have consequences for this important and understudied immune and tumor modulating cell population.

In human spontaneous colon cancers increased levels of circulating and tumor-derived MDSC—defined as CD11b + CD33 + HLA-DR-Lin-/low—were correlated to tumor stage and metastases but not tumor size [88]. The results of future studies will hopefully shed some light on the impact of biologics on the accumulation of MDSCs in UC patients and their relationship to dysplasia development.

4.1.6 Cytokines

TNFα It is an immediate gene product of NFκB-induced transcription [89, 90] and also induces NF-κB activation itself. TNFα induces its own gene expression as well as other potentially genetically destructive products including iNOS and reactive oxygen species (ROS) [91]. ROS is implicated in inducing base pair substitutions in DNA which left uncorrected by DNA repair mechanisms can lead to cancer initiation [92]. In colon cancer, elevated levels of TNFα have been observed and are an important link between inflammation and tumorigenesis. High levels of TNFα are associated with increased colon carcinoma [93]. And UC patients with the 308G→A SNP which results in increased TNFα production also have an increased risk of developing colon cancer [94].

TNF has 2 receptors TNFR1 and TNFR2. Loss of TNFR1 enhances colitis, while loss of TNFR2 protects against colitis [95]. Signaling through TNFR1 is responsible for most TNFα-induced apoptotic cell death. Deletion of *Tnfr1* (*Tnf-Rp55*) decreased the number of tumors in the AOM-DSS CAC model. *Tnfr1*$^{-/-}$ mice had fewer macrophages and neutrophils in their colonic lamina propria probably due to decreased expression of keratinocyte chemoattractant (KC/CXCL1) and CCL2, but T cells and dendritic cells were still present. Bone marrow chimeras identified the hematopoietic compartment as responsible for the protection from tumor growth, and administration of the anti-TNF antibody etanercept reduced tumor formation in WT mice even when given for a short period of time after the DSS cycles [96].

The epithelium is a source of TNFα and production of TNFα from the epithelium can induce a CD-like pathology in the intestine. However, recognition of TNFα by the epithelium was not required to induce this inflammation. When functional signaling through TNFR1 was present only in the epithelium, not on other stromal or hematopoietic cells, exogenous administration of TNFα resulted in epithelial apoptosis indicating some epithelial responsiveness to TNF signaling, but long-term overexpression of TNFα in this model did not lead to CD-like disease. In this setting overexpression of TNF from the epithelium activated local myofibroblasts, which drove the CD-like disease [97].

TNFα is also important in sporadic models of colon cancer. Mice with deletions in the *Apc* gene have increased circulating Tnfα and treatment of these mice with anti-TNFs reduced tumor growth [68]. TNFα is also important in angiogenesis and invasion promoting metastatic growth. By inducing the production of IL-8 and VEGF, TNFα signaling contributes to epithelial to mesenchymal transition (EMT)

[98], which is one of the first steps in cancer dissemination through destruction of the basement membrane at the invasive cancer front.

Suitably, the importance of TNFα in all aspects of CAC from initiation of the immune response, as an immediate gene product of TLR and NF-κB signaling, all the way through to EMT and the metastatic capability of tumor cells is mirrored by the observed role of its major producer—macrophages.

IL-10 IL-10 is an important regulatory cytokine and is induced by late phase TLR4 activation and TNFα signaling to promote immune resolution [99]. IL-10 is secreted by macrophages and dendritic cells to induce Tregs. It is also the major functional cytokine used by Tregs to suppress effector T cell responses. $Il\text{-}10^{-/-}$ deficient mice develop spontaneous colitis and adenocarcinomas [100, 101] in the presence of the microflora [101], which indicates the importance of dampening immune responses to maintain homeostasis and prevent tumor growth in the intestinal environment. Colitis induction is accelerated in $Il\text{-}10^{-/-}$ mice by administration of the NSAID piroxicam in food [102]. Recent studies have examined the role of IL-10 production by macrophages in the development of colitis. Zigmond, E et al demonstrated that loss of IL-10 secretion in macrophages does not lead to spontaneous colitis, but loss of recognition of IL-10 through its receptor IL-10Rα is required to develop colitis. Loss of IL-10Rα signaling in macrophages led to the increased T cell and neutrophil activation and accumulation. This study highlighted the importance of IL-10 in regulating both the adaptive T cells responses and the feedback control on the innate mediators required for immune restraint in the gut environment [103]. Human IBD has been associated with polymorphisms in the IL-10 [104] and IL-10RA gene [105], underscoring the importance of this cytokine in the pathogenesis of IBD and associated cancer.

IL-6 IL-6 is induced as an early product of NFκB activation induced by TLRs and TNFα. IL-6 has a complex signaling pathway, and can induce signaling through its receptor, IL-6Rα, in a heterocomplex with gp130 on the cell surface, but is also able to induce trans-signaling when the IL-6Rα is shed—then the soluble complex can bind to any cell expressing the gp130 protein—which may or may not express IL-6Rα [106, 107]. Activation of gp130 through either the membrane or soluble IL-6Rα complex activates JAK signaling and STAT3 translocation to the nucleus [108]. Serum IL-6 levels are elevated in sporadic CRC and correlate with prognosis, tumor size, metastases and survival [109]. IL-6 levels are also increased in IBD patients [110–112] and are predictive of time to relapse in CD patients. IL-6 is a pleiotropic cytokine and is involved in all branches of T helper subsets and B cell function.

IL-1β It is induced by a number of factors, but its secretion is tightly controlled. First, it is often transcribed after stimulation, but is maintained in large polyribosomes and is not translated. Second, it is maintained intracellularly in the pro- IL-1β form, and must be activated by cleavage by caspase-1 and the inflammasome. IL-1β is induced by TLRs as well as its own receptor, generating a feed-forward activation loop [113]. Loss of a repressor of IL-1β, Casp12, in mice increases colitis-associated tumor formation highlighting the role of IL-1β and inflammation in promoting tumor growth [114]. IL-1β is important in CAC and CRC for its ability to activate WNT/β-catenin signaling, and links immune activation to promotion of proliferation in cancer cells. Macrophage secretion of IL-1β activates WNT signaling in cancer cells by inactivating GSK3β [115] in an NFκB dependent manner. Epithelial to mesenchymal transition (EMT) has also been shown to be induced by IL-1β signaling in colon cancer cells [116]. IL-1β also prevents TRAIL-induced cell death in colon cancer cell lines with mutant β-catenin [117]. Its ability to induce EMT, activate of proliferation, and block apoptotic pathways make IL-1β a potent pro-tumorigenic cytokine in the progression of colon cancer.

4.2 Adaptive Immune Mediators

4.2.1 T Cells

T cells are adaptive immune effectors comprising several subsets and are either inflammatory or regulatory. T cells are antigen specific, and have the advantage of memory over the innate immune effectors. Inflammatory effector T cells (either CD4+ or CD8+) are required to control infections while regulatory (natural or induced) T cells are required to resolve inflammation and prevent inflammation-induced tissue destruction. Historically, CD was thought to be driven by Th1 T cell responses characterized by high levels of IFNγ production while UC was mainly driven by Th2 responses with high level production of IL-13. This paradigm has recently been re-examined with the discovery of Th17 T cells which secrete IL-17 [118–120]. In both CD and UC, Th17 cells are likely to supply pathogenic cytokine and effector cell functions. Moreover, there is a mixed population of Th cells which secrete both IFN-γ and IL-17A in IBD patients. Induced regulatory cells (iTreg) have an important role in the intestine to limit T cell effector functions and restore homeostasis after an insult. These cells are characterized as CD4+ CD25+ Foxp3+ cells that secrete IL-10. In IBD, there are increased numbers of iTregs in the mucosa and decreased circulating iTregs [121, 122] corresponding to increased recruitment to the inflamed area. Interestingly to the pathology of CD, the FoxP3+ T cells in the mucosa were found to secrete IL-17A [123, 124] but maintained CD4+ suppressive function.

In colitis-associated colon cancer iTregs are posited to have a dual role. In the early phase of tumorigenesis, control of inflammation is likely to prevent tumor growth, and in humans and mouse models the number of Tregs in the tumor stroma correlates with better overall survival [125, 126] and decreased tumor burden [36, 127, 38, 128]. As discussed above, this interaction may be based on iTreg interaction with other high cytokine producing cells like mast cells. However, as tumor growth progresses, iTregs will go on to prevent T cell effector responses capable of tumor killing. In a recent meta-analysis of studies examining the distribution of T cells and subsets in the tumor stroma, the number of CD45RO+ cells (effector memory) was the main T cell subset which was correlated to a positive overall survival [129]. This T cell subset was also important in predicting improved overall survival in MSI+ colon cancers. It was hypothesized that the frameshift mutations occurring in the proteins in these cancers increase antigenicity and improved adaptive T cell responses [17].

4.2.2 B Cells

Regulatory B cells producing IL-10 (B10 cells) are induced in mice given DSS and these cells were found to inhibit weight loss and improve histological scores. B10 cells transferred to *Il-10*$^{-/-}$ mice with spontaneous colitis or to *Rag2*$^{-/-}$ mice with CD4+ CD25-CD45RB^{high} T cells-induced colitis had reduced inflammatory scores [130]. Other earlier studies have also shown the regulatory role of B cells in colitis induced in TCR$\alpha^{-/-}$ and immunoglobulin (Ig) mu$^{-/-}$ (no T cells or B cells) mice. These mice have earlier onset of colitis which can be rescued by adoptive transfer of B cells from TCR$\alpha^{-/-}$ mice [131]. The mechanism of modulation was described as production of Il-10 and downregulation of Il-1 and Stat3 pathways rather than directly T cell suppression [132]. Consistent with the role of B cells as anti-inflammatory and protective in colitis is a reported case of an UC patient who had failed all previously therapy (including mesalamine, azathioprine, corticosteroids, adalimumab, infliximab etc.) and was administered rituximab (anti CD20 antibody) as a salvage therapy. Evidence from rheumatoid arthritis patients had shown that rituximab had clinical benefits for patients. However, the UC patient developed severe colitis symptoms, fever, and lost 4 kg of weight. The patient had nearly undetectable numbers of CD20+ B cells in the colonic mucosa and completely absent IL-10 production from lamina propria cells after a single treatment where previous to therapy lamina propria cells had produced considerable amounts [133]. In this case, virtually all of the IL-10 production in the colon was derived from B cells, and their loss removed the brake from severe inflammation. Spontaneous UC has also been noted in patients receiving rituximab for treatment of other disorders including nephrotic syndrome [134] and Graves' disease [135]. However, a small cohort of UC patients was treated with rituximab without significant deterioration, but with no improvement over placebo [136]. Collectively, both the animal models and the evidence from human studies underlines the importance of B cell in colitis in inducing regulatory responses and reducing inflammation.

A renewed interest in B cells in cancer has been founded on the observation that the numbers of tumor-infiltrating B cells correlate to better prognosis in several cancers including melanoma, breast and ovarian [137–140]. In sporadic colon cancer, tumor infiltrating B cells were shown to form follicular aggregates at the invasive front in close association with T cells and follicular dendritic cells. The B cells were oligoclonal, activated and antigen experienced as they had undergone VDJ rearrangement. These cells were reactive to tumor cells, but did not express CD27, a marker of memory B cells and expressed high levels of the T cell co-stimulatory molecule CD80 [141]. The authors postulated the importance of these B cells in driving protective T cells responses with their antigen presenting capabilities as they were much more numerous than dendritic cells in the tumor environment. However, B cells like T cells, can also function as regulatory B cells secreting IL-10 and TGFβ to reduce anti-tumor CTL responses. Additionally, they may produce low-affinity tumor reactive antibodies which contribute to inflammation, but are not effective enough to kill the tumor, which we have already demonstrated contributes to tumor growth. A small trial examined the anti-CD20 antibody rituximab in the treatment of colorectal cancer and found that 4/8 patients treated had their metastases size reduced [142]. In all, the role of B cells in spontaneous colon cancer and especially colitis-associated cancer has not be elaborately studied and may present a novel area of interest to promote anti-tumor T cell responses.

5 Molecular Pathways in Inflammation Associated Carcinogenesis

In CAC, clearly inflammation precedes tumorigenesis, suggesting that inflammation in addition to tumor promotion and progression plays a critical role in tumor initiation too. Several mechanisms involved in the initiation and progression of intestinal inflammation to cancer have been proposed. These inflammation associated oncogenic pathways can be grouped as follows: (i) oxidative stress induced DNA damage (ii) epigenetic alterations and (iii) cellular proliferative and survival pathways.

5.1 Oxidative Stress Induced DNA Damage

Stepwise analysis of genetic alterations in CAC has identified various early and late stage mutations in genes associated with regulation of intestinal cell proliferation and maintenance. Early oncogenic mutations in CAC are identified in p53 genes from dysplastic stages onwards. On the other hand, mutations in APC genes have been associated with later stages of CAC [39, 3, 11]. In humans it is not clear whether inflammation itself causes these mutations or environmental factors that predispose an individual to developing IBD play a role in these genetic alterations. However studies with chemical injury or infection induced chronic mouse models of colitis made it clear that prolonged inflammation by itself can initiate DNA damage and tumorigenesis [143, 144]. Infiltrated inflammatory cells such as macrophages and neutrophils in colitis can induce robust oxidative stress in its milieu. Moreover, epithelial cells and premalignant cells can also produce reactive species under the influence of inflammatory cytokines. The reactive oxygen species (ROS) and reactive nitrogen intermediates (RNIs) released by infiltrated immune cells are highly mutagenic and can cause direct DNA damage [145]. Mice deficient in *ataxia telangiectasia mutated* (*ATM*) gene, which encodes for DNA double strand break recognition and response protein manifested more 8-oxoguanine and nitrotyrosine, the markers of ROS mediated DNA damage as well as showed increased sensitivity towards chemical induced colitis [146]. Inhibition of reactive species or the genes involved in production of reactive species such as NO synthase decelerates DNA damage and tumor development [143]. In addition to direct DNA damage, hypoxia induced by inflammation can repress mismatch repair genes (MMR), increase epigenetic silencing and can induce the activity of DNA mutating activation-induced cytidine deaminase (AID) enzyme [147, 148]. AID protein is found at increased levels in IBD and CAC patients [149].

5.2 Epigenetic Changes

Epigenetic alterations via DNA methylation, histone modification or mi-RNA mediated silencing of key genes such as those involved in tumor suppression and DNA repair is an important mechanism of colon tumorigenesis [150]. Aberrant gene methylation is an early event in CAC. Hypermethylation of CpG islands in promoter regions of key cancer genes such as *P14, CDHI, P16, MYOD1* and *HPP1* is observed in UC patients [151, 152]. Current observations of UC patients suggest an increased methylation event in patients with high grade dysplasia and cancer compared to those with low grade dysplasia and without any cancer [153]. Hypermethylation and inactivation of *suppressor of cytokine signaling 3* (*SOCS3*) leads to aberrant activation of IL-6/STAT3 pathway in UC and CAC [154]. Cytokines such as IL-22 and IL-6 released during inflammation can activate DNA methyl transferases DNMT1 and DNMT3 which in turn may silence a number of target genes. In $Apc^{Min/+}$ mice, absence of *Dnmt1* reduces polyp formation and hypermethylation of CpG islands [155]. Methylation of Wnt pathway genes was found to be an early event in inflammation associated colon cancer with the number of methylation events increasing progressively from normal to colitic to neoplastic colon tissue [156]. Furthermore, microsatellite instability (MSI) observed in IBD associated cancer is associated with promoter hypermethylation of DNA mismatch repair gene *MLH1* [157]. Histone modifications are not well scrutinized in IBD associated cancer. In laboratory studies the repression of the promoter region of *MLH1* by histone deacetylation is reported to occur under hypoxic condition induced by the inflammatory environment of the cancer [148]. Moreover, histone deacetylases (HDACs) are considered as novel targets for prevention of cancer development in IBD [158].

Another mechanism by which inflammation can promote colon cancer involves action by miRNAs. To date, several miRNAs have been identified that are upregulated in CRC [159]. A recent study suggests differential expression of miRNAs in IBD patients from non-neoplasia to dysplasia to cancer [160]. Distinct miRNA expression patterns were found in epithelial and myeloid cells as well as before and after tumor induction in AOM-DSS model of CAC. The miRNA pool associated with inflammation and carcinogeneisis in this model was further shown to affect cell proliferative PI3K/Akt signaling pathway [161]. miRNAs can promote colon cancer either by regulating inflammatory mediators and/or by regulating the expression of oncogenes and tumor suppressor genes [162, 163]. For e.g. STAT3 activation induced transcription of miR-21 and miR-181b-1 inhibits tumor suppressor genes phosphatase and tensin homologue (PTEN) and cylindromatosis (CYLD). Inhibition of these tumor suppressors activates NF-κB pathway [162].

5.3 Cellular Proliferative and Survival Pathways

Inflammation coupled with tissue injury results in wound healing responses which include cell migration, increased cell proliferation and crypt regeneration [164]. During the same time as wound healing and regeneration, the cells in the inflamed mucosa start accumulating mutations and become premalignant. Thus repeated inflammation as observed in IBD patients disrupts the fine balance between cell proliferation and death. Moreover, release of inflammatory mediators enhances the resistance of normal and premalignant epithelial cells towards apoptosis. Several of the central signaling pathways that are required for normal intestinal maintenance are extensively studied in the context of cancer. These same pathways are deregulated in CAC either due to mutations or by under and/or over-activation of their components. The main signaling pathways involved in CAC are discussed below.

5.3.1 NF-κB Pathway

Proteins of the NF-κB (nuclear factor kappa B) pathway are transcriptional factors that regulate numerous cellular processes including maintenance of epithelial integrity and interactions between intestinal microbiota and the immune system [165]. NFκB regulates immune activation by inducing transcription of IL-6, iNOS, TNFα, COX1/2, IL-1β among other inflammatory effectors [166, 167]. Figure 2 depicts the multiple pathways regulated by NF-κB under inflammatory conditions. Under normal cellular conditions, NF-κB transcription factors are retained in an inactive state in the cytoplasm by protein inhibitors called IκB (inhibitors of NF-κB) [168]. In spite of the fact that aberrant expression of NF-κB proteins is commonly detected in patients with UC, CD and CRC; the mutations in genes encoding these proteins are rarely found [169, 170]. During inflammation, the NF-κB pathway is activated via the IκB kinase (IKK) complex comprising of IKKα, IKKβ and NEMO by various pro-inflammatory cytokines and pattern recognition receptors. Activated NF-κB translocates to the nucleus and induces the expression of other cytokines and chemokines, thereby sustaining the inflammatory insult in the intestinal epithelium. NK-κB activation in both epithelial and immune cells can promote CAC by different mechanisms. Mice that lack the *Ikkβ* gene in the intestinal epithelium show reduced tumor formation but not tumor growth after AOM-DSS treatment. However, these

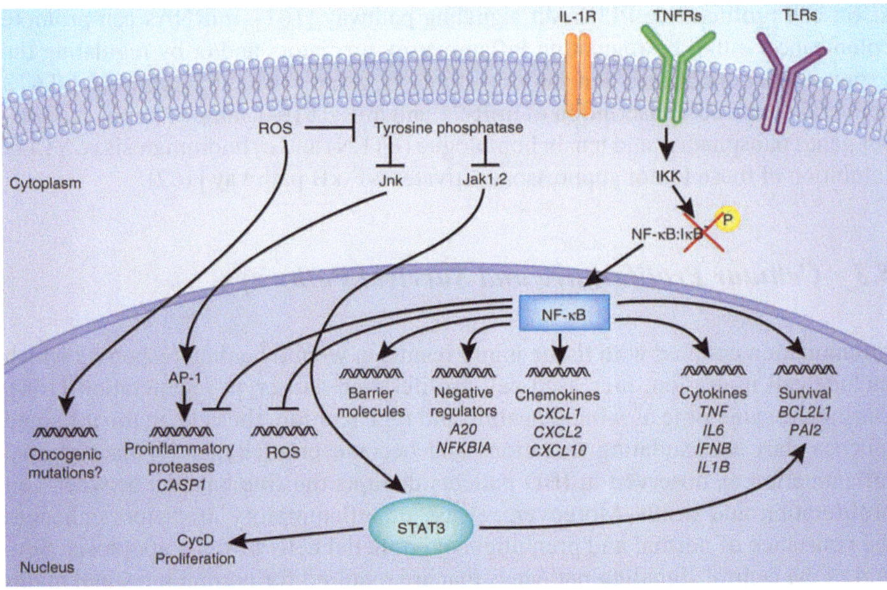

Fig. 2 Pathways regulated by NF-kB in inflammation and cancer. (Reprinted by permission from Macmillan Publishers Ltd: [289], copyright 2011)

mice manifested aggravated inflammation and injury after treatment with DSS [171]. It was further shown that the oncogenic potential of NF-κB in these mice is dependent on its anti-apoptotic function. Specifically, NF-κB prevents the apoptosis of pre-malignant cells by inducing the activity of anti-apoptotic protein such as BCL-xL. Similarly, inactivation of *Ikkβ* in myeloid cells in the AOM-DSS model of CAC reduced both tumor growth and development but without affecting apoptosis. The oncogenic effect of NF-κB in myeloid cells was exhibited by activation of tumor growth promoters such as IL-6. This study exemplifies the dynamic interplay between two of the compartments that make up the tumor microenvironment—and the different responses elicited by each using the same pathways. This study also demonstrated that NF-κB pathway plays a role in tumor initiation as well as promotion. Besides these classical pathways of NF-κB activation, the alternative pathway of NF-κB activation via constitutive β-catenin activation was also found in human colorectal carcinoma tissues [172]. Because of its multiple roles, targeting NF-κB in inflammation and tumorigenesis has not been practical.

5.3.2 STAT3 Pathway

The signal transducer and activator for transcription (STAT) proteins are other critical regulators of inflammation-associated cancer. STAT3 in particular is activated by a number of cytokines including IL-6, IL-17 and other growth factors. STAT3 target genes include proliferation and anti-apoptotic targets and may directly enhance cancer cell growth and survival. Indeed, overexpression of STAT3 in intestinal epithelial leads to increased tumor incidence and growth suggesting its role in tumor initiation and progression [173]. IL-6, derived from monocytes and macrophages, signaling through epithelial STAT3 has a profound impact on epithelial apoptosis following DSS as *Il-6*$^{-/-}$ and *Stat3*$^{-/-}$ mice have worse DSS induced colitis [174]. STAT3 however, is a pleiotropic transcription factor and is indispensable in hematopoietic development. STAT3 signaling downstream of IL-10Rα [175], is critical to the activity of NAPDH oxidases which limit bacterial invasion [176], and IL-12p40 induced Th1 activation [177]. Deletion of STAT3 in myeloid cells leads to a colitis-like disease [175–177]. STAT3 is also downstream of VEGFR signaling and promotes VEGF-induced proliferation and angiogenesis [178]. Additional evidence suggests that loss of STAT3 is also important in the *Apc*$^{Min/+}$ model of sporadic CRC, and loss of STAT3 in epithelial cells allowed tumors cells progression leading to larger tumors [179]. In fact, STAT3 and NF-κB play overlapping role in all the stages of cancer: initiation, progression and metastasis; by affecting similar cellular pathways [180]. In a pro-metastatic environment, NF-κB and STAT3 can directly target genes involved in EMT and angiogenesis including H1F1α, VEGF, cadherins, snail and chemokines.

5.3.3 Wnt/β-Catenin Pathway

Under normal conditions, β-catenin regulates cell-cell adhesion in a cadherin bound form after activation by Wnt ligands. Any unbound β-catenin is ubiquitinated and degraded after phosphorylation at N-terminal by constitutively active GSK-3β from the APC/Axin/GSK-3β complex [181]. In sporadic CRC, mutations in the Wnt/β-catenin pathway usually occur in very early stages of colon cancer. Inactivating mutation in the APC complex and/or impaired Wnt signaling leads to nuclear transport of stabilized β-catenin. Nuclear β-catenin activates T cell factor-lymphoid enhancing factor (TCF/LEF) mediated transcription of Wnt target genes such as c-Myc and Cyclin D1 that are involved in cell proliferation and survival [182]. Although the majority of CAC do not develop mutations in Wnt/β-catenin pathway until later stages of cancer development, recent studies suggest that Wnt/β-catenin pathway is activated in early phases of CAC [183]. Several animal models suggest that inflammatory pathways in CAC can directly activate Wnt/β-catenin pathways without any mutational changes in the genes associated with this pathway suggesting that there are pathways other than canonical Wnt signaling that can stabilize and activate β-catenin [115, 184]. The inflammatory mediator prostaglandin E (PGE_2) and TNFα activates Wnt/β-catenin pathway via PI3K/AKT pathway wherein AKT mediated C-terminal β-catenin phosphorylation at serine residue 552 leads to its release from APC/Axin/GSK-3β complex and stabilizes β-catenin [185, 98]. Constitutive activation of β-catenin possibly via PI3K/Akt pathway is observed in mice over-expressing TLR4 [86]. Indeed, β-catenin activity and NFκB activity are intricately connected and cross-regulated [186–190]. NFκB induces Wnt5a transcription and β-catenin activation by NFκB which in turn negatively regulates NFκB activation.

6 Role of Microbes and Innate Immunity in CAC

Recognition of microbes and microbial patterns by the innate immune system helps in maintaining intestinal homeostasis. In the past decades with the advent of next generation sequencing technologies it was found that in both IBD and CRC patients, the composition and diversity of the gut microbiota is different from healthy subjects [191–196]. This phenomenon of shifts in microbial populations in diseased patients is termed dysbiosis. The dysbiotic state in CRC patients is seen in both the luminal microbiota from stool and mucosa-associated bacteria from biopsy specimen of the patients. However the microbiota differences are not consistent across various studies. Moreover, these bacterial 16S based studies have not been able to link the disease with a particular pathogen or a spectrum of bacterial groups. Also from these studies it is difficult to decipher whether changes in the gut microbiota caused the disease or it was an aftermath of IBD or cancer. Additionally no study to date has described if human microbiota changes as inflammation progresses towards cancer. Thus although the role of commensal microbiota in the development of CAC is increasingly acknowledged, the mechanistic details are not well defined and are an intense focus of research.

Studies with germ-free animals and mono-association with specific microbes were the first ones to establish a link between microbiota, inflammation and colon cancer. Animals exposed to genotoxic agents and various genetic models of CAC failed to develop tumors under germ free conditions. For example spontaneously colitic *Il-10*$^{-/-}$ mice treated with AOM under germ free conditions do not develop tumors whereas conventionally raised mice do develop tumors [197]. Also, AOM treated *Il-10*$^{-/-}$ mice develop tumors when mono-associated with *Helicobacter hepaticus* but significantly less tumor formation occurs in the presence of *Bacteroides vulgatus,* suggesting that different microbial species can have varying effect on cancer outcome [198, 197]. In an effort to understand how microbes play a role in progression from inflammation to cancer, one study, mono-associated AOM treated *Il-10*$^{-/-}$ germ-free mice with an adherent-invasive *E. coli* (AIEC) or commensal *Enterococcus faecalis* strain [199]. Despite developing colitis after mono-association with both strains, only *E. coli* mono-associated mice developed invasive tumors, suggesting the role of specific bacterial virulence factors in inflammation-induced cancer. Deletion of this factor and in this case, colibactin encoded from a pathogenicity island from the *E. coli* strain reduced tumorigenesis without affecting colon inflammation [199]. It was also found that the colibactin could induce direct DNA damage. Similarly mono-association of multiple intestinal neoplasia (min), *Apc*$^{Min/+}$ mice, with *B. fragilis* toxin producing strain ETBF (enterotoxigenic *B. fragilis*) and not the non-toxic strain resulted in colonic tumor development. In min mice, ETFB induces pro-inflammatory Th17 immune responses by activating the STAT3 signaling pathway [200]. These studies were complemented by observations that both AIEC and ETBF are present in higher proportions in IBD and CRC patients. Altering the gut microbiota of ETBF colonized *Apc*$^{Min/+}$ mice by antibiotic treatment changed the tumor outcome in these mice without altering

ETBF colonization. These observations lead to the alpha bug hypothesis which suggests that in colon cancer certain bacteria act as pro-oncogenic agents [201]. These alpha bugs promote cancer by remodeling the colonic microbiota and surrounding epithelial cells. The alpha bugs with the help of remodeled helper bacteria reduce the population of protective bacteria in the intestine and eventually lead to cancer development.

Even though a plethora of insights is gained from germ free and mono-association studies, these studies do not entirely recapitulate human conditions wherein millions of bacteria reside in the intestine in a mutualistic, complex and fairly dynamic environment [202]. Thus, co-housing, fecal transplants with humanized microbiome, and antibiotic treatment of animal models are being increasingly used to tease apart the role of microbes in complex settings. The concept of dysbiotic gut microbiota affecting cancer outcomes was demonstrated in genetically engineered TRUC (*Tbet*$^{-/-}$ x *Rag2*$^{-/-}$ ulcerative colitis) mice using antibiotic treatment. Long-term exposure to broad-spectrum antibiotics reduced the incidence of neoplasia in TRUC mice [59]. Using a co-housing approach it has been shown that Nod2- deficient mice develop a CAC promoting bacterial community during AOM-DSS treatment. WT mice exposed to the CAC promoting bacterial community of Nod2 mutant mice developed an increased susceptibility towards colitis and CAC [203]. A recent study followed the microbiota of AOM-DSS murine model of CAC through different stages of inflammation driven tumorigenesis. This study showed that the basic microbial profile remains the same between inflamed and tumor stages of the mice. However the relative ratio of certain bacterial groups changed over the course of colitis associated cancer. Similar to step-wise mutations, a step-wise change in microbiota was observed in CAC wherein enrichment of different bacterial profiles was observed during inflammation, before cancer and after cancer stages of CAC development [204]. The temporal changes observed in microbiota during various stages of CAC supported the concept of driver-passenger microbes in colon cancer development. According to this concept, certain bacterial drivers in the intestine initiate tumorigenesis and once it is initiated, tumorigenesis further alters the microbiota by facilitating overgrowth of passenger bacteria and disappearance of driver bacteria. The passenger bacteria can either promote or suppress tumor formation depending on their microenvironment [205].

Nonpathogenic bacterial dysbiosis found in inflammatory states can deregulate immune system by modulating the expression of pattern recognition receptors (PRRs) such as Toll like receptors (TLRs) and Nod—like receptors. Polymorphism and mutations in genes encoding PRRs is another pathway by which genetic studies have associated innate immune system with IBD and colon cancer [206]. TLRs are expressed on both intestinal epithelial and innate immune cells. TLR4, a receptor for bacterial lipopolysaccharide, is expressed at increased levels in both the epithelium and the stroma of IBD and CAC patients [207, 208]. Even in animal models of CAC, Tlr4 promotes the development of colitis-associated cancer by releasing Cox2 and Peg2, and by activating NF-κB and Wnt/β-catenin pathway [209, 86]. *Tlr4*$^{-/-}$ mice are protected from AOM-DSS induced tumors whereas *Tlr2*$^{-/-}$ mice do develop tumors [210]. Likewise, myeloid differentiation factor 88 (Myd88) an adapter protein for TLRs and IL-1/IL-18 signaling is protective towards development of tumors in CAC [211]. The importance of bacterial signaling in colon tumorigenesis is well demonstrated in *Apc*$^{Min/+}$ mice by reduced tumor formation in the colon of these mice after loss of functional Myd88 [212]. Similarly deficiency of cytoplasmic PRR NOD2 has been associated with microbial dysbiosis, inflammation and cancer [203].

The studies above implicate that microbes can play a role in inflammation-associated cancer by producing genotoxic metabolites, aberrantly activating PRRs or by directly inducing inflammatory responses. Figure 3 outlines these three mechanisms by which microbes promote inflammation-associated cancer.

7 Influence of Dietary Factors on Inflammation and Cancer

Overwhelming evidence obtained from metanalyses, randomized controlled trials, animal models and in vitro studies, demonstrate the role of dietary components in colon cancer progression as well as its prevention [213, 214]. The Western diet rich in red meat and fat is a risk factor for colon cancer whereas a diet rich in fibers, fruits and vegetables is preventive against CRC [215]. However, it is not clear if a diet high in fat or its deficiency in other components is the root cause behind colon cancer. Other components of diet such as folic acid, calcium, vitamin D, unsaturated n-3 fatty acids, phytochemicals (curcumin, flavonoids, carotenoids, resveratrol) and certain types of carbohydrates are also protective towards colon cancer [214, 213]. Although the detailed mechanisms of these dietary factors are not well established, inflammation can be one of the links between diet and CAC. Dietary factors can promote or attenuate inflammation either by directly affecting inflammatory pathways or indirectly by altering the composition and metabolic output of host microbiota. For example people consuming high fat diet differ in their microbiota from those who are mainly on fiber rich diet with bacterial genes for methanogeneis and hydrogen sulphide production being abundant in the latter. Moreover a high fat diet is associated with increased levels of microbially converted secondary bile acid metabolites whereas short chain fatty acids such as butyrate are more abundant metabolites in fiber rich diets [216]. SCFAs the products of colonic bacteria, suppresses inflammation and neoplasia by affecting NF-κB and IFN-γ signaling pathways [217] whereas secondary bile acid metabolites possess proinflammatory and carcinogenic properties [218]. The dietary phytochemical astaxanthin, prevents colitis associated cancer in mice by modulating the expression of inflammatory cytokines and NF-κB [219]. Likewise curcumin and resveratrol suppress inflammation by downregulating Cox-2 and NF-κB expression [220]. Dietary unsaturated fatty acid reduces systemic inflammation and tumor incidence driven by AOM-DSS by modifying gut microbiota and the Notch1 signaling pathway [221]. The link between vitamin D and reduced risk for cancer development in IBD is demonstrated in animal models. *Smad3*$^{-/-}$ mice infected with *H. bilis* and

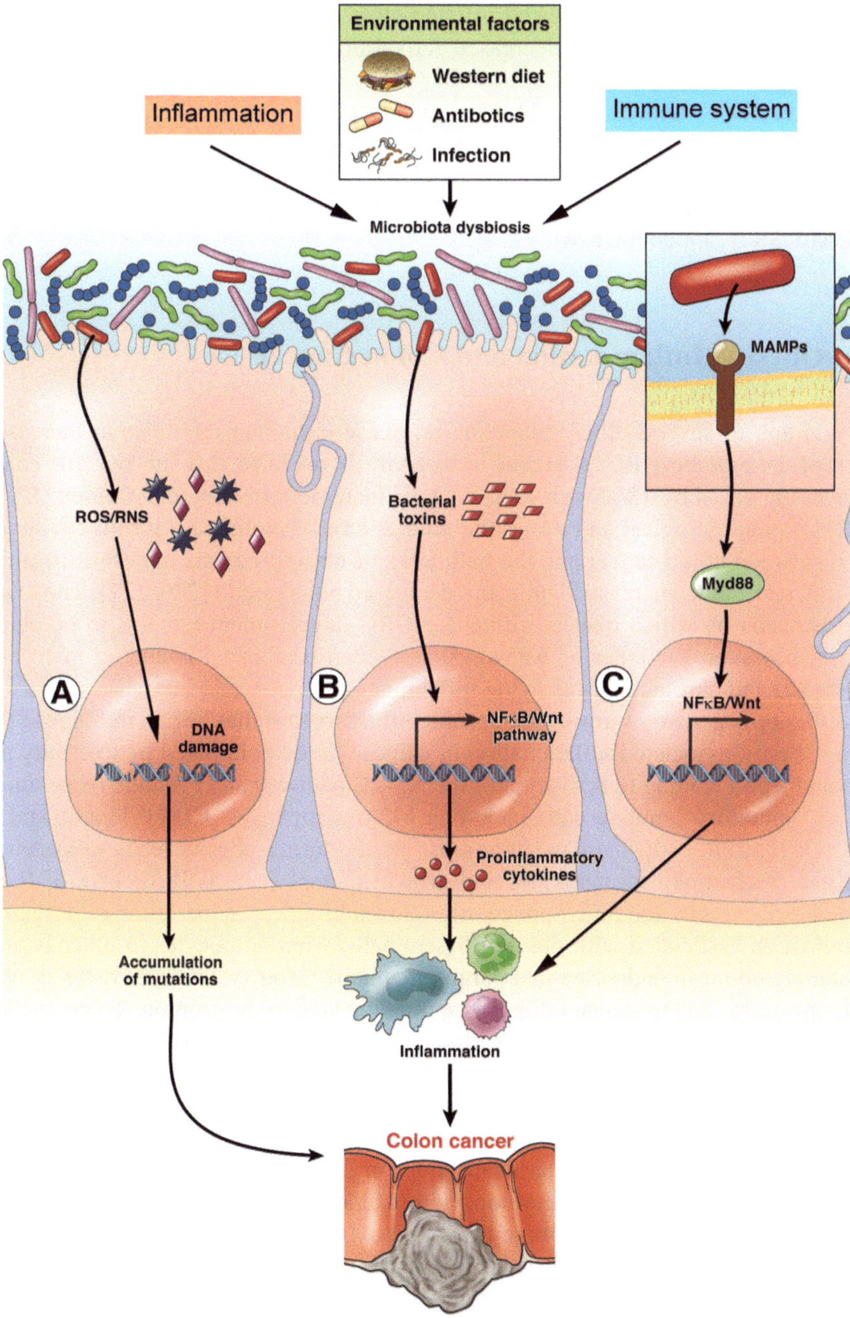

Fig. 3 Mechanisms by which microbes promote inflammation associated colon cancer. (Modified from Ref. [19], copyright 2014, with permission from Elsevier)

fed with vitamin D have reduced inflammatory cell infiltration, reduced inflammatory cytokine expression, decreased activation of MAPK and NF-κB pathways, and suppressed neoplasia [222]. Another study showed that the protective effect of vitamin D is mediated via TLR4 signaling [223]. The protective role of folic acid in colitis-associated cancer is adequately demonstrated in an IBD model using folate deficient mice that are heterozygous for *reduced folate carrier (Rfc1)* or *folate binding protein (Folbp1)*. These mice develop higher aberrant crypt foci after AOM-DSS treatment than WT [224].

8 Obesity, Inflammation and Colon Cancer

In addition to several other malignancies, obese individuals are also at increased risk of developing CRC. A recent meta-analysis revealed that the risk for colon cancer increases by 18 % with each 5 kg/m^2 increment in body mass index (BMI) [225]. Epidemiological and molecular studies have established a positive association between colon cancer and the hallmarks of obesity namely hyperinsulinemia, insulin resistance, increased leptin and decreased adiponectin [226, 227]. The exact mechanisms by which obesity influences CRC development remains to be elucidated, however a number of pathways have been implicated including microbial dysbiosis and chronic non-resolving inflammation.

Obesity is a state of low-grade chronic inflammation that begins in adipose tissue. Obesity-associated inflammation in adipose tissue is initiated by free fatty acids and leptin leading to activation of the NF-κB pathway and eventual recruitment of inflammatory immune cells including M2-macrophages, CD8+ T cells and B cells [226, 228–230]. Increase in adipose tissue during obesity leads to systemic release of chemokines, cytokines and growth factors such as IL-6, TNF-α, IL-8, C-reactive protein and macrophage migration inhibitory factor. Ultimately the inflammation associated with obesity disseminates from adipocytes to other organs including colon in endocrine and paracrine manner. The systemic increase in pro-inflammatory factors causes this dissemination and may promote colon cancer progression by activating NF-κB pathways [229, 230]. The colons of genetically or diet induced obese rodents recapitulate the inflammatory environment typically observed in CAC rodent models and manifest increased expression of these pro-inflammatory factors [231–233]. Also an increase in tumor incidences is observed in diet induced obese mice after AOM alone and AOM-DSS treatment [234, 235]. Similarly obese Zucker rats develop more adenomas than their lean counterparts after AOM treatment [236]. Obese KK-A^y mice (type II diabetic model with agouti gene) are also more susceptible to AOM and develop more tumors compared to normal C57BLl/6J mice [237]. Interestingly, other murine genetic models for obesity, the leptin deficient ob/ob mice and hyperleptinemic db/db mice, exhibit increased development of aberrant crypt foci in the colon yet do not develop any tumors in the colon after AOM treatment [238, 239]. However, the ob/ob mice develop higher colonic inflammation and tumors after treatment with AOM and DSS [240].

Development of increased aberrant crypt foci and tumors in the presence of obesity in colon cancer models suggests that obesity instigated inflammatory insult can be the link between obesity and colon cancer.

Another important mechanism by which obesity may promote inflammation-associated colon cancer is by inducing dysbiosis in host microbiota. Like CAC, germ free animals do not develop diet-induced obesity [241], suggesting that microbiota could play a link between obesity and inflammation associated cancer. The gut microbiota of obese humans and mice differs from healthy controls with an increase in the ratio of *Firmicutes* and *Bacteroidetes* [242–244]. Moreover, obesity is associated with increased gut permeability [245] and increased levels of circulating LPS [246, 247]. TLR4 activation by LPS induces activation of intracellular signaling cascades such as NF-κB and JNK—the pathways associated with CAC [248]. TLR4 signaling is also activated by fatty acids in adipocytes and macrophages [249]. Mice deficient in Tlr4 do not develop diet induced obesity and insulin resistance and neither do they develop colonic tumors [250, 209]. Interestingly, colonic inflammation leads to bacterial translocation into the adipose tissue [251]. All these findings suggest an inextricable link between microbiota, obesity, inflammation, and colon cancer.

9 Chemoprevention of CAC

The percentage of IBD patients developing CAC over time is at decline because of efficient disease management, development of anti-inflammatory therapeutics and increased accessibility to surveillance colonoscopies [7, 8]. However patients with longstanding and extensive colitis are still at high risk of developing cancer and warrants development of effective chemopreventive drugs. The first description of chemoprevention of colon cancer in IBD patients was reported with sulfasalazine exposure where it was found that UC patients with cancer had less exposure to this drug compared to UC patients without cancer [252]. Since then the potential of several chemopreventive agents has been described in IBD patients.

9.1 Mesalamines

Mesalamines are a derivative of the original sulfasalazine compound and are also known as mesalazine or 5-aminosalicylate (5-ASA). It is structurally similar to non-steroidal anti-inflammatory drug (NSAIDs) aspirin and is epidemiologically shown to reduce the incidence of cancer development in UC patients; although the findings are not consistent [253–257]. Furthermore its preventive role is supported by a large number of laboratory-based studies that demonstrate its ability to inhibit several pathways associated with colon cancer [253, 258–260]. So far, mesalamines have been shown to inhibit Cox2, NF-κB and Wnt-β catenin pathways [253,

259, 261]. It has also been shown to activate PPAR-γ, a transcription factor that inhibits the formation of aberrant crypt foci and cancer [260]. Improving replication fidelity thereby minimizing mutational frequency and modulating cell cycle proteins are other pathways by which mesalamines can exert their preventive effect [253, 258].

9.2 Thipourines

Azathioprine or 6-mercaptopurine have long been used as a maintenance therapy in IBD because of their anti-inflammatory properties [262]. However, only a few studies have evaluated the effect of these immunomodulators on the incidence of cancer development in IBD patients [263–265]. The results from these studies are inconclusive with regards to the effectiveness of these drugs in chemoprevention. In a prospective cohort of patients followed in France, the use of thiopurines decreased the incidence of CAC [266]. The protection against CAC should be balanced against the risk of the medications including hematopoietic malignancies [267, 268].

9.3 Other Biologics

Treatment using inhibitors for inflammatory cytokines such as TNF-α seems promising in animal models and as maintenance therapy in IBD but no clinical data is available yet on its effectiveness as a preventive medicine in CAC [269–271]. Infliximab, an inhibitor of TNF-α reduces the onset of colitis associated neoplasia in DSS caused chronic colitis only when administered in early stages of mild colitis [269]. Other cytokine inhibitors such as tocilizumab (anti-IL-6) have also been proposed as chemopreventive agents but no clinical data exists for their effectiveness [272].

Use of anti-inflammatory maintenance therapy is undisputedly a primary chemopreventive strategy in IBD patients. The use of anti-inflammatory agents described above have shown some conflicting results possibly due to confounding factors such as differences in study populations, drug dosage and duration, and the severity of the disease. But the need for safe and effective chemopreventive agents still remains.

10 Probiotics as Preventive and Treatment Therapy

As discussed before, microbes play a prominent role in inflammation and cancer. Based on that evidence it is logical to think that manipulation of host gut microbiota and consequently its metabolites can attenuate the carcinogenic potential of pathogenic gut microbiota associated with CAC. Commensal microbes play an important role in intestinal homeostasis by maintaining epithelial barrier and by regulating the immune system [273, 202]. Therefore, the use of beneficial live microbes as supplements, termed as probiotics, seems like a viable alternate therapeutic intervention for CAC patients. Epidemiologic and animal model-based studies have reported beneficial effects of probiotics for IBD [274]. However clinical data on the use of probiotics in CAC prevention is sporadic. Recently, the impact of the commercial probiotic VSL#3 has been studied in animal models of CAC [275–277]. VSL#3 is composed of eight different bacterial strains belonging to *Lactobacillus, Bifidobacterium* and *Streptococcus* species. Use of VSL#3 as preventive therapy before onset of inflammation, reduces adenoma formation in the AOM-DSS model of CAC, alters microbial community composition and delays the transition of inflammation to dysplasia in a rat model of CAC [275, 277]. However when VSL#3 is used after the onset of inflammation in AOM/*Il-10*$^{-/-}$ mice it no longer protects against CAC or inflammation [276]. In fact, an increase in tumorgenesis was observed in some mice and was associated with a loss of protective bacteria from the mucosa of host animals. Due to great variability in gut microbiota from person to person [278, 279], this study stresses the importance of a better understanding of individual host microbes and their metabolism before considering their modulation.

11 Summary

The mechanisms outlined in this chapter provide some insight about the pathogenesis of inflammation-associated cancer (Fig. 4). Inflammation associated cancer is a multifactorial and multi-step process involving complex interplay of host genetics, microbes, diet and immune system. The first step towards development of inflammation associated colon cancer involves increased permeability of the intestine in response to genetic susceptibility and external factors. The increased permeability leads to increased exposure of mucosa to microbial and dietary antigens and provides framework for tissue injury and inflammation. Inflammation is paralleled by microbial dysbiosis and release of pro-inflammatory cytokines that deregulates normal cell growth by inducing direct DNA damage, sustaining inflammation and by causing mutations and epigenetic changes in oncogenes and tumor suppressor genes leading to dysplasia. Further microbial dysbiosis, dysfunction of immune responses, cell signaling pathways and DNA damage promotes a dysplastic tissue towards malignancy.

Inflammation is crucial to tumor growth in a variety of ways. First, inflammation can damage cellular DNA and contribute to de novo mutations during the process of defending against invading organisms. Second, cytokines activate anti-apoptotic pathways in transformed cells, preventing cell death. Third, inflammatory cytokines induce the production of VEGF and allow the development of new blood

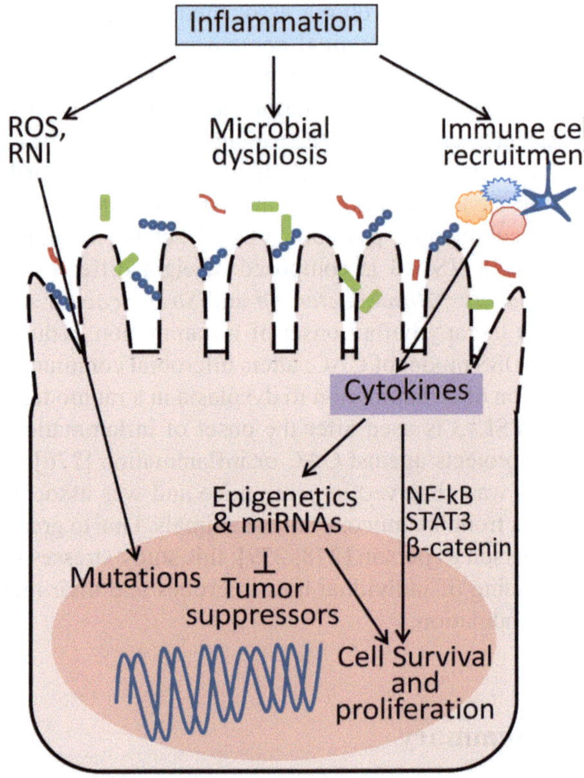

Fig. 4 Pathways involved in inflammation-associated tumor initiation

vessels, crucial to the continued growth of the tumor. Finally, aspects of inflammatory resolution, including development of regulatory responses by myeloid-derived suppressor cells and T cells and B cells, prevent immune destruction of the tumor. Therefore, any process that results in chronic inflammation will be prone to tumor development; but it also means that any tumor benefits from some low grade inflammation for the tumor to grow and progress and eventually metastasize [280, 281].

The effect of dietary and microbial metabolites on inflammation and cancer is complex and depends on host genetics, its microbiota and the internal immune environment. Identification and a deeper understanding of this complex network will be an important milestone for future therapeutic discoveries to treat not only CAC but other forms of CRC where inflammation plays a role in cancer progression. The priority should be towards developing individualized preventive and therapeutic strategies categorized according to specific microbial and genetic characteristics.

References

1. Haggar FA, Boushey RP (2009) Colorectal cancer epidemiology: incidence, mortality, survival, and risk factors. Clin Colon Rectal Surg 22(4):191–197. doi:10.1055/s-0029-1242458
2. Jiao S, Peters U, Berndt S, Brenner H, Butterbach K, Caan BJ, Carlson CS, Chan AT, Chang-Claude J, Chanock S, Curtis KR, Duggan D, Gong J, Harrison TA, Hayes RB, Henderson BE, Hoffmeister M, Kolonel LN, Marchand LL, Potter JD, Rudolph A, Schoen RE, Seminara D, Slattery ML, White E, Hsu L (2014) Estimating the heritability of colorectal cancer. Hum Mol Genet. doi:10.1093/hmg/ddu087
3. Terzic J, Grivennikov S, Karin E, Karin M (2010) Inflammation and colon cancer. Gastroenterology 138(6):2101–2114, e2105. doi:10.1053/j.gastro.2010.01.058
4. Canavan C, Abrams KR, Mayberry J (2006) Meta-analysis: colorectal and small bowel cancer risk in patients with Crohn's disease. Aliment Pharmacol Ther 23(8):1097–1104. doi:10.1111/j.1365-2036.2006.02854.x
5. Eaden JA, Abrams KR, Mayberry JF (2001) The risk of colorectal cancer in ulcerative colitis: a meta-analysis. Gut 48(4):526–535
6. Kappelman MD, Rifas-Shiman SL, Kleinman K, Ollendorf D, Bousvaros A, Grand RJ, Finkelstein JA (2007) The prevalence and geographic distribution of Crohn's disease and ulcerative colitis in the United States. Clin Gastroenterol Hepatol 5(12):1424–1429. doi:10.1016/j.cgh.2007.07.012
7. Andersen NN, Jess T (2013) Has the risk of colorectal cancer in inflammatory bowel disease decreased? World J Gastroenterol 19(43):7561–7568. doi:10.3748/wjg.v19.i43.7561
8. Lutgens MW, van Oijen MG, van der Heijden GJ, Vleggaar FP, Siersema PD, Oldenburg B (2013) Declining risk of colorectal cancer in inflammatory bowel disease: an updated meta-analysis of population-based cohort studies. Inflamm Bowel Dis 19(4):789–799. doi:10.1097/MIB.0b013e31828029c0
9. Yen EF, Pokhrel B, Bianchi LK, Roy HK, Du H, Patel A, Hall CR, Witt BL (2012) Decreased colorectal cancer and adenoma risk in patients with microscopic colitis. Dig Dis Sci 57(1):161–169. doi:10.1007/s10620-011-1852-2
10. Neufert C, Becker C, Neurath MF (2007) An inducible mouse model of colon carcinogenesis for the analysis of sporadic and inflammation-driven tumor progression. Nat Protoc 2(8):1998–2004. doi:10.1038/nprot.2007.279
11. Ullman TA, Itzkowitz SH (2011) Intestinal inflammation and cancer. Gastroenterology 140(6):1807–1816. doi:10.1053/j.gastro.2011.01.057
12. Fearon ER (2011) Molecular genetics of colorectal cancer. Annu Rev Pathol 6:479–507. doi:10.1146/annurev-pathol-011110-130235
13. Vogelstein B, Papadopoulos N, Velculescu VE, Zhou S, Diaz LA Jr, Kinzler KW (2013) Cancer genome landscapes. Science 339(6127):1546–1558. doi:10.1126/science.1235122
14. Vogelstein B, Kinzler KW (1993) The multistep nature of cancer. Trends Genet 9(4):138–141
15. Atreya I, Neurath MF (2008) Immune cells in colorectal cancer: prognostic relevance and therapeutic strategies. Expert Rev Anticancer Ther 8(4):561–572. doi:10.1586/14737140.8.4.561
16. Waldner MJ, Neurath MF (2008) Cytokines in colitis associated cancer: potential drug targets? Inflamm Allergy Drug Targets 7(3):187–194
17. Nosho K, Baba Y, Tanaka N, Shima K, Hayashi M, Meyerhardt JA, Giovannucci E, Dranoff G, Fuchs CS, Ogino S (2010) Tumour-infiltrating T-cell subsets, molecular changes in colorectal cancer, and prognosis: cohort study and literature review. J Pathol 222(4):350–366. doi:10.1002/path.2774
18. Sears CL, Garrett WS (2014) Microbes, microbiota, and colon cancer. Cell Host Microbe 15(3):317–328. doi:10.1016/j.chom.2014.02.007
19. Abreu MT, Peek RM Jr (2014) Gastrointestinal malignancy and the microbiome. Gastroenterology 146(6):1534–1546, e1533. doi:10.1053/j.gastro.2014.01.001
20. Greer JB, O'Keefe SJ (2011) Microbial induction of immunity, inflammation, and cancer. Front Physiol 1:168. doi:10.3389/fphys.2010.00168

21. Kato I, Majumdar AP, L and SJ, Barnholtz-Sloan JS, Severson RK (2010) Dietary fatty acids, luminal modifiers, and risk of colorectal cancer. Int J Cancer 127(4):942–951. doi:10.1002/ijc.25103
22. Sussman DA, Santaolalla R, Strobel S, Dheer R, Abreu MT (2012) Cancer in inflammatory bowel disease: lessons from animal models. Curr Opin Gastroenterol 28(4):327–333. doi:10.1097/MOG.0b013e328354cc36
23. Kanneganti M, Mino-Kenudson M, Mizoguchi E (2011) Animal models of colitis-associated carcinogenesis. J Biomed Biotechnol 2011:342637. doi:10.1155/2011/342637
24. Westbrook AM, Szakmary A, Schiestl RH (2010) Mechanisms of intestinal inflammation and development of associated cancers: lessons learned from mouse models. Mutat Res 705(1):40–59. doi:10.1016/j.mrrev.2010.03.001
25. Wirtz S, Neufert C, Weigmann B, Neurath MF (2007) Chemically induced mouse models of intestinal inflammation. Nat Protoc 2(3):541–546. doi:10.1038/nprot.2007.41
26. Cooper HS, Murthy S, Kido K, Yoshitake H, Flanigan A (2000) Dysplasia and cancer in the dextran sulfate sodium mouse colitis model. Relevance to colitis-associated neoplasia in the human: a study of histopathology, B-catenin and p53 expression and the role of inflammation. Carcinogenesis 21(4):757–768
27. Fukata M, Shang L, Santaolalla R, Sotolongo J, Pastorini C, Espana C, Ungaro R, Harpaz N, Cooper HS, Elson G, Kosco-Vilbois M, Zaias J, Perez MT, Mayer L, Vamadevan AS, Lira SA, Abreu MT (2011) Constitutive activation of epithelial TLR4 augments inflammatory responses to mucosal injury and drives colitis-associated tumorigenesis. Inflamm Bowel Dis 17(7):1464–1473. doi:10.1002/ibd.21527
28. Aust DE, Terdiman JP, Willenbucher RF, Chew K, Ferrell L, Florendo C, Molinaro-Clark A, Baretton GB, Lohrs U, Waldman FM (2001) Altered distribution of beta-catenin, and its binding proteins E-cadherin and APC, in ulcerative colitis-related colorectal cancers. Modern Pathol 14(1):29–39. doi:10.1038/modpathol.3880253
29. Suzuki R, Kohno H, Sugie S, Nakagama H, Tanaka T (2006) Strain differences in the susceptibility to azoxymethane and dextran sodium sulfate-induced colon carcinogenesis in mice. Carcinogenesis 27(1):162–169
30. Suzuki R, Kohno H, Sugie S, Tanaka T (2005) Dose-dependent promoting effect of dextran sodium sulfate on mouse colon carcinogenesis initiated with azoxymethane. Histol Histopathol 20(2):483–492
31. Bhan AK, Mizoguchi E, Smith RN, Mizoguchi A (1999) Colitis in transgenic and knockout animals as models of human inflammatory bowel disease. Immunol Rev 169:195–207
32. Moser AR, Pitot HC, Dove WF (1990) A dominant mutation that predisposes to multiple intestinal neoplasia in the mouse. Science 247(4940):322–324
33. Tanaka T, Kohno H, Suzuki R, Hata K, Sugie S, Niho N, Sakano K, Takahashi M, Wakabayashi K (2006) Dextran sodium sulfate strongly promotes colorectal carcinogenesis in Apc(Min/+) mice: inflammatory stimuli by dextran sodium sulfate results in development of multiple colonic neoplasms. Int J Cancer 118(1):25–34
34. Housseau F, Sears CL (2010) Enterotoxigenic Bacteroides fragilis (ETBF)-mediated colitis in Min (Apc+/–) mice: a human commensal-based murine model of colon carcinogenesis. Cell Cycle 9(1):3–5
35. Erdman SE, Poutahidis T, Tomczak M, Rogers AB, Cormier K, Plank B, Horwitz BH, Fox JG (2003) CD4+ CD25+ regulatory T lymphocytes inhibit microbially induced colon cancer in Rag2-deficient mice. Am J Pathol 162(2):691–702
36. Erdman SE, Sohn JJ, Rao VP, Nambiar PR, Ge Z, Fox JG, Schauer DB (2005) CD4+ CD25+ regulatory lymphocytes induce regression of intestinal tumors in ApcMin/+ mice. Cancer Res 65(10):3998–4004. doi:10.1158/0008-5472.CAN-04-3104
37. Blat D, Zigmond E, Alteber Z, Waks T, Eshhar Z (2014) Suppression of murine colitis and its associated cancer by carcinoembryonic antigen-specific regulatory T cells. Mol Ther 22(5):1018–1028. doi:10.1038/mt.2014.41
38. Erdman SE, Poutahidis T (2010) Roles for inflammation and regulatory T cells in colon cancer. Toxicol Pathol 38(1):76–87. doi:10.1177/0192623309354110

39. Grivennikov SI (2013) Inflammation and colorectal cancer: colitis-associated neoplasia. Semin Immunopathol 35(2):229–244. doi:10.1007/s00281-012-0352-6
40. Schulz O, Jaensson E, Persson EK, Liu X, Worbs T, Agace WW, Pabst O (2009) Intestinal CD103+, but not CX3CR1+, antigen sampling cells migrate in lymph and serve classical dendritic cell functions. J Exp Med 206(13):3101–3114. doi:10.1084/jem.20091925
41. Platt AM, Bain CC, Bordon Y, Sester DP, Mowat AM (2010) An independent subset of TLR expressing CCR2-dependent macrophages promotes colonic inflammation. J Immunol 184(12):6843–6854. doi:10.4049/jimmunol.0903987
42. Smythies LE, Shen R, Bimczok D, Novak L, Clements RH, Eckhoff DE, Bouchard P, George MD, Hu WK, Dandekar S, Smith PD (2010) Inflammation anergy in human intestinal macrophages is due to Smad-induced IkappaBalpha expression and NF-kappaB inactivation. J Biol Chem 285(25):19593–19604. doi:10.1074/jbc.M109.069955
43. Kamada N, Hisamatsu T, Okamoto S, Chinen H, Kobayashi T, Sato T, Sakuraba A, Kitazume MT, Sugita A, Koganei K, Akagawa KS, Hibi T (2008) Unique CD14 intestinal macrophages contribute to the pathogenesis of Crohn disease via IL-23/IFN-gamma axis. J Clin Invest 118(6):2269–2280. doi:10.1172/JCI34610
44. Cader MZ, Kaser A (2013) Recent advances in inflammatory bowel disease: mucosal immune cells in intestinal inflammation. Gut 62(11):1653–1664. doi:10.1136/gutjnl-2012-303955
45. Watanabe N, Ikuta K, Okazaki K, Nakase H, Tabata Y, Matsuura M, Tamaki H, Kawanami C, Honjo T, Chiba T (2003) Elimination of local macrophages in intestine prevents chronic colitis in interleukin-10-deficient mice. Dig Dis Sci 48(2):408–414
46. Zigmond E, Varol C, Farache J, Elmaliah E, Satpathy AT, Friedlander G, Mack M, Shpigel N, Boneca IG, Murphy KM, Shakhar G, Halpern Z, Jung S (2012) Ly6C hi monocytes in the inflamed colon give rise to proinflammatory effector cells and migratory antigen-presenting cells. Immunity 37(6):1076–1090. doi:10.1016/j.immuni.2012.08.026
47. Pollard JW (2004) Tumour-educated macrophages promote tumour progression and metastasis. Nat Rev Cancer 4(1):71–78. doi:10.1038/nrc1256
48. Zhou Q, Peng RQ, Wu XJ, Xia Q, Hou JH, Ding Y, Zhou QM, Zhang X, Pang ZZ, Wan DS, Zeng YX, Zhang XS (2010) The density of macrophages in the invasive front is inversely correlated to liver metastasis in colon cancer. J Transl Med 8:13. doi:10.1186/1479-5876-8-13
49. Forssell J, Oberg A, Henriksson ML, Stenling R, Jung A, Palmqvist R (2007) High macrophage infiltration along the tumor front correlates with improved survival in colon cancer. Clin Cancer Res 13(5):1472–1479. doi:10.1158/1078-0432.CCR-06-2073
50. Chaput N, Svrcek M, Auperin A, Locher C, Drusch F, Malka D, Taieb J, Goere D, Ducreux M, Boige V (2013) Tumour-infiltrating CD68+ and CD57+ cells predict patient outcome in stage II-III colorectal cancer. Br J Cancer 109(4):1013–1022. doi:10.1038/bjc.2013.362
51. Gulubova M, Ananiev J, Yovchev Y, Julianov A, Karashmalakov A, Vlaykova T (2013) The density of macrophages in colorectal cancer is inversely correlated to TGF-beta1 expression and patients' survival. J Mol Histol 44(6):679–692. doi:10.1007/s10735-013-9520-9
52. Funada Y, Noguchi T, Kikuchi R, Takeno S, Uchida Y, Gabbert HE (2003) Prognostic significance of CD8+ T cell and macrophage peritumoral infiltration in colorectal cancer. Oncol Rep 10(2):309–313
53. Ohno S, Inagawa H, Dhar DK, Fujii T, Ueda S, Tachibana M, Suzuki N, Inoue M, Soma G, Nagasue N (2003) The degree of macrophage infiltration into the cancer cell nest is a significant predictor of survival in gastric cancer patients. Anticancer Res 23(6D):5015–5022
54. Kinouchi M, Miura K, Mizoi T, Ishida K, Fujibuchi W, Ando T, Yazaki N, Saito K, Shiiba K, Sasaki I (2011) Infiltration of CD14-positive macrophages at the invasive front indicates a favorable prognosis in colorectal cancer patients with lymph node metastasis. Hepatogastroenterology 58(106):352–358
55. Edin S, Wikberg ML, Dahlin AM, Rutegard J, Oberg A, Oldenborg PA, Palmqvist R (2012) The distribution of macrophages with a M1 or M2 phenotype in relation to prognosis and the molecular characteristics of colorectal cancer. PLoS One 7(10):e47045. doi:10.1371/journal.pone.0047045

56. Algars A, Irjala H, Vaittinen S, Huhtinen H, Sundstrom J, Salmi M, Ristamaki R, Jalkanen S (2012) Type and location of tumor-infiltrating macrophages and lymphatic vessels predict survival of colorectal cancer patients. Int J Cancer 131(4):864–873. doi:10.1002/ijc.26457
57. Sugita J, Ohtani H, Mizoi T, Saito K, Shiiba K, Sasaki I, Matsuno S, Yagita H, Miyazawa M, Nagura H (2002) Close association between Fas ligand (FasL; CD95 L)-positive tumor-associated macrophages and apoptotic cancer cells along invasive margin of colorectal carcinoma: a proposal on tumor-host interactions. Jpn J Cancer Res 93(3):320–328
58. Gulubova MV, Ananiev JR, Vlaykova TI, Yovchev Y, Tsoneva V, Manolova IM (2012) Role of dendritic cells in progression and clinical outcome of colon cancer. Int J Colorectal Dis 27(2):159–169. doi:10.1007/s00384-011-1334-1
59. Garrett WS, Punit S, Gallini CA, Michaud M, Zhang D, Sigrist KS, Lord GM, Glickman JN, Glimcher LH (2009) Colitis-associated colorectal cancer driven by T-bet deficiency in dendritic cells. Cancer Cell 16(3):208–219. doi:10.1016/j.ccr.2009.07.015
60. Garrett WS, Lord GM, Punit S, Lugo-Villarino G, Mazmanian SK, Ito S, Glickman JN, Glimcher LH (2007) Communicable ulcerative colitis induced by T-bet deficiency in the innate immune system. Cell 131(1):33–45. doi:10.1016/j.cell.2007.08.017
61. Powell N, Walker AW, Stolarczyk E, Canavan JB, Gokmen MR, Marks E, Jackson I, Hashim A, Curtis MA, Jenner RG, Howard JK, Parkhill J, MacDonald TT, Lord GM (2012) The transcription factor T-bet regulates intestinal inflammation mediated by interleukin-7 receptor + innate lymphoid cells. Immunity 37(4):674–684. doi:10.1016/j.immuni.2012.09.008
62. Fridlender ZG, Albelda SM (2012) Tumor-associated neutrophils: friend or foe? Carcinogenesis 33(5):949–955. doi:10.1093/carcin/bgs123
63. Houghton AM (2010) The paradox of tumor-associated neutrophils: fueling tumor growth with cytotoxic substances. Cell Cycle 9(9):1732–1737
64. Houghton AM, Rzymkiewicz DM, Ji H, Gregory AD, Egea EE, Metz HE, Stolz DB, L and SR, Marconcini LA, Kliment CR, Jenkins KM, Beaulieu KA, Mouded M, Frank SJ, Wong KK, Shapiro SD (2010) Neutrophil elastase-mediated degradation of IRS-1 accelerates lung tumor growth. Nat Med 16(2):219–223. doi:10.1038/nm.2084
65. Nozawa H, Chiu C, Hanahan D (2006) Infiltrating neutrophils mediate the initial angiogenic switch in a mouse model of multistage carcinogenesis. Proc Natl Acad Sci U S A 103(33):12493–12498. doi:10.1073/pnas.0601807103
66. Rao HL, Chen JW, Li M, Xiao YB, Fu J, Zeng YX, Cai MY, Xie D (2012) Increased intratumoral neutrophil in colorectal carcinomas correlates closely with malignant phenotype and predicts patients' adverse prognosis. PLoS One 7(1):e30806. doi:10.1371/journal.pone.0030806
67. Shang K, Bai YP, Wang C, Wang Z, Gu HY, Du X, Zhou XY, Zheng CL, Chi YY, Mukaida N, Li YY (2012) Crucial involvement of tumor-associated neutrophils in the regulation of chronic colitis-associated carcinogenesis in mice. PLoS One 7(12):e51848. doi:10.1371/journal.pone.0051848
68. Gounaris E, Erdman SE, Restaino C, Gurish MF, Friend DS, Gounari F, Lee DM, Zhang G, Glickman JN, Shin K, Rao VP, Poutahidis T, Weissleder R, McNagny KM, Khazaie K (2007) Mast cells are an essential hematopoietic component for polyp development. Proc Natl Acad Sci U S A 104(50):19977–19982. doi:10.1073/pnas.0704620104
69. Gounaris E, Blatner NR, Dennis K, Magnusson F, Gurish MF, Strom TB, Beckhove P, Gounari F, Khazaie K (2009) T-regulatory cells shift from a protective anti-inflammatory to a cancer-promoting proinflammatory phenotype in polyposis. Cancer Res 69(13):5490–5497. doi:10.1158/0008-5472.CAN-09-0304
70. Egelston C, Kurko J, Besenyei T, Tryniszewska B, Rauch TA, Glant TT, Mikecz K (2012) Suppression of dendritic cell maturation and T cell proliferation by synovial fluid myeloid cells from mice with autoimmune arthritis. Arthritis Rheum 64(10):3179–3188. doi:10.1002/art.34494
71. Zhu B, Bando Y, Xiao S, Yang K, Anderson AC, Kuchroo VK, Khoury SJ (2007) CD11b+ Ly-6C(hi) suppressive monocytes in experimental autoimmune encephalomyelitis. J Immunol 179(8):5228–5237

72. Dilek N, Vuillefroy de Silly R, Blancho G, Vanhove B (2012) Myeloid-derived suppressor cells: mechanisms of action and recent advances in their role in transplant tolerance. Front Immunol 3:208. doi:10.3389/fimmu.2012.00208
73. Saiwai H, Kumamaru H, Ohkawa Y, Kubota K, Kobayakawa K, Yamada H, Yokomizo T, Iwamoto Y, Okada S (2013) Ly6C+ Ly6G– Myeloid-derived suppressor cells play a critical role in the resolution of acute inflammation and the subsequent tissue repair process after spinal cord injury. J Neurochem 125(1):74–88. doi:10.1111/jnc.12135
74. Nagaraj S, Schrum AG, Cho HI, Celis E, Gabrilovich DI (2010) Mechanism of T cell tolerance induced by myeloid-derived suppressor cells. J Immunol 184(6):3106–3116. doi:10.4049/jimmunol.0902661
75. Cheng P, Corzo CA, Luetteke N, Yu B, Nagaraj S, Bui MM, Ortiz M, Nacken W, Sorg C, Vogl T, Roth J, Gabrilovich DI (2008) Inhibition of dendritic cell differentiation and accumulation of myeloid-derived suppressor cells in cancer is regulated by S100A9 protein. J Exp Med 205(10):2235–2249. doi:10.1084/jem.20080132
76. Vogl T, Tenbrock K, Ludwig S, Leukert N, Ehrhardt C, van Zoelen MA, Nacken W, Foell D, van der Poll T, Sorg C, Roth J (2007) Mrp8 and Mrp14 are endogenous activators of Toll-like receptor 4, promoting lethal, endotoxin-induced shock. Nat Med 13(9):1042–1049. doi:10.1038/nm1638
77. Bunt SK, Clements VK, Hanson EM, Sinha P, Ostrand-Rosenberg S (2009) Inflammation enhances myeloid-derived suppressor cell cross-talk by signaling through Toll-like receptor 4. J Leukoc Biol 85(6):996–1004. doi:10.1189/jlb.0708446
78. Kallberg E, Vogl T, Liberg D, Olsson A, Bjork P, Wikstrom P, Bergh A, Roth J, Ivars F, Leanderson T (2012) S100A9 interaction with TLR4 promotes tumor growth. PLoS One 7(3):e34207. doi:10.1371/journal.pone.0034207
79. Hong EH, Chang SY, Lee BR, Kim YS, Lee JM, Kang CY, Kweon MN, Ko HJ (2013) Blockade of Myd88 signaling induces antitumor effects by skewing the immunosuppressive function of myeloid-derived suppressor cells. Int J Cancer 132(12):2839–2848. doi:10.1002/ijc.27974
80. Tartour E, Pere H, Maillere B, Terme M, Merillon N, Taieb J, Sandoval F, Quintin-Colonna F, Lacerda K, Karadimou A, Badoual C, Tedgui A, Fridman WH, Oudard S (2011) Angiogenesis and immunity: a bidirectional link potentially relevant for the monitoring of antiangiogenic therapy and the development of novel therapeutic combination with immunotherapy. Cancer Metastasis Rev 30(1):83–95. doi:10.1007/s10555-011-9281-4
81. Haile LA, von Wasielewski R, Gamrekelashvili J, Kruger C, Bachmann O, Westendorf AM, Buer J, Liblau R, Manns MP, Korangy F, Greten TF (2008) Myeloid-derived suppressor cells in inflammatory bowel disease: a new immunoregulatory pathway. Gastroenterology 135(3):871–881, 881, e871–e875. doi:10.1053/j.gastro.2008.06.032
82. Singh UP, Singh NP, Singh B, Hofseth LJ, Taub DD, Price RL, Nagarkatti M, Nagarkatti PS (2012) Role of resveratrol-induced CD11b(+) Gr-1(+) myeloid derived suppressor cells (MDSCs) in the reduction of CXCR3(+) T cells and amelioration of chronic colitis in IL-10$^{(-/-)}$ mice. Brain Behav Immun 26(1):72–82. doi:10.1016/j.bbi.2011.07.236
83. Zhang R, Ito S, Nishio N, Cheng Z, Suzuki H, Isobe KI (2011) Dextran sulphate sodium increases splenic Gr1(+)CD11b(+) cells which accelerate recovery from colitis following intravenous transplantation. Clin Exp Immunol 164(3):417–427. doi:10.1111/j.1365-2249.2011.04374.x
84. Turovskaya O, Foell D, Sinha P, Vogl T, Newlin R, Nayak J, Nguyen M, Olsson A, Nawroth PP, Bierhaus A, Varki N, Kronenberg M, Freeze HH, Srikrishna G (2008) RAGE, carboxylated glycans and S100A8/A9 play essential roles in colitis-associated carcinogenesis. Carcinogenesis 29(10):2035–2043. doi:10.1093/carcin/bgn188
85. Ichikawa M, Williams R, Wang L, Vogl T, Srikrishna G (2011) S100A8/A9 activate key genes and pathways in colon tumor progression. Mol Cancer Res 9(2):133–148. doi:10.1158/1541-7786.MCR-10-0394

86. Santaolalla R, Sussman DA, Ruiz JR, Davies JM, Pastorini C, Espana CL, Sotolongo J, Burlingame O, Bejarano PA, Philip S, Ahmed MM, Ko J, Dirisina R, Barrett TA, Shang L, Lira SA, Fukata M, Abreu MT (2013) TLR4 activates the beta-catenin pathway to cause intestinal neoplasia. PLoS One 8(5):e63298. doi:10.1371/journal.pone.0063298
87. Sade-Feldman M, Kanterman J, Ish-Shalom E, Elnekave M, Horwitz E, Baniyash M (2013) Tumor necrosis factor-alpha blocks differentiation and enhances suppressive activity of immature myeloid cells during chronic inflammation. Immunity 38(3):541–554. doi:10.1016/j.immuni.2013.02.007
88. Zhang B, Wang Z, Wu L, Zhang M, Li W, Ding J, Zhu J, Wei H, Zhao K (2013) Circulating and tumor-infiltrating myeloid-derived suppressor cells in patients with colorectal carcinoma. PLoS One 8(2):e57114. doi:10.1371/journal.pone.0057114
89. Collart MA, Baeuerle P, Vassalli P (1990) Regulation of tumor necrosis factor alpha transcription in macrophages: involvement of four kappa B-like motifs and of constitutive and inducible forms of NF-kappa B. Mol Cell Biol 10(4):1498–1506
90. Shakhov AN, Collart MA, Vassalli P, Nedospasov SA, Jongeneel CV (1990) Kappa B-type enhancers are involved in lipopolysaccharide-mediated transcriptional activation of the tumor necrosis factor alpha gene in primary macrophages. J Exp Med 171(1):35–47
91. Woo CH, Eom YW, Yoo MH, You HJ, Han HJ, Song WK, Yoo YJ, Chun JS, Kim JH (2000) Tumor necrosis factor-alpha generates reactive oxygen species via a cytosolic phospholipase A2-linked cascade. J Biol Chem 275(41):32357–32362. doi:10.1074/jbc.M005638200
92. Waris G, Ahsan H (2006) Reactive oxygen species: role in the development of cancer and various chronic conditions. J Carcinog 5:14. doi:10.1186/1477-3163-5-14
93. Kim S, Keku TO, Martin C, Galanko J, Woosley JT, Schroeder JC, Satia JA, Halabi S, Sandler RS (2008) Circulating levels of inflammatory cytokines and risk of colorectal adenomas. Cancer Res 68(1):323–328. doi:10.1158/0008-5472.CAN-07-2924
94. Garrity-Park MM, Loftus EV Jr, Bryant SC, Sandborn WJ, Smyrk TC (2008) Tumor necrosis factor-alpha polymorphisms in ulcerative colitis-associated colorectal cancer. Am J Gastroenterol 103(2):407–415. doi:10.1111/j.1572-0241.2007.01572.x
95. Wang K, Han G, Dou Y, Wang Y, Liu G, Wang R, Xiao H, Li X, Hou C, Shen B, Guo R, Li Y, Shi Y, Chen G (2012) Opposite role of tumor necrosis factor receptors in dextran sulfate sodium-induced colitis in mice. PLoS One 7(12):e52924. doi:10.1371/journal.pone.0052924
96. Popivanova BK, Kitamura K, Wu Y, Kondo T, Kagaya T, Kaneko S, Oshima M, Fujii C, Mukaida N (2008) Blocking TNF-alpha in mice reduces colorectal carcinogenesis associated with chronic colitis. J Clin Invest. doi:10.1172/JCI32453
97. Roulis M, Armaka M, Manoloukos M, Apostolaki M, Kollias G (2011) Intestinal epithelial cells as producers but not targets of chronic TNF suffice to cause murine Crohn-like pathology. Proc Natl Acad Sci U S A 108(13):5396–5401. doi:10.1073/pnas.1007811108
98. Wang H, Wang HS, Zhou BH, Li CL, Zhang F, Wang XF, Zhang G, Bu XZ, Cai SH, Du J (2013) Epithelial-mesenchymal transition (EMT) induced by TNF-alpha requires AKT/GSK-3beta-mediated stabilization of snail in colorectal cancer. PLoS One 8(2):e56664. doi:10.1371/journal.pone.0056664
99. Asadullah K, Sterry W, Volk HD (2003) Interleukin-10 therapy—review of a new approach. Pharmacol Rev 55(2):241–269. doi:10.1124/pr.55.2.4
100. Berg DJ, Davidson N, Kuhn R, Muller W, Menon S, Holland G, Thompson-Snipes L, Leach MW, Rennick D (1996) Enterocolitis and colon cancer in interleukin-10-deficient mice are associated with aberrant cytokine production and CD4(+) TH1-like responses. J Clin Invest 98(4):1010–1020
101. Sellon RK, Tonkonogy S, Schultz M, Dieleman LA, Grenther W, Balish E, Rennick DM, Sartor RB (1998) Resident enteric bacteria are necessary for development of spontaneous colitis and immune system activation in interleukin-10-deficient mice. Infect Immun 66(11):5224–5231
102. Narushima S, Spitz DR, Oberley LW, Toyokuni S, Miyata T, Gunnett CA, Buettner GR, Zhang J, Ismail H, Lynch RG, Berg DJ (2003) Evidence for oxidative stress in NSAID-induced colitis in $IL10^{-/-}$ mice. Free Radic Biol Med 34(9):1153–1166

103. Zigmond E, Bernshtein B, Friedlander G, Walker CR, Yona S, Kim KW, Brenner O, Krauthgamer R, Varol C, Muller W, Jung S (2014) Macrophage-restricted interleukin-10 receptor deficiency, but not IL-10 deficiency, causes severe spontaneous colitis. Immunity 40(5):720–733. doi:10.1016/j.immuni.2014.03.012
104. Jostins L, Ripke S, Weersma RK, Duerr RH, McGovern DP, Hui KY, Lee JC, Schumm LP, Sharma Y, Anderson CA, Essers J, Mitrovic M, Ning K, Cleynen I, Theatre E, Spain SL, Raychaudhuri S, Goyette P, Wei Z, Abraham C, Achkar JP, Ahmad T, Amininejad L, Ananthakrishnan AN, Andersen V, Andrews JM, Baidoo L, Balschun T, Bampton PA, Bitton A, Boucher G, Brand S, Buning C, Cohain A, Cichon S, D'Amato M, De Jong D, Devaney KL, Dubinsky M, Edwards C, Ellinghaus D, Ferguson LR, Franchimont D, Fransen K, Gearry R, Georges M, Gieger C, Glas J, Haritunians T, Hart A, Hawkey C, Hedl M, Hu X, Karlsen TH, Kupcinskas L, Kugathasan S, Latiano A, Laukens D, Lawrance IC, Lees CW, Louis E, Mahy G, Mansfield J, Morgan AR, Mowat C, Newman W, Palmieri O, Ponsioen CY, Potocnik U, Prescott NJ, Regueiro M, Rotter JI, Russell RK, Sanderson JD, Sans M, Satsangi J, Schreiber S, Simms LA, Sventoraityte J, Targan SR, Taylor KD, Tremelling M, Verspaget HW, De Vos M, Wijmenga C, Wilson DC, Winkelmann J, Xavier RJ, Zeissig S, Zhang B, Zhang CK, Zhao H, Silverberg MS, Annese V, Hakonarson H, Brant SR, Radford-Smith G, Mathew CG, Rioux JD, Schadt EE, Daly MJ, Franke A, Parkes M, Vermeire S, Barrett JC, Cho JH (2012) Host-microbe interactions have shaped the genetic architecture of inflammatory bowel disease. Nature 491(7422):119–124. doi:10.1038/nature11582
105. Moran CJ, Walters TD, Guo CH, Kugathasan S, Klein C, Turner D, Wolters VM, Bandsma RH, Mouzaki M, Zachos M, Langer JC, Cutz E, Benseler SM, Roifman CM, Silverberg MS, Griffiths AM, Snapper SB, Muise AM (2013) IL-10R polymorphisms are associated with very-early-onset ulcerative colitis. Inflamm Bowel Dis 19(1):115–123. doi:10.1002/ibd.22974
106. Naugler WE, Karin M (2008) The wolf in sheep's clothing: the role of interleukin-6 in immunity, inflammation and cancer. Trends Mol Med 14(3):109–119. doi:10.1016/j.molmed.2007.12.007
107. Matsumoto S, Hara T, Mitsuyama K, Yamamoto M, Tsuruta O, Sata M, Scheller J, Rose-John S, Kado S, Takada T (2010) Essential roles of IL-6 trans-signaling in colonic epithelial cells, induced by the IL-6/soluble-IL-6 receptor derived from lamina propria macrophages, on the development of colitis-associated premalignant cancer in a murine model. J Immunol 184(3):1543–1551. doi:10.4049/jimmunol.0801217
108. Waldner MJ, Foersch S, Neurath MF (2012) Interleukin-6—a key regulator of colorectal cancer development. Int J Biol Sci 8(9):1248–1253. doi:10.7150/ijbs.4614
109. Knupfer H, Preiss R (2010) Serum interleukin-6 levels in colorectal cancer patients—a summary of published results. Int J Colorectal Dis 25(2):135–140. doi:10.1007/s00384-009-0818-8
110. Louis E, Belaiche J, van Kemseke C, Franchimont D, de Groote D, Gueenen V, Mary JY (1997) A high serum concentration of interleukin-6 is predictive of relapse in quiescent Crohn's disease. Eur J Gastroenterol Hepatol 9(10):939–944
111. Van Kemseke C, Belaiche J, Louis E (2000) Frequently relapsing Crohn's disease is characterized by persistent elevation in interleukin-6 and soluble interleukin-2 receptor serum levels during remission. Int J Colorectal Dis 15(4):206–210
112. Atreya R, Neurath MF (2008) New therapeutic strategies for treatment of inflammatory bowel disease. Mucosal Immunol 1(3):175–182. doi:10.1038/mi.2008.7
113. Dinarello CA (2009) Immunological and inflammatory functions of the interleukin-1 family. Annu Rev Immunol 27:519–550. doi:10.1146/annurev.immunol.021908.132612
114. Dupaul-Chicoine J, Yeretssian G, Doiron K, Bergstrom KS, McIntire CR, LeBlanc PM, Meunier C, Turbide C, Gros P, Beauchemin N, Vallance BA, Saleh M (2010) Control of intestinal homeostasis, colitis, and colitis-associated colorectal cancer by the inflammatory caspases. Immunity 32(3):367–378. doi:10.1016/j.immuni.2010.02.012

115. Kaler P, Godasi BN, Augenlicht L, Klampfer L (2009) The NF-kappaB/AKT-dependent Induction of Wnt signaling in colon cancer cells by macrophages and IL-1beta. Cancer Microenviron. doi:10.1007/s12307-009-0030-y
116. Li Y, Wang L, Pappan L, Galliher-Beckley A, Shi J (2012) IL-1beta promotes stemness and invasiveness of colon cancer cells through Zeb1 activation. Mol Cancer 11:87. doi:10.1186/1476-4598-11-87
117. Kaler P, Galea V, Augenlicht L, Klampfer L (2010) Tumor associated macrophages protect colon cancer cells from TRAIL-induced apoptosis through IL-1beta-dependent stabilization of Snail in tumor cells. PLoS One 5(7):e11700. doi:10.1371/journal.pone.0011700
118. Kobayashi T, Okamoto S, Hisamatsu T, Kamada N, Chinen H, Saito R, Kitazume MT, Nakazawa A, Sugita A, Koganei K, Isobe K, Hibi T (2008) IL23 differentially regulates the Th1/Th17 balance in ulcerative colitis and Crohn's disease. Gut 57 (12):1682–1689. doi:10.1136/gut.2007.135053
119. Sugihara T, Kobori A, Imaeda H, Tsujikawa T, Amagase K, Takeuchi K, Fujiyama Y, Andoh A (2010) The increased mucosal mRNA expressions of complement C3 and interleukin-17 in inflammatory bowel disease. Clin Exp Immunol 160(3):386–393. doi:10.1111/j.1365-2249.2010.04093.x
120. Fujino S, Andoh A, Bamba S, Ogawa A, Hata K, Araki Y, Bamba T, Fujiyama Y (2003) Increased expression of interleukin 17 in inflammatory bowel disease. Gut 52(1):65–70
121. Wang Y, Liu XP, Zhao ZB, Chen JH, Yu CG (2011) Expression of CD4+ forkhead box P3 (FOXP3)+ regulatory T cells in inflammatory bowel disease. J Dig Dis 12(4):286–294. doi:10.1111/j.1751-2980.2011.00505.x
122. Maul J, Loddenkemper C, Mundt P, Berg E, Giese T, Stallmach A, Zeitz M, Duchmann R (2005) Peripheral and intestinal regulatory CD4+ CD25(high) T cells in inflammatory bowel disease. Gastroenterology 128(7):1868–1878
123. Hovhannisyan Z, Treatman J, Littman DR, Mayer L (2011) Characterization of interleukin-17-producing regulatory T cells in inflamed intestinal mucosa from patients with inflammatory bowel diseases. Gastroenterology 140(3):957–965. doi:10.1053/j.gastro.2010.12.002
124. Kryczek I, Wu K, Zhao E, Wei S, Vatan L, Szeliga W, Huang E, Greenson J, Chang A, Rolinski J, Radwan P, Fang J, Wang G, Zou W (2011) IL-17+ regulatory T cells in the microenvironments of chronic inflammation and cancer. J Immunol 186(7):4388–4395. doi:10.4049/jimmunol.1003251
125. Salama P, Phillips M, Grieu F, Morris M, Zeps N, Joseph D, Platell C, Iacopetta B (2009) Tumor-infiltrating FOXP3+ T regulatory cells show strong prognostic significance in colorectal cancer. J Clin Oncol 27(2):186–192. doi:10.1200/JCO.2008.18.7229
126. Frey DM, Droeser RA, Viehl CT, Zlobec I, Lugli A, Zingg U, Oertli D, Kettelhack C, Terracciano L, Tornillo L (2010) High frequency of tumor-infiltrating FOXP3(+) regulatory T cells predicts improved survival in mismatch repair-proficient colorectal cancer patients. Int J Cancer 126(11):2635–2643. doi:10.1002/ijc.24989
127. Erdman SE, Poutahidis T (2010) Cancer inflammation and regulatory T cells. Int J Cancer 127(4):768–779. doi:10.1002/ijc.25430
128. Erdman SE, Rao VP, Poutahidis T, Ihrig MM, Ge Z, Feng Y, Tomczak M, Rogers AB, Horwitz BH, Fox JG (2003) CD4(+)CD25(+) regulatory lymphocytes require interleukin 10 to interrupt colon carcinogenesis in mice. Cancer Res 63(18):6042–6050
129. Mei Z, Liu Y, Liu C, Cui A, Liang Z, Wang G, Peng H, Cui L, Li C (2014) Tumour-infiltrating inflammation and prognosis in colorectal cancer: systematic review and meta-analysis. Br J Cancer 110(6):1595–1605. doi:10.1038/bjc.2014.46
130. Maseda D, Candando KM, Smith SH, Kalampokis I, Weaver CT, Plevy SE, Poe JC, Tedder TF (2013) Peritoneal cavity regulatory B cells (B10 cells) modulate IFN-gamma+CD4+ T cell numbers during colitis development in mice. J Immunol 191(5):2780–2795. doi:10.4049/jimmunol.1300649
131. Mizoguchi A, Mizoguchi E, Smith RN, Preffer FI, Bhan AK (1997) Suppressive role of B cells in chronic colitis of T cell receptor alpha mutant mice. J Exp Med 186(10):1749–1756

132. Mizoguchi A, Mizoguchi E, Takedatsu H, Blumberg RS, Bhan AK (2002) Chronic intestinal inflammatory condition generates IL-10-producing regulatory B cell subset characterized by CD1d upregulation. Immunity 16(2):219–230
133. Goetz M, Atreya R, Ghalibafian M, Galle PR, Neurath MF (2007) Exacerbation of ulcerative colitis after rituximab salvage therapy. Inflamm Bowel Dis 13(11):1365–1368. doi:10.1002/ibd.20215
134. Ardelean DS, Gonska T, Wires S, Cutz E, Griffiths A, Harvey E, Tse SM, Benseler SM (2010) Severe ulcerative colitis after rituximab therapy. Pediatrics 126(1):e243–e246. doi:10.1542/peds.2009-3395
135. El Fassi D, Nielsen CH, Kjeldsen J, Clemmensen O, Hegedus L (2008) Ulcerative colitis following B lymphocyte depletion with rituximab in a patient with Graves' disease. Gut 57(5):714–715. doi:10.1136/gut.2007.138305
136. Leiper K, Martin K, Ellis A, Subramanian S, Watson AJ, Christmas SE, Howarth D, Campbell F, Rhodes JM (2011) Randomised placebo-controlled trial of rituximab (anti-CD20) in active ulcerative colitis. Gut 60(11):1520–1526. doi:10.1136/gut.2010.225482
137. Erdag G, Schaefer JT, Smolkin ME, Deacon DH, Shea SM, Dengel LT, Patterson JW, Slingluff CL Jr (2012) Immunotype and immunohistologic characteristics of tumor-infiltrating immune cells are associated with clinical outcome in metastatic melanoma. Cancer Res 72(5):1070–1080. doi:10.1158/0008-5472.CAN-11-3218
138. Schmidt M, Bohm D, von Torne C, Steiner E, Puhl A, Pilch H, Lehr HA, Hengstler JG, Kolbl H, Gehrmann M (2008) The humoral immune system has a key prognostic impact in node-negative breast cancer. Cancer Res 68(13):5405–5413. doi:10.1158/0008-5472.CAN-07-5206
139. Milne K, Kobel M, Kalloger SE, Barnes RO, Gao D, Gilks CB, Watson PH, Nelson BH (2009) Systematic analysis of immune infiltrates in high-grade serous ovarian cancer reveals CD20, FoxP3 and TIA-1 as positive prognostic factors. PLoS One 4(7):e6412. doi:10.1371/journal.pone.0006412
140. Linnebacher M (2013) Tumor-infiltrating B cells come into vogue. World J Gastroenterol 19(1):8–11. doi:10.3748/wjg.v19.i1.8
141. Maletzki C, Jahnke A, Ostwald C, Klar E, Prall F, Linnebacher M (2012) Ex-vivo clonally expanded B lymphocytes infiltrating colorectal carcinoma are of mature immunophenotype and produce functional IgG. PLoS One 7(2):e32639. doi:10.1371/journal.pone.0032639
142. Barbera-Guillem E, Nelson MB, Barr B, Nyhus JK, May KF Jr, Feng L, Sampsel JW (2000) B lymphocyte pathology in human colorectal cancer. Experimental and clinical therapeutic effects of partial B cell depletion. Cancer Immunol Immunother 48(10):541–549
143. Erdman SE, Rao VP, Poutahidis T, Rogers AB, Taylor CL, Jackson EA, Ge Z, Lee CW, Schauer DB, Wogan GN, Tannenbaum SR, Fox JG (2009) Nitric oxide and TNF-alpha trigger colonic inflammation and carcinogenesis in Helicobacter hepaticus-infected, Rag2-deficient mice. Proc Natl Acad Sci U S A 106(4):1027–1032. pii:0812347106. doi:10.1073/pnas.0812347106
144. Meira LB, Bugni JM, Green SL, Lee CW, Pang B, Borenshtein D, Rickman BH, Rogers AB, Moroski-Erkul CA, McFaline JL, Schauer DB, Dedon PC, Fox JG, Samson LD (2008) DNA damage induced by chronic inflammation contributes to colon carcinogenesis in mice. J Clin Investig 118(7):2516–2525. doi:10.1172/JCI35073
145. Halliwell B (1999) Oxygen and nitrogen are pro-carcinogens. Damage to DNA by reactive oxygen, chlorine and nitrogen species: measurement, mechanism and the effects of nutrition. Mutat Res 443(1–2):37–52
146. Westbrook AM, Schiestl RH (2010) Atm-deficient mice exhibit increased sensitivity to dextran sulfate sodium-induced colitis characterized by elevated DNA damage and persistent immune activation. Cancer Res 70(5):1875–1884. doi:10.1158/0008-5472.CAN-09-2584
147. Takai A, Marusawa H, Minaki Y, Watanabe T, Nakase H, Kinoshita K, Tsujimoto G, Chiba T (2012) Targeting activation-induced cytidine deaminase prevents colon cancer development despite persistent colonic inflammation. Oncogene 31(13):1733–1742. doi:10.1038/onc.2011.352

148. Edwards RA, Witherspoon M, Wang K, Afrasiabi K, Pham T, Birnbaumer L, Lipkin SM (2009) Epigenetic repression of DNA mismatch repair by inflammation and hypoxia in inflammatory bowel disease-associated colorectal cancer. Cancer Res 69(16):6423–6429. doi:10.1158/0008-5472.CAN-09-1285
149. Endo Y, Marusawa H, Kou T, Nakase H, Fujii S, Fujimori T, Kinoshita K, Honjo T, Chiba T (2008) Activation-induced cytidine deaminase links between inflammation and the development of colitis-associated colorectal cancers. Gastroenterology 135(3):889–898, 898, e881–e883. doi:10.1053/j.gastro.2008.06.091
150. Hammoud SS, Cairns BR, Jones DA (2013) Epigenetic regulation of colon cancer and intestinal stem cells. Curr Opin Cell Biol 25(2):177–183. doi:10.1016/j.ceb.2013.01.007
151. Saito S, Kato J, Hiraoka S, Horii J, Suzuki H, Higashi R, Kaji E, Kondo Y, Yamamoto K (2011) DNA methylation of colon mucosa in ulcerative colitis patients: correlation with inflammatory status. Inflamm Bowel Dis 17(9):1955–1965. doi:10.1002/ibd.21573
152. Wang FY, Arisawa T, Tahara T, Takahama K, Watanabe M, Hirata I, Nakano H (2008) Aberrant DNA methylation in ulcerative colitis without neoplasia. Hepatogastroenterology 55(81):62–65
153. Issa JP, Ahuja N, Toyota M, Bronner MP, Brentnall TA (2001) Accelerated age-related CpG island methylation in ulcerative colitis. Cancer Res 61(9):3573–3577
154. Li Y, Deuring J, Peppelenbosch MP, Kuipers EJ, de Haar C, van der Woude CJ (2012) IL-6-induced DNMT1 activity mediates SOCS3 promoter hypermethylation in ulcerative colitis-related colorectal cancer. Carcinogenesis 33(10):1889–1896. doi:10.1093/carcin/bgs214
155. Eads CA, Nickel AE, Laird PW (2002) Complete genetic suppression of polyp formation and reduction of CpG-island hypermethylation in Apc(Min/+) Dnmt1-hypomorphic Mice. Cancer Res 62(5):1296–1299
156. Dhir M, Montgomery EA, Glockner SC, Schuebel KE, Hooker CM, Herman JG, Baylin SB, Gearhart SL, Ahuja N (2008) Epigenetic regulation of WNT signaling pathway genes in inflammatory bowel disease (IBD) associated neoplasia. J Gastrointest Surg 12(10):1745–1753. doi:10.1007/s11605-008-0633-5
157. Fleisher AS, Esteller M, Harpaz N, Leytin A, Rashid A, Xu Y, Liang J, Stine OC, Yin J, Zou TT, Abraham JM, Kong D, Wilson KT, James SP, Herman JG, Meltzer SJ (2000) Microsatellite instability in inflammatory bowel disease-associated neoplastic lesions is associated with hypermethylation and diminished expression of the DNA mismatch repair gene, hMLH1. Cancer Res 60(17):4864–4868
158. Glauben R, Batra A, Stroh T, Erben U, Fedke I, Lehr HA, Leoni F, Mascagni P, Dinarello CA, Zeitz M, Siegmund B (2008) Histone deacetylases: novel targets for prevention of colitis-associated cancer in mice. Gut. pii:gut.2007.134650. doi:10.1136/gut.2007.134650
159. Motoyama K, Inoue H, Takatsuno Y, Tanaka F, Mimori K, Uetake H, Sugihara K, Mori M (2009) Over- and under-expressed microRNAs in human colorectal cancer. Int J Oncol 34(4):1069–1075
160. Kanaan Z, Rai SN, Eichenberger MR, Barnes C, Dworkin AM, Weller C, Cohen E, Roberts H, Keskey B, Petras RE, Crawford NP, Galandiuk S (2012) Differential microRNA expression tracks neoplastic progression in inflammatory bowel disease-associated colorectal cancer. Hum Mutat 33(3):551–560. doi:10.1002/humu.22021
161. Josse C, Bouznad N, Geurts P, Irrthum A, Huynh-Thu VA, Servais L, Hego A, Delvenne P, Bours V, Oury C (2014) Identification of a microRNA landscape targeting the PI3K/Akt signaling pathway in inflammation-induced colorectal carcinogenesis. Am J Physiol Gastrointest Liver Physiol 306(3):G229–G243. doi:10.1152/ajpgi.00484.2012
162. Iliopoulos D, Jaeger SA, Hirsch HA, Bulyk ML, Struhl K (2010) STAT3 activation of miR-21 and miR-181b-1 via PTEN and CYLD are part of the epigenetic switch linking inflammation to cancer. Mol Cell 39(4):493–506. doi:10.1016/j.molcel.2010.07.023
163. Okayama H, Schetter AJ, Harris CC (2012) MicroRNAs and inflammation in the pathogenesis and progression of colon cancer. Dig Dis 30(2):9–15. doi:10.1159/000341882
164. Sturm A, Dignass AU (2008) Epithelial restitution and wound healing in inflammatory bowel disease. World J Gastroenterol 14(3):348–353

165. Nenci A, Becker C, Wullaert A, Gareus R, van Loo G, Danese S, Huth M, Nikolaev A, Neufert C, Madison B, Gumucio D, Neurath MF, Pasparakis M (2007) Epithelial NEMO links innate immunity to chronic intestinal inflammation. Nature 446(7135):557–561. doi:10.1038/nature05698
166. Hayden MS, West AP, Ghosh S (2006) NF-kappaB and the immune response. Oncogene 25(51):6758–6780. doi:10.1038/sj.onc.1209943
167. Kawai T, Akira S (2007) Signaling to NF-kappaB by Toll-like receptors. Trends Mol Med 13(11):460–469. doi:10.1016/j.molmed.2007.09.002
168. Vallabhapurapu S, Karin M (2009) Regulation and function of NF-kappaB transcription factors in the immune system. Annu Rev Immunol 27:693–733. doi:10.1146/annurev.immunol.021908.132641
169. Rogler G, Brand K, Vogl D, Page S, Hofmeister R, Andus T, Knuechel R, Baeuerle PA, Scholmerich J, Gross V (1998) Nuclear factor kappaB is activated in macrophages and epithelial cells of inflamed intestinal mucosa. Gastroenterology 115(2):357–369
170. Sakamoto K, Maeda S, Hikiba Y, Nakagawa H, Hayakawa Y, Shibata W, Yanai A, Ogura K, Omata M (2009) Constitutive NF-kappaB activation in colorectal carcinoma plays a key role in angiogenesis, promoting tumor growth. Clin Cancer Res 15(7):2248–2258. doi:10.1158/1078-0432.CCR-08-1383
171. Greten FR, Eckmann L, Greten TF, Park JM, Li ZW, Egan LJ, Kagnoff MF, Karin M (2004) IKKbeta links inflammation and tumorigenesis in a mouse model of colitis-associated cancer. Cell 118(3):285–296
172. Lauscher JC, Grone J, Dullat S, Hotz B, Ritz JP, Steinhoff U, Buhr HJ, Visekruna A (2010) Association between activation of atypical NF-kappaB1 p105 signaling pathway and nuclear beta-catenin accumulation in colorectal carcinoma. Mol Carcinog 49(2):121–129. doi:10.1002/mc.20606
173. Bollrath J, Phesse TJ, von Burstin VA, Putoczki T, Bennecke M, Bateman T, Nebelsiek T, Lundgren-May T, Canli O, Schwitalla S, Matthews V, Schmid RM, Kirchner T, Arkan MC, Ernst M, Greten FR (2009) gp130-mediated Stat3 activation in enterocytes regulates cell survival and cell-cycle progression during colitis-associated tumorigenesis. Cancer Cell 15(2):91–102. doi:10.1016/j.ccr.2009.01.002
174. Grivennikov S, Karin E, Terzic J, Mucida D, Yu GY, Vallabhapurapu S, Scheller J, Rose-John S, Cheroutre H, Eckmann L, Karin M (2009) IL-6 and Stat3 are required for survival of intestinal epithelial cells and development of colitis-associated cancer. Cancer Cell 15(2):103–113. doi:10.1016/j.ccr.2009.01.001
175. Takeda K, Clausen BE, Kaisho T, Tsujimura T, Terada N, Forster I, Akira S (1999) Enhanced Th1 activity and development of chronic enterocolitis in mice devoid of Stat3 in macrophages and neutrophils. Immunity 10(1):39–49
176. Welte T, Zhang SS, Wang T, Zhang Z, Hesslein DG, Yin Z, Kano A, Iwamoto Y, Li E, Craft JE, Bothwell AL, Fikrig E, Koni PA, Flavell RA, Fu XY (2003) STAT3 deletion during hematopoiesis causes Crohn's disease-like pathogenesis and lethality: a critical role of STAT3 in innate immunity. Proc Natl Acad Sci U S A 100(4):1879–1884. doi:10.1073/pnas.0237137100
177. Kobayashi M, Kweon MN, Kuwata H, Schreiber RD, Kiyono H, Takeda K, Akira S (2003) Toll-like receptor-dependent production of IL-12p40 causes chronic enterocolitis in myeloid cell-specific Stat3-deficient mice. J Clin Invest 111(9):1297–1308. doi:10.1172/JCI17085
178. Waldner MJ, Wirtz S, Jefremow A, Warntjen M, Neufert C, Atreya R, Becker C, Weigmann B, Vieth M, Rose-John S, Neurath MF (2010) VEGF receptor signaling links inflammation and tumorigenesis in colitis-associated cancer. J Exp Med 207(13):2855–2868. doi:10.1084/jem.20100438
179. Musteanu M, Blaas L, Mair M, Schlederer M, Bilban M, Tauber S, Esterbauer H, Mueller M, Casanova E, Kenner L, Poli V, Eferl R (2010) Stat3 is a negative regulator of intestinal tumor progression in Apc(Min) mice. Gastroenterology 138(3):1003–1011, e1001–e1005. doi:10.1053/j.gastro.2009.11.049

180. Bollrath J, Greten FR (2009) IKK/NF-kappaB and STAT3 pathways: central signalling hubs in inflammation-mediated tumour promotion and metastasis. EMBO Rep 10(12):1314–1319. doi:10.1038/embor.2009.243
181. de Lau W, Barker N, Clevers H (2007) WNT signaling in the normal intestine and colorectal cancer. Front Biosci 12:471–491
182. Kanwar SS, Yu Y, Nautiyal J, Patel BB, Majumdar AP (2010) The Wnt/beta-catenin pathway regulates growth and maintenance of colonospheres. Mol Cancer 9:212. doi:10.1186/1476-4598-9-212
183. Claessen MM, Schipper ME, Oldenburg B, Siersema PD, Offerhaus GJ, Vleggaar FP (2010) WNT-pathway activation in IBD-associated colorectal carcinogenesis: potential biomarkers for colonic surveillance. Cell Oncol 32(4):303–310. pii:F4784644K33268J1. doi:10.3233/CLO-2009-0503
184. van Dekken H, Wink JC, Vissers KJ, Franken PF, Ruud Schouten W, WC JH, Kuipers EJ, Fodde R, Janneke van der Woude C (2007) Wnt pathway-related gene expression during malignant progression in ulcerative colitis. Acta Histochem 109(4):266–272. doi:10.1016/j.acthis.2007.02.007
185. Castellone MD, Teramoto H, Williams BO, Druey KM, Gutkind JS (2005) Prostaglandin E2 promotes colon cancer cell growth through a Gs-axin-beta-catenin signaling axis. Science 310(5753):1504–1510
186. Duan Y, Liao AP, Kuppireddi S, Ye Z, Ciancio MJ, Sun J (2007) Beta-catenin activity negatively regulates bacteria-induced inflammation. Lab Invest 87(6):613–624. doi:10.1038/labinvest.3700545
187. Deng J, Xia W, Miller SA, Wen Y, Wang HY, Hung MC (2004) Crossregulation of NF-kappaB by the APC/GSK-3beta/beta-catenin pathway. Mol Carcinog 39(3):139–146. doi:10.1002/mc.10169
188. Du Q, Zhang X, Cardinal J, Cao Z, Guo Z, Shao L, Geller DA (2009) Wnt/beta-catenin signaling regulates cytokine-induced human inducible nitric oxide synthase expression by inhibiting nuclear factor-kappaB activation in cancer cells. Cancer Res 69(9):3764–3771. doi:10.1158/0008-5472.CAN-09-0014
189. Schon S, Flierman I, Ofner A, Stahringer A, Holdt LM, Kolligs FT, Herbst A (2014) Beta-catenin regulates NF-kappaB activity via TNFRSF19 in colorectal cancer cells. Int J Cancer. doi:10.1002/ijc.28839
190. Chandrakesan P, Jakkula LU, Ahmed I, Roy B, Anant S, Umar S (2013) Differential effects of beta-catenin and NF-kappaB interplay in the regulation of cell proliferation, inflammation and tumorigenesis in response to bacterial infection. PLoS One 8(11):e79432. doi:10.1371/journal.pone.0079432
191. Chen W, Liu F, Ling Z, Tong X, Xiang C (2012) Human intestinal lumen and mucosa-associated microbiota in patients with colorectal cancer. PLoS One 7(6):e39743. doi:10.1371/journal.pone.0039743
192. Frank DN, Robertson CE, Hamm CM, Kpadeh Z, Zhang T, Chen H, Zhu W, Sartor RB, Boedeker EC, Harpaz N, Pace NR, Li E (2011) Disease phenotype and genotype are associated with shifts in intestinal-associated microbiota in inflammatory bowel diseases. Inflamm Bowel Dis 17(1):179–184. doi:10.1002/ibd.21339
193. Shen XJ, Rawls JF, Randall T, Burcal L, Mpande CN, Jenkins N, Jovov B, Abdo Z, Sandler RS, Keku TO (2010) Molecular characterization of mucosal adherent bacteria and associations with colorectal adenomas. Gut Microbes 1(3):138–147. doi:10.4161/gmic.1.3.12360
194. Sobhani I, Tap J, Roudot-Thoraval F, Roperch JP, Letulle S, Langella P, Corthier G, Tran Van Nhieu J, Furet JP (2011) Microbial dysbiosis in colorectal cancer (CRC) patients. PLoS One 6(1):e16393. doi:10.1371/journal.pone.0016393
195. Wang Y, Devkota S, Musch MW, Jabri B, Nagler C, Antonopoulos DA, Chervonsky A, Chang EB (2010) Regional mucosa-associated microbiota determine physiological expression of TLR2 and TLR4 in murine colon. PLoS One 5(10):e13607. doi:10.1371/journal.pone.0013607

196. Willing BP, Dicksved J, Halfvarson J, Andersson AF, Lucio M, Zheng Z, Jarnerot G, Tysk C, Jansson JK, Engstrand L (2010) A pyrosequencing study in twins shows that gastrointestinal microbial profiles vary with inflammatory bowel disease phenotypes. Gastroenterology 139(6):1844–1854, e1841. doi:10.1053/j.gastro.2010.08.049
197. Uronis JM, Muhlbauer M, Herfarth HH, Rubinas TC, Jones GS, Jobin C (2009) Modulation of the intestinal microbiota alters colitis-associated colorectal cancer susceptibility. PLoS One 4(6):e6026. doi:10.1371/journal.pone.0006026
198. Nagamine CM, Rogers AB, Fox JG, Schauer DB (2008) Helicobacter hepaticus promotes azoxymethane-initiated colon tumorigenesis in BALB/c-IL10-deficient mice. Int J Cancer 122(4):832–838. doi:10.1002/ijc.23175
199. Arthur JC, Perez-Chanona E, Muhlbauer M, Tomkovich S, Uronis JM, Fan TJ, Campbell BJ, Abujamel T, Dogan B, Rogers AB, Rhodes JM, Stintzi A, Simpson KW, Hansen JJ, Keku TO, Fodor AA, Jobin C (2012) Intestinal inflammation targets cancer-inducing activity of the microbiota. Science 338(6103):120–123. doi:10.1126/science.1224820
200. Wu S, Rhee KJ, Albesiano E, Rabizadeh S, Wu X, Yen HR, Huso DL, Brancati FL, Wick E, McAllister F, Housseau F, Pardoll DM, Sears CL (2009) A human colonic commensal promotes colon tumorigenesis via activation of T helper type 17 T cell responses. Nat Med 15(9):1016–1022. doi:10.1038/nm.2015
201. Sears CL, Pardoll DM (2011) Perspective: alpha-bugs, their microbial partners, and the link to colon cancer. J Infect Dis 203(3):306–311. doi:10.1093/jinfdis/jiq061
202. Backhed F, Ley RE, Sonnenburg JL, Peterson DA, Gordon JI (2005) Host-bacterial mutualism in the human intestine. Science 307(5717):1915–1920
203. Couturier-Maillard A, Secher T, Rehman A, Normand S, De Arcangelis A, Haesler R, Huot L, Grandjean T, Bressenot A, Delanoye-Crespin A, Gaillot O, Schreiber S, Lemoine Y, Ryffel B, Hot D, Nunez G, Chen G, Rosenstiel P, Chamaillard M (2013) NOD2-mediated dysbiosis predisposes mice to transmissible colitis and colorectal cancer. J Clin Invest 123(2):700–711. doi:10.1172/JCI62236
204. Liang X, Li H, Tian G, Li S (2014) Dynamic microbe and molecule networks in a mouse model of colitis-associated colorectal cancer. Sci Rep 4:4985. doi:10.1038/srep04985
205. Tjalsma H, Boleij A, Marchesi JR, Dutilh BE (2012) A bacterial driver-passenger model for colorectal cancer: beyond the usual suspects. Nat Rev Microbiol 10(8):575–582. doi:10.1038/nrmicro2819
206. Moossavi S, Rezaei N (2013) Toll-like receptor signalling and their therapeutic targeting in colorectal cancer. Int Immunopharmacol 16(2):199–209. doi:10.1016/j.intimp.2013.03.017
207. Cario E, Podolsky DK (2000) Differential alteration in intestinal epithelial cell expression of toll-like receptor 3 (TLR3) and TLR4 in inflammatory bowel disease. Infect Immun 68(12):7010–7017
208. Hausmann M, Kiessling S, Mestermann S, Webb G, Spottl T, Andus T, Scholmerich J, Herfarth H, Ray K, Falk W, Rogler G (2002) Toll-like receptors 2 and 4 are up-regulated during intestinal inflammation. Gastroenterology 122(7):1987–2000
209. Fukata M, Chen A, Vamadevan AS, Cohen J, Breglio K, Krishnareddy S, Hsu D, Xu R, Harpaz N, Dannenberg AJ, Subbaramaiah K, Cooper HS, Itzkowitz SH, Abreu MT (2007) Toll-like receptor-4 promotes the development of colitis-associated colorectal tumors. Gastroenterology 133(6):1869–1881. doi:10.1053/j.gastro.2007.09.008
210. Lowe EL, Crother TR, Rabizadeh S, Hu B, Wang H, Chen S, Shimada K, Wong MH, Michelsen KS, Arditi M (2010) Toll-like receptor 2 signaling protects mice from tumor development in a mouse model of colitis-induced cancer. PLoS One 5(9):e13027. doi:10.1371/journal.pone.0013027
211. Salcedo R, Worschech A, Cardone M, Jones Y, Gyulai Z, Dai RM, Wang E, Ma W, Haines D, O'HUigin C, Marincola FM, Trinchieri G (2010) MyD88-mediated signaling prevents development of adenocarcinomas of the colon: role of interleukin 18. J Exp Med 207(8):1625–1636. doi:10.1084/jem.20100199

212. Rakoff-Nahoum S, Medzhitov R (2007) Regulation of spontaneous intestinal tumorigenesis through the adaptor protein MyD88. Science 317(5834):124–127. doi:10.1126/science.1140488
213. Kim YS, Milner JA (2007) Dietary modulation of colon cancer risk. J Nutr 137(11 Suppl):2576S–2579S
214. Pericleous M, Mandair D, Caplin ME (2013) Diet and supplements and their impact on colorectal cancer. J Gastrointest Oncol 4(4):409–423. doi:10.3978/j.issn.2078-6891.2013.003
215. Magalhaes B, Peleteiro B, Lunet N (2012) Dietary patterns and colorectal cancer: systematic review and meta-analysis. Eur J Cancer Prevent 21(1):15–23. doi:10.1097/CEJ.0b013e3283472241
216. Ou J, Carbonero F, Zoetendal EG, DeLany JP, Wang M, Newton K, Gaskins HR, O'Keefe SJ (2013) Diet, microbiota, and microbial metabolites in colon cancer risk in rural Africans and African Americans. Am J Clin Nutr 98(1):111–120. doi:10.3945/ajcn.112.056689
217. Hofmanova J, Strakova N, Vaculova AH, Tylichova Z, Safarikova B, Skender B, Kozubik A (2014) Interaction of dietary fatty acids with tumour necrosis factor family cytokines during colon inflammation and cancer. Mediators Inflamm 2014:848632. doi:10.1155/2014/848632
218. Bernstein C, Holubec H, Bhattacharyya AK, Nguyen H, Payne CM, Zaitlin B, Bernstein H (2011) Carcinogenicity of deoxycholate, a secondary bile acid. Arch Toxicol 85 (8):863–871. doi:10.1007/s00204-011-0648-7
219. Yasui Y, Hosokawa M, Mikami N, Miyashita K, Tanaka T (2011) Dietary astaxanthin inhibits colitis and colitis-associated colon carcinogenesis in mice via modulation of the inflammatory cytokines. Chem Biol Interact 193(1):79–87. doi:10.1016/j.cbi.2011.05.006
220. Chung MY, Lim TG, Lee KW (2013) Molecular mechanisms of chemopreventive phytochemicals against gastroenterological cancer development. World J Gastroenterol 19(7):984–993. doi:10.3748/wjg.v19.i7.984
221. Piazzi G, D'Argenio G, Prossomariti A, Lembo V, Mazzone G, Candela M, Biagi E, Brigidi P, Vitaglione P, Fogliano V, D'Angelo L, Fazio C, Munarini A, Belluzzi A, Ceccarelli C, Chieco P, Balbi T, Loadman PM, Hull MA, Romano M, Bazzoli F, Ricciardiello L (2014) Eicosapentaenoic acid free fatty acid prevents and suppresses colonic neoplasia in colitis-associated colorectal cancer acting on Notch signaling and gut microbiota. Int J Cancer. doi:10.1002/ijc.28853
222. Meeker S, Seamons A, Paik J, Treuting PM, Brabb T, Grady WM, Maggio-Price L (2014) Increased dietary vitamin D suppresses MAPK signaling, colitis and colon cancer. Cancer Res. doi:10.1158/0008-5472.CAN-13-2820
223. Murillo G, Nagpal V, Tiwari N, Benya RV, Mehta RG (2010) Actions of vitamin D are mediated by the TLR4 pathway in inflammation-induced colon cancer. J Steroid Biochem Mol Biol 121(1–2):403–407. doi:10.1016/j.jsbmb.2010.03.009
224. Chapkin RS, Kamen BA, Callaway ES, Davidson LA, George NI, Wang N, Lupton JR, Finnell RH (2009) Use of a novel genetic mouse model to investigate the role of folate in colitis-associated colon cancer. J Nutr Biochem 20(8):649–655. doi:10.1016/j.jnutbio.2008.07.001
225. Ning Y, Wang L, Giovannucci EL (2010) A quantitative analysis of body mass index and colorectal cancer: findings from 56 observational studies. Obes Rev 11(1):19–30. doi:10.1111/j.1467-789X.2009.00613.x
226. Aleksandrova K, Nimptsch K, Pischon T (2013) Obesity and colorectal cancer. Front Biosci 5:61–77
227. Chen J, Huang XF (2009) Adiponectin in obesity-associated colon cancer and its preventive implications: comment on "Association of visceral fat accumulation and adiponectin levels with colorectal neoplasia". Dig Dis Sci 54(8):1810–1811. doi:10.1007/s10620-009-0857-6
228. Yehuda-Shnaidman E, Schwartz B (2012) Mechanisms linking obesity, inflammation and altered metabolism to colon carcinogenesis. Obes Rev 13(12):1083–1095. doi:10.1111/j.1467-789X.2012.01024.x

229. Arkan MC, Hevener AL, Greten FR, Maeda S, Li ZW, Long JM, Wynshaw-Boris A, Poli G, Olefsky J, Karin M (2005) IKK-beta links inflammation to obesity-induced insulin resistance. Nat Med 11(2):191–198. pii:nm1185. doi:10.1038/nm1185
230. Suganami T, Tanimoto-Koyama K, Nishida J, Itoh M, Yuan X, Mizuarai S, Kotani H, Yamaoka S, Miyake K, Aoe S, Kamei Y, Ogawa Y (2007) Role of the Toll-like receptor 4/NF-kappaB pathway in saturated fatty acid-induced inflammatory changes in the interaction between adipocytes and macrophages. Arterioscler Thromb Vasc Biol 27(1):84–91. doi:10.1161/01.ATV.0000251608.09329.9a
231. de La Serre CB, Ellis CL, Lee J, Hartman AL, Rutledge JC, Raybould HE (2010) Propensity to high-fat diet-induced obesity in rats is associated with changes in the gut microbiota and gut inflammation. Am J Physiol Gastrointest Liver Physiol 299(2):G440–G448. doi:10.1152/ajpgi.00098.2010
232. Fenton JI, Nunez NP, Yakar S, Perkins SN, Hord NG, Hursting SD (2009) Diet-induced adiposity alters the serum profile of inflammation in C57BL/6N mice as measured by antibody array. Diabetes Obes Metab 11(4):343–354. doi:10.1111/j.1463-1326.2008.00974.x
233. Siegmund B, Lehr HA, Fantuzzi G (2002) Leptin: a pivotal mediator of intestinal inflammation in mice. Gastroenterology 122(7):2011–2025
234. Tuominen I, Al-Rabadi L, Stavrakis D, Karagiannides I, Pothoulakis C, Bugni JM (2013) Diet-induced obesity promotes colon tumor development in azoxymethane-treated mice. PLoS One 8(4):e60939. doi:10.1371/journal.pone.0060939
235. Park SY, Kim JS, Seo YR, Sung MK (2012) Effects of diet-induced obesity on colitis-associated colon tumor formation in A/J mice. Int J Obes (Lond) 36(2):273–280. doi:10.1038/ijo.2011.83
236. Jain SS, Bird RP (2010) Elevated expression of tumor necrosis factor-alpha signaling molecules in colonic tumors of Zucker obese (fa/fa) rats. Int J Cancer 127(9):2042–2050. doi:10.1002/ijc.25232
237. Teraoka N, Mutoh M, Takasu S, Ueno T, Nakano K, Takahashi M, Imai T, Masuda S, Sugimura T, Wakabayashi K (2011) High susceptibility to azoxymethane-induced colorectal carcinogenesis in obese KK-Ay mice. Int J Cancer 129(3):528–535. doi:10.1002/ijc.25711
238. Ealey KN, Lu S, Archer MC (2008) Development of aberrant crypt foci in the colons of ob/ob and db/db mice: evidence that leptin is not a promoter. Mol Carcinog 47(9):667–677. doi:10.1002/mc.20419
239. Hirose Y, Hata K, Kuno T, Yoshida K, Sakata K, Yamada Y, Tanaka T, Reddy BS, Mori H (2004) Enhancement of development of azoxymethane-induced colonic premalignant lesions in C57BL/KsJ-db/db mice. Carcinogenesis 25(5):821–825. doi:10.1093/carcin/bgh059
240. Flores MB, Rocha GZ, Damas-Souza DM, Osorio-Costa F, Dias MM, Ropelle ER, Camargo JA, de Carvalho RB, Carvalho HF, Saad MJ, Carvalheira JB (2012) Obesity-induced increase in tumor necrosis factor-alpha leads to development of colon cancer in mice. Gastroenterology 143(3):741–753, e741–e744. doi:10.1053/j.gastro.2012.05.045
241. Backhed F, Manchester JK, Semenkovich CF, Gordon JI (2007) Mechanisms underlying the resistance to diet-induced obesity in germ-free mice. Proc Natl Acad Sci U S A 104(3):979–984. doi:10.1073/pnas.0605374104
242. Ley RE, Backhed F, Turnbaugh P, Lozupone CA, Knight RD, Gordon JI (2005) Obesity alters gut microbial ecology. Proc Natl Acad S U S A 102(31):11070–11075. doi:10.1073/pnas.0504978102
243. Ley RE, Turnbaugh PJ, Klein S, Gordon JI (2006) Microbial ecology: human gut microbes associated with obesity. Nature 444(7122):1022–1023. doi:10.1038/4441022a
244. Turnbaugh PJ, Backhed F, Fulton L, Gordon JI (2008) Diet-induced obesity is linked to marked but reversible alterations in the mouse distal gut microbiome. Cell Host Microbe 3(4):213–223. doi:10.1016/j.chom.2008.02.015
245. Brun P, Castagliuolo I, Di Leo V, Buda A, Pinzani M, Palu G, Martines D (2007) Increased intestinal permeability in obese mice: new evidence in the pathogenesis of nonalcoholic steatohepatitis. Am J Physiol Gastrointest Liver Physiol 292(2):G518–G525. doi:10.1152/ajpgi.00024.2006

246. Erridge C, Attina T, Spickett CM, Webb DJ (2007) A high-fat meal induces low-grade endotoxemia: evidence of a novel mechanism of postprandial inflammation. Am J Clinical Nutr 86(5):1286–1292
247. Lassenius MI, Pietilainen KH, Kaartinen K, Pussinen PJ, Syrjanen J, Forsblom C, Porsti I, Rissanen A, Kaprio J, Mustonen J, Groop PH, Lehto M (2011) Bacterial endotoxin activity in human serum is associated with dyslipidemia, insulin resistance, obesity, and chronic inflammation. Diabetes Care 34(8):1809–1815. doi:10.2337/dc10-2197
248. Kim SJ, Choi Y, Choi YH, Park T (2011) Obesity activates toll-like receptor-mediated proinflammatory signaling cascades in the adipose tissue of mice. J Nutr Biochem. doi:10.1016/j.jnutbio.2010.10.012
249. Nakarai H, Yamashita A, Nagayasu S, Iwashita M, Kumamoto S, Ohyama H, Hata M, Soga Y, Kushiyama A, Asano T, Abiko Y, Nishimura F (2011) Adipocyte-macrophage interaction may mediate LPS-induced low-grade inflammation: potential link with metabolic complications. Innate Immun. doi:10.1177/1753425910393370
250. Shi H, Kokoeva MV, Inouye K, Tzameli I, Yin H, Flier JS (2006) TLR4 links innate immunity and fatty acid-induced insulin resistance. J Clinical Invest 116(11):3015–3025. doi:10.1172/JCI28898
251. Batra A, Heimesaat MM, Bereswill S, Fischer A, Glauben R, Kunkel D, Scheffold A, Erben U, Kuhl A, Loddenkemper C, Lehr HA, Schumann M, Schulzke JD, Zeitz M, Siegmund B (2012) Mesenteric fat—control site for bacterial translocation in colitis? Mucosal Immunol 5(5):580–591. doi:10.1038/mi.2012.33
252. Pinczowski D, Ekbom A, Baron J, Yuen J, Adami HO (1994) Risk factors for colorectal cancer in patients with ulcerative colitis: a case-control study. Gastroenterology 107(1):117–120
253. Stolfi C, De Simone V, Pallone F, Monteleone G (2013) Mechanisms of action of non-steroidal anti-inflammatory drugs (NSAIDs) and mesalazine in the chemoprevention of colorectal cancer. Int J Mol Sci 14(9):17972–17985. doi:10.3390/ijms140917972
254. Bernstein CN, Nugent Z, Blanchard JF (2011) 5-aminosalicylate is not chemoprophylactic for colorectal cancer in IBD: a population based study. A J Gastroenterol 106(4):731–736. doi:10.1038/ajg.2011.50
255. Tang J, Sharif O, Pai C, Silverman AL (2010) Mesalamine protects against colorectal cancer in inflammatory bowel disease. Dig Dis Sci 55(6):1696–1703. doi:10.1007/s10620-009-0942-x
256. Terdiman JP, Steinbuch M, Blumentals WA, Ullman TA, Rubin DT (2007) 5-Aminosalicylic acid therapy and the risk of colorectal cancer among patients with inflammatory bowel disease. Inflamm Bowel Dis 13(4):367–371. doi:10.1002/ibd.20074
257. van Staa TP, Card T, Logan RF, Leufkens HG (2005) 5-Aminosalicylate use and colorectal cancer risk in inflammatory bowel disease: a large epidemiological study. Gut 54(11):1573–1578. doi:10.1136/gut.2005.070896
258. Gasche C, Goel A, Natarajan L, Boland CR (2005) Mesalazine improves replication fidelity in cultured colorectal cells. Cancer Res 65(10):3993–3997
259. Koelink PJ, Robanus-Maandag EC, Devilee P, Hommes DW, Lamers CB, Verspaget HW (2009) 5-Aminosalicylic acid inhibits colitis-associated but not sporadic colorectal neoplasia in a novel conditional Apc mouse model. Carcinogenesis 30(7):1217–1224. pii:bgp113. doi:10.1093/carcin/bgp113
260. Lu D, Cottam HB, Corr M, Carson DA (2005) Repression of beta-catenin function in malignant cells by nonsteroidal antiinflammatory drugs. Proc Natl Acad Sci U S A 102(51):18567–18571. doi:10.1073/pnas.0509316102
261. Stolfi C, Pellegrini R, Franze E, Pallone F, Monteleone G (2008) Molecular basis of the potential of mesalazine to prevent colorectal cancer. World J Gastroenterol 14(28):4434–4439
262. Konidari A, Matary WE (2014) Use of thiopurines in inflammatory bowel disease: Safety issues. World J Gastrointest Pharmacol Ther 5(2):63–76. doi:10.4292/wjgpt.v5.i2.63
263. Matula S, Croog V, Itzkowitz S, Harpaz N, Bodian C, Hossain S, Ullman T (2005) Chemoprevention of colorectal neoplasia in ulcerative colitis: the effect of 6-mercaptopurine. Clin Gastroenterol Hepatol 3(10):1015–1021

264. van Schaik FD, van Oijen MG, Smeets HM, van der Heijden GJ, Siersema PD, Oldenburg B (2012) Thiopurines prevent advanced colorectal neoplasia in patients with inflammatory bowel disease. Gut 61(2):235–240. doi:10.1136/gut.2011.237412
265. Subramanian V, Logan RF (2011) Chemoprevention of colorectal cancer in inflammatory bowel disease. Best Pract Res Clin Gastroenterol 25(4–5):593–606. doi:10.1016/j.bpg.2011.09.003
266. Beaugerie L, Svrcek M, Seksik P, Bouvier AM, Simon T, Allez M, Brixi H, Gornet JM, Altwegg R, Beau P, Duclos B, Bourreille A, Faivre J, Peyrin-Biroulet L, Flejou JF, Carrat F, Group CS (2013) Risk of colorectal high-grade dysplasia and cancer in a prospective observational cohort of patients with inflammatory bowel disease. Gastroenterology 145(1):166–175, e168. doi:10.1053/j.gastro.2013.03.044
267. Lopez A, Mounier M, Bouvier AM, Carrat F, Maynadie M, Beaugerie L, Peyrin-Biroulet L, Group CS (2014) Increased risk of acute myeloid leukemias and myelodysplastic syndromes in patients who received thiopurine treatment for inflammatory bowel disease. Clin Gastroenterol Hepatol. doi:10.1016/j.cgh.2014.02.026
268. Sokol H, Beaugerie L, Maynadie M, Laharie D, Dupas JL, Flourie B, Lerebours E, Peyrin-Biroulet L, Allez M, Simon T, Carrat F, Brousse N, Group CS (2012) Excess primary intestinal lymphoproliferative disorders in patients with inflammatory bowel disease. Inflamm Bowel Dis 18(11):2063–2071. doi:10.1002/ibd.22889
269. Kim YJ, Hong KS, Chung JW, Kim JH, Hahm KB (2010) Prevention of colitis-associated carcinogenesis with infliximab. Cancer Prev Res 3(10):1314–1333. doi:10.1158/1940-6207.CAPR-09-0272
270. Csillag C, Borup R, Olsen J, Nielsen FC, Nielsen OH (2007) Treatment response and colonic gene expression in patients with Crohn's disease. Scand J Gastroenterol 42(7):834–840. doi:10.1080/00365520601127166
271. Rutgeerts P, Sandborn WJ, Feagan BG, Reinisch W, Olson A, Johanns J, Travers S, Rachmilewitz D, Hanauer SB, Lichtenstein GR, de Villiers WJ, Present D, Sands BE, Colombel JF (2005) Infliximab for induction and maintenance therapy for ulcerative colitis. N Engl J Med 353(23):2462–2476. pii:353/23/2462. doi:10.1056/NEJMoa050516
272. Yao X, Huang J, Zhong H, Shen N, Faggioni R, Fung M, Yao Y (2014) Targeting interleukin-6 in inflammatory autoimmune diseases and cancers. Pharmacol Ther 141(2):125–139. doi:10.1016/j.pharmthera.2013.09.004
273. Abraham C, Medzhitov R (2011) Interactions between the host innate immune system and microbes in inflammatory bowel disease. Gastroenterology 140(6):1729–1737. doi:10.1053/j.gastro.2011.02.012
274. Isaacs K, Herfarth H (2008) Role of probiotic therapy in IBD. Inflamm Bowel Dis 14(11):1597–1605. doi:10.1002/ibd.20465
275. Appleyard CB, Cruz ML, Isidro AA, Arthur JC, Jobin C, De Simone C (2011) Pretreatment with the probiotic VSL#3 delays transition from inflammation to dysplasia in a rat model of colitis-associated cancer. Am J Physiol Gastrointest Liver Physiol 301(6):G1004–G1013. doi:10.1152/ajpgi.00167.2011
276. Arthur JC, Gharaibeh RZ, Uronis JM, Perez-Chanona E, Sha W, Tomkovich S, Muhlbauer M, Fodor AA, Jobin C (2013) VSL#3 probiotic modifies mucosal microbial composition but does not reduce colitis-associated colorectal cancer. Sci Rep 3:2868. doi:10.1038/srep02868
277. Bassaganya-Riera J, Viladomiu M, Pedragosa M, De Simone C, Hontecillas R (2012) Immunoregulatory mechanisms underlying prevention of colitis-associated colorectal cancer by probiotic bacteria. PLoS One 7(4):e34676. doi:10.1371/journal.pone.0034676
278. Ley RE, Hamady M, Lozupone C, Turnbaugh PJ, Ramey RR, Bircher JS, Schlegel ML, Tucker TA, Schrenzel MD, Knight R, Gordon JI (2008) Evolution of mammals and their gut microbes. Science 320(5883):1647–1651. pii:1155725. doi:10.1126/science.1155725
279. Lozupone CA, Stombaugh JI, Gordon JI, Jansson JK, Knight R (2012) Diversity, stability and resilience of the human gut microbiota. Nature 489(7415):220–230. doi:10.1038/nature11550

280. Mueller MM, Fusenig NE (2004) Friends or foes—bipolar effects of the tumour stroma in cancer. Nat Rev Cancer 4(11):839–849. doi:10.1038/nrc1477
281. Coussens LM, Werb Z (2002) Inflammation and cancer. Nature 420(6917):860–867
282. Oohashi Y, Ishioka T, Wakabayashi K, Kuwabara N (1981) A study on carcinogenesis induced by degraded carrageenan arising from squamous metaplasia of the rat colorectum. Cancer Lett 14(3):267–272
283. Wang JG, Wang DF, Lv BJ, Si JM (2004) A novel mouse model for colitis-associated colon carcinogenesis induced by 1,2-dimethylhydrazine and dextran sulfate sodium. World J Gastroenterol 10(20):2958–2962
284. Shattuck-Brandt RL, Varilek GW, Radhika A, Yang F, Washington MK, DuBois RN (2000) Cyclooxygenase 2 expression is increased in the stroma of colon carcinomas from IL-$10^{(-/-)}$ mice. Gastroenterology 118(2):337–345
285. Rudolph U, Finegold MJ, Rich SS, Harriman GR, Srinivasan Y, Brabet P, Boulay G, Bradley A, Birnbaumer L (1995) Ulcerative colitis and adenocarcinoma of the colon in G alpha i2-deficient mice. Nat Genet 10(2):143–150. doi:10.1038/ng0695-143
286. Chu FF, Esworthy RS, Chu PG, Longmate JA, Huycke MM, Wilczynski S, Doroshow JH (2004) Bacteria-induced intestinal cancer in mice with disrupted Gpx1 and Gpx2 genes. Cancer Res 64(3):962–968
287. Kado S, Uchida K, Funabashi H, Iwata S, Nagata Y, Ando M, Onoue M, Matsuoka Y, Ohwaki M, Morotomi M (2001) Intestinal microflora are necessary for development of spontaneous adenocarcinoma of the large intestine in T-cell receptor beta chain and p53 double-knockout mice. Cancer Res 61(6):2395–2398
288. Sohn KJ, Shah SA, Reid S, Choi M, Carrier J, Comiskey M, Terhorst C, Kim YI (2001) Molecular genetics of ulcerative colitis-associated colon cancer in the interleukin 2- and beta(2)-microglobulin-deficient mouse. Cancer Res 61(18):6912–6917
289. Ben-Neriah Y, Karin M (2011) Inflammation meets cancer, with NF-[kappa]B as the matchmake. Nat Immunol 12(8):715–723

Sphingolipids in Intestinal Inflammation and Tumorigenesis

Mónica García-Barros, Nicolas Coant and Ashley J. Snider

Abstract Sphingolipids, and their metabolizing enzymes, have emerged as significant players in cell signaling and biology. These bioactive lipids function not only as structural lipids in cell, but mediate significant biologic functions as well. Sphingolipid enzymes function in the intestinal lumen to degrade exogenous dietary sphingolipids and in the enterocytes themselves to generate endogenous sphingolipids. Among the most studied bioactive sphingolipids are ceramide, sphingosine, and sphingosine-1-phosphate. These sphingolipids and their synthetic enzymes have been extensively implicated in intestinal inflammation and tumorigenesis and will be the focus of this chapter.

Keywords Sphingolipid · Sphingosine kinase · Sphingosine-1-phosphate · Ceramide · Sphingosine · Sphingomyelinase · Colitis · Colon cancer

Abbreviations

SPT	Serine palmitoyltransferase
CerS	Ceramide synthase
SM	Sphingomyelin
SMS	Sphingomyelin synthase
C1P	Ceramide-1-phosphate
SMase	Sphingomyelinase
CDase	Ceramidase
SK	Sphingosine kinase
S1P	Sphingosine-1-phosphate

A. J. Snider (✉) · M. García-Barros · N. Coant
Department of Medicine, Cancer Center, Stony Brook University, 100 Nicolls Road, T15, 023, Stony Brook, NY 11794, USA
e-mail: ashley.snider@stonybrook.edu

A. J. Snider
VAMC Northport, 79 Middleville Road, Northport, NY, USA

M. García-Barros
e-mail: Monica.Garcia-Barros@stonybrookmedicine.edu

N. Coant
e-mail: Nicolas.Coant@stonybrookmedicine.edu

V. W. Yang, A. B. Bialkowska (eds.), *Intestinal Tumorigenesis*,
DOI 10.1007/978-3-319-19986-3_9

SPP	S1P phosphatase
SPL	S1P lyase
MAPK	Mitogen activated protein kinase
AKT	Protein kinase B
TNFα	Tumor necrosis factor alpha
ERM	Ezrin, radixin and moesin
COX-2	Cyclooxygenase 2
PKC	Protein kinase C
TRAF2	TNF receptor-associated factor 2
HDAC	Histone deacetylase
S1PR	S1P receptor
PI3K	Phosphatidylinositol-4,5-bisphosphate 3-kinase
JNK	c-Jun N-terminal kinases
STAT3	Signal transducer and activator of transcription 3
VCAM	Vascular cell adhesion molecule
ICAM	Intracellular adhesion molecule
CAC	Colitis associated cancer
Alk-SMase	Alkaline sphingomyelinase
nCDase	Neutral ceramidase
FAP	Familial adenomatous poylposis
GCS	Glucosylceramide synthase
GBA	Glucocerebrosidase
DSS	Dextran sulfate sodium
MDR	Multi-drug resistance
aSMase	Acid sphingomyelinase
nSMase	Neutral sphingomyelinase
IL10	Interleukin 10
UDCA	Ursodeoxycholic acid
UA	Ursolic acid
AOM	Azoxymethane
ICR	Imprinting control region
APC	Adenomatous polyposis coli
TIMP1	Tissue inhibitor of metalloproteinase
aCDase	Acid ceramidase
ACF	Aberrant crypt foci
TNBS	Trinitrobenzene sulfonate
EGF	Epidermal growth factor
BRDU	5-bromo-2′-deoxyuridine
5FU	Fluorouracil
HIF1α	Hypoxia-inducible factor 1 alpha
TCF4	Transcription factor 4
Glu/Gal-Cer Synthase	Glucosyl/glalactosylceramide

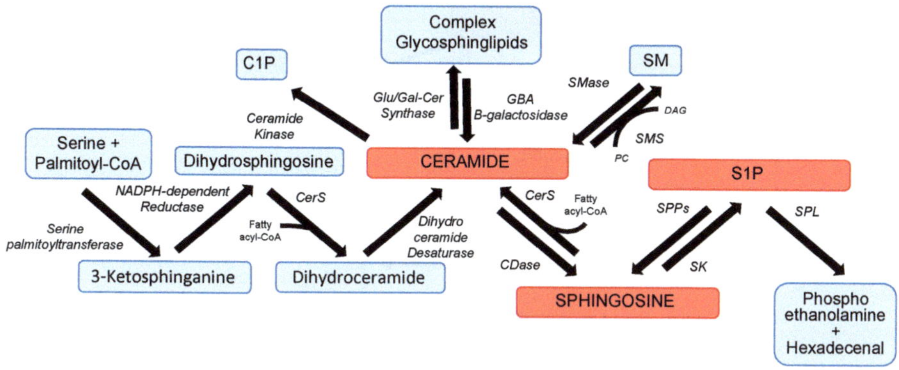

Fig. 1 Sphingolipid metabolism. De novo sphingolipid production begins with the action of serine palmitoyltransferase and the condensation of serine and palmitoyl Co-A. This is rapidly processed to dihydro-sphingosine, which is then acylated by ceramide synthases (CerS) to form dihydroceramide. Ceramide is formed by desaturation of dihydroceramide or by the degradation of complex sphingolipids such as SM or glycosphingolipids. At this point ceramide has many fates. Ceramide kinase can phosphorylate ceramide to form C1P, deaclyation of ceramide can occur via one of CDases, resulting in the generation of sphingosine. CerS can then reacylate sphingosine to ceramide or SKs can rapidly phosphorylate sphingosine to S1P. S1P can be dephosphorylated by S1P phosphatases or terminally degraded by S1P lyase to form hexadecanal and phosphoethanolamine. (Glu/Gal-Cer Synthase: glucosyl/glalactosylceramide synthase; GBA: glucocerebrocosidase.)

1 Sphingolipid Metabolism

Sphingolipids, first described as structural components of cell membranes, have rapidly moved to the foreground of lipid research as potent bioactive lipids. The nomenclature for this family of lipids is derived from the Sphinx due to their enigmatic properties. Sphingolipids consist of an amino-alcohol backbone, and upon modification results in a vast array of sphingolipids being formed with distinct biologic functions.

Sphingolipids can be generated via two different pathways, the de novo pathway and the salvage pathway. These two pathways lead to the formation of ceramides considered as the central hub of sphingolipid metabolism (Fig. 1). De novo sphingolipid synthesis begins in the ER with the condensation of serine and palmitoyl-CoA catalyzed by serine palmitoyltransferase (SPT) to generate 3-keto-dihydrosphingosine [1]. Recent studies have also implicated that other amino acids and fatty acyl-CoAs as substrates for SPTs [2]. 3-keto-dihydrosphingosine is then reduced to form dihydrosphingosine (sphinganine), which is then N-acylated by one of six (dihydro) ceramide synthases (CerS) to produce dihydroceramide. Ceramide is then generated as the product of dihydroceramide desaturase. From this point ceramide has many fates. Sphingomyelin synthases (SMS) can then convert ceramide to sphingomyelin (SM). Ceramide can be phosphorylated by ceramide kinase to form ceramide-1-phosphate (C1P), or glycosylated by either glucosyl or galactosyl ceramide forming the backbone for complex glycosphingolipids [3]. In the salvage pathway,

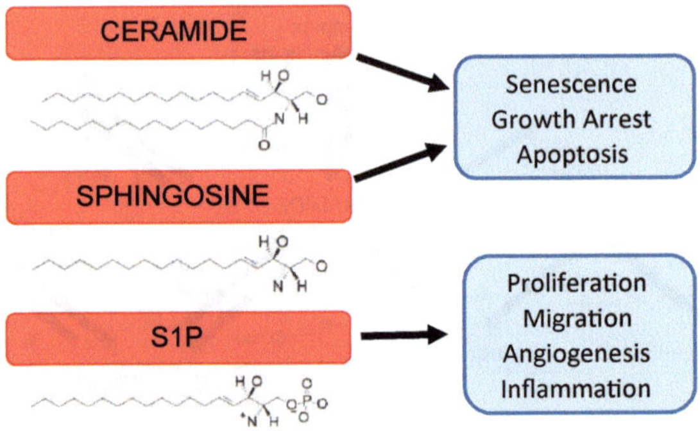

Fig. 2 Sphingolipid structure and function in biology. Sphingolipids consist of a sphingoid backbone containing an amino-alcohol (sphingosine), which is modified to form all sphingolipids including ceramide and S1P (shown here). These bioactive sphingolipids have distinct cellular and biologic functions. Ceramide and sphingosine are involved in growth arrest, senescence, apoptosis and apoptosis; while S1P regulates proliferation, migration, angiogenesis and inflammation

SM can be hydrolyzed into ceramide by the activity of several sphingomyelinases (SMases) characterized by the pH required for optimal enzymatic activity [4–6]. In addition, exit from the sphingolipid metabolic pathway occurs through the action of ceramidases (CDases), which hydrolyze ceramide to produce sphingosine. There are currently five known CDases which are characterized by the pH required for optimal activity (acid, neutral, and alkaline) [7]. Sphingosine is phosphorylated by Sphingosine kinases 1 and 2 (SK1 and 2) to form Sphingosine-1-phosphate (S1P) or reacylated back to ceramide via the CerS, which demonstrate high specificity with regards to the acyl chain length [8]. S1P can be metabolized back to sphingosine by S1P phosphatases (SPPs), or irreversibly broken down by S1P lyase (SPL) to generate ethanolamine phosphate and hexadecenal resulting in the only exit pathway from sphingolipid metabolism [9, 10]. Hexadecenal can also be reduced further allowing for formation of palmitoyl-CoA to potentially re-enter From this network of interconnected enzymes with both temporal and spatial regulation a number of potent bioactive sphingolipids have emerged. Ceramide, sphingosine, and S1P are among the most studied bioactive sphingolipids (Fig. 2). This chapter will focus of the role of sphingolipid enzymes and their bioactive lipid products in intestinal inflammation and tumorigenesis the sphingolipid metabolism pathway [11].

2 Bioactive Sphingolipids

2.1 Ceramide and Sphingosine

The lipid ceramide actually encompasses many distinct molecular species characterized by specific fatty-acyl chain lengths and, as recently identified, different backbones [12, 13] (Fig. 2). The identification and classification for the numerous species of ceramide is a fairly recent development made possible through the use of advanced mass spectrometry [14, 15]. Ceramide is the central lipid in sphingolipid metabolism serving as a precursor for C1P, SM, and glycosphingolipids, moreover the breakdown of ceramide is the only known mechanism to generate sphingosine and S1P.

Ceramide has been implicated in cell death [16, 17], cellular stress responses [18, 19], senescence [20] and cell cycle arrest [20, 21, 19, 16]. Since these seminal discoveries for the biologies exerted by ceramide there have been few direct cellular targets identified. The action of ceramide activation of protein phosphatases has been implicated in dephosphorylation of cell signaling proteins including protein kinase C alpha [22], protein Rb [23], p38 MAPK [24] and AKT [25]. Ceramide activated protein phosphatases has also been implicated in the blockade of insulin signaling mediated by tumor necrosis factor alpha (TNFα) [26]. One of the recent implications of ceramide activated protein phosphatase, of particular interest to cancer, is the dephosphorylation of ERM proteins (ezrin, radixin and moesin) [27], which play a role in actin-cytoskeletal rearrangement and migration. This study implicated that ceramide generated by bacterial SMase at the plasma membrane dose dependently decreased the phosphorylation state of ERM proteins.

Ceramide generation by CerS has also been implicated in colon cancer in vitro. HCT116 cells treated with the cyclooxygenase 2 (COX-2) inhibitor celecoxib demonstrated growth inhibition, partially attributed to ceramide generation by CerS6. Generation of ceramide by overexpression of Cers4 and CerS6 in HCT116 cells has also been shown to inhibit proliferation and induce apoptosis [28].

Sphingosine, the metabolite of ceramide breakdown, has been associated with growth arrest and cell death [29–31]. Specifically, sphingosine has been shown to inhibit PKC [32], and analogues of sphingosine have since been used a molecular inhibitors for PKC [33]. Casein kinase II has also been implicated as a target for sphingosine [34].

Together the studies examining the generation or accumulation of sphingosine and ceramide have implicated these lipids in anti-proliferative, senescent and pro-apoptotic pathways. Future studies should yield insight into the functions for specific ceramide species in apoptosis and growth arrest.

2.2 *Sphingosine-1-Phosphate*

Sphingosine is generated through the actions of ceramidases and is rapidly phosphorylated by SKs to S1P, a potent bioactive lipid with roles in cellular proliferation, migration, invasion and inflammation (reviewed in [35]). S1P has both intracellular and extracellular targets. Intracellular S1P has been implicated in the E3-ubiquitin ligase activity of TRAF2 [36], and in the inhibition of histone deacetylases 1 and 2 (HDAC1 and HDAC2) [37]. S1P also exerts its effects extracellularly and is thought to be primarily generated inside the cell and exported out by several possible transporters, such as the ABCC1 transporter [38, 39]. However, there is also evidence that SK1 can be secreted from monocytic cells and vascular endothelial cells to generate S1P [40, 41]. Extracellular S1P can then bind to one of 5 G-protein coupled receptors (S1PR1–5) acting in an autocrine or paracrine manner [42, 43].

S1PRs bind different G-protein subunits and have specific cell-type distribution, thus permitting S1P to exert these vast biological effects including activation of key cellular signaling pathways such as MAPK, PI3K, and JNK [44, 45, 42]. S1PR1 has been shown to activate MAPK, and inhibit adenylyl cyclase in a pertussis toxin-dependent manner. A major role for S1PRs, and specifically S1PR1, has recently emerged with the development of the S1PR modulator and immunosuppressant drug, FTY720 (recently approved to treat multiple sclerosis and marketed as Gilyena™ [46]. When phosphorylated, FTY720 binds to and downregulates S1PRs (except S1PR2) preventing lymphocyte egress from secondary lymphoid organs [47, 48]. S1PR1 has also been implicated in the phosphorylation of signal transducer and activator of transcription 3 (STAT3) in colon carcinogenesis [49]. S1PR2 stimulates calcium mobilization [50], and has recently been implicated in lamellipodia formation in response to S1P, through the phosphorylation of ERM proteins [51, 27, 52]. S1PR3 signaling induces calcium mobilization in a pertussis toxin-independent manner and inhibits adenylyl cyclase in a pertussis toxin-dependent manner [50]. While the role of S1PR3 has been clearly defined in cancer it has been implicated in lung [53] and breast cancer [54]. S1PR4 activates growth-related signaling pathways [55], and is highly expressed in lymphoid tissues [56]. S1PR5 has not been extensively studied, but has been shown to function primarily in oligodendrocytes [57]. Therefore, through activation of the receptors, S1P can exert diverse biologic functions.

S1P, and its synthetic enzyme SK, are of primary interest to intestinal tumorigenesis due to their functions as a mitogentic and angiogenic factors [58–60], and their involvement in cytokine-mediated inflammation [61, 62]. Blocking S1P actions with a monoclonal antibody reduced angiogenesis and tumor development in a xenograft model using multiple cell lines, including HT29 colon cancer cells [63]. Furthermore, activation of SK1, and generation of S1P, by TNFα induces vascular cell adhesion molecule (VCAM) and intracellular adhesion molecule (ICAM) in endothelial cells [64]. Perhaps most importantly for intestinal tumorigenesis, SK1 and S1P are necessary for COX-2 expression in colon cancer cells [61], inflammatory models of colitis [65, 66], and in models of colitis associated cancer (CAC) [67, 61].

3 Metabolism of Dietary Sphingolipids

Sphingolipids are part of a normal diet. Dairy products, together with fruit and vegetables, are particularly enriched in sphingolipids. Even though the daily dietary intake of sphingolipids is about 300–400 mg the nutritional requirements for them is unknown. While sphingolipids are abundant in adult diets, human milk is the only source of sphingolipids for neonates, and it seems that feeding neonates with commercial bovine milk leads to abnormal content of sphingolipids in the intestine.

This can result in deficiencies the immune system and defects in the development of central nervous system, suggesting an important role of sphingolipids in development in neonates [68].

SM and ceramide cannot be absorbed directly by the intestine and SM digestion is slow in the small intestine allowing the colon to be exposed to high amounts. During digestion, cholecystokinin is released from the intestinal endocrine cells to stimulate the contraction of gallbladder allowing the release of alkalkine sphingomyelinase (alk-SMase) together with its natural activator, bile salt, into the intestinal lumen. Cholecystokinin will also stimulate the secretion of trypsin from the pancreas which will cleave the alk-SMase bound on the intestinal mucosa releasing it into the lumen and enhancing its activity. Alk-SMase will catalyze the generation of ceramide within the intestinal lumen, which can then be further hydrolyzed into sphingosine by neutral ceramidase (nCDase). Sphingosine can be absorbed by the enterocytes via a passive diffusion mechanism where it can be converted back to ceramide and other complex sphingolipids or phosphorylated by SKs to generate S1P. Exit from the sphingolipid pathway occurs in the enterocyte via SPL, which is highly expressed in the intestine.

4 Sphingolipid Enzymes and Intestinal Inflammation and Tumorigenesis

Alteration of sphingolipid metabolism has been suggested to play a critical role in intestinal inflammation (Table 1), CAC (Table 2), as well as susceptibility to colon cancer. Evidence suggests that colon cancer may be associated with a decrease in the levels of ceramide and an overall increase in the levels of S1P (Fig. 3, adapted from [69]). This may be due to distinct changes in enzymes implicated in sphingolipid metabolism, including glycosphingolipid enzymes, alk-SMase, SKs, SPL, or SPPs or through the action of S1P on its receptors.

5 Glycosphingolipid Enzymes

Complex glycosphingolipids encompass a vast array of sphingolipids, many of which have not been well studied. Glucosylceramide synthase (GCS) is the first enzyme in the complex glycosphingolipid pathway. Mice null for *Ugcg* (gene encoding GCS) were embryonic lethal and demonstrated an essential role for complex glycosphingolipids in embryonic development. Embryos were able to differentiate into endodermal, mesodermal and ectodermal layers, but unable to form well differentiated tissues [70]. With the development of the Cre-Lox system *Gcs* has been deleted selectively in specific tissues with much of the work focused on neural development. However, the enterocyte-specific deletion has recently been reported [71]. Just after birth, there were no distinct differences between mice

Table 1 Sphingolipids in intestinal inflammation

Enzyme	Model	Modulation	Response	Ref.
GBA				
	DSS-induced Colitis	GBA2$^{-/-}$ mice	No difference in disease from WT; Increased JNK and STAT3 phosphorylation	[71]
Alk-SMase				
	Patient samples	Ulcerative colitis	Expression decreased 25 %	[76]
	DSS-induced Colitis	Yeast alk-SMase administration	Decreased damage score and TNFα expression in colon	[77]
	IL10$^{-/-}$ mice	VSL#3 administration	Increased alk-SMase activity and decreased intestinal inflammation	[78]
	Patient samples	VSL#3 administration	Increased alk-SMase activity and decreased intestinal inflammation	[78]
ASMase				
	DSS-induced colitis	SMA-7 (aSMase inhibitor)	Decreased pathology and TNFα expression in colon	[79]
NCDase				
	DSS-induced colitis	nCDase$^{-/-}$ mice	Increased systemic inflammation and COX-2 expression in colon	[65]
SK1				
	Patient Samples	Ulcerative colitis	Increased expression	[66]
	DSS-induced colitis	SK1$^{-/-}$ mice	Decreased systemic inflammation and decreased pathology and Cox-2 and granulocytes in colon	[66]
SK2				
	DSS-induced colitis	SK2$^{-/-}$ mice	Enhanced SK1 expression, S1P production, and severity of colitis	[49]
	DSS & TNBS-induced colitis	ABC294640 SK2 inhibitor	Decreased disease index and cytokine induction in colon	[103, 104]
S1PRs				
	DSS-induced colitis	S1PR1$^{-/-}$ mice	Increased colonic bleeding	[123]
	Oxazolone-induced colitis	FTY720 S1PR agonist	Reduced cytokines, histologic damage, and myeleoperoxidase activity in colon	[128]
	T-cell transfer colitis	FTY720 S1PR agonist	Decreased pathology	[129]
	TNBS-induced colitis	FTY720 S1PR agonist	No protection from disease	[130]
	DSS-induced colitis	W-061 S1PR1 agonist	Decreased Th17-cell infiltrate into lamina propria; no change in Tregs	[125]
	IL10$^{-/-}$ mice	KRP-203 S1PR agonist	Reduced lymphocytes in circulation and in colon	[126]
	IL10$^{-/-}$ mice	SEW2871 S1PR agonist	Decreased established colitis and STA3 phosphorylation in colon	[127]
	DSS-induced colitis	S1PR4$^{-/-}$ mice	Prevented weight loss and IL-6 production	[124]

Table 2 Sphingolipids in intestinal tumorigenesis

Enzyme	Model	Modulation	Response	Ref.
SMase				
	DMH	WT rats	Decreased nSMase activity	[80]
	AOM	Sprague-Dawley rats	Decreased alk-SMase activities	[83]
	AOM	Sprague-Dawley rats with UCDA	Increased aSMase and nSMase activities	[83]
	Patient samples	Colon cancer	Decreased alk-SMase activity, as well as aSMase and nSMase	[88]
	Patient samples	FAP	Decreased alk-SMase activity, as well as aSMase and nSMase	[91]
	AOM/DSS	VSL3# administration	Increased alk-SMase activity, complete inhibition of tumor formation	[86]
SKs				
	$APC^{Min/+}$ mice	$SK1^{-/-}$ mice	Decreased tumor size	[105]
	AOM	F344 rats	Increased SK1 expression	[61]
	AOM/DSS	$SK1^{-/-}$ mice	Reduced ACF and tumor incidence	[67]
	AOM/DSS	$SK2^{-/-}$ mice	Enhanced tumor number and size	[49]
	Patient samples	Colon cancer	SK1 expression correlated with tumor stage and mortality	[67, 106, 107]
SPL				
	$APC^{Min/+}$ mice	No modulation	Decreased SPL activity in adenomas	[111]
	Patient samples	Colon cancer	Decreased SPL and SPP expression	[111, 112]
S1PRs				
	AOM/DSS	WT and $SK2^{-/-}$ mice	Increased S1PR1 expression	[49]

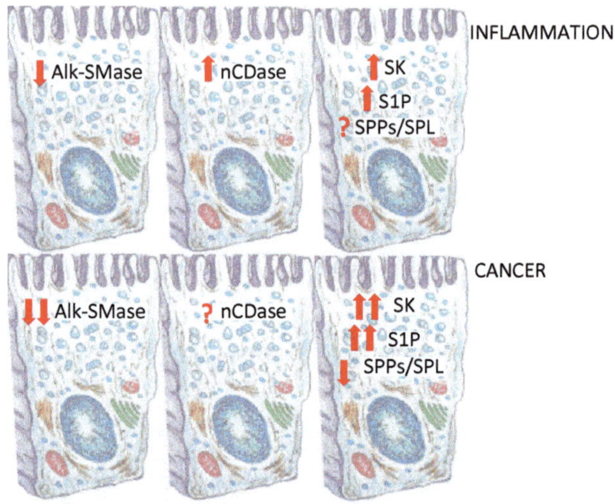

Fig. 3 Sphingolipid enzymes in intestinal inflammation and tumorigenesis. Sphingolipid enzyme levels in intestinal inflammation and tumorigenesis and associated sphingolipid changes

lacking *Gcs* in intestinal villi and their wild type littermates. Between days 5 and 7 mice demonstrated impaired villi structure and an inability to absorb nutrients, which resulted in early death [71].

5.1 Glycosphingolipid Enzymes and Intestinal Inflammation

Breakdown of complex glycosphingolipids occur through the action of glucocerebrosidase (GBA) (also referred to as beta-glucosidase or glucosylceramidase), and mutations in this gene result in Gaucher's Disease. In the dextran sulfate sodium (DSS) model for ulcerative colitis mice deficient in Gba2 were not protected from disease. In fact, their tissue response to DSS did not differ from their wild type controls. However, in *Gba2*$^{-/-}$ mice signaling proteins involved in regeneration were up-regulated, potentially suggesting that loss of GBA could influence the recovery from DSS [72].

5.2 Glycosphingolipid Enzymes and Intestinal Tumorigenesis

Much of what is known about glycosphingolipids in intestinal tumorigenesis stems from work down in cell culture, with no published work to date in animal models. GCS as mentioned above is the initial step in glycosphingolipid metabolism and thus can serve as a shunt for ceramide, potentially decreasing ceramide-induced apoptosis. This increase in GCS has been implicated in multi-drug resistance (MDR). In fact, suppression of GCS restores the sensitivity of MDR colon cancer cells to drug treatment, and complimentary studies demonstrated that the over-expression of GCS is associated with MDR in HCT-8 colon cancer cells [73]. The lack of in vivo data on the role of complex glycosphingolipids or their metabolizing enzymes, coupled with the few key studies presented here suggest the potential importance of these enzymes in intestinal homeostasis, inflammation and potentially tumorigenesis.

6 Sphingomyelinases

SMases serve to breakdown sphingomyelin to ceramide, and as mentioned previously are named based on the pH optima at which they function, alkaline, neutral and acid. Alk-SMase is the most studied SMase in the intestinal tract (with very few studies on either neutral or acid) and is the major enzyme responsible for hydrolysis of sphingomyelin in the intestine. Its longitudinal distribution results in the highest expression at the jejunum and lowest in the colon. Alk-SMase, which shares no structural similarity to other members of the SMase family, is classified as a nucleotide pyrophosphatase/phosphodiesteras (NPP). The extracellular domain of alk-SMase contains the catalytic site and is localized in the intestinal lumen. This domain can be cleaved and activated by pancreatic trypsin or dissociated from the membrane by bile salts. Thus alk-SMase may plays a major role in the initial steps of sphingolipid breakdown in the intestinal tract. Mice null for *alk-SMase* did not exhibit changes in the length of the intestine, and presented no gross disorganization of the mucosal structure. However, these mice present with hypertrophy of the intestinal epithelium and widening of the villi, crypts and the thickening of the muscle layer under the crypts with no spontaneous tumorigenesis in the intestine up to 10 months of age. The changes in the ability to digest SM showed that digestion, ceramide formation, and fatty acid absorption are severely affected [74].

Diet has been shown to regulate alk-SMase activity, in fact high fat diet has been shown to significantly decrease activity of alk-SMase, as well as neutral and acid SMases [75]. In a similar study by Yang et al. adding beef fat to a high fat diet did not alter SMase activities. Interestingly, adding fiber to the high fat diet increased both acid and neutral SMase activities, but did not alter alk-SMase activity [76]. These studies suggest that SMases in the intestinal tract, specifically alk-SMase, are important to dietary sphingolipid metabolism, and reciprocally that SMase activities can be altered by diet.

6.1 *SMase and Intestinal Inflammation*

Expression of alk-SMase is decreased 25 % in patients with ulcerative colitis [77]. In the DSS model for ulcerative colitis, administration of yeast alk-SMase decreased colitis damage in Sprague-Dawley Rats [78]. Treatment with the probiotic

VSL#3 resulted in increased alk-SMase activity in Interleukin-10 (*IL-10*) knockout mice, which spontaneously develop colitis, as well as in humans. Furthermore this increase in alk-SMase activity resulted in decreased intestinal inflammation in both the murine model and human patients [79]. Together these studies suggest that alk-SMase activity may function to protect against colitis.

The acid SMase (aSMase) inhibitor (SMA-7) has been shown to decrease lipopolysaccharide (LPS)-induced macrophage production of cytokines, and to increase cell viability in response to LPS in HT29 colon cancer cells [80]. In addition, SMA-7 administered to mice with DSS-induced colitis decreased pro-inflammatory cytokine production in the colon and partially protected against DSS-induced colitis in both an acute and chronic model [80].

6.2 SMase in Intestinal Tumorigenesis Models

One of the early studies examining sphingomyelinase activity in relation to colon cancer demonstrated that rats treated with 1,2-dimethylhydrazine (DMH) exhibited increases in colonic SM; which was attributed to decreased neutral SMase activity [81]. Additionally, early studies implicated a role for SMases in colon carcinogenesis by examining SMase activities in tissues following administration of potential novel therapeutics. Ursodeoxycholic acid (UDCA) and ursolic acid (UA) have both been shown to exert chemopreventive effects in colon cancer in animal models; and UA has recently been shown to decrease aSMase activity in intestinal cells [82]. UDCA alone was administered to rats and SMase activities were assessed in the colonic mucosa with alk-SMase activity increasing dose-dependently, but with only slight increases in aSMase and no changes in nSMase. Interestingly in this study, caspase-3 activity was also increased in the colonic mucosa following UCDA administration [83]. Administration of AOM to Sprague-Dawley rats significantly reduced alkSMase activity in colonic mucosa. Addition of UA as a potential therapeutic did not affect alk-SMase activity, but significantly increased both nSMase and aSMase activities in vehicle and AOM treated rats [84]. Furthermore, CD1 mice treated with azoxymethane (AOM) followed with 1 week of 1 % DSS administered different amounts of UDCA for 17 weeks showed a significant inhibition in the multiplicity of colonic adenocarcinomas [85].

Dietary sphingomyelin has also been utilized to examine alterations in intestinal tumorigenesis. In imprinting control region (ICR) mice were maintained on a diet with or without dietary SM for 22 weeks. These mice were subjected to the DMH model for colon carcinogenesis. In these mice dietary SM reduced tumor incidence and increased alk-SMase activity in the colon (no changes in aSMase or nSMase were observed) [86]. Moreover, the pro-biotic VSL#3 which increased in alk-SMase activity in an animal model of colitis were used in a CAC, model, namely AOM/DSS. In this model administration of VSL#3 completely inhibited tumor formation [87]. Alk-SMase activity was also increased in the colonic mucosa upon VSL3# administration, suggesting that alk-SMase may play anti-cancer role in the colon. Moreover, inactivating mutations in alk-SMase have been identified in colon cancer cell lines [88] and its activity is decreased in colon cancer [89]. Interestingly, alk-SMase has not been linked to Adenomatous Polyposis Coli (APC) gene mutations, as its activity was very similar in $Apc^{Min/+}$ and wild type animals [90].

Metastasis of colon cancer often involves the liver and in an animal model of liver metastasis, aSMase deficient mice demonstrated enhanced tumor growth and abolished in mice where aSMase was overexpressed in the liver [91]. Interestingly, this was attributed to increased production of S1P, macrophage migration and expression of TIMP1 (tissue inhibitor of metalloproteinase).

6.3 Clinical Implications for SMase in Intestinal Tumorigenesis

In human colorectal adenocarcinoma samples alk-SMase activity is significantly decreased (75 %) when compared to normal adjacent tissues [89]. Furthermore, this group also determined that alk-SMase activity is decreased by 90 % in tumors from patients with familial adenomatous polyposis (FAP) when compared to normal adjacent tissues [92]. Interestingly, these studies also determined that both acid and neutral SMase activities were decreased in colorectal carcinoma samples when compared to normal adjacent tissues. Presence of alk-SMase in stool has been suggested as a potential biomarker, as its activity has been shown to be down-regulated in colorectal cancer in human patients and reduced expression has also been detected in stool of patients with colorectal cancer [93].

The studies implicating SMases in intestinal tumorigenesis suggest that alk-SMase activity is suppressed, with somewhat conflicting evidence for the roles of aSMase or nSMase. The studies examining liver metastasis, begin to indicate complex roles for sphingolipid enzymes in intestinal tumorigenesis, suggesting that these enzymes may serve distinct roles in the tumor microenvironment when compared to the tumor itself.

7 Ceramidases

Acid, alkaline, and neutral CDases have been identified in the intestinal mucosa. These enzymes catalyze the hydrolysis of ceramide into sphingosine. Acid and alkaline ceramidases have not been extensively studied in the intestinal tract; however, neutral ceramidase (nCDase) is the main enzyme responsible for ceramide digestion in the intestinal tract and has a longitudinal distribution similar to that of alk-SMase. NCDase is a type II integral membrane protein found at the brush border of the intestinal epithelium and it can be cleaved to generate a soluble secreted protein, thus functioning as an ectoenzyme. Mice null for *nCDase* are viable and appear without severe developmental defects. Tissue ceramide levels were not significantly altered by the loss of nCDase; however, in the jejunum levels of C16:0

ceramide were increased concomitant with a decrease in the levels of sphingosine (S1P remained unchanged). *nCDase* null mice demonstrated impaired degradation of sphingolipids in the colon, in addition to increased ceramide levels in the colon and feces. The use of null mice demonstrate that nCDase plays an essential role in the digestion of dietary sphingolipids regulating the levels of ceramide (and potentially sphingosine/S1P).

7.1 Ceramidases in Intestinal Inflammation

nCDase activity has been demonstrated to increase in the colonic epithelium, but not in the sub-mucosa, in the DSS-induced colitis model [65]. *nCDase*$^{-/-}$ mice exhibited significant systemic inflammation, as well as increased Cox-2 expression in the colon following acute DSS treatment (5 % wt/vol for 5 days). This study also demonstrated slight increases in nCDase expression in tissues from human patients. These data suggested that nCDase may play a protective role in intestinal inflammation. Similarly, nCDase has been implicated to protect from cytokine-induced cell death in pancreatic beta cells [94]. However, gemcitabine reduced nCDase expression in transformed endothelial cells and knockdown of *nCDase* alone was sufficient to induce cell cycle arrest [95]. These studies suggest that nCDase may play a key role in inflammation, as well as proliferation. As yet, no studies have been published on either alkaline or acid ceramidases in intestinal inflammation; however, aCDase has been shown to be required for Tnfα-induced Cox-2 expression in L929 fibroblasts [96]. There are also no published studies on CDases in colon cancer, but data from cell lines and other animal models suggest that CDases may play key roles in both intestinal inflammation and tumorigenesis.

8 Sphingosine Kinases

Two isoforms of SK have been cloned, SK1 and SK2. Both SK enzymes phosphorylate sphingosine to form S1P. SK2 activation has been implicated in apoptosis [97] and has been shown to regulate gene expression by associating with HDACs [37]. SK1, which appears be highly regulated, has been implicated in far different cellular responses including proliferation and migration. SK1 has also been implicated in inflammation, as it can be activated by and mediate downstream responses of pro-inflammatory cytokines, including TNFα and interleukin 1 beta (IL1β), leading to induction of COX-2 and production of prostaglandins [62].

Mice null for *Sk1*$^{-/-}$ present no phenotype, are viable, fertile and with no obvious abnormalities. Serum and plasma levels of S1P are reduced by a 50 % compared to wild type mice; however, there is little alteration in tissue S1P levels. This is thought to be due to high tissue expression of SPL. Immune cells, including macrophages and neutrophils, from *Sk1*$^{-/-}$ mice have been shown to function normally

[98, 99]; though *Sk1*$^{-/-}$ mice are rendered lymphopenic in presence of FTY720 [100]. Mice deficient in Sk2 present no phenotype, but exhibit increased circulating S1P levels [101]. Homozygotes are viable, fertile and exhibit a mild anaphylaxis when challenged [102].

8.1 Sphingosine Kinases in Intestinal Inflammation

Several studies have implicated a role for SKs in intestinal inflammation utilizing mice deficient in the SKs or SK inhibitors. Expression of SK1 has been shown to be elevated in patients with ulcerative colitis [66]. Moreover, acute DSS administration increased SK1 message levels and activity in the colon, as well as increased circulating S1P levels. In this study, *Sk1*$^{-/-}$ mice were partially protected from the systemic leukocytosis, as well as tissue histopathologic damage including Cox-2 induction and neutrophil infiltration [66]. In contrast, mice deficient in Sk2 exhibited increased disease severity in acute DSS-induced colitis and were unable to recover following removal of DSS [49]. The increased severity of disease in *Sk2*$^{-/-}$ mice was suggested to be due to increased SK1 and S1P. The results in the *Sk2*$^{-/-}$ mice also contradict studies using the purported SK2 inhibitor, ABC294640. This compound, which has recently been shown to bind and inhibit the estrogen receptor [103] (indicating potential off-target effects), has been shown to decrease DSS-induced colitis in both an acute and chronic DSS-induced [104], as well as Trinitrobenzene sulfonate (TNBS)-induced colitis [105]. These studies suggest critical roles for SKs in intestinal inflammation and strongly indicate the necessity for specific (and effective in vivo) SK inhibitors.

8.2 Sphingosine Kinases in Intestinal Tumorigenesis Models

It is important to note that sphingolipid metabolism is abnormal in Min mouse polyps since both S1P and sphingosine are high when compared to surrounding tissues. Consistent with this, *Apc*$^{Min/+}$ mice polyps also exhibit decreased levels and activity of SPL (an enzyme that irreversibly degrades S1P). In addition, SK1, one of two enzymes implicated in S1P synthesis, is required for enlargement of adenomas in *Apc*$^{Min/+}$ mouse, since loss of SK1 resulted in smaller tumor size in the *Apc*$^{Min/+}$ model [106].

In the AOM model for colon carcinogenesis, F344 rats treated with AOM (15 mg/kg body weight once a week for two weeks) demonstrated an upregulation in SK1 expression that correlated with tumor progression [61]. Furthermore, in the AOM/DSS murine model for colon cancer wild-type mice demonstrated an increase in Sk1 expression in the colon, as well as increased S1P levels in the colon and in circulation. In this study mice null for *Sk1* subjected to AOM/DSS presented significantly less aberrant crypt foci (ACF) and tumor incidence when compared to wild-type [67]. Interestingly, *Sk2*$^{-/-}$ mice have been shown to develop more tumors when compared to WT mice upon administration of AOM/DSS suggesting that SK1, may play a more critical role in colon cancer formation [49]. This study implicated increased SK1, S1PR1 and phosphorylation of STAT3 in the increased susceptibility of *Sk2*$^{-/-}$ mice to tumorigenesis. It is important remark that deletion of S1P receptors 2 and 3, and the loss of an allele in receptor 1 did not recapitulate the SK1 deletion.

8.3 Clinical Implications for Sphingosine Kinases in Intestinal Tumorigenesis

SKs, specifically SK1, have been implicated in numerous tumors types including non–small-cell lung, ovarian, and colon cancers [107]. Kawamori et al. showed that SK1 expression is upregulated in human colon cancer and suggested that SK1 expression may be associated with colon cancer progression and metastasis. Indeed, a human colorectal tissue microarray showed that SK1 is expressed to a very limited extent in normal tissues and in normal epithelium surrounding colon carcinoma; however, SK1 expression in adenomas was higher, and about 89 % of adenocarcinoma samples and 100 % of metastatic lesions showed high levels of SK1 [67]. Increased SK1 expression in colon cancer patients was confirmed in a study of 303 colon cancer cases. Interestingly, in this study high SK1 expression correlated with advanced tumor stage and increased mortality [108].

Together the studies examining the role of SKs and S1P in intestinal tumorigenesis suggest that SK1, and not SK2, plays a significant role in colitis, CAC and associated is with tumor progression in human patients. Thus suggesting that SK1 may be a novel potential therapeutic target in intestinal tumorigenesis.

9 S1P Lyase and S1P Phosphatases

SPL and SPPs contribute to control the levels of S1P. Therefore, it is important to note that expression of both SPL and SPPs are high in normal epithelial cells. In particular, SPL is highly expressed in the apex of the intestinal villi, controlling the levels of S1P and modulating mitogenic signaling, thus promoting the fast turn over characteristic of this tissue. Mice null for SPL survived only up to 15 weeks of life, demonstrating an impaired capacity to degrade S1P. Deletion of SPL significantly increased circulating S1P levels and resulted in a pro-inflammatory response but impaired neutrophil migration from blood into tissues [109]. The intestinal mucosa of these mice showed that cell type and distribution differed from the wild type mice, and that the *lamina propria* of the small intestinal contained less lymphocytes and more granulocytes than did wild type mice [110, 109]. Deletion of *Spp1* in mice results in stunted growth during the first week of life and multiple epidermal abnormalities resulting usually in death before weaning [111].

9.1 S1P Lyase and S1P Phosphatase in Intestinal Tumorigenesis

In the *Apc*$^{Min/+}$ model for intestinal tumorigenesis SPL expression and activity were decreased in adenomas when compared to normal surrounding tissues. This decreased breakdown of sphingolipids resulted in increased sphingosine and S1P, potentially contributing to tumorigenesis [112]. Further implicating a role for S1P in tumorigenesis, expression of SPL and SPPs have been shown to be downregulated in samples from colon cancer patients [112, 113].

As S1P has been implicated in lymphocyte trafficking, colitis and colitis associated cancer, the regulation of this bioactive lipid is critical. The enzyme that generates S1P, SK1, is up in colon cancer and the proteins involved in S1P are decreased, suggesting that S1P levels may go un-checked and significantly contribute to intestinal tumorigenesis.

10 Sphingosine-1-Phosphate Receptors

S1PRs are ubiquitously expressed and induce a myriad of cellular responses. S1PR1 has been demonstrated to function in lymphocyte egress from the thymus and secondary lymphoid organs [114], and has been demonstrated to regulate basal and inflammation-induced vascular leak in vivo [115]. Homozygous *S1pr1*$^{-/-}$ mice are embryonic lethal due to defects in angiogenesis [116]. S1PR2 has recently been shown to function in lamellipodia formation in response to epidermal growth factor (EGF) and S1P [51], which has implications for migration and metastasis. Mice deficient in S1PR2 have been shown to develop seizures and become deaf by one month of age [117–119]. S1pr3 has been implicated in gastric cancer cell migration [120]. Mice deficient in S1pr3 display no obvious phenotype, but knockout mice for both *S1pr2* and *S1pr3* demonstrated decreased survivability and litter size [121]. Expression of S1PR4 has been demonstrated in lymphoid tissues, and lymphocytes, and mice deficient in this receptor display decreased megakaryocyte differentiation [122]. S1PR5 has been shown to be necessary for monocyte homeostasis in circulation [123]. The cell type distribution, as well as multiple G-protein subunits, contribute to the diverse processes regulated by S1PRs.

10.1 S1PRs and Intestinal Inflammation

In order to investigate the role of S1PR1 in intestinal inflammation, conditional knockouts were generated. Deletion of S1PR1 resulted in increased bleeding and decreased survival in a chronic DSS model for intestinal inflammation [124], demonstrating a critical role for S1PR1 in colonic vascular integrity. On the other hand, deletion of *S1pr4* in the DSS-induced colitis model resulted in protection from disease, including decreased weight loss and serum Il-6 [125].

In a similar study with W-061, a S1PR1/3 agonist protected against weight loss and pathology induced by DSS. This study also demonstrated that W-061 prevented migration of Th1 and Th17 cells into the *lamina propria* [126]. An additional S1PR1 agonist, KRP-203, demonstrated protection from colitis in the *Il-10*$^{-/-}$ mouse model. KRP-203 administration decreased lymphocytes in peripheral blood and in the colonic *lamina propria* [127]. A recent study with SEW2871, also in *Il-10*$^{-/-}$ mice, diminished already established colitis, decreased pro-inflammatory cytokines and STAT3 phosphorylation [128].

FTY720 has been studied extensively in animal models of intestinal inflammation. This S1PR modulator has been shown to reduce cytokine production and pathology associated with oxazolone-induced colitis [129]. In the of CD4+/CD62 L+ T adaptive transfer model colitis, migration of CD4+ T cells into the colon tissue was prevented by FTY720 administration [130]. Conversely, in the TNBS model, which is commonly used to mimic Crohn's Disease, administration of FTY720 did not prevent TNBS-associated weight loss [131]; nor was it able to protect against pathology associated with infection with *Citrobacter rodentium* [132]. Similarly, FTY720 has been associated with increased diarrhea in approximately 10 % of patients [133].

Together the studies on S1PR modulation in intestinal inflammation suggest that these novel molecules may serve a potential therapeutics. As the modulators become more specific for receptor sub-type, perhaps the side-effects will decrease.

10.2 S1PRs in Intestinal Tumorigenesis

S1PR1 expression is increased in tissues from human patients with ulcerative colitis [124, 134] and in animal models of colitis associated cancer [49]. FTY720 (at high doses) decreased proliferation of colon cancer cell lines [135], and has recently been demonstrated to decrease CAC in a mouse model. In this study, FTY720 decreased epithelia cell proliferation, STAT3 activation and S1PR1 expression in WT and *Sk2*$^{-/-}$ mice [49].

There is very little data on the other S1PRs in colon cancer. However, these receptors may play a significant role in the inflammatory aspect of CAC. With the recent publications on FTY720 and phospho-STAT3, the effects of more specific S1PR modulators may prove interesting.

11 Effect of Dietary Sphingolipids in Colon Cancer Models

One of the first studies to remark the importance of sphingolipids and cancer in the digestive system was presented by Dudeja et al., where they found differences in both sphingolipid composition and sphingolipid enzyme activities between tumor and normal rat tissues [81]. This study demonstrated an increase of SM levels accompanied by a decrease of SMase activity after treatment with the pro-carcinogen DMH. In human colon cancer; however, it is been described that while levels of SM and ceramide are decreased [136, 137] and S1P levels are increased [67].

As dietary sphingolipids are metabolized in the intestinal lumen, their potential as therapeutics have been assessed. In the $Apc^{Min/+}$ model for intestinal tumorigenesis evaluated by Schmelz et al., mice were fed diets supplemented with either ceramide or a mixture of complex sphingolipids (sphingomyelin, glucosylceramide, lactosylceramide and ganglioside G_{D3}). This study indicated that diets enriched with sphingolipids reduced tumors in all regions of the intestine and corresponded to a redistribution of β-catenin from the cytosol towards the junctions between intestinal epithelial cells [138].

The effect of SM in colon cancer has been studied in several animal models. CF1 female mice treated with DMH and fed with AIN76 A (considered a diet with very low sphingolipid content <0.005 %) supplemented with milk SM presented a significant reduction in the number of ACF by 70 % and aberrant crypts per focus by 30 %. Interestingly, after 40 weeks there were no differences in tumor incidence; however, mice fed with SM presented up to 31 % of adenomas, while the animals fed with the control diet presented only carcinomas. This study suggests that dietary SM may play a role in the appearance or advancement to more malignant tumors [139, 140]. To investigate whether chemically synthesized SM could recapitulate the effects of this compound in the milk, CF1 mice were treated with DMH and fed AIN76A alone or supplemented with SM isolated from skim milk, synthetic N-palmitoylsphingomyelin or N-palmitoyldihydrosphingomyelin for four weeks. The results demonstrated that these synthetic compounds induced a decrease in the number of ACF in all three groups when compared with control by 54, 52 and 70 % respectively [141]. These data established that SM, and not other constituents in SM isolated from milk, was responsible for the suppressive response in colon cancer. Since SM from milk demonstrated a suppressive effect in ACF formation and tumor advancement, the effects of glycosphingolipids (the other major sphingolipid group in milk) were also evaluated. DMH treated CF1 mice were fed with AIN76A supplemented with glucosylceramide, lactosylceramide or ganglioside G_{D3} during four weeks. All treatment groups exhibited a decrease in the number of ACF by >40 % when compared to controls. This decreased tumor growth was accompanied by a reduction in the 5-bromo-2′-deoxyuridine (BrdU)-positive cells with an 80 % decrease in proliferation in the upper half of the crypts [142].

In addition, the effects of oral SM were also evaluated in rats treated with AOM. This study demonstrated that SM decreased the number of ACFs present and proliferation in the base of the crypt [143]. This effect was p53 independent since $p53^{+/-}$ mice treated with AOM and fed with SM presented a decrease in tumor incidence and size similar to wild type littermates [144]. The effects of dietary SM were also assessed in a HT29 xenograft model, where 5-fluorouracil (5FU) nor SM alone exerted no effects, but in combination these agents decreased tumor size [145]. Similarly, combination of exogenous SM with irinotecan increased individual responses in several xenograft models (HT29, HCT15 and CW-39) [146].

From these studies it was clear that dietary SM and specifically milk sphingolipids inhibited intestinal tumorigenesis when supplemented in the diet in a DMH colon cancer model; however, the effect of plant sphingolipids, structurally different from those in mammals, had yet to be determined. To address this question CF1 mice treated with DMH were fed with a diet supplemented with soy glucosylceramide. As with sphingolipids from animals, plant sphingolipids suppressed proliferation in the upper-half of the crypt by at least 50 %, reduced the number of ACF by 52 % and resulted in 37 % fewer adenomas when used at 0.1 % [147]. This study demonstrated that, like milk (animal based) sphingolipids, plant sphingolipids possessed anti-tumor effects in the colon. The effect of plant sphingolipids was also studied in the Min mouse model. Soy glucosylceramide administration to *Apc*$^{Min/+}$ mice resulted in reduction in the number of adenomas by 70 %, and also affected the expression of 96 genes (more than 2-fold) in intestinal mucosal cells. Two of these genes were important transcription factors associated with cancer, as hypoxia-inducible factor 1α (HIF1α) and transcription factor 4 (TCF4) [148].

Sphingadienes, a family of endogenous sphingolipids recently identified in *Drosophila melanogaster* characterized by conjugated trans double bonds at C4,5 [also found in the tobacco hornworm *Manduca sexta* (*Lepidoptera*) [149] and in the larvae of the silk moth *Bombyx mori* [150]], induced colon cancer cell death in vitro. Sphingadienes also prevented intestinal tumorigenesis in *Apc*$^{Min/+}$ by blocking AKT translocation to the membrane, inhibiting protein transcription and promoting apoptosis and autophagy [151]. Also, blocking of AKT with sphingadienes decreased activation of WNT/β-catenin pathway [152]. To investigate the hypothesis that sphingolipids could be used as both chemotherapeutic and chemopreventive agents, CF1 mice were treated with SM isolated from whey prior to DMH, throughout DMH treatment or after tumor initiation. Both groups demonstrated a reduction in proliferation at the bottom of the crypt, and similar decrease in tumor formation when compared with control, with no obvious differences between the treated groups, confirming that sphingolipids can act as both, chemotherapeutic and chemopreventive compounds [153].

Dietary sphingolipids have been extensively studied in colon carcinogenesis which lead to studies on synthetized sphingolipids. Schmelz and colleagues designed a sphingolipid that would be specifically delivered to the colon, a ceramide analogue with a D-glucuronic acid attached at the primary hydroxyl of N-palmitoyl-D-sphingosine (Ceramide-β-glucuronide) [154]. CF1 female mice treated with 30 mg/kg DMH in 1mM in EDTA once a week during 6 weeks fed with a diet control induced 30.2±4.2 ACF per animal, however mice fed with ceramide-β-glucuronide demonstrated reduced number of ACF in a dose manner. In addition, mice treated with the ceramide analogue presented a reduction in the crypt multiplicity of ACF (number of aberrant crypt/focus that indicates the probability of adenoma development). Treatment with this synthetic ceramide also suppressed BrdU-positive cells, both in the bottom half (by 20 % at a dose of 1 %) and the upper half (by 70 % at the same dose).

Another sphingolipid analog that has been utilized in colon cancer models is Enigmol, a sphingoid base that cannot be phosphorylated by sphingosine kinases, thus avoiding the formation of S1P. Oral administration of Enigmol showed antitumor efficacy in $Apc^{Min/+}$ mice and decreased the number of tumors by half with no evidence of toxicity at the doses utilized [155]. The effect of synthetized sphingolipids in colon cancer was also evaluated in a subcutaneous mouse model showing that LCL-30 (C16-pyridinium ceramide) reduces proliferation and tumor growth [156].

The studies examining dietary and synthesized sphingolipids suggest that dietary SM, and other synthetic sphingolipids, may have therapeutic and/or chemopreventive properties. These sphingolipids may also improve the efficacy of current therapeutics utilized in patients, which would be an important future direction for the field.

12 Conclusion

Sphingolipids are an integral part not only of cellular membranes but possess bioactive properties and are key regulators in important biologic processes. Sphingolipids enzymes are a vast set of interacting enzymes that function to balance the levels (and actions) of these bioactive lipids. This is possible in part to sub-cellular localization, as well as cell-type specific expression of these enzymes. Several key enzymes in sphingolipid generation and breakdown are highly expressed in the intestinal tract, in both basal and tumorigenic states. These enzymes and their lipid products have the ability to effect key pathways in intestinal homeostasis and in the promotion of cancer, including proliferative pathways involving MAPK and AKT, inflammatory pathways involving cytokines and downstream mediators such as COX-2, and the Wnt/β-catenin pathway; all of which have been extensively studied in intestinal tumorigenesis. Additionally, loss of sphingolipid enzymes, such as SK1, alk-SMase and nCDase have dramatic effects on intestinal inflammation and tumorigenesis. Furthermore, dietary and synthetic sphingolipids demonstrate both chemopreventive and chemotherapeutic properties. Therefore, the studies examining sphingolipids and their metabolizing enzymes may yield novel potential mechanisms, interventions and novel therapeutic targets for intestinal inflammation and tumorigenesis.

Acknowledgements This work was supported by the National Institutes of Health grant CA097132. Thanks to J-P Laigneau for the artwork in Fig. 3.

References

1. Saddoughi SA, Ogretmen B (2013) Diverse functions of ceramide in cancer cell death and proliferation. Adv Cancer Res 117:37–58. doi:10.1016/B978-0-12-394274-6.00002-9
2. Merrill AH Jr (2011) Sphingolipid and glycosphingolipid metabolic pathways in the era of sphingolipidomics. Chem Rev 111(10):6387–6422. doi:10.1021/cr2002917
3. Liu YY, Hill RA, Li YT (2013) Ceramide glycosylation catalyzed by glucosylceramide synthase and cancer drug resistance. Adv Cancer Res 117:59–89. doi:10.1016/B978-0-12-394274-6.00003-0

4. Henry B, Ziobro R, Becker KA, Kolesnick R, Gulbins E (2013) Acid sphingomyelinase. Handb Exp Pharmacol 215:77–88. doi:10.1007/978-3-7091-1368-4_4
5. Airola MV, Hannun YA (2013) Sphingolipid metabolism and neutral sphingomyelinases. Handb Exp Pharmacol 215:57–76. doi:10.1007/978-3-7091-1368-4_3
6. Holthuis JC, Luberto C (2010) Tales and mysteries of the enigmatic sphingomyelin synthase family. Adv Exp Med Biol 688:72–85
7. Mao C, Obeid LM (2008) Ceramidases: regulators of cellular responses mediated by ceramide, sphingosine, and sphingosine-1-phosphate. Biochim Biophys Acta 1781(9):424–434. pii:S1388-1981(08)00118-2. doi:10.1016/j.bbalip.2008.06.002
8. Heffernan-Stroud LA, Obeid LM (2013) Sphingosine kinase 1 in cancer. Adv Cancer Res 117:201–235. doi:10.1016/B978-0-12-394274-6.00007-8
9. Bartke N, Hannun YA (2009) Bioactive sphingolipids: metabolism and function. J Lipid Res 50(Suppl):S91–S96. doi:10.1194/jlr.R800080-JLR200
10. Gault CR, Obeid LM, Hannun YA (2010) An overview of sphingolipid metabolism: from synthesis to breakdown. Adv Exp Med Biol 688:1–23
11. Bandhuvula P, Saba JD (2007) Sphingosine-1-phosphate lyase in immunity and cancer: silencing the siren. Trends Mol Med 13(5):210–217. doi:10.1016/j.molmed.2007.03.005
12. Hornemann T, Penno A, Rutti MF, Ernst D, Kivrak-Pfiffner F, Rohrer L, von Eckardstein A (2009) The SPTLC3 subunit of serine palmitoyltransferase generates short chain sphingoid bases. J Biol Chem 284(39):26322–26330. doi:10.1074/jbc.M109.023192
13. Han G, Gupta SD, Gable K, Niranjanakumari S, Moitra P, Eichler F, Brown RH Jr, Harmon JM, Dunn TM (2009) Identification of small subunits of mammalian serine palmitoyltransferase that confer distinct acyl-CoA substrate specificities. Proc Natl Acad Sci U S A 106(20):8186–8191. doi:10.1073/pnas.0811269106
14. Merrill AH Jr, Stokes TH, Momin A, Park H, Portz BJ, Kelly S, Wang E, Sullards MC, Wang MD (2009) Sphingolipidomics: a valuable tool for understanding the roles of sphingolipids in biology and disease. J Lipid Res 50(Suppl):S97–S102. doi:10.1194/jlr.R800073-JLR200
15. Merrill AH Jr, Sullards MC, Allegood JC, Kelly S, Wang E (2005) Sphingolipidomics: high-throughput, structure-specific, and quantitative analysis of sphingolipids by liquid chromatography tandem mass spectrometry. Methods (San Diego, Calif) 36(2):207–224
16. Obeid LM, Linardic CM, Karolak LA, Hannun YA (1993) Programmed cell death induced by ceramide. Science (New York, NY) 259(5102):1769–1771
17. Mullen TD, Jenkins RW, Clarke CJ, Bielawski J, Hannun YA, Obeid LM (2011) Ceramide synthase-dependent ceramide generation and programmed cell death: involvement of salvage pathway in regulating postmitochondrial events. J Biol Chem 286(18):15929–15942. doi:10.1074/jbc.M111.230870
18. Hannun YA, Obeid LM (1997) Ceramide and the eukaryotic stress response. BiochemSocTrans 25:1171–1175
19. Hannun YA, Obeid LM (1995) Ceramide: an intracellular signal for apoptosis. Trends Biochem Sci 20(2):73–77
20. Venable ME, Lee JY, Smyth MJ, Bielawska A, Obeid LM (1995) Role of ceramide in cellular senescence. J Biol Chem 270(51):30701–30708
21. Jayadev S, Liu B, Bielawska AE, Lee JY, Nazaire F, Pushkareva M, Obeid LM, Hannun YA (1995) Role for ceramide in cell cycle arrest. J Biol Chem 270(5):2047–2052
22. Lee JY, Hannun YA, Obeid LM (1996) Ceramide inactivates cellular protein kinase Calpha. J Biol Chem 271 (22):13169–13174
23. Kishikawa K, Chalfant CE, Perry DK, Bielawska A, Hannun YA (1999) Phosphatidic acid is a potent and selective inhibitor of protein phosphatase 1 and an inhibitor of ceramide-mediated responses. J Biol Chem 274(30):21335–21341
24. Perry DM, Kitatani K, Roddy P, El-Osta M, Hannun YA (2012) Identification and characterization of protein phosphatase 2C activation by ceramide. J Lipid Res 53(8):1513–1521. doi:10.1194/jlr.M025395
25. Bourbon NA, Sandirasegarane L, Kester M (2002) Ceramide-induced inhibition of Akt is mediated through protein kinase Czeta: implications for growth arrest. J Biol Chem 277(5):3286–3292. doi:10.1074/jbc.M110541200

26. Ghosh N, Patel N, Jiang K, Watson JE, Cheng J, Chalfant CE, Cooper DR (2007) Ceramide-activated protein phosphatase involvement in insulin resistance via Akt, serine/arginine-rich protein 40, and ribonucleic acid splicing in L6 skeletal muscle cells. Endocrinology 148(3):1359–1366. doi:10.1210/en.2006-0750
27. Canals D, Roddy P, Hannun YA (2012) Protein phosphatase 1alpha mediates ceramide-induced ERM protein dephosphorylation: a novel mechanism independent of phosphatidylinositol 4,5-biphosphate (PIP2) and myosin/ERM phosphatase. J Biol Chem 287(13):10145–10155. doi:10.1074/jbc.M111.306456
28. Hartmann D, Lucks J, Fuchs S, Schiffmann S, Schreiber Y, Ferreiros N, Merkens J, Marschalek R, Geisslinger G, Grosch S (2012) Long chain ceramides and very long chain ceramides have opposite effects on human breast and colon cancer cell growth. Int J Biochem Cell Biol 44(4):620–628. doi:10.1016/j.biocel.2011.12.019
29. Johnson KR, Becker KP, Facchinetti MM, Hannun YA, Obeid LM (2002) PKC-dependent activation of sphingosine kinase 1 and translocation to the plasma membrane. Extracellular release of sphingosine-1-phosphate induced by phorbol 12-myristate 13-acetate (PMA). J Biol Chem 277(38):35257–35262
30. Seufferlein T, Rozengurt E (1994) Sphingosine induces p125FAK and paxillin tyrosine phosphorylation, actin stress fiber formation, and focal contact assembly in Swiss 3T3 cells. J Biol Chem 269(44):27610–27617
31. Sweeney EA, Inokuchi J, Igarashi Y (1998) Inhibition of sphingolipid induced apoptosis by caspase inhibitors indicates that sphingosine acts in an earlier part of the apoptotic pathway than ceramide. FEBS Lett 425(1):61–65
32. Hannun YA, Loomis CR, Merrill AH Jr, Bell RM (1986) Sphingosine inhibition of protein kinase C activity and of phorbol dibutyrate binding in vitro and in human platelets. J Biol Chem 261(27):12604–12609
33. Khan WA, Mascarella SW, Lewin AH, Wyrick CD, Carroll FI, Hannun YA (1991) Use of D-erythro-sphingosine as a pharmacological inhibitor of protein kinase C in human platelets. Biochem J 278(Pt 2):387–392
34. McDonald OB, Hannun YA, Reynolds CH, Sahyoun N (1991) Activation of casein kinase II by sphingosine. J Biol Chem 266(32):21773–21776
35. Snider AJ, Orr Gandy KA, Obeid LM (2010) Sphingosine kinase: role in regulation of bioactive sphingolipid mediators in inflammation. Biochimie 92(6):707–715. pii:S0300-9084(10)00052-0. doi:10.1016/j.biochi.2010.02.008
36. Alvarez SE, Harikumar KB, Hait NC, Allegood J, Strub GM, Kim EY, Maceyka M, Jiang H, Luo C, Kordula T, Milstien S, Spiegel S (2010) Sphingosine-1-phosphate is a missing cofactor for the E3 ubiquitin ligase TRAF2. Nature 465(7301):1084–1088. doi:10.1038/nature09128
37. Hait NC, Allegood J, Maceyka M, Strub GM, Harikumar KB, Singh SK, Luo C, Marmorstein R, Kordula T, Milstien S, Spiegel S (2009) Regulation of histone acetylation in the nucleus by sphingosine-1-phosphate. Science (New York, NY) 325 (5945):1254–1257
38. Tanfin Z, Serrano-Sanchez M, Leiber D (2011) ATP-binding cassette ABCC1 is involved in the release of sphingosine 1-phosphate from rat uterine leiomyoma ELT3 cells and late pregnant rat myometrium. Cell Signal 23(12):1997–2004. doi:10.1016/j.cellsig.2011.07.010
39. Mitra P, Oskeritzian CA, Payne SG, Beaven MA, Milstien S, Spiegel S (2006) Role of ABCC1 in export of sphingosine-1-phosphate from mast cells. Proc Natl Acad Sci U S A 103(44):16394–16399
40. Hammad SM, Taha TA, Nareika A, Johnson KR, Lopes-Virella MF, Obeid LM (2006) Oxidized LDL immune complexes induce release of sphingosine kinase in human U937 monocytic cells. Prostaglandins Other Lipid Mediat 79(1–2):126–140. doi:10.1016/j.prostaglandins.2005.12.004
41. Venkataraman K, Thangada S, Michaud J, Oo ML, Ai Y, Lee YM, Wu M, Parikh NS, Khan F, Proia RL, Hla T (2006) Extracellular export of sphingosine kinase-1a contributes to the vascular S1P gradient. Biochem J 397(3):461–471
42. Spiegel S, Milstien S (2003) Sphingosine-1-phosphate: an enigmatic signalling lipid. Nat Rev Mol Cell Biol 4(5):397–407

43. Kluk MJ, Hla T (2002) Signaling of sphingosine-1-phosphate via the S1P/EDG-family of G-protein-coupled receptors. Biochim Biophys Acta 1582(1–3):72–80
44. Taha TA, Argraves KM, Obeid LM (2004) Sphingosine-1-phosphate receptors: receptor specificity versus functional redundancy. Biochim Biophys Acta 1682(1–3):48–55
45. Hla T (2004) Physiological and pathological actions of sphingosine 1-phosphate. Semin Cell Dev Biol 15(5):513–520
46. Brinkmann V (2009) FTY720 (fingolimod) in multiple sclerosis: therapeutic effects in the immune and the central nervous system. Br J Pharmacol 158(5):1173–1182. doi:10.1111/j.1476-5381.2009.00451.x
47. Matloubian M, Lo CG, Cinamon G, Lesneski MJ, Xu Y, Brinkmann V, Allende ML, Proia RL, Cyster JG (2004) Lymphocyte egress from thymus and peripheral lymphoid organs is dependent on S1P receptor 1. Nature 427(6972):355–360
48. Brinkmann V, Cyster JG, Hla T (2004) FTY720: sphingosine 1-phosphate receptor-1 in the control of lymphocyte egress and endothelial barrier function. Am J Transplant 4(7):1019–1025. doi:10.1111/j.1600-6143.2004.00476.x. pii:AJT476
49. Liang J, Nagahashi M, Kim EY, Harikumar KB, Yamada A, Huang WC, Hait NC, Allegood JC, Price MM, Avni D, Takabe K, Kordula T, Milstien S, Spiegel S (2013) Sphingosine-1-phosphate links persistent STAT3 activation, chronic intestinal inflammation, and development of colitis-associated cancer. Cancer Cell 23(1):107–120. doi:10.1016/j.ccr.2012.11.013
50. An S, Goetzl EJ, Lee H (1998) Signaling mechanisms and molecular characteristics of G protein-coupled receptors for lysophosphatidic acid and sphingosine 1-phosphate. J Cell Biochem Suppl 31:147–157
51. Orr Gandy KA, Adada M, Canals D, Carroll B, Roddy P, Hannun YA, Obeid LM (2013) Epidermal growth factor-induced cellular invasion requires sphingosine-1-phosphate/sphingosine-1-phosphate 2 receptor-mediated ezrin activation. FASEB J. doi:10.1096/fj.13-228460
52. Gandy KA, Canals D, Adada M, Wada M, Roddy P, Snider AJ, Hannun YA, Obeid LM (2013) Sphingosine 1-phosphate induces filopodia formation through S1PR2 activation of ERM proteins. Biochem J 449(3):661–672. doi:10.1042/BJ20120213
53. Hsu A, Zhang W, Lee JF, An J, Ekambaram P, Liu J, Honn KV, Klinge CM, Lee MJ (2012) Sphingosine-1-phosphate receptor-3 signaling up-regulates epidermal growth factor receptor and enhances epidermal growth factor receptor-mediated carcinogenic activities in cultured lung adenocarcinoma cells. Int J Oncol 40(5):1619–1626. doi:10.3892/ijo.2012.1379
54. Sukocheva O, Wadham C, Holmes A, Albanese N, Verrier E, Feng F, Bernal A, Derian CK, Ullrich A, Vadas MA, Xia P (2006) Estrogen transactivates EGFR via the sphingosine 1-phosphate receptor Edg-3: the role of sphingosine kinase-1. J Cell Biol 173(2):301–310. doi:10.1083/jcb.200506033
55. Van Brocklyn JR, Graler MH, Bernhardt G, Hobson JP, Lipp M, Spiegel S (2000) Sphingosine-1-phosphate is a ligand for the G protein-coupled receptor EDG-6. Blood 95(8):2624–2629.
56. Wang W, Graeler MH, Goetzl EJ (2005) Type 4 sphingosine 1-phosphate G protein-coupled receptor (S1P4) transduces S1P effects on T cell proliferation and cytokine secretion without signaling migration. FASEB J 19(12):1731–1733
57. Novgorodov AS, El-Alwani M, Bielawski J, Obeid LM, Gudz TI (2007) Activation of sphingosine-1-phosphate receptor S1P5 inhibits oligodendrocyte progenitor migration. FASEB J 21(7):1503–1514. pii:fj.06-7420com. doi:10.1096/fj.06-7420com
58. Anelli V, Gault CR, Snider AJ, Obeid LM (2010) Role of sphingosine kinase-1 in paracrine/transcellular angiogenesis and lymphangiogenesis in vitro. FASEB J 24(8):2727–2738. doi:10.1096/fj.09-150540
59. Anelli V, Gault CR, Cheng AB, Obeid LM (2008) Sphingosine Kinase 1 Is Up-regulated during Hypoxia in U87MG Glioma Cells: ROLE OF HYPOXIA-INDUCIBLE FACTORS 1 AND 2. J Biolog Chem 283(6):3365–3375
60. Xia P, Gamble JR, Wang L, Pitson SM, Moretti PA, Wattenberg BW, D'Andrea RJ, Vadas MA (2000) An oncogenic role of sphingosine kinase. Curr Biol 10(23):1527–1530.
61. Kawamori T, Osta W, Johnson KR, Pettus BJ, Bielawski J, Tanaka T, Wargovich MJ, Reddy BS, Hannun YA, Obeid LM, Zhou D (2006) Sphingosine kinase 1 is up-regulated in colon carcinogenesis. FASEB J 20(2):386–388

62. Pettus BJ, Bielawski J, Porcelli AM, Reames DL, Johnson KR, Morrow J, Chalfant CE, Obeid LM, Hannun YA (2003) The sphingosine kinase 1/sphingosine-1-phosphate pathway mediates COX-2 induction and PGE2 production in response to TNF-alpha. FASEB J 17(11):1411–1421
63. Visentin B, Vekich JA, Sibbald BJ, Cavalli AL, Moreno KM, Matteo RG, Garland WA, Lu Y, Yu S, Hall HS, Kundra V, Mills GB, Sabbadini RA (2006) Validation of an anti-sphingosine-1-phosphate antibody as a potential therapeutic in reducing growth, invasion, and angiogenesis in multiple tumor lineages. Cancer Cell 9(3):225–238. doi:10.1016/j.ccr.2006.02.023
64. Xia P, Gamble JR, Rye KA, Wang L, Hii CS, Cockerill P, Khew-Goodall Y, Bert AG, Barter PJ, Vadas MA (1998) Tumor necrosis factor-alpha induces adhesion molecule expression through the sphingosine kinase pathway. Proc Natl Acad Sci U S A 95(24):14196–14201
65. Snider AJ, Wu BX, Jenkins RW, Sticca JA, Kawamori T, Hannun YA, Obeid LM (2012) Loss of neutral ceramidase increases inflammation in a mouse model of inflammatory bowel disease. Prostaglandins Other Lipid Mediat 99(3–4):124–130. doi:10.1016/j.prostaglandins.2012.08.003
66. Snider AJ, Kawamori T, Bradshaw SG, Orr KA, Gilkeson GS, Hannun YA, Obeid LM (2009) A role for sphingosine kinase 1 in dextran sulfate sodium-induced colitis. FASEB J 23(1):143–152
67. Kawamori T, Kaneshiro T, Okumura M, Maalouf S, Uflacker A, Bielawski J, Hannun YA, Obeid LM (2009) Role for sphingosine kinase 1 in colon carcinogenesis. FASEB J 23(2):405–414
68. Rueda R, Sabatel JL, Maldonado J, Molina-Font JA, Gil A (1998) Addition of gangliosides to an adapted milk formula modifies levels of fecal *Escherichia coli* in preterm newborn infants. J Pediatrics 133(1):90–94
69. Garcia-Barros M, Coant N, Truman JP, Snider AJ, Hannun YA (2013) Sphingolipids in colon cancer. Biochim Biophys Acta. doi:10.1016/j.bbalip.2013.09.007
70. Yamashita T, Wada R, Sasaki T, Deng C, Bierfreund U, Sandhoff K, Proia RL (1999) A vital role for glycosphingolipid synthesis during development and differentiation. Proc Natl Acad Sci U S A 96(16):9142–9147
71. Jennemann R, Kaden S, Sandhoff R, Nordstrom V, Wang S, Volz M, Robine S, Amen N, Rothermel U, Wiegandt H, Grone HJ (2012) Glycosphingolipids are essential for intestinal endocytic function. J Biol Chem 287(39):32598–32616. doi:10.1074/jbc.M112.371005
72. Scharl M, Leucht K, Frey-Wagner I, Zeitz J, Hausmann M, Fischbeck A, Liebisch G, Kellermeier S, Pesch T, Arikkat J, Schmitz G, Fried M, Yildiz Y, Rogler G (2011) Knock-out of beta-glucosidase 2 has no influence on dextran sulfate sodium-induced colitis. Digestion 84(2):156–167. doi:10.1159/000327380
73. Song M, Zang W, Zhang B, Cao J, Yang G (2012) GCS overexpression is associated with multidrug resistance of human HCT-8 colon cancer cells. J Exp Clin Cancer Res 31:23. doi:10.1186/1756-9966-31-23
74. Zhang Y, Cheng Y, Hansen GH, Niels-Christiansen LL, Koentgen F, Ohlsson L, Nilsson A, Duan RD (2011) Crucial role of alkaline sphingomyelinase in sphingomyelin digestion: a study on enzyme knockout mice. J Lipid Res 52(4):771–781. doi:10.1194/jlr.M012880
75. Yang L, Mutanen M, Cheng Y, Duan RD (2002) Effects of fat, beef and fiber in diets on activities of sphingomyelinase, ceramidase and caspase-3 in rat colonic mucosa. Med Princ Pract 11(3):150–156. doi:63247
76. Yang L, Mutanen M, Cheng Y, Duan R (2002) Effects of red meat and fiber in high fat diet on activities of sphingomyelinase, ceramidase and caspase-3 in rat colonic mucosa. J Nutr Biochem 13(8):499
77. Sjoqvist U, Hertervig E, Nilsson A, Duan RD, Ost A, Tribukait B, Lofberg R (2002) Chronic colitis is associated with a reduction of mucosal alkaline sphingomyelinase activity. Inflamm Bowel Dis 8(4):258–263
78. Andersson D, Kotarsky K, Wu J, Agace W, Duan RD (2009) Expression of alkaline sphingomyelinase in yeast cells and anti-inflammatory effects of the expressed enzyme in a rat colitis model. Dig Dis Sci 54(7):1440–1448. doi:10.1007/s10620-008-0509-2

79. Soo I, Madsen KL, Tejpar Q, Sydora BC, Sherbaniuk R, Cinque B, Di Marzio L, Cifone MG, Desimone C, Fedorak RN (2008) VSL#3 probiotic upregulates intestinal mucosal alkaline sphingomyelinase and reduces inflammation. Can J Gastroenterol 22(3):237–242
80. Sakata A, Ochiai T, Shimeno H, Hikishima S, Yokomatsu T, Shibuya S, Toda A, Eyanagi R, Soeda S (2007) Acid sphingomyelinase inhibition suppresses lipopolysaccharide-mediated release of inflammatory cytokines from macrophages and protects against disease pathology in dextran sulphate sodium-induced colitis in mice. Immunology 122(1):54–64. doi:10.1111/j.1365-2567.2007.02612.x
81. Dudeja PK, Dahiya R, Brasitus TA (1986) The role of sphingomyelin synthetase and sphingomyelinase in 1,2-dimethylhydrazine-induced lipid alterations of rat colonic plasma membranes. Biochim Biophy Acta 863(2):309–312
82. Zhang P, Cheng Y, Duan RD (2013) Ursolic acid inhibits acid sphingomyelinase in intestinal cells. Phytother Res 27(2):173–178. doi:10.1002/ptr.4709
83. Cheng Y, Tauschel HD, Nilsson A, Duan RD (1999) Ursodeoxycholic acid increases the activities of alkaline sphingomyelinase and caspase-3 in the rat colon. Scand J Gastroenterol 34(9):915–920
84. Andersson D, Cheng Y, Duan RD (2008) Ursolic acid inhibits the formation of aberrant crypt foci and affects colonic sphingomyelin hydrolyzing enzymes in azoxymethane-treated rats. J Cancer Res Clin Oncol 134(1):101–107. doi:10.1007/s00432-007-0255-4
85. Kohno H, Suzuki R, Yasui Y, Miyamoto S, Wakabayashi K, Tanaka T (2007) Ursodeoxycholic acid versus sulfasalazine in colitis-related colon carcinogenesis in mice. Clin Cancer Res 13(8):2519–2525. doi:10.1158/1078-0432.CCR-06-2727
86. Zhang P, Li B, Gao S, Duan RD (2008) Dietary sphingomyelin inhibits colonic tumorigenesis with an up-regulation of alkaline sphingomyelinase expression in ICR mice. Anticancer Res 28(6A):3631–3635
87. Appleyard CB, Cruz ML, Isidro AA, Arthur JC, Jobin C, De Simone C (2011) Pretreatment with the probiotic VSL#3 delays transition from inflammation to dysplasia in a rat model of colitis-associated cancer. Am J Physiol Gastrointest Liver Physiol 301(6):G1004–G1013. doi:10.1152/ajpgi.00167.2011
88. Wu J, Cheng Y, Nilsson A, Duan RD (2004) Identification of one exon deletion of intestinal alkaline sphingomyelinase in colon cancer HT-29 cells and a differentiation-related expression of the wild-type enzyme in Caco-2 cells. Carcinogenesis 25(8):1327–1333. doi:10.1093/carcin/bgh140
89. Hertervig E, Nilsson A, Nyberg L, Duan RD (1997) Alkaline sphingomyelinase activity is decreased in human colorectal carcinoma. Cancer 79(3):448–453
90. Hertervig E, Nilsson A, Nilbert M, Duan RD (2003) Reduction in alkaline sphingomyelinase in colorectal tumorigenesis is not related to the APC gene mutation. Int J Colorectal Dis 18(4):309–313. doi:10.1007/s00384-002-0471-y
91. Osawa Y, Suetsugu A, Matsushima-Nishiwaki R, Yasuda I, Saibara T, Moriwaki H, Seishima M, Kozawa O (2013) Liver acid sphingomyelinase inhibits growth of metastatic colon cancer. J Clin Invest 123(2):834–843. doi:10.1172/JCI65188
92. Hertervig E, Nilsson A, Bjork J, Hultkrantz R, Duan RD (1999) Familial adenomatous polyposis is associated with a marked decrease in alkaline sphingomyelinase activity: a key factor to the unrestrained cell proliferation? Br J Cancer 81(2):232–236. doi:10.1038/sj.bjc.6690682
93. Di Marzio L, Di Leo A, Cinque B, Fanini D, Agnifili A, Berloco P, Linsalata M, Lorusso D, Barone M, De Simone C, Cifone MG (2005) Detection of alkaline sphingomyelinase activity in human stool: proposed role as a new diagnostic and prognostic marker of colorectal cancer. Cancer Epidemiol Biomarkers Prev 14(4):856–862. doi:10.1158/1055-9965.EPI-04-0434
94. Zhu Q, Jin JF, Shan XH, Liu CP, Mao XD, Xu KF, Liu C (2008) Chronic activation of neutral ceramidase protects beta-cells against cytokine-induced apoptosis. Acta Pharmacol Sin 29(5):593–599. doi:10.1111/j.1745-7254.2008.00781.x
95. Wu BX, Zeidan YH, Hannun YA (2009) Downregulation of neutral ceramidase by gemcitabine: implications for cell cycle regulation. Biochim Biophys Acta 1791(8):730–739. pii:S1388-1981(09)00086-9. doi:10.1016/j.bbalip.2009.03.012

96. Zeidan YH, Pettus BJ, Elojeimy S, Taha T, Obeid LM, Kawamori T, Norris JS, Hannun YA (2006) Acid ceramidase but not acid sphingomyelinase is required for tumor necrosis factor-{alpha}-induced PGE2 production. J Biol Chem 281(34):24695–24703
97. Liu H, Toman RE, Goparaju SK, Maceyka M, Nava VE, Sankala H, Payne SG, Bektas M, Ishii I, Chun J, Milstien S, Spiegel S (2003) Sphingosine kinase type 2 is a putative BH3-only protein that induces apoptosis. J Biol Chem 278 (41):40330–40336
98. Zemann B, Urtz N, Reuschel R, Mechtcheriakova D, Bornancin F, Badegruber R, Baumruker T, Billich A (2007) Normal neutrophil functions in sphingosine kinase type 1 and 2 knockout mice. Immunol Lett 109(1):56–63
99. Xiong Y, Lee HJ, Mariko B, Lu YC, Dannenberg AJ, Haka AS, Maxfield FR, Camerer E, Proia RL, Hla T (2013) Sphingosine kinases are not required for inflammatory responses in macrophages. J Biol Chem 288(45):32563–32573. doi:10.1074/jbc.M113.483750
100. Allende ML, Sasaki T, Kawai H, Olivera A, Mi Y, van Echten-Deckert G, Hajdu R, Rosenbach M, Keohane CA, Mandala S, Spiegel S, Proia RL (2004) Mice deficient in sphingosine kinase 1 are rendered lymphopenic by FTY720. J Biol Chem 279(50):52487–52492
101. Kharel Y, Raje M, Gao M, Gellett AM, Tomsig JL, Lynch KR, Santos WL (2012) Sphingosine kinase type 2 inhibition elevates circulating sphingosine 1-phosphate. Biochem J 447(1):149–157. doi:10.1042/BJ20120609
102. Mizugishi K, Yamashita T, Olivera A, Miller GF, Spiegel S, Proia RL (2005) Essential role for sphingosine kinases in neural and vascular development. Mol Cell Biol 25(24):11113–11121
103. Antoon JW, White MD, Meacham WD, Slaughter EM, Muir SE, Elliott S, Rhodes LV, Ashe HB, Wiese TE, Smith CD, Burow ME, Beckman BS (2010) Antiestrogenic effects of the novel sphingosine kinase-2 inhibitor ABC294640. Endocrinology 151(11):5124–5135. doi:10.1210/en.2010-0420
104. Maines LW, Fitzpatrick LR, French KJ, Zhuang Y, Xia Z, Keller SN, Upson JJ, Smith CD (2008) Suppression of ulcerative colitis in mice by orally available inhibitors of sphingosine kinase. Dig Dis Sci 53(4):997–1012
105. Maines LW, Fitzpatrick LR, Green CL, Zhuang Y, Smith CD (2013) Efficacy of a novel sphingosine kinase inhibitor in experimental Crohn's disease. Inflammopharmacology 18(2):73–85. doi:10.1007/s10787-010-0032-x
106. Kohno M, Momoi M, Oo ML, Paik JH, Lee YM, Venkataraman K, Ai Y, Ristimaki AP, Fyrst H, Sano H, Rosenberg D, Saba JD, Proia RL, Hla T (2006) Intracellular role for sphingosine kinase 1 in intestinal adenoma cell proliferation. Mol Cell Biol 26(19):7211–7223
107. Johnson KR, Johnson KY, Crellin HG, Ogretmen B, Boylan AM, Harley RA, Obeid LM (2005) Immunohistochemical distribution of sphingosine kinase 1 in normal and tumor lung tissue. J Histochem Cytochem 53(9):1159–1166
108. Tan SS, Khin LW, Wong L, Yan B, Ong CW, Datta A, Salto-Tellez M, Lam Y, Yap CT (2014) Sphingosine kinase 1 promotes malignant progression in colon cancer and independently predicts survival of patients with colon cancer by competing risk approach in South asian population. Clin Transl Gastroenterol 5:e51. doi:10.1038/ctg.2013.21
109. Allende ML, Bektas M, Lee BG, Bonifacino E, Kang J, Tuymetova G, Chen W, Saba JD, Proia RL (2011) Sphingosine-1-phosphate lyase deficiency produces a pro-inflammatory response while impairing neutrophil trafficking. J Biol Chem 286(9):7348–7358. doi:10.1074/jbc.M110.171819
110. Vogel P, Donoviel MS, Read R, Hansen GM, Hazlewood J, Anderson SJ, Sun W, Swaffield J, Oravecz T (2009) Incomplete inhibition of sphingosine 1-phosphate lyase modulates immune system function yet prevents early lethality and non-lymphoid lesions. PLoS ONE 4(1):e4112. doi:10.1371/journal.pone.0004112
111. Allende ML, Sipe LM, Tuymetova G, Wilson-Henjum KL, Chen W, Proia RL (2013) Sphingosine-1-phosphate phosphatase 1 regulates keratinocyte differentiation and epidermal homeostasis. J Biol Chem 288(25):18381–18391. doi:10.1074/jbc.M113.478420
112. Oskouian B, Sooriyakumaran P, Borowsky AD, Crans A, Dillard-Telm L, Tam YY, Bandhuvula P, Saba JD (2006) Sphingosine-1-phosphate lyase potentiates apoptosis via p53- and p38-dependent pathways and is down-regulated in colon cancer. Proc Natl Acad Sci U S A 103(46):17384–17389. doi:10.1073/pnas.0600050103

113. Oskouian B, Saba J (2007) Sphingosine-1-phosphate metabolism and intestinal tumorigenesis: lipid signaling strikes again. Cell Cycle 6(5):522–527
114. Pappu R, Schwab SR, Cornelissen I, Pereira JP, Regard JB, Xu Y, Camerer E, Zheng YW, Huang Y, Cyster JG, Coughlin SR (2007) Promotion of lymphocyte egress into blood and lymph by distinct sources of sphingosine-1-phosphate. Science (New York, NY) 316(5822):295–298
115. Camerer E, Regard JB, Cornelissen I, Srinivasan Y, Duong DN, Palmer D, Pham TH, Wong JS, Pappu R, Coughlin SR (2009) Sphingosine-1-phosphate in the plasma compartment regulates basal and inflammation-induced vascular leak in mice. J Clin Invest 119(7):1871–1879
116. Liu Y, Wada R, Yamashita T, Mi Y, Deng CX, Hobson JP, Rosenfeldt HM, Nava VE, Chae SS, Lee MJ, Liu CH, Hla T, Spiegel S, Proia RL (2000) Edg-1, the G protein-coupled receptor for sphingosine-1-phosphate, is essential for vascular maturation. J Clin Invest 106(8):951–961
117. Lorenz JN, Arend LJ, Robitz R, Paul RJ, MacLennan AJ (2007) Vascular dysfunction in S1P2 sphingosine 1-phosphate receptor knockout mice. Am J Physiol Regul Integr Comp Physiol 292(1):R440–R446. doi:10.1152/ajpregu.00085.2006
118. Kono M, Belyantseva IA, Skoura A, Frolenkov GI, Starost MF, Dreier JL, Lidington D, Bolz SS, Friedman TB, Hla T, Proia RL (2007) Deafness and stria vascularis defects in S1P2 receptor-null mice. J Biol Chem 282(14):10690–10696. pii:M700370200. doi:10.1074/jbc.M700370200
119. MacLennan AJ, Carney PR, Zhu WJ, Chaves AH, Garcia J, Grimes JR, Anderson KJ, Roper SN, Lee N (2001) An essential role for the H218/AGR16/Edg-5/LP(B2) sphingosine 1-phosphate receptor in neuronal excitability. Eur J Neurosci 14(2):203–209
120. Yamashita H, Kitayama J, Shida D, Yamaguchi H, Mori K, Osada M, Aoki S, Yatomi Y, Takuwa Y, Nagawa H (2006) Sphingosine 1-phosphate receptor expression profile in human gastric cancer cells: differential regulation on the migration and proliferation. J Surg Res 130(1):80–87. doi:10.1016/j.jss.2005.08.004
121. Ishii I, Ye X, Friedman B, Kawamura S, Contos JJ, Kingsbury MA, Yang AH, Zhang G, Brown JH, Chun J (2002) Marked perinatal lethality and cellular signaling deficits in mice null for the two sphingosine 1-phosphate (S1P) receptors, S1P(2)/LP(B2)/EDG-5 and S1P(3)/LP(B3)/EDG-3. J Biol Chem 277(28):25152–25159
122. Golfier S, Kondo S, Schulze T, Takeuchi T, Vassileva G, Achtman AH, Graler MH, Abbondanzo SJ, Wiekowski M, Kremmer E, Endo Y, Lira SA, Bacon KB, Lipp M (2010) Shaping of terminal megakaryocyte differentiation and proplatelet development by sphingosine-1-phosphate receptor S1P4. FASEB J 24(12):4701–4710. doi:10.1096/fj.09-141473
123. Debien E, Mayol K, Biajoux V, Daussy C, De Aguero MG, Taillardet M, Dagany N, Brinza L, Henry T, Dubois B, Kaiserlian D, Marvel J, Balabanian K, Walzer T (2013) S1PR5 is pivotal for the homeostasis of patrolling monocytes. Eur J Immunol 43(6):1667–1675. doi:10.1002/eji.201343312
124. Montrose DC, Scherl EJ, Bosworth BP, Zhou XK, Jung B, Dannenberg AJ, Hla T (2013) S1P(1) localizes to the colonic vasculature in ulcerative colitis and maintains blood vessel integrity. J Lipid Res 54(3):843–851. doi:10.1194/jlr.M034108
125. Schulze T, Golfier S, Tabeling C, Rabel K, Graler MH, Witzenrath M, Lipp M (2011) Sphingosine-1-phospate receptor 4 (S1P(4)) deficiency profoundly affects dendritic cell function and TH17-cell differentiation in a murine model. FASEB J 25(11):4024–4036. doi:10.1096/fj.10-179028
126. Sanada Y, Mizushima T, Kai Y, Nishimura J, Hagiya H, Kurata H, Mizuno H, Uejima E, Ito T (2011) Therapeutic effects of novel sphingosine-1-phosphate receptor agonist W-061 in murine DSS colitis. PLoS ONE 6(9):e23933. doi:10.1371/journal.pone.0023933
127. Song J, Matsuda C, Kai Y, Nishida T, Nakajima K, Mizushima T, Kinoshita M, Yasue T, Sawa Y, Ito T (2008) A novel sphingosine 1-phosphate receptor agonist, 2-amino-2-propanediol hydrochloride (KRP-203), regulates chronic colitis in interleukin-10 gene-deficient mice. J Pharmacol Exp Ther 324(1):276–283

128. Dong J, Wang H, Wu G, Zhao J, Zhang L, Zuo L, Zhu W, Gong J, Li Y, Gu L, Li J (2014) Oral treatment with SEW2871, a sphingosine-1-phosphate type 1 receptor agonist, ameliorates experimental colitis in Interleukin-10 gene deficient mice. Clin Exp Immunol. doi:10.1111/cei.12304
129. Daniel C, Sartory NA, Zahn N, Schmidt R, Geisslinger G, Radeke HH, Stein JM (2007) FTY720 ameliorates oxazolone colitis in mice by directly affecting T helper type 2 functions. Mol Immunol 44(13):3305–3316
130. Deguchi Y, Andoh A, Yagi Y, Bamba S, Inatomi O, Tsujikawa T, Fujiyama Y (2006) The S1P receptor modulator FTY720 prevents the development of experimental colitis in mice. Oncol Rep 16(4):699–703
131. Radi ZA, Heuvelman DM, Masferrer JL, Benson EL (2011) Pharmacologic evaluation of sulfasalazine, FTY720, and anti-IL-12/23p40 in a TNBS-induced Crohn's disease model. Dig Dis Sci 56(8):2283–2291. doi:10.1007/s10620-011-1628-8
132. Murphy CT, Hall LJ, Hurley G, Quinlan A, MacSharry J, Shanahan F, Nally K, Melgar S (2012) The sphingosine-1-phosphate analogue FTY720 impairs mucosal immunity and clearance of the enteric pathogen Citrobacter rodentium. Infect Immun 80(8):2712–2723. doi:10.1128/IAI.06319-11
133. Tedesco-Silva H, Mourad G, Kahan BD, Boira JG, Weimar W, Mulgaonkar S, Nashan B, Madsen S, Charpentier B, Pellet P, Vanrenterghem Y (2004) FTY720, a novel immunomodulator: efficacy and safety results from the first phase 2A study in de novo renal transplantation. Transplantation 77(12):1826–1833
134. Nguyen AV, Wu YY, Liu Q, Wang D, Nguyen S, Loh R, Pang J, Friedman K, Orlofsky A, Augenlicht L, Pollard JW, Lin EY (2013) STAT3 in epithelial cells regulates inflammation and tumor progression to malignant state in colon. Neoplasia 15(9):998–1008
135. Nagaoka Y, Otsuki K, Fujita T, Uesato S (2008) Effects of phosphorylation of immunomodulatory agent FTY720 (fingolimod) on antiproliferative activity against breast and colon cancer cells. Biol Pharm Bull 31(6):1177–1181
136. Selzner M, Bielawska A, Morse MA, Rudiger HA, Sindram D, Hannun YA, Clavien PA (2001) Induction of apoptotic cell death and prevention of tumor growth by ceramide analogues in metastatic human colon cancer. Cancer Res 61(3):1233–1240
137. Merchant TE, Kasimos JN, de Graaf PW, Minsky BD, Gierke LW, Glonek T (1991) Phospholipid profiles of human colon cancer using 31P magnetic resonance spectroscopy. Int J Colorectal Dis 6(2):121–126
138. Schmelz EM, Roberts PC, Kustin EM, Lemonnier LA, Sullards MC, Dillehay DL, Merrill AH Jr (2001) Modulation of intracellular beta-catenin localization and intestinal tumorigenesis in vivo and in vitro by sphingolipids. Cancer Res 61(18):6723–6729
139. Schmelz EM, Dillehay DL, Webb SK, Reiter A, Adams J, Merrill AH Jr (1996) Sphingomyelin consumption suppresses aberrant colonic crypt foci and increases the proportion of adenomas versus adenocarcinomas in CF1 mice treated with 1,2-dimethylhydrazine: implications for dietary sphingolipids and colon carcinogenesis. Cancer Res 56(21):4936–4941
140. Dillehay DL, Webb SK, Schmelz EM, Merrill AH Jr (1994) Dietary sphingomyelin inhibits 1,2-dimethylhydrazine-induced colon cancer in CF1 mice. J Nutr 124(5):615–620
141. Merrill AH Jr, Schmelz EM, Wang E, Dillehay DL, Rice LG, Meredith F, Riley RT (1997) Importance of sphingolipids and inhibitors of sphingolipid metabolism as components of animal diets. J Nutr 127(5 Suppl):830S–833S
142. Schmelz EM, Sullards MC, Dillehay DL, Merrill AH Jr (2000) Colonic cell proliferation and aberrant crypt foci formation are inhibited by dairy glycosphingolipids in 1,2-dimethylhydrazine-treated CF1 mice. J Nutr 130(3):522–527
143. Exon JH, South EH (2003) Effects of sphingomyelin on aberrant colonic crypt foci development, colon crypt cell proliferation and immune function in an aging rat tumor model. Food Chem Toxicol 41(4):471–476
144. Hu Y, Le Leu RK, Belobrajdic D, Young GP (2008) The potential of sphingomyelin as a chemopreventive agent in AOM-induced colon cancer model: wild-type and p53+ /– mice. Mol Nutr Food Res 52(5):558–566. doi:10.1002/mnfr.200700258

145. Modrak DE, Lew W, Goldenberg DM, Blumenthal R (2000) Sphingomyelin potentiates chemotherapy of human cancer xenografts. Biochem Biophys Res Commun 268(2):603–606. doi:10.1006/bbrc.2000.2178
146. Modrak DE, Rodriguez MD, Goldenberg DM, Lew W, Blumenthal RD (2002) Sphingomyelin enhances chemotherapy efficacy and increases apoptosis in human colonic tumor xenografts. Int J Oncol 20(2):379–384
147. Symolon H, Schmelz EM, Dillehay DL, Merrill AH Jr (2004) Dietary soy sphingolipids suppress tumorigenesis and gene expression in 1,2-dimethylhydrazine-treated CF1 mice and ApcMin/+ mice. J Nutr 134(5):1157–1161
148. Symolon H, Schmelz EM, Dillehay DL, Merrill AH Jr (2004) Dietary soy sphingolipids suppress tumorigenesis and gene expression in 1,2-dimethylhydrazine-treated CF1 mice and ApcMin/+ mice. J Nutr 134(5):1157–1161
149. Abeytunga DT, Glick JJ, Gibson NJ, Oland LA, Somogyi A, Wysocki VH, Polt R (2004) Presence of unsaturated sphingomyelins and changes in their composition during the life cycle of the moth Manduca sexta. J Lipid Res 45(7):1221–1231. doi:10.1194/jlr.M300392-JLR200
150. Kwon HC, Lee KC, Cho OR, Jung IY, Cho SY, Kim SY, Lee KR (2003) Sphingolipids from Bombycis Corpus 101A and their neurotrophic effects. J Nat Prod 66(4):466–469. doi:10.1021/np0204491
151. Fyrst H, Oskouian B, Bandhuvula P, Gong Y, Byun HS, Bittman R, Lee AR, Saba JD (2009) Natural sphingadienes inhibit Akt-dependent signaling and prevent intestinal tumorigenesis. Cancer Res 69(24):9457–9464. doi:10.1158/0008-5472.CAN-09-2341
152. Kumar A, Pandurangan AK, Lu F, Fyrst H, Zhang M, Byun HS, Bittman R, Saba JD (2012) Chemopreventive sphingadienes downregulate Wnt signaling via a PP2A/Akt/GSK3beta pathway in colon cancer. Carcinogenesis 33(9):1726–1735. doi:10.1093/carcin/bgs174
153. Lemonnier LA, Dillehay DL, Vespremi MJ, Abrams J, Brody E, Schmelz EM (2003) Sphingomyelin in the suppression of colon tumors: prevention versus intervention. Arch Biochem Biophys 419(2):129–138
154. Schmelz EM, Bushnev AS, Dillehay DL, Sullards MC, Liotta DC, Merrill AH Jr (1999) Ceramide-beta-D-glucuronide: synthesis, digestion, and suppression of early markers of colon carcinogenesis. Cancer Res 59(22):5768–5772
155. Symolon H, Bushnev A, Peng Q, Ramaraju H, Mays SG, Allegood JC, Pruett ST, Sullards MC, Dillehay DL, Liotta DC, Merrill AH Jr (2011) Enigmol: a novel sphingolipid analogue with anticancer activity against cancer cell lines and in vivo models for intestinal and prostate cancer. Mol Cancer Ther 10(4):648–657. doi:10.1158/1535-7163.MCT-10-0754
156. Dahm F, Bielawska A, Nocito A, Georgiev P, Szulc ZM, Bielawski J, Jochum W, Dindo D, Hannun YA, Clavien PA (2008) Mitochondrially targeted ceramide LCL-30 inhibits colorectal cancer in mice. Br J Cancer 98(1):98–105. doi:10.1038/sj.bjc.6604099

Aging: An Etiological Factor in The Development of Intestinal Tumorigenesis

Jyoti Nautiyal, Krystyn Purvis and Adhip P. N. Majumdar

Abstract Sporadic colorectal cancer (CRC) occurs in people who have essentially no family history of the disease. Many probable reasons have been suggested for the age-related rise in sporadic CRC, including altered carcinogen metabolism and the cumulative effects of long-term exposure to cancer-causing agents. In this review article we have discussed the current evidence that implicates aging as an etiological factor in the initiation and progression of CRCs, with particular reference to the involvement of colon cancer stem/stem-like cells ***(CSCs/CSLCs)***, a small sub-population of self-renewing tumor cells, in these processes. In addition, the role of EGFR and as well as miRNAs, specifically microRNA 21 in regulating CSCs/CSLCs in the colon during aging is discussed.

Keywords Colorectal cancer · Aging · Mutations · Epidermal growth factor receptor(s) EGFRs · MicroRNA · Cancer stem cells(CSCs)

1 Epidemiology of CRC (Global and USA)

Cancer is a global concern, being the leading cause of death in developed countries and second leading cause of deaths in developing countries. According to GLOBOCAN 2008, about 12.7 million cancer incidences and 7.6 million cancer-related deaths are estimated to have occurred globally in 2008 [1]. While prostate, colorectal, female breast and lung cancers top the list in developed countries, cancers related to infections that occur in the stomach, liver and cervical were among leading cancers in developing countries. There are evident disparities between incidence

A. P. N. Majumdar (✉)
Department of Internal Medicine, Veterans Affairs Medical Center, Research Service, Karmanos Cancer Institute, Wayne State University, Room-B4238, 4646 John R, Detroit, MI 48201, USA
e-mail: a.majumdar@wayne.edu

J. Nautiyal · K. Purvis
Department of Biological Sciences, College of Liberal Arts and Science, Wayne State University, Detroit, MI 48201, USA

V. W. Yang, A. B. Bialkowska (eds.), *Intestinal Tumorigenesis*,
DOI 10.1007/978-3-319-19986-3_10

and mortality rates between developed and developing economies. About 56 % of the reported incidences and 64 % of cancer related deaths are reported to be in developing economies. In the Unites States of America, cancer is the cause of 1 in 4 deaths. Based on data from several cancer registries, the projections for year 2012 are approximately 1.6 million new cancer cases and 0.5 million related deaths in the US. Among these, cancers of the digestive tract collectively account for 17.4 % of all cancers. Digestive cancers include cancers of esophagus, stomach, small intestine, colon, rectum, anus, liver, gall bladder and pancreas. Colorectal cancer (CRC) that comprises 50 % of estimated GI cancers remains a high mortality malignancy [2]. It is estimated that CRC will result in 9 % of all cancer related deaths. CRC is the third leading malignancy affecting both males and females equally; comprising 9 % of estimated new cancer cases and 9 % cancer related deaths [2].

2 Biology of the GI Tract

The detail of intestinal histology has been provided elsewhere in the textbook. We will briefly cover the dynamics of GI tract in this chapter. The intestinal epithelium is one of the most fascinating organ systems in terms of maintenance of its architecture and functional and structural integrity. Different lineages of cells in the small intestine are arranged in proliferative crypts and differentiated villi (Fig. 1). The actively dividing cells at the base of crypt migrate upwards within the crypts and villi and are finally shed into the lumen. As these cells migrate upward, they undergo differentiation to give rise to different types of lineages. Interestingly, villi and Paneth cells are not found in colorectal epithelium [3, 4]. Through the fine tuning of various pathways exercising their morphogenic and mitogenic effects, the structure and function of the GI tract is maintained during the lifetime. The GI tract has one of the highest turn-over rates. The self-renewal of the epithelial cells of the GI tract takes 2–7 days under normal conditions. Proliferative cells are known to occupy the bottom two-thirds of the intestinal crypts, while differentiated cells constitute the upper third and the surface epithelium. The proliferative zones of the intestine are sustained by repeated divisions of precursor stem cells at the base of crypts [5, 6]. In the small intestine, base of the crypt, just superior to the paneth cells, is known to contain these stem cells. However in the large intestine, stem cells are known to be located in the mid-crypt of ascending colon and crypt base of the descending colon [7].

Due to salient features of self-renewal and mitosis, GI tract epithelium is quite prone to carcinogenesis. The short lifespan of gastrointestinal epithelial cells does not allow accumulation of enough mutations to progress into cancer. However, intestinal stem cells are considered to accumulate multiple mutations throughout their lifetime and thus are more susceptible to neoplastic events. CRC being one of the most extensively studied multi-step carcinogenesis provides valuable insight into different signaling events involved in normal function as well as carcinogenesis [8–10]. Dysregulation of one or more of these pathways can lead to neoplastic

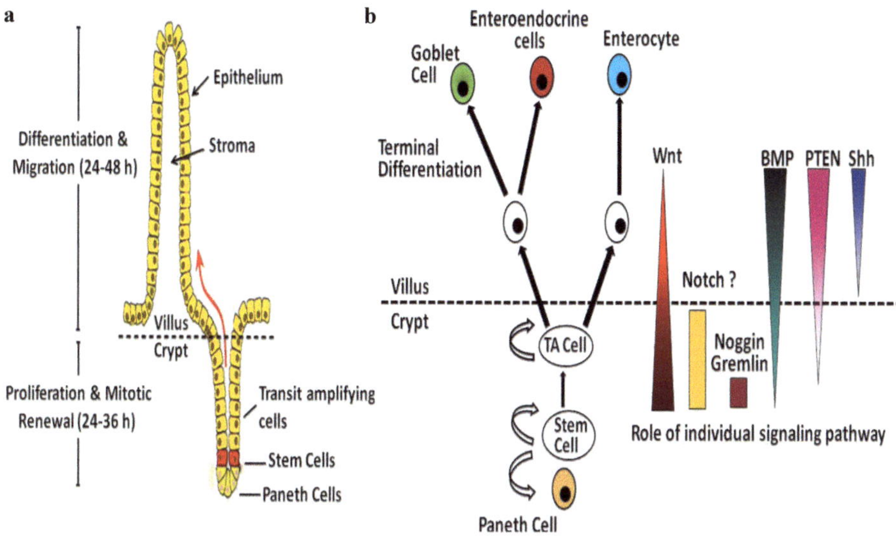

Fig. 1 Schematic representation of architecture and signaling in the small intestinal mucosa. **a** In the epithelial lining of the normal intestinal mucosa, stem cells are located at the bottom of crypts and divide asymmetrically to give rise to different lineage of cells. The daughter cells undergoing differentiation migrate upwards to give rise to transit amplifying and terminally differentiated cells. The latter are programmed for cell death and are shed into the lumen. **b** Intestinal stem cells give rise to a lineage of differentiated epithelial cells. The proliferation and differentiation of stem cells is tightly regulated by distinct signaling pathways. (**a** and **b** with permission from [18])

transformation of GI epithelium. Stem cells renewal and their differentiation state is maintained by various pathways, deregulation of which can lead to neoplasia. Some of these pathways are AKT/PKB, Wnt, Bone Morphogenic protein (BMP), sonic HedgeHog (HH) and Notch [11, 12] (Fig. 1). Few of the prominent pathways involved in sequential initiation and progression of CRC are Wnt/β-catenin pathway, EGFR pathway, TGF-β response inactivation, loss of p53 function. These pathways are known to regulate various processes of carcinogenesis like proliferation, survival and favor metastatic spread to secondary organs [8].

3 Aging and CRC

3.1 Epidemiology

This century has witnessed global phenomenon of accelerated growth of aging population (persons 60 years old and over). Population aging is a phenomenon associated with shift toward an increased proportion of older persons in the population. According to Madrid International Plan of Action on Ageing (MIPAA) in 2002 there were 632 million people aged 60 years or over in the world. This elderly population rose by 28 % and reached 810 million by year 2012 [13]. While this trend is observed in both developed and developing countries, growth of aging population is far more rapid in developing economies due to decreased rates of fertility coupled with reductions in mortality and increased longevity. In developed countries, aging population is estimated to grow by 1.9 % annually and rise from 264 million in 2009–416 million in 2050 [14]. While in developing countries this growth is expected to surpass 3 % annually and population age 60 and over projected to rise from 475 million in 2009 to 1.6 billion in 2050 [15]. Data on the sex and age composition of the United States derived from the 2010 Census provides similar statistics for aging population. While overall population showed a growth of 9.7 % between 2000 and 2010, the elderly population (62 years and above) demonstrated a growth of 21 %. The growth rates have been almost similar for elderly men and women (9.9 and 9.5 %) [16]. Further examination of Census data shows that the oldest old (85 years and over) had the largest growth increase. The percentage point increase for the oldest-old population over the previous two decades was concentrated in the 90 to 94 year old age group, (increased from 25.0 % in 1990 to 26.2 % in 2000, and 26.4 % in 2010) [17].

This increasing elderly population provides opportunity to investigate and gain a better understanding of various mechanisms of aging and their association with age-associated diseases. Aging of the body goes handin hand with structural and functional changes in various organ- systems and is thought to be a causative factor for various diseases in the elderly population. Cancer is a disease of aging where various mutations are accumulated in cells resulting in neoplasia and its spread to other organs. Among the top cancers affected by aging are prostate, lung & bronchus, and colorectal cancer (CRC) in males and in females, cancers of the breast, lung & bronchus and colorectum are most predominant. Each of these cancers shows progressive increase in susceptibility to developing invasive cancer with increasing age. For CRC in particular, a male in age group 70 years and above has 54 times higher risk of getting cancer compared to younger male (birth to 39 years). For females this risk increases 49 times for age 70 years and over. By the time an individual is 70 years or older, every 1 in 23 male and 1 in 25 female will be affected by invasive CRC [2].

3.2 *Mechanisms of Age Related Carcinogenesis*

3.2.1 Increased Susceptibility to Carcinogens and Mutational Events

Accumulation of mutations during lifetime is one of the major causative factors for initiation of cancer. Mutagenic chemicals are not restricted to environmental origin. Several mutagens have been found naturally in foods and are also generated during cooking. Sometimes, the compound itself may not be mutagenic but may be metabolized by liver enzymes and/or colon resident bacteria into active mutagens and/or carcinogens. Therefore, gastrointestinal tract is often exposed to plethora of such food ingredients which are active mutagens themselves or will give rise to some. Few such examples include piperine in black pepper, agaritine in common mushroom, theobromine in cocoa powder, caffeic acid in coffee, allyl isothiocyanate in mustard, horseradish, food preservatives such as nitrates and burnt materials like bread crust, burnt toast and burnt surfaces of meat cooked at high temperatures [18, 19].

Aging is associated with declined cellular functions and altered metabolism leading to enhanced susceptibility of the GI tract to various carcinogens and predisposition to carcinogenesis. Aberrant crypt foci (ACF) are microscopic lesions, considered pre-malignant lesions and are found to be more frequent with increasing age [20, 21]. ACF are shown to demonstrate similar spectrum of mutations as those observed in adenomatous polyps, which are precursors of adenomas and carcinomas [18, 22–34]). Sprague Dawley [35] and Fisher-344 rats [36] showed increase in formation of ACF in response to colonic carcinogen azoxymethane. Our more recent data confirms to this contention in regards to cancer precursor cells that the aged rats showed a greater susceptibility to the carcinogen dimethylhydrazine (DMH), compared to their younger counterparts [37]. Intestinal mucosa of aged rats demonstrated higher population of cancer stem cells which are hypothesized to be involved in the development and progression of intestinal malignancies [38]. These precursor cells are discussed in more details in later sections. We observed that intestinal mucosa of older rats had higher proportion of cancer stem cells to begin with, which increased further in response to DMH exposure. This DMH-induced increase in cancer stem cell population was higher in older rats than their younger counterparts [37, 39].

3.2.2 Accumulation of Mutations During Lifetime

Accumulation of a constellation of mutations during lifetime is considered a major contributing factor to various neoplasia (reviewed in [8]). Carcinogenesis is a multistep process, resulting from accumulation of mutations during progression from normal epithelium to carcinoma [9]. Colon cancer is one of the extensively studied malignancies that shows the loss or inactivation of the tumor suppressor gene adenomatous polyposis coli (*APC*) as precursor of genomic instability. According to genetic model of colorectal tumorigenesis, neoplasia develops as a result of accumulation of mutations involving loss or inactivation of tumor suppressor genes and aberrant activation of oncogenes over the period of 20–40 years [9, 40]. According to Vogelstein model and its recent modified version by Fearon et al. (2011), commonly affected gatekeeper genes are *RAS*, *EGFR*, *PIK3CA*, *BRAF*, *cMYC*, *p53*, *APC*, *PTEN*, *SMADS* (*2,3,4*), *TGFβ-2* etc. [8, 10]. We have examined age-associated changes in few of these critical genes in the gastric mucosa of healthy subjects and found the tumor-suppressor genes *APC*, *p53*, and *DCC* (deleted in colorectal cancer) to be inactivated in the gastric mucosa in older subjects [41]. We reported that the loss of heterozygosity (LOH) is one of the mechanisms for this loss-of function of genes [41].

3.2.3 Failure of Immune Response

The process of aging is associated with progressive deterioration of multiple cellular and molecular functions that eventually translate to various ailments and malignancies. Progression of age has profound impact on the immune system, leaving the elderly subject highly susceptible to various diseases. Immune system is severely compromised in both aging human population and experimental animals. This is associated with decrease in innate immunity (immunosurveillance) as well as the adaptive immunity (type and amount of T- and B-lymphocyte produced including the cytokines produced by them) [42–44]. Several studies show that the compromised immune system with several other cellular processes are strongly associated with development and progression of cancers [45]. Immune system prevents tumorigensis as well as spread of cancer by recognizing and targeting cancer cells. Cancer immunosurveillance is known to decline with progressing age leading to immunosenescence [46, 47]. Another model associates the failure of immune system to effectively regulate various cellular mechanisms and alleviate inflammation and increased incidences of cancer. Of all human cancers in adults, around 20 % are associated with inflammation [48–50]. There is abundance of epidemiologic data showing higher rates of colorectal cancer among individuals harboring *Helicobacter pylori* infections and among people diagnosed with Crohn's disease or ulcerative colitis [51–53]. It is proposed that within inflammatory milieu cells secrete various growth factors that provide cancer initiated cells survival and growth advantage [54].

3.2.4 Dysregulation of Growth as an Early Event in Initiation of Colorectal Carcinogenesis

The tissue architecture and functional integrity of intestinal mucosa is maintained by a fine balance between proliferation of precursor cells at the bottom and apoptosis of cells at the luminal surface [6, 55]. Therefore, any deregulation of processes of proliferation and apoptosis leads to hyperplasia or atrophy of the GI tract. Better understanding of these processes during aging would help us to understand the age-associated etiology of CRC.

There is plethora of evidence that shows the loss of balance in replacing the apoptotic cells in the aging GI tract. The mucosa of aging barrier-reared Fischer-344 rats demonstrated increased proliferative activity in the stomach and the small and large intestines [56–65]. The increased susceptibility of the aging intestinal mucosa to azoxymethane was also associated with increased ACF and proliferative activity [36]. Additionally, carcinogen exposure in aging rats leads to a change in distribution of proliferating cells in the colonic mucosa. While most of the proliferating cells were located at the bottom half of the crypt in young rats, exposure of aging rats to aozxymetahne, caused proliferative cells to be distributed throughout the crypt with the exception of 10–20 % of the upper region. These findings have been paralleled in the colon, in patients with precancerous lesions of the colon and/or rectum [64]. Similarly, the colonic mucosa of human volunteers revealed that actively dividing cells are restricted to lower two-thirds of the crypt in young subjects, whereas aging is associated with spread of these cells base to the middle and upper-third of the gland [66, 67]. Age-related increase in proliferative activity as well as expansion of proliferative region signifies deregulation of major pathways leading to increased incidence of CRC [57, 60, 63, 68, 69].

Conversely, decreased rates of apoptosis with aging can also lead to deregulation of GI tissue dynamics pushing the hyperplastic tissue towards neoplasia [70, 71]. Similar to observations in a variety of cancers, decreased apoptosis has been implicated in processes of initiation and progression of colorectal neoplasia [70–73]. We have investigated age-related changes in apoptotic events in the gastric and colonic mucosa of Fischer-344 rats. Number of apoptotic cells, as well as various biochemical events associated with apoptosis were found to be reduced in aging rats when compared to their younger counterparts [65]. Although there is not a lot of evidence implicating reduced apoptosis to colorectal neoplasia, our observation of the age-related increase in proliferation and decrease in apoptosis suggests that these events play critical role in the development and progression of CRC during the later part of life.

3.2.5 EGFRs in Gastrointestinal Carcinogenesis

Vogelstein's model, with some improvisations, has been the widely accepted for understanding the multi-step process of colorectal carcinogenesis [8–10, 74–79]. Though there is ample evidence to associate various oncogenes and tumor suppressors to colorectal neoplasia, tyrosine kinases form a significant minor class of contributing factors. Among this class of oncogenes, prominent genes to be associated with GI malignancies are pp60 c-Src, a non-receptor protein, EGFR(s), hepatocyte growth factor receptor (c-MET), vascular endothelial growth factor receptor (VEGFR) as well as focal adhesion kinases (FAK) (reviewed in [18]). These are pleiotropic genes that play an indispensable role in regulating cell survival and growth (Fig. 2). Deregulation of these signaling pathways have been associated with multiple malignancies, including cancers of digestive system [10, 80–83, 84–89]. Since EGFR(s) are a major family of receptor tyrosine kinases, in this chapter we will lay special emphasis on EGFR signaling pathways and various ways to target them to successfully treat GI neoplasia.

EGFR Family Members and Gastrointestinal Neoplasia

Normal cellular homeostasis is maintained by growth factor receptors by regulating multiple processes that include growth, proliferation, differentiation, migration, angiogenesis, and cell death. Deregulation of tyrosine kinase signaling pathways, therefore, has serious consequences for the organ and the organism. EGFR(s) drive multiple processes of carcinogenesis which among others include cellular growth, migration, differentiation, invasion, angiogenesis and metastasis (Fig. 2) [90–93]. EGFR family consists of four members: EGFR/ErbB1/HER1, ErbB2/HER2,

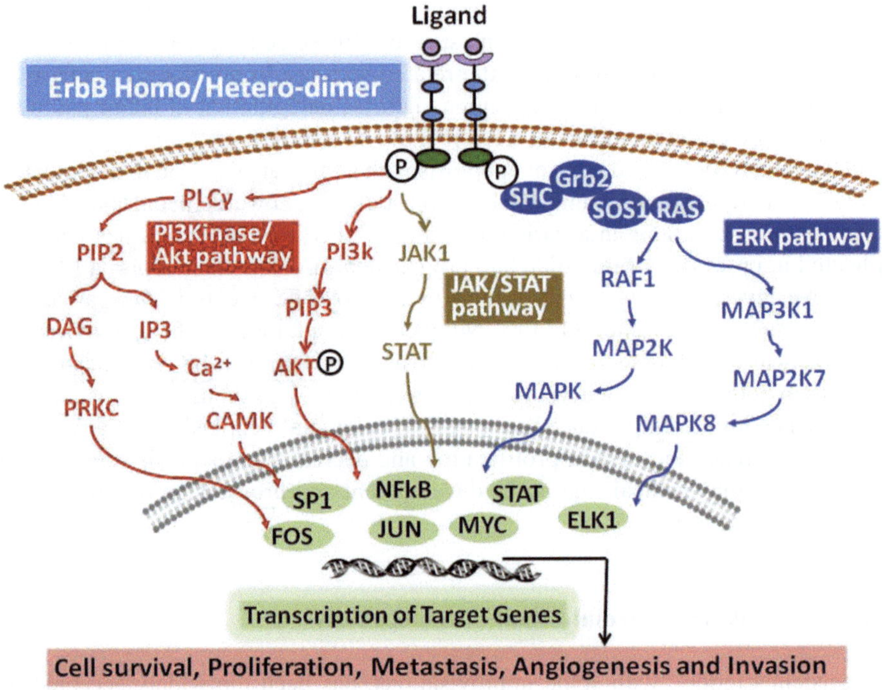

Fig. 2 Schematic representation of ErbBs/EGFRs signaling pathways. Ligand binding to ectodomain of EGFR(s) leads to dimerization of the receptor and subsequent activation of tyrosine kinase domain. Complexity of ErbB signaling is regulated by ligand specific binding, leading to different combinations of homo/-hetero dimerization of the receptors. Depending on the type of homo/-hetero dimer formed, a distinct intracellular pathway is activated. These pathways are involved in several processes of cellular homeostasis. *PI3K* Phosphoinositide-3-kinase, *Akt* phospho kinase B (PKB), *JAK* Janus kinase, *STAT* signal transducer and activator of transcription, *ERK* Extracellular signal regulated protein kinase, *MAPK* Mitogen activated protein kinase (with permission from [18])

ErbB3/HER3, and ErbB4/HER4. With the exception of HER2, each receptor has its own specific ligands which results in a highly regulated sequence of intracellular signal transduction. Upon binding of a specific ligand, the monomer receptor forms homo-/hetero- dimers with other members of the family. The dimer is enzymatically activated by auto-/trans-phosphorylation of the intracellular tyrosine kinase domain. HER3 containing a kinase-dead domain, must form hetero-dimers to relay intracellular signals (reviewed in [94]). The EGFR signaling is complex and involves crosstalk with other signaling pathways. This complexity is a result of set of receptor specific ligands, homo-/hetero dimers and depending on specific tyrosine phosphorylation, activation of unique downstream effectors (reviewed in [94]). EGFRs crosstalk with other pathways, including MAPK signaling, PI3/AKT pathway, Janus kinase/signal transducer and activator of transcription (JAK/STAT) pathways and also modulate the expression of VEGF and interleukin-8 Fig. 2 [95, 96].

Aberrant activation of EGFRs can result from wide range of factors that among others include (a) over expression of either receptor and/ or their ligands (b) mutations leading to constitutive activation (c) inefficient mechanisms of inactivation and (d) heterodimerization leading to transactivation [92, 97, 98]. While there is ample evidence to implicate EGFR, HER2 and HER3 in various processes of carcinogenesis, role of HER4 is not clearly delineated (reviewed in [18]). While some reports suggest tumor–promoting role for HER4 [99, 100] others have proposed its role as a tumor-suppressor [101, 102].

Aging and EGR Signaling

Aberrant activation of EGFR signaling which in turn leads to overstimulation of downstream pathways like Ras/Raf/MAPK kinase and PI3K/Akt signaling provide proliferative and survival signals [98, 103–105]. Age related deregulation of proliferation and apoptosis in the GI mucosa is one of the pre-disposing factors for cancers of the digestive system [56, 60, 61, 69]. Considering the pleiotropic role of EGFRs signaling, one is inclined to postulate a significant role of EGFRs in the age-related increase in processes of carcinogenesis. Our investigations on these lines show increased expression and activation of EGFR and its family members ErbB2/ErbB3 the GI mucosa during aging [36, 39, 106–108]. Additionally, one of the most common ligands of EGFR, TGF-α, is also implicated in the development of colorectal neoplasia through autocrine/paracrine signaling of cancer cells. Several cancers, including those of colon exhibit increased expression of not only EGFR but also its ligand TGF-α in the GI mucosa of CRC [109–113].

Though the precise mechanism of the age-related deregulation of proliferation and apoptosis is not known, we postulate that EGFR signaling plays a significant role in regulating these processes. This contention is supported by our findings that aging is associated with increased activation of EGFR(s) and its signaling events [106, 107], which could partly be attributed to increased amount of membrane associated precursor forms of TGF-α, one of the primary ligand of EGFR [107]. Others have also reported similar findings in pre-malignant and neolpastic lesions [64, 112, 114, 115]. We thus postulated that the age-related increase in susceptibility of the GI mucosa to carcinogens would lead to enhanced EGFR signaling. Support for this postulation comes from our observation that the DMH-induced neoplasia in the colon of rats is associated with a concomitant increase in the expression of EGFR [39]. In response to DMH, ACF formation in rats was increased 2-folds, with a 12-fold increase in expression of EGFR compared to controls [39]. Additionally, in humans, aging was associated with increased expression of EGFR in macroscopically normal mucosa in patients with adenomatous polyps [108]. Recently, we have reported that that the age-related rise in EGFR signaling could be further increased by DMH [37]. Taken together, the results a potential regulatory role of EGFRs in the development and progression of GI malignancies, specifically CRC.

4 Aging and Recurrent Colorectal Cancer

CRC remains a huge health concern worldwide and in the US. Although with recent advances in medicine the 5-year survival of patients with CRC has gone up to 90 %, nearly half of them will show recurrence of the disease [116, 117]. In the final stage of primary CRC, cancerous cells migrate to blood/lymph initiating secondary tumor in another organ. This process defined as metastasis, is considered a poor prognostic indicator as most of CRC associated deaths are due to recurrence of the disease [118, 119].

According to elderly population projections cancer incidence for people over 65 years of age is expected to double over the next decades [120]. Colorectal cancer has median age at diagnosis as 71 years and that of death associated with metastatic cancer as 75 years [121]. Considering the fact that CRC is a malignancy of older population and recurrent cancer is the major cause for deaths underscores the need to gain a better understanding of the mechanisms of aging and cancer recurrence.

4.1 Recurrent Colorectal Cancer and Chemoresistance

Cancer cells modulate various signaling pathways to evade killing by anti-tumor therapies or develop de novo chemoresistance. One such prominent pathway exploited by cancer cells is EGFR signaling. Overexpression of EGFR signaling is often associated with development of resistance to many known therapies including chemotherapy, hormone therapy, radiotherapy in a variety of cancers, including breast, ovarian, colo-rectal, head and neck and non-small cell lung cancers (NSCLC) and cancer recurrence (reviewed in [18]). Cancer cells can exploit various mechanisms to evade killing by therapies and relapse with more aggressive and invasive phenotype. These among others include: (1) overexpression of growth factor receptors or their ligands and (2) overexpression of other members of the growth factor receptor family. In our studies utilizing chemo-surviving and chemo-resistant human colon cancer cells we observed 5-fluorouracil (5-FU) or 5-FU plus oxaliplatin (FOLFOX), the mainstay of colorectal cancer chemotherapeutics, to produce only a partial response. Chemo-surviving cells were generated by growing HCT-116 and HT-29 human colon cancer cells in monolayer and incubating with 50 μM 5-FU and 1.25 μM oxaliplatin [123]. The cells surviving at the end of treatment period (48 h FOLFOX exposure) were collected and used for various analyses. Chemoresistant cells on the other hand were generated by exposing colon cancer cells to increasing concentrations of 5-FU and oxaliplatin [122]. Once these cells could survive at the highest dose of FOLFOX, they were maintained in normal cell culture medium containing FOLFOX [122]. We found EGFR expression to increase considerably in cells that survived exposure to FOLFOX [122, 123]. The increased expression of EGFR could partly be attributed to hypomethylation of the *EGFR* promoter [122]. Additional evidence suggests that development of enhanced drug resistance is often associated with expression of more than one member of the EGFR family [124, 125]. We observed a similar phenomenon in our FOLFOX surviving as well as

FOLFOX resistant colon cancer cells. These cells not only showed higher expression but also activation (as gaused by tyrosine phosphorylation) of EGFR and its family members [122, 123]. Furthermore, their downstream effectors were also found to be aberrantly activated in these chemo-resistant cells [122, 123]. The fact that EGFR plays a role in development of resistance to anti-cancer therapies and that EGFR expression in colonic mucosa increases with advancing age suggests that this growth factor receptor exerts a critical role in the age-related rise CRC.

4.2 *miRNAs, Colorectal Carcinogenesis and Aging*

More recently microRNAs (miRNA/miR) have been studied extensively and accumulating data associate miRNAs to various processes like proliferation, apoptosis, differentiation, metabolism, and invasion [126]. miRs are short RNAs 18–25 nucleotides in length that regulate functions of various genes by targeting their translation [127]. These endogenous small noncoding RNAs bind to sequence at the 3′ untranslated region (UTR) of target mRNAs, resulting in either translational repression or degradation [131]. Nearly 30 % of human protein coding genes are known to be regulated by miRNAs [126, 128]. Due to their pleiotropic effects in cells, miRNAs are also implicated in processes of cancer initiation and progression [126]. Specific to colorectal cancer two classes of miRNA have been identified. One class is "oncomiR" (a miRNA with oncogenic properties) that are known to be upregulated with progression of cancer. Examples of this class are let-7b, miR-9, miR-18, miR-21, miR-200 (reviewed in [129, 130]). Among miRNA associated with tumor suppressor function are let-7a, miR-143, miR-145, which are downregulated during carcinogenesis (reviewed in [129, 130]). Among proposed targets of these miRNAs are APC, PI3K, PTEN, KRAS, p53 and SMADs [130, 133].

miR-21 is one of the earliest microRNAs to be discovered and has been investigated in great details. miR-21, acting as an "Oncomir", is found to be up regulated in all cancers including colorectal neoplasia and plays a key role in several processes of tumor promotion, invasion, and metastasis [131–133]. Due to its pleitotropic regulatory role, overexpression of miR-21 has also been linked to poor prognosis of colorectal cancer and also shows a strong association with poor survival [134–135]. miR-21 has been associated with anti-cancer chemoresistance in pancreatic, prostate, breast and glioblastoma cancers (reviewed in [132]). Targets of miR-21 have been found to be a critical regulator of tumors suppressor genes such as programmed cell death gene-4 (*PDCD4*), phosphates and tensin homolog (*PTEN*) and transforming growth factor βR-2 (*TGFR-β2*) [136–137]. Recent findings from our laboratory also support this contention. Xenografts of miR-21 overexpressing human colon cancer cells in SCID mice showed down regulation of PTEN and up regulation of Akt activity [138]. We further demonstrated that downregulation of miR-21 leads to differentiation of chemoresistant human colon cancer cells leaving them susceptible to anti-cancer therapies [139]. This is in agreement with other evidence showing that downregulation or inhibition of miR-21 leads to inhibition of tumor growth, induce apoptosis and inhibit processes of invasion and metastasis [132, 137]. These observations suggest miR-21 to be a promising target for anti-tumor therapies.

While the mechanistic details of miRNA induced carcinogenesis are still being unraveled, there are data suggesting a role for miRNAs in aging and suggest that they are critically involved in regulating the age-related progression of cancer [37, 130, 140]. Noren Hooten et al. have demonstrated an age-related change in miRNA expression in human subjects [141]. Out of 800 miRNA investigated, 9 miRNAs showed drastic reduction with aging; they are miR-24, miR-103, miR-107, miR-128, miR-130a, miR-155, miR-221, miR-496, and miR-1538. Among them, five miRNAs: miR-24, miR-128, miR-130a, miR-155, and miR-221 have been implicated in carcinogenesis with predicted targets such as PI3K, c-kit, H2AX showing elevated expression with advancing age [141]. Our recent findings also support the contention that miRNAs play a pivotal role in aging and age-related increase in colon carcinogenesis [37]. We have conducted a similar miRNA expression profile and observed differential expression of miRNAs in aging rats. Among the up regulated miRNAs, miR-21 invoked further interest. In Fisher-344 rats, the expression of miR-21 in colonic mucosa was found to be considerably higher than in their younger counterparts [37]. The expression of miR-21 increased in both young and aging rats in response to colonic carcinogen DMH. However the increase was further augmented in aging rats in response to DMH [37].

There is also evidence that shows regulation of miR-21 by EGFR [142], which in turn is up regulated during aging [36, 105, 106]. Our observation that in aging rats, administration of cetuximab, monoclonal antibodies to EGFR, not only inhibits EGFR but also miR-21 suggests that EGFR regulates miR-21 in the aged [37]. Interestingly, this regulation of miR-21 by EGFR was evident only in aging rats [37]. These observations further suggest that miR-21 in association with EGFR signaling pathways play a key role in the age-related rise in colorectal neoplasia. Furthermore, miRNAs are known to regulate stemness or differentiation of cancer stem cells [37, 139]. These cancer stem cells represent specialized populations of cells which have been recently implicated in driving processes of carcinogenesis and chemo-resistance in variety of cancers. These cells and their regulatory function in initiation and progression in cancer are discussed in next section.

5 Aging and Colo-rectal CSCs

5.1 Cancer Stem Cells in Colorectal Cancers

The cancer stem-cell concept was first proposed by John Dick group in late 1990s in hematological malignancies [143]. This concept explained the presence of phenotypically heterogenous cells in tumors and a relatively smaller fraction of cells being tumorigenic [144]. This concept of cancer stem cells or cancer stem-like cells CSCs/CSLCs in the initiation, progression and recurrence of hematological malignancies as well as solid cancers is now well accepted [18, 145]. The CSC/CSLC model proposes that a small proportion of tumor cells defined as CSCs/CSLCs exists within the tumor which is able to self-renew and differentiate into heterogenous lineage of cells that form the bulk of tumor [146]. There is abundant evidence for the presence of CSCs in GI cancers [147–149]. Currently, CSCs/CSLCs are identified by expression of specific cell-surface epitopes and have the ability to form tumors at extremely diluted concentration in immunodeficient mice [148]. These cells are also associated with resistance to number of conventional chemotherapies which target bulk of non-CSC cells but fail to eliminate CSCs [38, 150]. Currently, in CRC, CSCs are recognized by expression of cell-surface epitopes CD133, CD44, CD166 and ALDH [147]. Our recent data show that FOLFOX resistant human colon cancer cells HCT-116 and HT-29, contain higher proportion of CSCs as identified by expression of these cell-surface markers [151]. Additionally, these chemoresistant cells contained higher proportion of CD44 and CD166 co-expressing cells [151]. CSCs with dual expression of CD44 and CD166 are known to be more tumorigenic [147].

5.2 CSCs During Aging and Susceptibility to Carcinogens

Our recent findings demonstrate that the macroscopically normal mucosa of patients with adenomatous polyps shows an age-related increase in CSCs/CSLCs (Fig. 3) [107]. The number of polyps also showed a positive correlation with advancing age ($r^2=0.92, p<0.02$) [107]. CSCs/CSLCs were identified by the expression of surface markers CD166, CD44, Ep-CAM. The cells expressing CSC markers also showed a marked increased in EGFR expression. However, the age-related increase in EGFR and possible its role in CRC was discussed in Sect. 3.2.5. We further investigated role of CSCs in rendering aging colon more susceptible to carcinogens. We observed that DMH exposure lead to increase in CSC population in the rat colonic crypts [39]. Cells within normal colonic crypt epithelium showed higher expression of CD166, CD44, Ep-CAM [39]. This with a concomitant increase in EGFR expression suggests that effects of carcinogens are more pronounced in normal epithelium of colonic crypts of aging rats [39]. Similar finding were confirmed by our recent studies showing increased expression of CD44, CD166 and ALDH in colonic mucosa of aging rats [37]. DMH induced the expression of these markers in both colonic mucosa as well as single cells isolated and cultures in stem-cell media. This increase was observed in both young and old rats, though the effect was more pronounced in aging rats [37]. These findings along with increased expression and activation of EGFR in the colonic mucosa supports the contention that aging leads to increased susceptibility to tumorigenesis, which in part is associated with increase CSC population.

5.3 Mechanisms of CSCs Regulation in Age Associated CRC

To examine mechanisms of regulation of CSCs with regards to growth factor receptors, specifically EGFR and its family members, we investigated the levels of EGFR in normal tissue as well as adenomatous polyps from patients and carcinogen induced rat colonic crypts. Expression of EGFR increases with advancing age in macroscopically normal mucosa of patients with adenomatous polyps [107] as well in carcinogen-induced rat colonic crypts [39]. This carcinogen induced increase in CSC population is significantly higher in aging rat colonic mucosa than their younger counterparts [39]. We have also reported that the FOLFOX resistant colon cancer cells that show increased expression of several CSC markers, including CD133, CD44, CD166 and ALDH-1 also show a concomitant rise in EGFR [122]. Additionally we reported that aging is associated with concomitant increase in miR-21, which plays a significant role in colon carcinogenesis [37]. In order to understand the mechanisms of CSC regulation in aging colon, we employed targeted therapy, cetuximab to inhibit EGFR activation and its downstream signaling. Targeted inhibition of EGFR was associated with significant reduction in CSCS surface markers CD166 and ALDH-1 as well as miR-21 [37]. Taken together, our results support the hypothesis that EGFR plays a significant role in age-related rise in CRC, which in part is attributed to maintenance of CSCs and miR regulation.

The exact mechanisms of EGFR regulation of CSC in during aging and tumorigenesis remain to be determined. A better understanding of the role of different signaling pathways during aging will enable us to develop novel therapeutics that target CSC and eliminate age-associated primary and secondary colorectal tumors (Fig. 3).

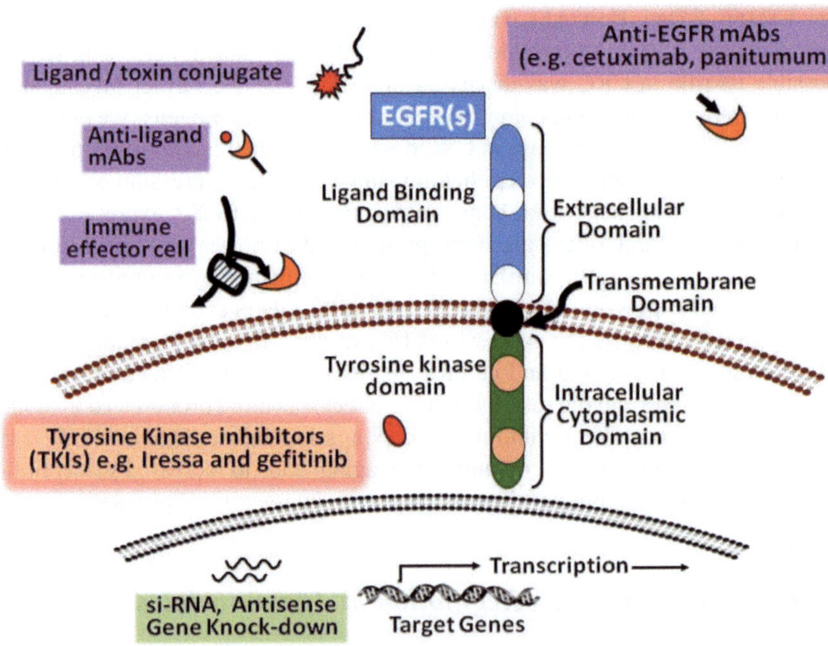

Fig. 3 Schematic representation of different strategies to inhibit EGFR(s). Inhibition of EGFRs can be achieved by targeting extracellular ligand binding domain and/or intracellular tyrosine kinase domain of the receptor. Among the available strategies, mAbs that target the ligand binding domain and TKIs that target intracellular tyrosine kinase activity have been shown to be more successful. mAbs: monoclonal antibodies; siRNA: small interfering ribonucleic acid (with permission from [18])

Acknowledgements A part of the work presented in this communication has been supported by grants to Dr Majumdar by NIH/NIA (AG 014343) and the Department of Veterans Affairs (VA Merit Review)

References

1. Jemal A, Bray F, Center MM, Ferlay J, Ward E, Forman D (2011) Global cancer statistics. CA Cancer J Clin 61(2):69–90. doi:10.3322/caac.20107
2. Siegel R, Naishadham D, Jemal A (2012) Cancer statistics, 2012. CA Cancer J Clin 62(1):10–29. doi:10.3322/caac.20138
3. Scoville DH, Sato T, He XC, Li L (2008) Current view: intestinal stem cells and signaling. Gastroenterology 134(3):849–864. doi:S0016-5085(08)00185-6 [pii] 10.1053/j.gastro.2008.01.079
4. Brabletz S, Schmalhofer O, Brabletz T (2009) Gastrointestinal stem cells in development and cancer. J Pathol 217(2):307–317. doi:10.1002/path.2475
5. Leblond CP, Messier B (1958) Renewal of chief cells and goblet cells in the small intestine as shown by radioautography after injection of thymidine-H3 into mice. Anat Rec 132(3):247–259
6. Leblond CP, Stevens CE (1948) The constant renewal of the intestinal epithelium in the albino rat. Anat Rec 100(3):357–377
7. Wright NA (2000) Epithelial stem cell repertoire in the gut: clues to the origin of cell lineages, proliferative units and cancer. Int J Exp Pathol 81(2):117–143. doi:iep146 [pii]
8. Fearon ER (2011) Molecular genetics of colorectal cancer. Ann Rev Pathol 6:479–507. doi:10.1146/annurev-pathol-011110-130235
9. Fearon ER, Vogelstein B (1990) A genetic model for colorectal tumorigenesis. Cell 61(5):759–767. doi:http://dx.doi.org/10.1016/0092-8674(90)90186-I
10. Vogelstein B, Fearon ER, Hamilton SR, Kern SE, Preisinger AC, Leppert M, Nakamura Y, White R, Smits AM, Bos JL (1988) Genetic alterations during colorectal-tumor development. N Engl J Med 319(9):525–532. doi:10.1056/NEJM198809013190901
11. Radtke F, Clevers H (2005) Self-renewal and cancer of the gut: two sides of a coin. Science 307(5717):1904–1909. doi:307/5717/1904 [pii]10.1126/science.1104815
12. Roy S, Majumdar AP (2012) Signaling in colon cancer stem cells. J Mol Signal 7(1):11. doi:10.1186/1750-2187-7-11
13. United Nations Department of Economic and Social Affairs, Population Division (2012) Population ageing and development: ten years after Madrid. http://www.un.org/esa/population/publications/popfacts/popfacts_2012–4.pdf
14. United Nations Department of Economic and Social Affairs, Population Division (2009) World population to exceed 9 billion by 2050. http://www.un.org/esa/population/publications/wpp2008/pressrelease.pdf
15. http://www.census.gov/population/www/pop-profile/elderpop.html. Accessed Sept 2009
16. Howden LM, Meyer JA (2011) Age and Sex Composition: 2010. United States Census Bureau
17. Werner CA (2011) The Older Population: 2010. US Census Bureau
18. Nautiyal J, Kanwar SS, Majumdar AP (2010) EGFR(s) in aging and carcinogenesis of the gastrointestinal tract. Curr Protein Pept Sci 11(6):436–450
19. Weinberg RA (2013) The biology of cancer, 2nd edn. Garland Science Taylor and Francis Group, New York
20. Banerjee S, Hussain M, Wang Z, Saliganan A, Che M, Bonfil D, Cher M, Sarkar FH (2007) In vitro and in vivo molecular evidence for better therapeutic efficacy of ABT-627 and taxotere combination in prostate cancer. Cancer Res 67(8):3818–3826. doi:67/8/3818 [pii]10.1158/0008-5472.CAN-06–3879

21. Bird RP (1987) Observation and quantification of aberrant crypts in the murine colon treated with a colon carcinogen: preliminary findings. Cancer Lett 37(2):147–151
22. Pretlow TP, Brasitus TA, Fulton NC, Cheyer C, Kaplan EL (1993) K-ras mutations in putative preneoplastic lesions in human colon. J Natl Cancer Inst 85(24):2004–2007
23. Pretlow TP, O'Riordan MA, Pretlow TG, Stellato TA (1992) Aberrant crypts in human colonic mucosa: putative preneoplastic lesions. J Cell Biochem Suppl 16G:55–62
24. Yamashita N, Minamoto T, Ochiai A, Onda M, Esumi H (1995) Frequent and characteristic K-ras activation in aberrant crypt foci of colon. Is there preference among K-ras mutants for malignant progression? Cancer 75(6 Suppl):1527–1533
25. Yamashita N, Minamoto T, Ochiai A, Onda M, Esumi H (1995) Frequent and characteristic K-ras activation and absence of p53 protein accumulation in aberrant crypt foci of the colon. Gastroenterology 108(2):434–440. doi:S0016508595000503 [pii]
26. Heinen CD, Shivapurkar N, Tang Z, Groden J, Alabaster O (1996) Microsatellite instability in aberrant crypt foci from human colons. Cancer Res 56(23):5339–5341
27. Nucci MR, Robinson CR, Longo P, Campbell P, Hamilton SR (1997) Phenotypic and genotypic characteristics of aberrant crypt foci in human colorectal mucosa. Hum Pathol 28(12):1396–1407
28. Pedroni M, Sala E, Scarselli A, Borghi F, Menigatti M, Benatti P, Percesepe A, Rossi G, Foroni M, Losi L, Di Gregorio C, De Pol A, Nascimbeni R, Di Betta E, Salerni B, de Leon MP, Roncucci L (2001) Microsatellite instability and mismatch-repair protein expression in hereditary and sporadic colorectal carcinogenesis. Cancer Res 61(3):896–899
29. Shivapurkar N, Huang L, Ruggeri B, Swalsky PA, Bakker A, Finkelstein S, Frost A, Silverberg S (1997) K-ras and p53 mutations in aberrant crypt foci and colonic tumors from colon cancer patients. Cancer Lett 115(1):39–46. doi:S0304-3835(97)04709-5 [pii]
30. Shpitz B, Bomstein Y, Shalev M, Liverant S, Kaufman Z, Klein E, Mekori Y, Bernheim J (1999) Oncoprotein coexpression in human aberrant crypt foci and minute polypoid lesions of the large bowel. Anticancer Res 19(4B):3361–3366
31. Smith AJ, Stern HS, Penner M, Hay K, Mitri A, Bapat BV, Gallinger S (1994) Somatic APC and K-ras codon 12 mutations in aberrant crypt foci from human colons. Cancer Res 54(21):5527–5530
32. Suzuki H, Watkins DN, Jair KW, Schuebel KE, Markowitz SD, Chen WD, Pretlow TP, Yang B, Akiyama Y, Van Engeland M, Toyota M, Tokino T, Hinoda Y, Imai K, Herman JG, Baylin SB (2004) Epigenetic inactivation of SFRP genes allows constitutive WNT signaling in colorectal cancer. Nat Genet 36(4):417–422. doi:10.1038/ng1330ng1330 [pii]
33. Takayama T, Katsuki S, Takahashi Y, Ohi M, Nojiri S, Sakamaki S, Kato J, Kogawa K, Miyake H, Niitsu Y (1998) Aberrant crypt foci of the colon as precursors of adenoma and cancer. N Engl J Med 339(18):1277–1284
34. Takayama T, Ohi M, Hayashi T, Miyanishi K, Nobuoka A, Nakajima T, Satoh T, Takimoto R, Kato J, Sakamaki S, Niitsu Y (2001) Analysis of K-ras, APC, and beta-catenin in aberrant crypt foci in sporadic adenoma, cancer, and familial adenomatous polyposis. Gastroenterology 121(3):599–611. doi:S0016508501387322 [pii]
35. Magnuson BA, South EH, Exon JH, Dashwood RH, Xu M, Hendrix K, Hubele S (2000) Increased susceptibility of adult rats to azoxymethane-induced aberrant crypt foci. Cancer Lett 161(2):185–193. doi:S0304383500006091 [pii]
36. Schmelz EM, Levi E, Du J, Xu H, Majumdar AP (2004) Age-related loss of EGF-receptor related protein (ERRP) in the aging colon is a potential risk factor for colon cancer. Mech Ageing Dev 125(12):917–922. doi:S0047-6374(04)00185-X [pii]10.1016/j.mad.2004.03.010
37. Nautiyal J, Du J, Yu Y, Kanwar SS, Levi E, Majumdar APN (2012) EGFR regulation of colon cancer stem-like cells during aging and in response to the colonic carcinogen dimethylhydrazine. Am J Physiol Gastrointestinal Liver Physiol 302(7):G655–G663. doi:10.1152/ajpgi.00323.2011
38. Jordan CT, Guzman ML, Noble M (2006) Cancer stem cells. New Engl J Med 355(12):1253–1261. doi:10.1056/NEJMra061808
39. Levi E, Misra S, Du J, Patel BB, Majumdar APN (2009) Combination of aging and dimethylhydrazine treatment causes an increase in cancer—stem cell population of rat colonic

crypts. Biochem Biophys Res Commun 385(3):430–433. doi:http://dx.doi.org/10.1016/j.bbrc.2009.05.080
40. Yokozaki H, Yasui W, Tahara E (2001) Genetic and epigenetic changes in stomach cancer. Int Rev Cytol 204:49–95
41. Moragoda L, Jaszewski R, Kulkarni P, Majumdar AP (2002) Age-associated loss of heterozygosity of tumor suppressor genes in the gastric mucosa of humans. Am J Physiol Gastrointest Liver Physiol 282(6):G932–G936. doi:10.1152/ajpgi.00312.2001
42. Pawelec G, Derhovanessian E, Larbi A (2010) Immunosenescence and cancer. Crit Rev Oncol/Hematol 75(2):165–172. doi:10.1016/j.critrevonc.2010.06.012
43. Frasca D, Landin AM, Riley RL, Blomberg BB (2008) Mechanisms for decreased function of B cells in aged mice and humans. J Immunol 180(5):2741–2746
44. LeMaoult J, Szabo P, Weksler ME (1997) Effect of age on humoral immunity, selection of the B-cell repertoire and B-cell development. Immunologic Rev 160:115–126
45. Ahmad A, Banerjee S, Wang Z, Kong D, Majumdar AP, Sarkar FH (2009) Aging and inflammation: etiological culprits of cancer. Curr Aging Sci 2(3):174–186
46. Fulop T, Larbi A, Hirokawa K, Mocchegiani E, Lesourds B, Castle S, Wikby A, Franceschi C, Pawelec G (2007) Immunosupportive therapies in aging. Clin Interv Aging 2(1):33–54
47. Larbi A, Fulop T, Pawelec G (2008) Immune receptor signaling, aging and autoimmunity. Adv Exp Med Biol 640:312–324. doi:10.1007/978-0-387-09789-3_21
48. Balkwill F, Coussens LM (2004) Cancer: an inflammatory link. Nature 431(7007):405–406. doi:10.1038/431405a
49. Coussens LM, Werb Z (2002) Inflammation and cancer. Nature 420(6917):860–867. doi:10.1038/nature01322
50. Lu H, Ouyang W, Huang C (2006) Inflammation, a key event in cancer development. Mol Cancer Res: MCR 4(4):221–233. doi:10.1158/1541-7786.MCR-05-0261
51. Peek RM Jr, Crabtree JE (2006) Helicobacter infection and gastric neoplasia. J Pathol 208(2):233–248. doi:10.1002/path.1868
52. Thun MJ, Henley SJ, Gansler T (2004) Inflammation and cancer: an epidemiological perspective. Novartis Found Symp 256:6–21; discussion 22–28, 49–52, 266–269
53. Vagefi PA, Longo WE (2005) Colorectal cancer in patients with inflammatory bowel disease. Clin Colorectal Cancer 4(5):313–319
54. Marx J (2004) Cancer research. Inflammation and cancer: the link grows stronger. Science 306(5698):966–968. doi:10.1126/science.306.5698.966
55. Lipkin M (1987) Proliferation and differentiation of normal and diseased gastrointyestinal cells. Physiology of gastrointestinal tract. Raven Press, Newyork
56. Atillasoy E, Holt PR (1993) Gastrointestinal proliferation and aging. J Gerontol 48(2):B43–B49
57. Holt PR, Yeh KY (1988) Colonic proliferation is increased in senescent rats. Gastroenterology 95(6):1556–1563. doi:S0016508588003853 [pii]
58. Holt PR, Yeh KY (1989) Small intestinal crypt cell proliferation rates are increased in senescent rats. J Gerontol 44(1):B9–B14
59. Holt PR, Yeh KY, Kotler DP (1988) Altered controls of proliferation in proximal small intestine of the senescent rat. Proc Natl Acad Sci U S A 85(8):2771–2775
60. Jaszewski R, Ehrinpreis MN, Majumdar AP (1999) Aging and cancer of the stomach and colon. Front Biosci 4:D322–D328
61. Majumdar AP, Edgerton EA, Arlow FL (1988) Gastric mucosal tyrosine kinase activity during aging and its relationship to cell proliferation in rats. Biochim Biophys Acta 965(2–3):97–105
62. Majumdar AP, Jasti S, Hatfield JS, Tureaud J, Fligiel SE (1990) Morphological and biochemical changes in gastric mucosa of aging rats. Dig Dis Sci 35(11):1364–1370
63. Majumdar AP, Jaszewski R, Dubick MA (1997) Effect of aging on the gastrointestinal tract and the pancreas. Proc Soc Exp Biol Med 215(2):134–144
64. Malecka-Panas E, Kordek R, Biernat W, Tureaud J, Liberski PP, Majumdar AP (1997) Differential activation of total and EGF receptor (EGF-R) tyrosine kinase (tyr-k) in the rectal

mucosa in patients with adenomatous polyps, ulcerative colitis and colon cancer. Hepatogastroenterology 44(14):435–440

65. Xiao ZQ, Moragoda L, Jaszewski R, Hatfield JA, Fligiel SE, Majumdar AP (2001) Aging is associated with increased proliferation and decreased apoptosis in the colonic mucosa. Mech Ageing Dev 122(15):1849–1864. doi:S0047-6374(01)00323-2 [pii]
66. Roncucci L, Ponz de Leon M, Scalmati A, Malagoli G, Pratissoli S, Perini M, Chahin NJ (1988) The influence of age on colonic epithelial cell proliferation. Cancer 62(11):2373–2377
67. Stadler J, Stern HS, Yeung KS, McGuire V, Furrer R, Marcon N, Bruce WR (1988) Effect of high fat consumption on cell proliferation activity of colorectal mucosa and on soluble faecal bile acids. Gut 29(10):1326–1331
68. Majumdar AP (2003) Regulation of gastrointestinal mucosal growth during aging. J Physiol Pharmacol 54(Suppl 4):143–154
69. Majumdar APN, Jaszewski R (2003) Aging of the esophagus, and stomach. Aging and the gastrointestinal tract. Krager, New York
70. Carson DA, Ribeiro JM (1993) Apoptosis and disease. Lancet 341 (8855):1251–1254. doi:0140-6736(93)91154-E [pii]
71. Schwartzman RA, Cidlowski JA (1993) Apoptosis: the biochemistry and molecular biology of programmed cell death. Endocr Rev 14(2):133–151
72. Bedi A, Pasricha PJ, Akhtar AJ, Barber JP, Bedi GC, Giardiello FM, Zehnbauer BA, Hamilton SR, Jones RJ (1995) Inhibition of apoptosis during development of colorectal cancer. Cancer Res 55(9):1811–1816
73. Thompson CB (1995) Apoptosis in the pathogenesis and treatment of disease. Science 267(5203):1456–1462
74. Hruban RH, Adsay NV, Albores-Saavedra J, Compton C, Garrett ES, Goodman SN, Kern SE, Klimstra DS, Kloppel G, Longnecker DS, Luttges J, Offerhaus GJ (2001) Pancreatic intraepithelial neoplasia: a new nomenclature and classification system for pancreatic duct lesions. Am J Surg Pathol 25(5):579–586
75. Hruban RH, Takaori K, Klimstra DS, Adsay NV, Albores-Saavedra J, Biankin AV, Biankin SA, Compton C, Fukushima N, Furukawa T, Goggins M, Kato Y, Kloppel G, Longnecker DS, Luttges J, Maitra A, Offerhaus GJ, Shimizu M, Yonezawa S (2004) An illustrated consensus on the classification of pancreatic intraepithelial neoplasia and intraductal papillary mucinous neoplasms. Am J Surg Pathol 28(8):977–987. doi:00000478-200408000-00001 [pii]
76. Lin J, Beerm DG (2004) Molecular biology of upper gastrointestinal malignancies. Semin Oncol 31(4):476–486. doi:S0093775404002404 [pii]
77. Nowell PC (1976) The clonal evolution of tumor cell populations. Science 194(4260):23–28
78. Vogelstein B, Kinzler KW (2004) Cancer genes and the pathways they control. Nat Med 10(8):789–799. doi:10.1038/nm1087nm1087 [pii]
79. Yuasa Y (2003) Control of gut differentiation and intestinal-type gastric carcinogenesis. Nat Rev Cancer 3(8):592–600. doi:10.1038/nrc1141nrc1141 [pii]
80. Glenney JR Jr (1992) Tyrosine-phosphorylated proteins: mediators of signal transduction from the tyrosine kinases. Biochim Biophys Acta 1134(2):113–127. doi:0167-4889(92)90034-9 [pii]
81. Hunter T, Cooper JA (1985) Protein-tyrosine kinases. Annu Rev Biochem 54:897–930. doi:10.1146/annurev.bi.54.070185.004341
82. Joensuu H, Roberts PJ, Sarlomo-Rikala M, Andersson LC, Tervahartiala P, Tuveson D, Silberman S, Capdeville R, Dimitrijevic S, Druker B, Demetri GD (2001) Effect of the tyrosine kinase inhibitor STI571 in a patient with a metastatic gastrointestinal stromal tumor. N Engl J Med 344(14):1052–1056
83. Yarden Y, Ullrich A (1988) Growth factor receptor tyrosine kinases. Annu Rev Biochem 57:443–478. doi:10.1146/annurev.bi.57.070188.002303
84. Brunton VG, Ozanne BW, Paraskeva C, Frame MC (1997) A role for epidermal growth factor receptor, c-Src and focal adhesion kinase in an in vitro model for the progression of colon cancer. Oncogene 14(3):283–293. doi:10.1038/sj.onc.1200827
85. Cartwright CA, Kamps MP, Meisler AI, Pipas JM, Eckhart W (1989) pp60c-src activation in human colon carcinoma. J Clin Invest 83(6):2025–2033. doi:10.1172/JCI114113

86. Cartwright CA, Coad CA, Egbert BM (1994) Elevated c-Src tyrosine kinase activity in premalignant epithelia of ulcerative colitis. J Clin Invest 93(2):509–515. doi:10.1172/JCI117000
87. Craven RJ, Xu LH, Weiner TM, Fridell YW, Dent GA, Srivastava S, Varnum B, Liu ET, Cance WG (1995) Receptor tyrosine kinases expressed in metastatic colon cancer. Int J Cancer 60(6):791–797
88. Pena SV, Melhem MF, Meisler AI, Cartwright CA (1995) Elevated c-yes tyrosine kinase activity in premalignant lesions of the colon. Gastroenterology 108(1):117–124. doi:S0016508595000060 [pii]
89. Takahashi Y, Kitadai Y, Bucana CD, Cleary KR, Ellis LM (1995) Expression of vascular endothelial growth factor and its receptor, KDR, correlates with vascularity, metastasis, and proliferation of human colon cancer. Cancer Res 55(18):3964–3968
90. Baselga J (2002) Why the epidermal growth factor receptor? The rationale for cancer therapy. Oncologist 7(90004):2–8. doi:10.1634/theoncologist.7-suppl_4-2
91. Bundred NJ, Chan K, Anderson NG (2001) Studies of epidermal growth factor receptor inhibition in breast cancer. Endocr Relat Cancer 8(3):183–189
92. Prenzel N, Fischer OM, Streit S, Hart S, Ullrich A (2001) The epidermal growth factor receptor family as a central element for cellular signal transduction and diversification. Endocr Relat Cancer 8(1):11–31
93. Wells A (1999) EGF receptor. Int J Biochem Cell Biol 31(6):637–643. doi:S1357272599000151 [pii]
94. Nautiyal J PBaMA (2009) Tyrosine kinases in colorectal cancer. Eur Gastroenterol Hepatol 4(2):64–69
95. Rak J, Yu JL, Klement G, Kerbel RS (2000) Oncogenes and angiogenesis: signaling three-dimensional tumor growth. J Invest Dermatol Symp Proc 5(1):24–33. doi:10.1046/j.1087-0024.2000.00012.x
96. Schlessinger J (2000) Cell signaling by receptor tyrosine kinases. Cell 103(2):211–225
97. Hackel PO, Zwick E, Prenzel N, Ullrich A (1999) Epidermal growth factor receptors: critical mediators of multiple receptor pathways. Curr Opin Cell Biol 11(2):184–189. doi:S0955-0674(99)80024-6 [pii]
98. Moghal N, Sternberg PW (1999) Multiple positive and negative regulators of signaling by the EGF-receptor. Curr Opin Cell Biol 11(2):190–196. doi:S0955-0674(99)80025-8 [pii]
99. Gilbertson RJ, Perry RH, Kelly PJ, Pearson AD, Lunec J (1997) Prognostic significance of HER2 and HER4 coexpression in childhood medulloblastoma. Cancer Res 57(15):3272–3280
100. Kataoka H, Joh T, Kasugai K, Okayama N, Moriyama A, Asai K, Kato T (1998) Expression of mRNA for heregulin and its receptor, ErbB-3 and ErbB-4, in human upper gastrointestinal mucosa. Life Sci 63(7):553–564. doi:S0024320598003063 [pii]
101. Lyne JC, Melhem MF, Finley GG, Wen D, Liu N, Deng DH, Salup R (1997) Tissue expression of Neu differentiation factor/heregulin and its receptor complex in prostate cancer and its biologic effects on prostate cancer cells in vitro. Cancer J Sci Am 3(1):21–30
102. Srinivasan R, Poulsom R, Hurst HC, Gullick WJ (1998) Expression of the c-erbB-4/HER4 protein and mRNA in normal human fetal and adult tissues and in a survey of nine solid tumour types. J Pathol 185(3):236–245
103. Khwaja A, Rodriguez-Viciana P, Wennstrom S, Warne PH, Downward J (1997) Matrix adhesion and Ras transformation both activate a phosphoinositide 3-OH kinase and protein kinase B/Akt cellular survival pathway. EMBO J 16(10):2783–2793. doi:10.1093/emboj/16.10.2783
104. Majumdar AP, Du J (2006) Phosphatidylinositol 3-kinase/Akt signaling stimulates colonic mucosal cell survival during aging. Am J Physiol Gastrointest Liver Physiol 290 (1):G49–55. doi:00106.2005 [pii]10.1152/ajpgi.00106.2005
105. Majumdar APN, Basson MD (2006) Effect of aging on the gastrointestinal tract. Physiology of gastrointestinal tract. Academic, New York
106. Tureaud J, Sarkar FH, Fligiel SE, Kulkarni S, Jaszewski R, Reddy K, Yu Y, Majumdar AP (1997) Increased expression of EGFR in gastric mucosa of aged rats. Am J Physiol 273(2 Pt 1):G389–G398

107. Xiao ZQ, Majumdar AP (2001) Increased in vitro activation of EGFR by membrane-bound TGF-alpha from gastric and colonic mucosa of aged rats. Am J Physiol Gastrointest Liver Physiol 281(1):G111–G116
108. Patel BB, Yu Y, Du J, Levi E, Phillip PA, Majumdar AP (2009) Age-related increase in colorectal cancer stem cells in macroscopically normal mucosa of patients with adenomas: a risk factor for colon cancer. Biochem Biophys Res Commun 378(3):344–347. doi:S0006-291X(08)02083-4 [pii]10.1016/j.bbrc.2008.10.179
109. Anzano MA, Rieman D, Prichett W, Bowen-Pope DF, Greig R (1989) Growth factor production by human colon carcinoma cell lines. Cancer Res 49(11):2898–2904
110. Barnard JA, Beauchamp RD, Russell WE, Dubois RN, Coffey RJ (1995) Epidermal growth factor-related peptides and their relevance to gastrointestinal pathophysiology. Gastroenterology 108(2):564–580
111. Coffey RJ Jr, Shipley GD, Moses HL (1986) Production of transforming growth factors by human colon cancer lines. Cancer Res 46(3):1164–1169
112. Podolsky D (1994) Peptide growth factors in the gastrointestinal tract. In: Leonard RJ, Fayez KG, Jonathan DK, Juanita LM, Hamid MS, Jackie DW (eds) Physiology of the gastrointestinal tract, Two Volume Set, 5th edn. Raven Press
113. Saeki T, Salomon DS, Johnson GR, Gullick WJ, Mandai K, Yamagami K, Moriwaki S, Tanada M, Takashima S, Tahara E (1995) Association of epidermal growth factor-related peptides and type I receptor tyrosine kinase receptors with prognosis of human colorectal carcinomas. Jpn J Clin Oncol 25(6):240–249
114. Relan NK, Saeed A, Ponduri K, Fligiel SE, Dutta S, Majumdar AP (1995) Identification and evaluation of the role of endogenous tyrosine kinases in azoxymethane induction of proliferative processes in the colonic mucosa of rats. Biochim Biophys Acta 1244(2–3):368–376
115. Yoshida K, Kyo E, Tsuda T, Tsujino T, Ito M, Niimoto M, Tahara E (1990) EGF and TGF-alpha, the ligands of hyperproduced EGFR in human esophageal carcinoma cells, act as autocrine growth factors. Int J Cancer 45(1):131–135
116. André T, Boni C, Mounedji-Boudiaf L, Navarro M, Tabernero J, Hickish T, Topham C, Zaninelli M, Clingan P, Bridgewater J, Tabah-Fisch I, de Gramont A (2004) Oxaliplatin, fluorouracil, and leucovorin as adjuvant treatment for colon cancer. N Engl J Med 350(23):2343–2351. doi:10.1056/NEJMoa032709
117. Jemal A, Center MM, Ward E, Thun MJ (2009) Cancer occurrence. Methods Mol Biol 471:3–29. doi:10.1007/978-1-59745-416-2_1
118. Neugut AI, Lautenbach E, Abi-Rached B, Forde KA (1996) Incidence of adenomas after curative resection for colorectal cancer. Am J Gastroenterol 91(10):2096–2098
119. Welch JP, Donaldson GA (1979) The clinical correlation of an autopsy study of recurrent colorectal cancer. Ann Surg 189(4):496–502
120. Edwards BK, Howe HL, Ries LA, Thun MJ, Rosenberg HM, Yancik R, Wingo PA, Jemal A, Feigal EG (2002) Annual report to the nation on the status of cancer, 1973–1999, featuring implications of age and aging on U.S. cancer burden. Cancer 94(10):2766–2792
121. Howlader NNA, Krapcho M, Garshell J, Neyman N, Altekruse SF, Kosary CL, Yu M, Ruhl J, Tatalovich Z, Cho H, Mariotto A, Lewis DR, Chen HS, Feuer EJ, Cronin KA (eds) (2013) SEER Cancer statistics review, 1975–2010. National Cancer Institute, Bethesda (based on November 2012 SEER data submission, posted to the SEER web site, April 2013)
122. Yu Y, Kanwar SS, Patel BB, Nautiyal J, Sarkar FH, Majumdar AP (2009) Elimination of colon cancer stem-like cells by the combination of curcumin and FOLFOX. Transl Oncol 2(4):321–328
123. Patel BB, Gupta D, Elliott AA, Sengupta V, Yu Y, Majumdar AP (2010) Curcumin targets FOLFOX-surviving colon cancer cells via inhibition of EGFRs and IGF-1R. Anticancer Res 30(2):319–325
124. Chen X, Yeung TK, Wang Z (2000) Enhanced drug resistance in cells coexpressing ErbB2 with EGF receptor or ErbB3. Biochem Biophys Res Commun 277(3):757–763. doi:10.1006/bbrc.2000.3731

125. Pinkas-Kramarski R, Soussan L, Waterman H, Levkowitz G, Alroy I, Klapper L, Lavi S, Seger R, Ratzkin BJ, Sela M, Yarden Y (1996) Diversification of Neu differentiation factor and epidermal growth factor signaling by combinatorial receptor interactions. EMBO J 15(10):2452–2467
126. Bartel DP (2004) MicroRNAs: genomics, biogenesis, mechanism, and function. Cell 116(2):281–297
127. Slaby O, Svoboda M, Michalek J, Vyzula R (2009) MicroRNAs in colorectal cancer: translation of molecular biology into clinical application. Mol Cancer 8:102. doi:10.1186/1476-4598-8-102
128. Engels BM, Hutvagner G (2006) Principles and effects of microRNA-mediated post-transcriptional gene regulation. Oncogene 25(46):6163–6169. doi:10.1038/sj.onc.1209909
129. Dong Y, Wu WK, Wu CW, Sung JJ, Yu J, Ng SS (2011) MicroRNA dysregulation in colorectal cancer: a clinical perspective. Br J Cancer 104(6):893–898. doi:10.1038/bjc.2011.57
130. Yang L, Belaguli N, Berger DH (2009) MicroRNA and colorectal cancer. World J Surg 33(4):638–646. doi:10.1007/s00268-008-9865-5
131. Medina PP, Nolde M, Slack FJ (2010) OncomiR addiction in an in vivo model of microRNA-21-induced pre-B-cell lymphoma. Nature 467(7311):86–90. doi:10.1038/nature09284
132. Pan X, Wang Z-X, Wang R (2010) MicroRNA-21: A novel therapeutic target in human cancer. Cancer Biol Ther 10(12):1224–1232
133. Ziyan W, Shuhua Y, Xiufang W, Xiaoyun L (2011) MicroRNA-21 is involved in osteosarcoma cell invasion and migration. Med Oncology 28(4):1469–1474. doi:10.1007/s12032-010-9563-7
134. Schetter AJ, Leung SY, Sohn JJ, Zanetti KA, Bowman ED, Yanaihara N, Yuen ST, Chan TL, Kwong DL, Au GK, Liu CG, Calin GA, Croce CM, Harris CC (2008) MicroRNA expression profiles associated with prognosis and therapeutic outcome in colon adenocarcinoma. JAMA 299(4):425–436. doi:10.1001/jama.299.4.425
135. Slaby O, Svoboda M, Fabian P, Smerdova T, Knoflickova D, Bednarikova M, Nenutil R, Vyzula R (2007) Altered expression of miR-21, miR-31, miR-143 and miR-145 is related to clinicopathologic features of colorectal cancer. Oncology 72(5–6):397–402. doi:10.1159/000113489
136. Jazbutyte V, Thum T (2010) MicroRNA-21: from cancer to cardiovascular disease. Curr Drug Target 11(8):926–935
137. Lu Z, Liu M, Stribinskis V, Klinge CM, Ramos KS, Colburn NH, Li Y (2008) MicroRNA-21 promotes cell transformation by targeting the programmed cell death 4 gene. Oncogene 27(31):4373–4379. doi:10.1038/onc.2008.72
138. Roy S, Yu Y, Padhye SB, Sarkar FH, Majumdar AP (2013) Difluorinated-curcumin (CDF) restores PTEN expression in colon cancer cells by down-regulating miR-21. PloS ONE 8(7):e68543. doi:10.1371/journal.pone.0068543
139. Yu Y, Sarkar FH, Majumdar AP (2013) Down-regulation of miR-21 Induces differentiation of chemoresistant colon cancer cells and enhances susceptibility to therapeutic regimens. Transl Oncol 6(2):180–186
140. Kashyap L (2011) Can microRNAs act as biomarkers of aging? Bioinformation 5(9):396–397
141. Noren HN, Abdelmohsen K, Gorospe M, Ejiogu N, Zonderman AB, Evans MK (2010) microRNA expression patterns reveal differential expression of target genes with age. PloS ONE 5(5):e10724. doi:10.1371/journal.pone.0010724
142. Seike M, Goto A, Okano T, Bowman ED, Schetter AJ, Horikawa I, Mathe EA, Jen J, Yang P, Sugimura H, Gemma A, Kudoh S, Croce CM, Harris CC (2009) MiR-21 is an EGFR-regulated anti-apoptotic factor in lung cancer in never-smokers. Proc Natl Acad Sci U S A 106(29):12085–12090. doi:10.1073/pnas.0905234106
143. Bonnet D, Dick JE (1997) Human acute myeloid leukemia is organized as a hierarchy that originates from a primitive hematopoietic cell. Nat Med 3(7):730–737
144. Rasheed ZA, Kowalski J, Smith BD, Matsui W (2011) Concise review: emerging concepts in clinical targeting of cancer stem cells. Stem Cells 29(6):883–887. doi:10.1002/stem.648

145. Sanders MA, Majumdar AP (2011) Colon cancer stem cells: implications in carcinogenesis. Front Biosci 16:1651–1662
146. Clevers H (2011) The cancer stem cell: premises, promises and challenges. Nat Med 17(3):313–319. doi:10.1038/nm.2304
147. Dalerba P, Dylla SJ, Park IK, Liu R, Wang X, Cho RW, Hoey T, Gurney A, Huang EH, Simeone DM, Shelton AA, Parmiani G, Castelli C, Clarke MF (2007) Phenotypic characterization of human colorectal cancer stem cells. Proc Natl Acad Sci U S A 104(24):10158–10163. doi:0703478104 [pii]10.1073/pnas.0703478104
148. O'Brien CA, Pollett A, Gallinger S, Dick JE (2007) A human colon cancer cell capable of initiating tumour growth in immunodeficient mice. Nature 445(7123):106–110. doi:nature05372 [pii]10.1038/nature05372
149. Ricci-Vitiani L, Lombardi DG, Pilozzi E, Biffoni M, Todaro M, Peschle C, De Maria R (2007) Identification and expansion of human colon-cancer-initiating cells. Nature 445(7123):111–115. doi:nature05384 [pii]10.1038/nature05384
150. Dean M, Fojo T, Bates S (2005) Tumour stem cells and drug resistance. Nat Rev Cancer 5(4):275–284. doi:nrc1590 [pii]10.1038/nrc1590
151. Nautiyal J, Kanwar SS, Yu Y, Majumdar AP (2011) Combination of dasatinib and curcumin eliminates chemo-resistant colon cancer cells. J Mol Signal 6:7. doi:10.1186/1750-2187-6-7

Epithelial to Mesenchymal Transition (EMT) and Intestinal Tumorigenesis

Ashleigh Pulkoski-Gross, Xi E. Zheng, Deborah Kim, Jillian Cathcart and Jian Cao

Abstract Colorectal cancer is the world's third most common cancer and has one of the highest rates of morbidity. From studying familial occurrences of colon cancer, significant advances were made in understanding the molecular basis of the disease, even those of a sporadic nature. While there have been great strides in understanding the generation of colorectal cancer, the largest determinant of morbidity, metastasis, is far less understood. Metastasis is primarily the cause of death of patients with colorectal cancer and more attention is being paid to the molecular determinants of cell invasion and migration, as these are critical steps in the metastatic cascade. The conversion of epithelial cells to mesenchymal-like ones that are capable of leaving the primary tumor site requires the coordination of many signal transduction pathways and changes in transcriptional activity. These reversible changes are referred to as epithelial-to-mesenchymal transition (EMT). As EMT is a crucial program that determines metastatic capacity, recent efforts have been made to identify inhibitors of EMT in order to augment current chemotherapies and improve patient outcome.

Keywords Epithelial-to-mesenchymal transition (EMT) · Mesenchymal-to-epithelial transition (MET) · Colorectal cancer (CRC) · Metastasis · Wnt · TGF-β · Cancer stem cells

---

J. Cao (✉) · A. Pulkoski-Gross · D. Kim · J. Cathcart
Department of Medicine/Division of Cancer Prevention, Stony Brook University,
Stony Brook, New York, NY 11794, USA
e-mail: jian.cao@sunysb.edu

A. Pulkoski-Gross
e-mail: ashleigh.pulkoski-gross@stonybrook.edu

X. E. Zheng
Department of Public Health, Weill Cornell Medical College, New York, NY 11065, USA
e-mail: xez9001@nyp.org

D. Kim
e-mail: deborahkim@hotmail.com

J. Cathcart
e-mail: jillian.cathcart@gmail.com

V. W. Yang, A. B. Bialkowska (eds.), *Intestinal Tumorigenesis*,
DOI 10.1007/978-3-319-19986-3_11

1 Introduction

Metastasis is the most common cause of cancer related death in patients with solid tumors, including intestinal tumors [1, 2]. Tumor invasion is one of the major factors contributing to early stages of cancer metastasis [3]. In order to invade, cancer cells must gain the ability to cleave cell-cell and cell-matrix adhesion molecules, digest the basement membrane and extracellular matrices, and enhance cell motility [4]. Since most human solid tumors are epithelial in origin, transition from an epithelial phenotype to a migratory mesenchymal-like phenotype is thought to be a key step in gaining invasive ability. This transition is called epithelial-to-mesenchymal transition (EMT) [5–7]. Based on the biological context, EMT has been categorized into three types—Type I EMT refers to the physiologic EMT that occurs during development, Type II EMT refers to wound healing and chronic inflammation leading to tissue fibrosis, and Type III EMT (oncogenic EMT) refers to the transdifferentiation of epithelial cancer cells to more invasive mesenchymal cancer cells during oncogenesis [7].

EMT is a biological process by which epithelial cells lose their cell polarity and cell-cell adhesion and gain migratory ability, becoming mesenchymal-like cells during embryonic development, tissue remodeling, and wound repair [8, 9]. Mounting experimental studies have demonstrated that EMT serves as a critical step in the conversion of early stage tumors to invasive malignancies that achieve local invasion and dissemination into the systemic circulation [10, 11]. Emerging evidence of a link between EMT and cancer stem cells (CSCs) has highlighted the importance of EMT in cancer molecular pathobiology [12]. Distinctions between epithelial and mesenchymal phenotypes are summarized in Fig. 1.

Although the accumulating evidence supporting EMT as a crucial component of disease progression, histological examination of metastatic carcinomas has revealed organized epithelial structures similar to the primary tumor, despite the obvious phenotypic switch that occurs at the invasive front of the primary tumor [13]. Such observations suggest that once metastasis has occurred, tumor cells are able to regain the epithelial phenotype of the primary tumor. This observation is consistent with the reversibility of EMT (mesenchymal-to-epithelial transition, MET) and the concept that epithelial function needs to be regained to facilitate metastatic growth [14, 15]. Thus dedifferentiated phenotypes, which cancer cells demonstrate during the processes of dissemination, may not be fixed irrevocably by genetic alterations in the genome of these cells. A major problem in proving the EMT concept is the transient, reversible nature of EMT during carcinoma invasion and metastasis. Therefore, not all pathologists currently agree on the relevance of EMT in human cancer progression.

Nevertheless, there is some consensus about the role of EMT in cancer dissemination. It is thought that small numbers of differentiated carcinoma cells in a tumor transiently undergo EMT resulting in enhanced invasion and intravasation into blood and lymphatic vessels. Following extravasation at distant sites, the mesenchymal-like cells revert back (MET) toward their original epithelial phenotype (see hypothetical model, Fig. 2).

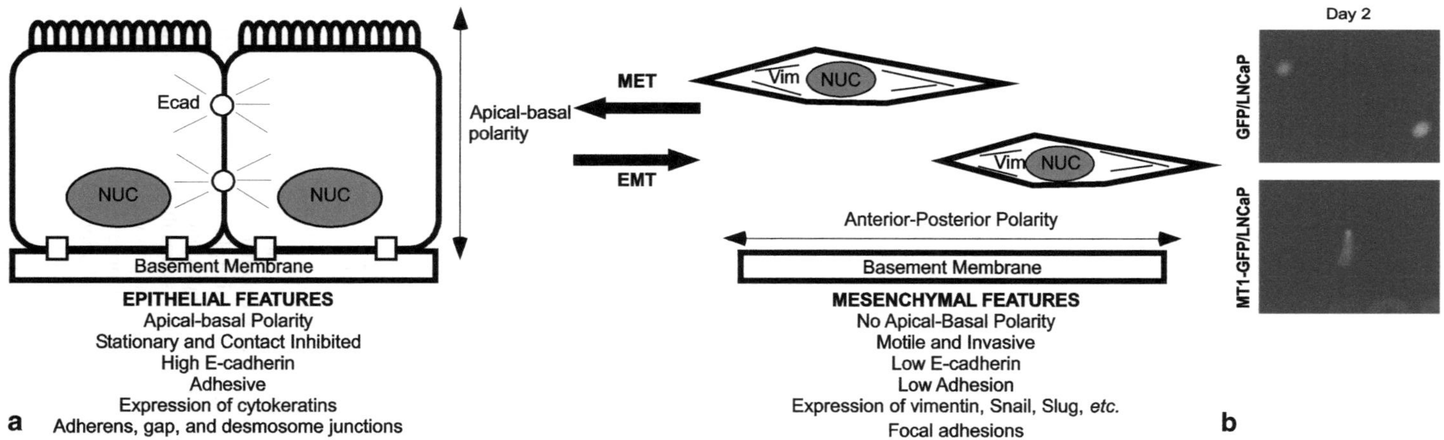

Fig. 1 Distinctions between epithelial and mesenchymal cell phenotype. (**a**) Epithelial cells display contact inhibition and adhere both to neighboring cells and the basement membrane. They express epithelial markers such as cytokeratins and E-cadherin (*E-cad*). Transitioning to a mesenchymal phenotype results in decreased adhesion and increased motility. E-cad expression is lost and EMT effectors are increased, including Snail and Slug, in addition to increases in mesenchymal markers such as vimentin (*vim*). (**b**) LNCaP cells expressing GFP or MMP-14-GFP were mixed with a type I collagen 3D matrix and cells were examined under fluorescent microscopy. LNCaP cells expressing MMP-14-GFP display a more invasive phenotype, as demonstrated by the extension of the cell through the collagen matrix. This is indicative of EMT

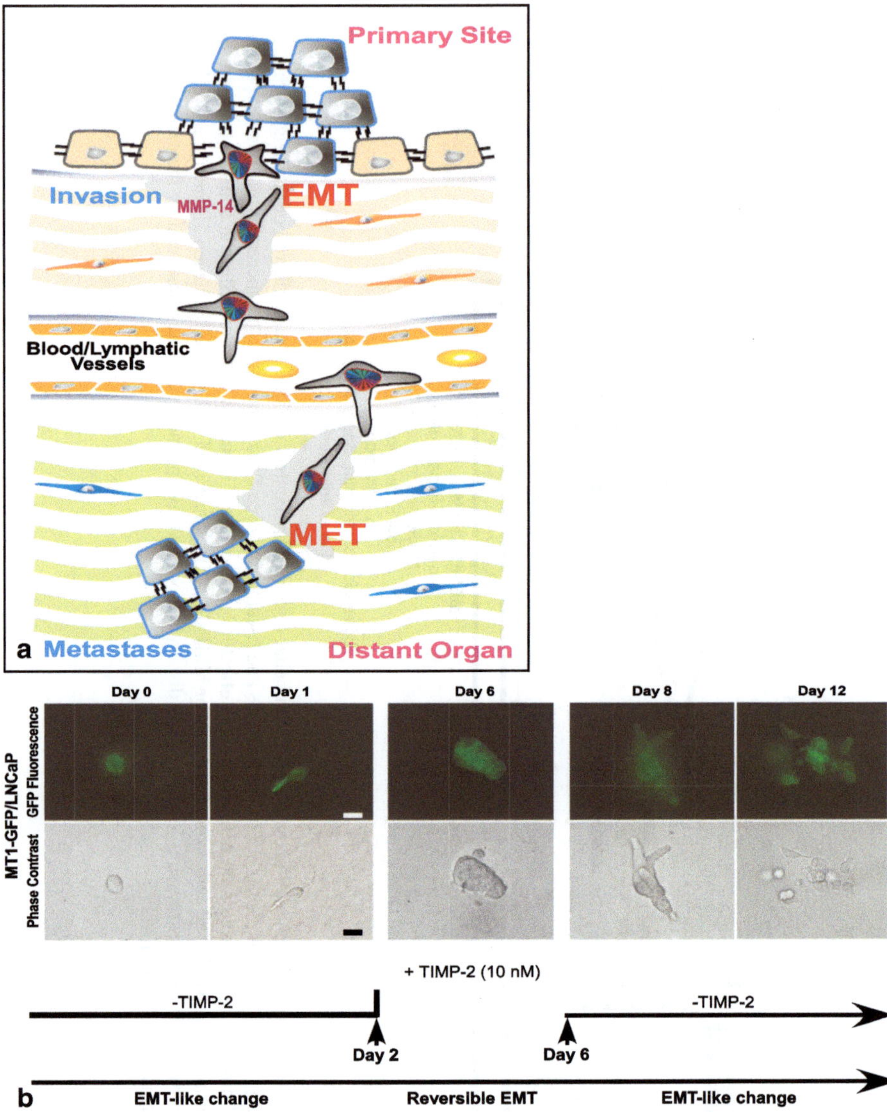

Fig. 2 Hypothetical model of EMT/MET and the reversibility of EMT. (**a**) Early stage carcinoma cells maintain epithelial properties. Various stimuli can induce reduction in Ecad, an increase in EMT-associated transcription factors, and a conversion to a mesenchymal transcriptional program. Cancer cells undergoing EMT display a more invasive phenotype characterized by proteolytic degradation of basement membrane and other ECM components, followed by intravasation and then extravasation from blood and lymphatic vessels. As a result of cues generated in the distant organ, the disseminated mesenchymal-like cells revert back toward an epithelial phenotype (MET) and can proliferate in the new environment to form a secondary lesion. (**b**) MMP-14 induced EMT can be reversed by inhibition of proteolytic activity. Incubation of LNCaP cells expressing MMP-14 in a 3D collagen matrix results in a change to a more mesenchymal phenotype, determined based on morphology (EMT). Introduction of TIMP-2, an endogenous MMP-14 inhibitor, results in reversal of the mesenchymal phenotype to a more epithelial-like one (MET). Removal of the inhibitor results in reversion to mesenchymal behavior (EMT)

In this chapter, we will explore the role of EMT in intestinal tumorigenesis. EMT has been implicated in the progression of colorectal cancer (CRC) and as such we will use this pathology to discuss the complex conditions that are currently thought to invoke EMT and MET, including signaling pathways and the resultant activation of EMT markers. In addition to the induction of EMT, we will also discuss the various mechanisms of control of EMT as it is currently understood. Finally, this chapter endeavours to look toward the future of cancer treatments, discussing novel therapeutic strategies to prevent the progression of colorectal cancer based on the current knowledge of EMT.

2 EMT and Colorectal Cancer

Since a tumor is composed of multiple components, including activated stroma and cancer cells, influence over EMT signaling can be derived from a variety of sources. Several factors have been implicated in induction of EMT, including transforming growth factor-β (TGF-β), Wnt, Sonic Hedgehog (Shh), interleukin-6 (IL-6), Notch, and receptor tyrosine kinase (RTK) signaling, implicating ligands such as epidermal growth factor (EGF), platelet derived growth factor (PDGF), hepatocyte growth factor (HGF), insulin-like growth factor (IGF), tumor necrosis factor-α (TNF-α), keratinocyte growth factor (KGF), and fibroblast growth factor (FGF) [16, 17]. The effectors of the EMT process that are up-regulated in response to these signaling pathways include transcription factors/repressors such as Snail (Snail1), Slug (Snail2), Smuc (Snail3), Zeb1 and Zeb2 (SIP1), Twist1 and Twist2, E12/E47 (TCF3), E2-2 (TCF4), Goosecoid, FOXC2, KLF8, Brachyury (Tbx1), Tbx2, Tbx3, Lbx1, Six1, and Prrx1 resulting in changes in gene expression and the classical loss of epithelial characteristics [18]. These characteristics include upregulation of vimentin and matrix metalloproteinases (MMPs). Micro-RNAs (miRs) have also been increasingly associated with various stages of cancer progression, including EMT [19]. The bulk of these components and effectors have been implicated in the progression of CRC.

CRC is the third most common malignancy and is one of the cancers with the highest morbidity in the Western world [20, 21]. The pathogenesis of CRC was defined in the early 1990s by Fearon and Vogelstein as a progression from adenoma to carcinoma in a process termed the adenoma-carcinoma sequence [21]. Aberrant crypt foci (ACF) form as the earliest lesions and development of the ACF to an increasingly hyperplastic and dysplastic mass may lead to carcinoma and invasive adenocarcinoma [22]. Fearon et al. proposed that the first genetic hit is one to the Wnt pathway; generally sporadic cases of CRC harbor either adenomatous polyposis coli (APC), β-catenin, or Axin mutations such that negative regulation of this pathway is impossible [22]. The adenoma-carcinoma sequence continues when other genetic hits are incurred, such as an activating mutation of Ras or a deactivation of tumor suppressor p53 [23, 24]. CRC is thought to develop as a result of an accumulation of mutations during this sequence. However, the definition of the pathology has been refined to include the idea that normal intestinal stem cells (ISCs) can be the cell of origin for the tumor [25, 26]. Support for this idea lies in the facts that the long-lived nature of the ISCs allow for genetic mutation accumulation, the Wnt pathway that is commonly dysregulated in CRC is a crucial pathway for the maintenance of ISCs, and CRC presents with heterogeneity that can easily be acquired by having an ISC as the cell of origin [27].

The normal physiology of the colon is layered, with an outer layer of sheets of smooth muscle that control peristalsis and the center layer that is a stromal layer consisting of vasculature and nerve fibers, in addition to immune components [28]. The luminal surface of the intestine is a one cell-layer thick epithelium known as the mucosa. This layer plays a critical role in nutritional absorption and the proper production of stool [28]. The intestinal system consists of both the small intestine and large intestine, or colon. These two portions are physiologically different, with the small intestine containing villi, or protrusions into the mucosa, which serve to increase the surface area for absorption, while the colon have intussusceptions known as the crypts of Lieberkühn [28]. The intestines collectively contain at least four types of cells: enterocytes, enteroendocrine, mucosecreting (goblet cells), and Paneth cells, however Paneth cells are primarily found in the small intestine [29]. Following birth, the colonic crypts become deeper and monoclonal during development of the mature colon and as the organ grows larger toward adulthood, crypt fission occurs to provide new crypts [28].

The epithelia of both intestines are subject to rapid turnover and can only be sustained by the ISC population at the base of the crypts [28]. ISCs are capable of undergoing asymmetric division to self-renew and to eventually provide the crypt with the appropriate population of differentiated cells [29]. Terminally differentiated cells reside at the top one-third of the crypt until they are shed to the lumen and transit-amplifying cells have a middling residence above the ISCs at the base (Fig. 3) [29]. The ISCs maintain the integrity of the intestinal crypt based on signals from their descendants and the microenvironment, which is crucial considering the high turnover rate of the organ [24]. The ISCs residence at the base of the crypt give it an advantageous position to collect signals from the environment which include Wnt, Notch, bone morphogenetic proteins (BMPs), and Hh which serve to maintain the stem cell compartment [24].

Despite the highly regulated nature of the stem cells by these signaling pathways, environmental and genetic factors may play a role in deregulating these pathways and thus lead to CRC. Western diets, inflammatory bowel disease, and genetic predisposition (*i.e.* familial adenomatous polyposis [FAP] and hereditary non-polyposis CRC [HNPCC]), are crucial determinants of CRC development [24]. The development of CRC begins with adenoma formation in which the cells of the crypt increase in number and in tandem are more immature in nature. Fission of the adenomatous crypt occurs and stromal cells are recruited that influence disease progression by secreting factors such as IL-6 and HGF [24]. The transition all the way to carcinoma is accomplished by the accumulation of other mutations, as discussed. The end result is a colorectal carcinoma that includes intestinal epithelium on a spectrum of differentiation from differentiated to CSCs, in addition to recruited fibroblasts and myeloid cells [24].

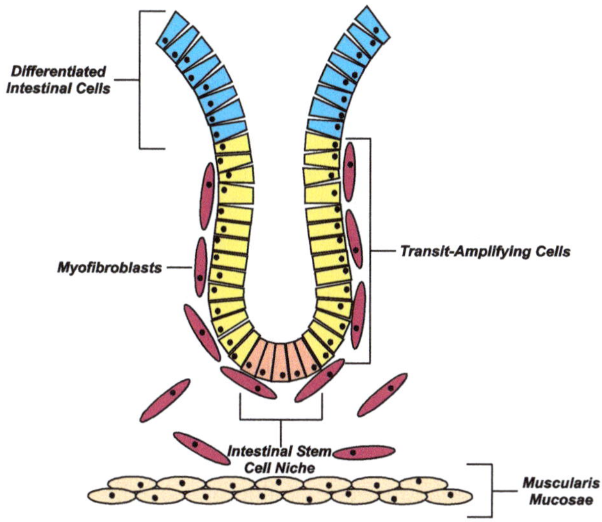

Fig. 3 Schematic representation of the colonic crypt (crypt of Lieberkühn). The colonic mucosal epithelium is organized into crypts where the intestinal stem cells (*ISCs*) reside at the base. The proliferative and trans-differentiating cells derived from asymmetrical division of the stem cells migrate upward as they differentiate. Completely differentiated cells are exposed to the lumen of the colon, as they are the final layer of the crypt. The terminally differentiated cells reside there until they are shed to the lumen. Myofibroblasts in the stromal compartment aid in maintaining the microenvironment required to support the normal ISC function

The role of EMT has been documented to play an important role in CRC progression. EMT can influence proliferation of CRC cells, but also correlates to invasion and distant metastasis. The ability of a cell to move away from the primary tumor is dictated by whether or not they have undergone EMT. EMT derived production of cells with the CSC status is also a contributor to disease progression. CSCs are powerful in that they can be motile and can seed a secondary site well; they are able not only to populate the new site with CSC clones, but also recapitulate the non-CSC population of the tumor. They also tend to have several potential mechanisms for drug resistance, making it clinically difficult to treat. CRC development has been attributed to the overpopulation of CSCs [30], a population that can increase due to the occurrence of EMT.

Pathways associated with EMT are also those that are associated with oncogenesis, including Wnt pathway, Notch signaling, and RTK signaling [31]. The transition from epithelial to mesenchymal phenotype correlates with poor prognosis in CRC and several of the pathways we will discuss here are directly implicated [31]. One of the foremost studied pathways implicated in CRC is the Wnt signaling pathway. The majority of CRC cases involve some mutation leading to Wnt activation; as such there is great interest in understanding its role in disease progression. While Wnt is a crucial pathway we will address, there are several additional pathways that play a major role in EMT in colorectal cancer.

3 EMT Induction and Mechanisms of Control

3.1 Matrix Metalloproteinase (MMP) Induced EMT

Matrix metalloproteinases (MMPs) are a family of zinc-dependent endopeptidases that are responsible for extracellular matrix (ECM) remodeling, in addition to cleavage of growth factor receptors, adhesion molecules, and cytokines. These enzymes can either be secreted to the extracellular milieu or membrane bound and for the most part are kept in their latent states by their pro-domains. Once the pro-domains are removed, these proteins can be inhibited by their endogenous inhibitors, tissue inhibitor of matrix metalloproteinases (TIMPs). A primary tumor consists not only of the malignant cells, but also of activated stroma which has been revealed to be a great contributor to the invasive potential of the cancerous cells [32]. Cancer associated fibroblasts and immune cells can both contribute MMPs to the tumor cell environment. Extracellular matrix metalloproteinase inducer (EMMPRIN) and transforming growth factor-β signaling (TGF-β) can stimulate the production of enzymes by tumor associated fibroblasts and these proteases contribute to invasion of tumor cells. Extracellular proteases are generally found in higher concentrations at the invasive fronts of tumors, implicating them in the EMT process. Not only do the MMPs contribute to cell invasion by degradation of the extracellular matrix for clearance, but also this degradation uncovers what are now termed "matrikins" [32]. These are cryptic sites in extracellular matrix molecules that can elicit a response from the cell, including stimulation of invasion. MMPs also cleave cell adhesion molecules; a hallmark of EMT is the loss of cell adhesion molecules and therefore their normal anchoring. For example, MMP-14 is often upregulated in cancer and, accordingly, expression of MMP-14 leads to shedding of E-cadherin (Ecad) from the cell surface (Fig. 4). This activity is important in that it disrupts cell adhesion by reducing the functional Ecad at the cell membrane and studies have shown the

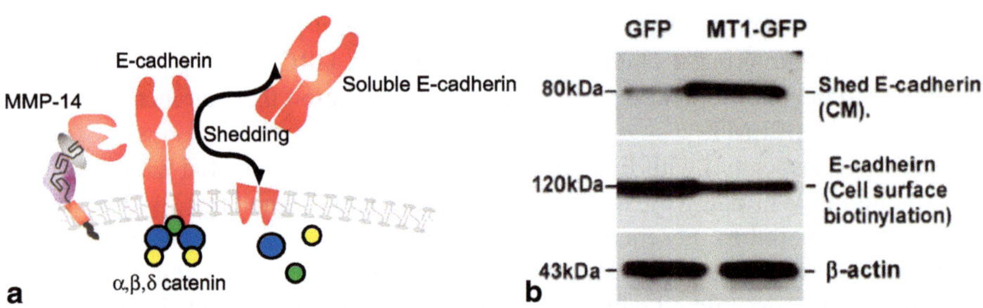

Fig. 4 MMP-14 cleaves E-cadherin (*Ecad*) from the cell membrane. **a** MMP-14 that is embedded in the cell membrane can cleave neighboring Ecad. Releasing the soluble fragment of Ecad to the ECM effectively reduces the ability of the cell to adhere to neighboring cells and may release β-catenin from its reserve and this allows it to translocate for transcription. **b** Expression of MMP-14 in vitro results in a decrease of full-length Ecad at the cell surface and a reciprocal increase in soluble Ecad in conditioned media

soluble fragments of Ecad disrupt cell aggregation [33]. Importantly, the soluble Ecad can further induce expression of MMPs, specifically MMP-2, MMP-9, and MMP-14 [34]. The ability of MMPs to cleave proteins in the ECM allows them to destroy proteins that restrict growth factors, including those that sequester fibroblast growth factor (FGF), insulin-like growth factor (IGF) and TGF-β [32, 35–37]. The influence of EMT by MMPs is therefore greater, as signaling that results from engagement of the receptors for these ligands result in switching to a mesenchymal-like program and changing cell phenotype from an epithelial cell to a mesenchymal one. Overexpression of MMPs by tumor cells themselves also contributes to EMT (Figs. 1 and 3). Loss of Ecad due to shedding and upregulation of EMT-associated transcription factors such as Snail contribute to the mesenchymal phenotype of MMP expressing or exposed cells [38, 39].

MMP expression in CRC have been established and MMP-2 has been consistently found to be overexpressed and MMP-9 has been found to be overexpressed in patient samples in which metastatic disease was found at the mRNA and protein level [40]. Not only are the proteins overexpressed, but they are also active; this is a relevant point as these enzymes are usually secreted in a latent form with little ability to cleave their targets. Activation of the proteins must be accomplished in order for them to exert their effects on cells, ECM, and EMT. Work in colon cell lines demonstrate that MMP-2 and MMP-9 are capable of increasing migratory and invasive behavior [41]. MMP-7 has also been reported to be expressed in the case of CRC and it correlates to state of the disease [42]. Not only is this MMP expressed, but it can be found in its active form at the site of liver metastases [42].

3.2 EMT-Inducing Signaling Pathways

3.2.1 Transforming Growth Factor-β (TGF-β) Signaling

TGF-β was identified in 1994 as a major factor that influences EMT [43]. It can induce opposing effects, dependent on the context in which it is active, as it is a tumor suppressor in early stages but contributes to EMT in later stages of cancer [44]. TGF-β signaling pathways have been associated with cells derived from breast carcinomas that have a mesenchymal and stem-like profile [45]. TGF-β stimulation results in induction of Snail1, Zeb2, and Slug, influencing EMT as these are a subset of EMT associated transcription factors [46–49]. TGF-β binds its receptors (TβR-I, TβR-II, or TβR-III) and induces canonical signaling through the Smad pathway or non-canonical signaling independent of the Smad pathway [50]. TβRI and TβRII are bound by its ligand and these serine-threonine kinase receptors serve to activate receptor-regulated Smads (R-Smads), Smad2 and Smad3, which then complex with Smad4. Once the Smad complex has assembled, translocation to the nucleus occurs and transcription is mediated by Smads in cooperation with other transcriptional regulators [51]. TGF-β signaling through Smads results in expression of high mobility group A2 (HMGA2) which influences the upregulation of Snail, Slug, and

Twist, leading to the hallmark reduction in Ecad during EMT [52, 53]. Not only does the activation of canonical TGF-β signaling influence transcription levels of EMT effectors, it has also been shown that Smads can interact directly with Snail1, to promote EMT further by co-localizing to the promoter of genes such as Ecad and occludin and preventing transcription [54]. The EMT associated transcription factors Zeb1 and Zeb2 have also been documented to differentially modulate TGF-β signaling effects by their interaction with Smads [55–58].

Non-canonical pathways coupled to TGF-β signaling (Fig. 5) have been found to be augmented in metastatic cancer cells, potentially dominating any anti-tumoral signaling derived from TGF-β associated Smad signaling [59]. One non-canonical TGF-β driven pathway is the PI3K/Akt cascade and it is required for some of the changes in the mesenchymal transition, including junction disassembly and is independent of the Smad signaling pathway [60]. The activated Akt can initiate signaling through mammalian target of rapamycin (mTOR) or phosphorylate GSK-3β to allow EMT transcription factor accumulation, ultimately leading to EMT [61]. Additionally, activation of PI3K by TGF-β signaling can result in activation of p21-

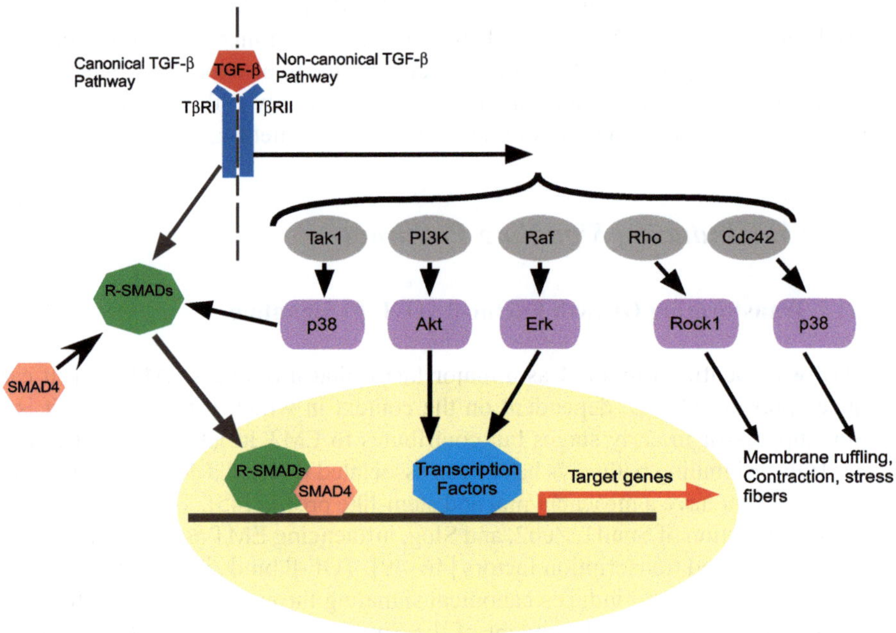

Fig. 5 TGF-β signaling is a critical regulator of EMT. The canonical TGF-β signaling pathway requires binding of the TGF-β ligand to TβRI and TβRII heterodimer. Once the ligand has bound, the SMAD complex (R-SMADs) can join with SMAD4 and translocate to the nucleus and act as a transcription factor in conjunction with other transcriptional regulators. The non-canonical pathway involves several downstream effectors. Several of the pathways that are briefly listed result in transcriptional activation of EMT associated components. In addition, activation of Rho and Cdc43 result in stress fiber formation, contraction, and membrane ruffling, all of which are physical necessities for cell migration and invasion

activated kinase-2 (PAK2), leading to classic morphological changes associated with EMT [62]. The GTPase RhoA has also been suggested to be upstream of PI3K/Akt, as RhoA activity increases with TGF-β treatment and expression of a dominant negative RhoA (N19-RhoA) disallows the typical morphology change associated with EMT [63]. Furthermore, TGF-β receptor activation has been connected to activity of Cdc42 and its downstream effector $p38^{MAPK}$ in addition to RhoA and its downstream effector ROCK1, resulting in membrane ruffling [64]. More recently, it was found that TGF-β activates TGF-β-associated kinase 1 (TAK1) via activation of TNF-receptor associated factor 6 (TRAF-6), which allows Tak1 to activate an intermediate kinase, MAP kinase kinase 3/6 (MKK3/6), that phosphorylates $p38^{MAPK}$ [65]. Yamashita, et al. specifically showed that TRAF6 is required for TGF-β induced EMT [66]. An additional pathway that was identified related to TGF-β and RhoA is the Par6-Smurf1-RhoA pathway; this pathway results in dissolution of tight junctions by degradation of RhoA after ubiquitination by activation of Smurf1 by Par6 association with the TβRI/TβRII complex after TGF-β binding [67]. This particular example highlights the function of TGF-β outside of any gene regulation function. Furthermore, TGF-β stimulation results in activation of the JNK pathway, resulting in increased fibronectin seemingly independent of Smad4, however separate work suggests the Smad pathway and the JNK pathway are synergistic [68, 69], as is the case with a variety of the non-canonical signaling pathways activated upon TGF-β stimulation [70]. TGF-β receptor activation leads to phosphorylation of ShcA which allows for association of Grb2 and Sos; the complexing of these proteins results in rapid activation of Raf and downstream kinases Erk1 and Erk2 [71]. Importantly, the Ras-Raf pathway was determined to be necessary for induction of EMT via TGF-β [72]. Interestingly, this work implied that the Ras-Raf pathway was required for TGF-β induced EMT, but the PI3K/Akt signaling activity was only necessary for evasion of TGF-β induced apoptosis, conflicting with results of studies in separate cell lines that imply PI3K is a major mediator of TGF-β induced EMT [72, 73].

TGF-β plays a role in the progression of colorectal cancer in particular; TGF-β1 is overexpressed in colon cancer and circulating levels of this particular cytokine correlate with liver metastases [74, 75]. Recent studies indicate that human colorectal cancer samples demonstrate expression of Twist1 and Snail1, EMT transcription factors, and their expression is associated with TGF-β1 expression [76]. Cell based studies indicate that expression of dominant-negative TβRII reduces mesenchymal phenotype and invasiveness [77]. Additionally, HNPCC samples that demonstrate mutated TβRII show impaired ability to respond to TGF-β and have reduced invasive capacity; restoration of low levels of wild-type TβRII allow these colon cancer cells to invade matrix and respond to TGF-β [77]. The liver is a primary site for metastasis in CRC progression, with few treatment options, as is the case with most metastatic cancers. TGF-β promotion of CRC liver metastases can be reversed in an in vivo model of cancer by use of TGF-β blocking peptides resulting in a reversal of EMT cell phenotype and decreased liver metastasis. This work implies that TGF-β plays a critical role in EMT because blocking it results in reversal of metastasis in CRC [78].

3.2.2 Interleukin-6

Interleukin-6 (IL-6) is a cytokine that normally functions as a mediator of the immune response; however it has long been known to alter the phenotype of epithelial cells. In a report from 1989, treatment of epithelial cells with IL-6 resulted in loss of adherens junctions and desmosomes and increased motility [79]. The significance in this particular phenomenon lies in the fact that many different types of human tumors stain positively for IL-6, including colon cancer, and is also a clinically relevant prognostic factor [80, 81]. In the years since this discovery, many reports of IL-6 influence on EMT have been published. Treatment with IL-6 results in loss of cell adhesion and a concomitant loss of Ecad staining at the membrane, a prominent characteristic of EMT [82]. In a head and neck model of cancer, IL-6 treatment led to decreased Ecad and increased vimentin and the EMT associated transcription factor, Snail, in vitro [83]. Interestingly, Snail transcription factors are associated with increased expression of cytokines, including IL-6, establishing a feed forward mechanism that potentiates EMT [84, 85]. Importantly, this expression of IL-6 led to increased tumor growth, EMT, and metastatic lesions in an in vivo mouse model [83]. This work concluded that IL-6 is exerting its effect on cells via the JAK/STAT3 pathway [83]. STAT3 has been identified not only as an inducer of Snail, but also as a direct transcriptional regulator of the EMT transcription factor Twist [86]. In particular to CRC, STAT3 is a direct regulator of another transcription factor associated with EMT, ZEB1. STAT3 upregulates ZEB1 which in turn results in a decrease in Ecad, a hallmark of EMT [87]. IL-6 is responsible for EMT in cancer cells and accordingly it has been found to induce development of $CD44^+$ cancer stem cell (CSC) populations that are usually associated with resistance to current therapies [88].

3.2.3 Wnt Signaling

Increased Wnt signaling is often found in cancer and this lipid-modified, soluble glycoprotein can either signal through the canonical β-catenin pathway or through non-canonical pathways, both of which share dishevelled (Dsh) as a cytoplasmic signal transducer [89]. The canonical pathway entails Wnt ligand binding to the Frizzled (Fzd)/lipoprotein receptor-related protein (LRP)-5/6 receptor complex; resultant Dsh phosphorylation leads to glycogen synthase kinase-3β (GSK3β) inhibition, leading to dissolution of the multiprotein complex that includes GSK3β, Axin, adenomatous polyposis coli (APC), and β-catenin. This complex is responsible for β-catenin sequestration and loss of this complex allows for β-catenin to accumulate in the cytoplasm and eventually translocate to the nucleus to participate in transcription in conjunction with the TCF/LEF family of transcription factors [90, 91]. β-catenin activity has been documented to result in a decrease in occludin protein and transcription factors such as *c-jun*, in addition to an indirect increase in proteases [92]. Furthermore, β-catenin, in conjunction with LEF-1, has been found responsible for fibronectin expression [93]. β-catenin signaling has also been documented to upregulate another effector of EMT, Slug [94]. Interestingly, Snail has also been documented to physically interact with β-catenin to induce expression of Wnt target genes, suggesting that EMT associated transcription factor interplay with signaling pathways can potentially increase invasiveness of cancer cells [95]. All of these components contribute to the invasive phenotype of transitioned mesenchymal cells.

Non-canonical pathways for Wnt signaling includes the planar cell polarity pathway (PCP) or the Wnt/Ca^{2+} pathway [89]. PCP signal transduction begins with a Wnt ligand binding Fzd resulting in phosphorylation of Dsh; once Dsh is phosphorylated in this case, Dishevelled associated activator of morphogenesis 1 (Daam1) is activated and can induce Rho-ROCK activity to mediate actin polymerization [89]. Not only does Dsh induce Daam1 activity, but also induces profilin and Rac-JNK activity to moderate the actin cytoskeleton and activation of JNK also results in AP-1 induction of target genes [89, 96]. Mediation of the cytoskeleton is crucial to the invasive behavior of cells that undergo EMT, as this type of reorganization is necessary for cell movement. One co-receptor for Fzd that has been identified as a player in this non-canonical pathway includes the receptor tyrosine kinase Ror2. It can act as a receptor for particular Wnt ligands, activating a β-catenin independent signaling pathway that includes Dsh, JNK, Src kinases, and eventually results in expression of vimentin and MMP-13 [97]. Studies in several cell lines have shown that the result of this activity is a gain of invasive behavior relating this pathway to EMT via Snail [97–101], however Ror2 involvement in EMT is still ambiguous as a later report indicated that Ror2 can suppress EMT [102]. The Wnt/Ca^{2+} pathway seems to be a modulator of both the canonical pathway and the PCP; activation of Fzd by Wnt ligands activates Dsh by G-proteins in this case [89]. Dsh activation via the G-proteins results in phosphodiesterase activity that inactivates PKG to inhibit Ca^{2+} release. Alternatively, Dsh can activate PLC, which leads to production of IP_3 and DAG, leading to Ca^{2+} release and PKC activity, respectively. The Ca^{2+} release by IP_3 can negatively influence β-catenin/TCF transcriptional activity [89]. The key players in this pathway are summarized in Fig. 6.

Wnt ligands and Fzd receptors have traditionally been classified by whether or not they activate canonical or non-canonical pathways, but this is not completely accurate as the ability of a certain Wnt or Fzd to activate a particular pathway is contextual [96]. The effects of the activation of the non-canonical pathways are still ambiguous. For example, while Wnt signaling is associated with EMT and cell invasion, Wnt5a can activate non-canonical signaling pathways through Dsh and phospholipase C (PLC) and result in increased intracellular calcium [103, 104]. Recent work indicates that Wnt5a acts to inhibit EMT and cellular invasiveness in colon cancer cells as well as in an in vivo model [104]. Analysis of patient samples revealed that most colon cancers have downregulated Wnt5a expression, but expression of Wnt5a correlated with better survival rates in patients [104]. Wnt5a expression results in Ecad expression and a downregulation of Twist and Zeb1, both EMT transcription factors, and interestingly, components of the canonical Wnt signaling pathway [104]. In vitro studies have shown that EMT can be induced by Wnt-1 and Wnt-5a because of the associated increase in Snail (an EMT associated

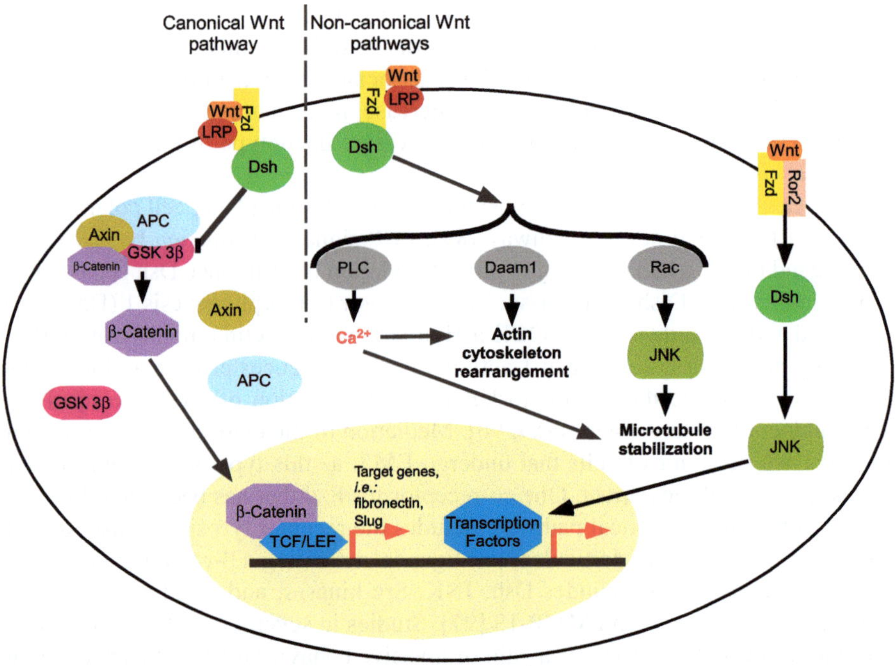

Fig. 6 Simplified summary of canonical and non-canonical Wnt signaling pathways. In the canonical pathway, Wnt ligand binding to Fzd/LRP complex results in phosphorylation of Dishevelled (*Dsh*), which results in inhibition of GSK3β. This inhibition leads to dissolution of the APC complex that keeps β-catenin contained. β-catenin is released and can accumulate in the nucleus to act in conjunction with TCF/LEF proteins and results in increased transcription of EMT-associated genes such as fibronectin and Slug. In cases of CRC, this pathway is often dysregulated. The non-canonical pathway can also lead to gene transcription, but some of the downstream events also include actin cytoskeletal rearrangement and microtubule stabilization, both of which are necessary for effective migration and invasion. The non-canonical pathway also operates through Dsh, but Daam1, PLC, Rac, or JNK may be activated

transcription factor) protein levels and Brachyury [105–107]. It has been purported that cancer cells initially function through the canonical Wnt pathway, but tumor associated cells induce a switch to the non-canonical pathways, leading to invasion and eventual metastasis [108].

The Wnt pathway has been heavily linked to CRC. Frequently, sporadic cases of CRC generate because of a mutation in APC (most common), Axin, or β-catenin, which in any case, causes an inability to negatively regulate the Wnt signaling pathway. Further, familial adenomatous polyposis provided evidence that loss of APC as a negative regulator of the Wnt pathway is a crucial factor in CRC development. Wnt signaling is required for maintenance of the normal intestinal crypt that includes intestinal stem cells; once this signaling pathway is turned on constitutively, the stem cell population rises and an accumulation of progenitor cells ensues, resulting in an adenoma [22]. This adenoma formation is an initial step in the progression to carcinoma and follows the idea that CRC is a consequence of stem cell overpopulation [30, 96]. In CRC, nuclear β-catenin was noted to be found primarily at the invasive front of the tumor, thus implicating Wnt signaling in EMT [109]. It is postulated that the cells at the invasive front of a colon tumor undergo EMT differentially because of their exposure to a separate microenvironment as compared to the cells central in the tumor [31]. Nonetheless, Wnt signaling is capable of inducing matrix proteases, reducing adhesion molecules, and increasing cell invasion and migration characteristic of EMT and crucial to the development of metastases [110].

The importance of Wnt signaling to EMT and CRC progression is demonstrated by the ability of soluble Wnt inhibitors to impede tumor growth and their ability to decrease EMT. Secreted Frizzled-related protein 3 (sFRP3) expression results in a decrease in tumorigenicity and a reversal of EMT [111]. The expression of this soluble Wnt inhibitor results in a decrease in the EMT transcription factors Slug and Twist [111]. Furthermore, there is a decrease in N-cadherin, vimentin, MMPs, and fibronectin and an increase in Ecad, along with a decrease in invasion and migration [111].

3.2.4 Hedgehog Signaling

The Hedgehog (Hh) family of proteins are secreted ligands that tend to act by autocrine and paracrine means and include the 3 ligands, Sonic (Shh), Indian (Ihh), and Desert (Dhh) [112]. Interaction of these ligands with the transmembrane protein Patched (PTCH) results in the release of Smoothened (SMO) inhibition and allows for hyperphosphorylation of this protein. This ultimately leads to transcription of target genes with the cooperation by the Gli family of zinc-finger transcription factors after activation by SMO [112, 113]. Shh has been implicated in cancer, as mutations in this pathway have been related to high rates of cellular proliferation and transformation and finally, cancer. The active Shh pathway has been associated with EMT and resultant metastasis; inhibition of Shh has been shown to inhibit cancer progression [112]. One piece of evidence that the Hh signaling pathway is responsible for determining epithelial plasticity comes from developmental biology studies. Overexpression of Shh during development of murine pancreata results in organs that have reduced epithelial cells and an increase in mesenchymal cells [113]. In a prostate cancer model, overexpression of the downstream transcription factor Gli1 results in rapid expression of the EMT associated transcription factor Snail and, accordingly, have lower Ecad expression [114, 115]. Furthermore, additional evidence has been presented that supports the idea that the Hh pathway is required and active in early and late stages of human CRC [116]. Importantly, the Hh signaling pathway was found to be required for CSC proliferation and renewal and therefore tumor progression [116]. Overall, Shh signaling can directly participate in EMT and has been implicated in mediating other EMT associated signaling pathways, such as TGF-β signaling, Wnt signaling, PI3K/Akt pathways, and Notch signaling [51, 117].

3.2.5 Notch Signaling

Notch signaling involvement in EMT is not as well understood as some of the other signaling pathways. Notch receptors are transmembrane receptors that have an extracellular domain with epidermal growth factor-like repeats while the intracellular domain (NIC) contains nuclear localization signals [118]. Cells harboring membrane bound ligands (Delta-like-1, -3, -4, and Jagged-1, -2) can interact with neighboring cells that express the membrane embedded Notch receptors (juxtacrine signaling) [118]. This interaction results in the cleavage of Notch by sequential proteolysis first by A Disintegrin and Metalloprotease-10 or -17 (ADAMs), followed by γ-secretase, leading to accumulation of NIC in the nucleus where it can bind to transcription factors (CBF1/RBPjk or nuclear factor-kappa B [NFκB]) and induce expression of its target genes, including those associated with EMT such as Snail, Akt, and NFκB (Fig. 7) [119, 120]. Overexpression of Notch-1 has been shown to increase mesenchymal markers, including vimentin and the EMT transcription factors Zeb1, Zeb2, and Slug, resulting in increased invasive behavior [121, 122]. Importantly, inhibition of the Notch signaling pathway results in a decrease in the ability of CSCs to proliferate and produce tumorspheres, indicating the crucial nature of Notch signaling in EMT and tumor progression [123, 124].

Normal colonic crypts express Notch and its signaling is necessary for maintenance of the progenitors of the intestine, however, Notch signaling pathway activity is found upregulated in primary CRC and corresponding liver metastases [125, 126]. From this it is purported that patients may benefit from inhibition of Notch signaling as a mechanism of preventing CRC progression [127]. γ-secretase inhibitors (GSIs), antibodies targeting Notch and other members of the pathway, and small peptides targeting the Notch pathway constituents are in preclinical and clinical development [for review, see [128]. Notch inhibition via γ-secretase inhibition results in an increase in KLF4, a tumor suppressor in CRC, and effectively impedes tumorigenicity [129, 130]. Also, in vitro and in vivo models using γ-secretase inhibitors demonstrate that Notch signaling is downregulated upon treatment and can reduce the CSC phenotype, including reducing CSC phenotype induced by chemotherapeutics in CRC, implying a role for Notch in EMT [131].

3.2.6 Receptor Tyrosine Kinase Signaling

Several receptor tyrosine kinase ligands have been associated with EMT, including EGF, FGF, PDGF, HGF, IGF-II, and TNF-α [16, 17]. Several ligand generated signaling pathways have been implicated in EMT by modulating expression of Zeb1, Zeb2, Snail1, and Twist [19]. With the exception of IGF-II, these ligands can induce morphology changes and cell scattering and these activities are associated with downstream effectors including Src family kinases and Ras [16]. IGF-II alternatively induces EMT by redistribution of Ecad and β-catenin from the cell membrane complexes where they are normally sequestered into the intracellular compartment. The loss of Ecad at the membrane reduces cell-cell association and the movement

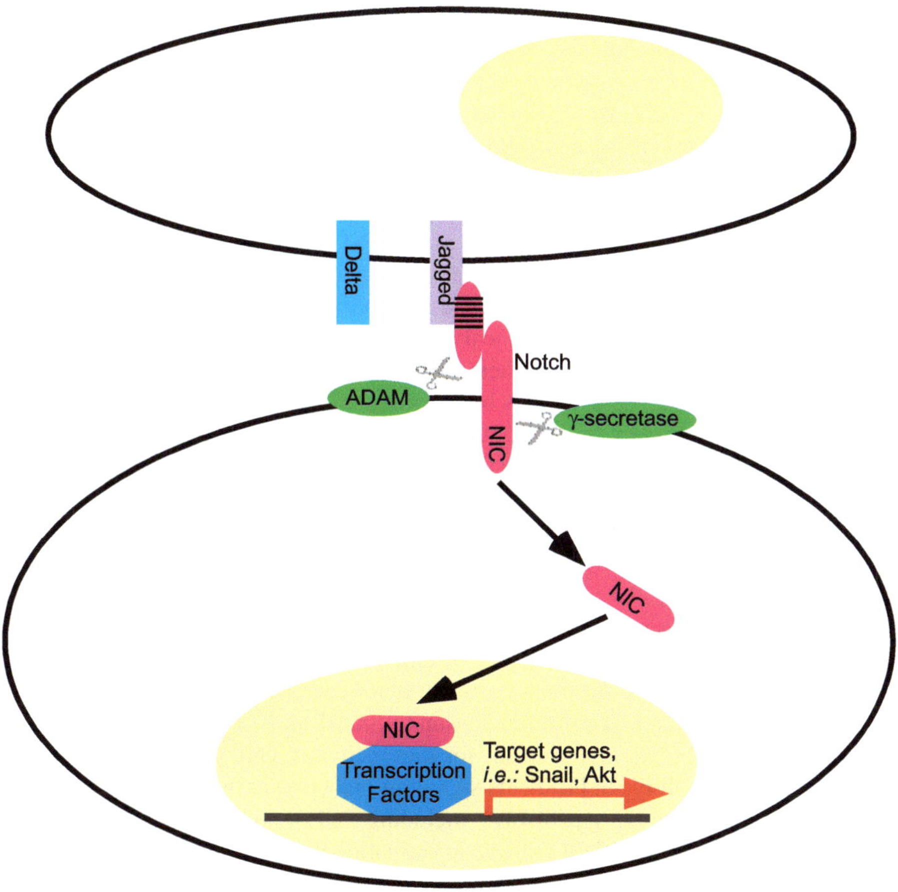

Fig. 7 Notch signaling induces EMT associated transcription. Notch signaling is unique in that the ligand and receptor are membrane bound; juxtaposition is required for ligand binding and activation of the receptor. Once a neighboring cell engages the Notch receptor, it is sequentially cleaved by A Disintegrin and Metalloproteinase (*ADAM*) and then by γ-secretase. Once the Notch membrane bound intermediate is cleaved by the secretase, the Notch intracellular fragmenet (*NIC*) translocates to the nucleus, as it contains several nuclear localization signals. This NIC functions as a transcriptional factor and its target genes include those related to EMT such as Snail and Akt

of β-catenin from the cell surface to the nucleus allows for β-catenin mediated transcription [132], which can account for transcription of EMT associated genes such as vimentin, fibronectin, and the transcriptional repressor, Slug [133].

EGF receptor (EGFR), a transmembrane receptor tyrosine kinase that can activate Ras and PI3K signaling pathways, is often dysregulated in cases of carcinoma, including CRC [134, 135]. Exogenous EGF treatment results in increased motility and invasiveness of cancer cells and several inhibitors for this pathway are clinically available and provide benefit to patients [136, 137]. EGF treatment has been linked to increased expression of extracellular proteases, including MMPs, that contribute to invasiveness [138]. Not only does EGF/EGFR signaling contribute in this way, but EGF-induced changes in morphology have been associated with changes in focal adhesion kinase phosphorylation status, which can influence a cell's ability to adhere to substrate and promote motility [139]. Furthermore, EGFR signaling results in decreased Ecad at the cell membrane by caveolin-1 mediated internalization of the junction protein and transcriptional downregulation of Ecad is accomplished by an increase in Snail after EGF treatment [140]. Crucially, constitutively active EGFR expression results in increased motility, probably due to the increase in invasion promoting gene expression such as MMP-1 and MMP-13 [141]. However, despite the impact of EGFR signaling in several cancer lines and the ability to detect EGFR in CRC cell lines and primary and metastatic lesions, no clear clinical correlations have been made with prognosis or metastasis in cases of CRC [142–144].

FGF, a heparin-binding growth factor that interacts with fibroblast growth factor receptors (FGFR-1, -2, -3, or -4) has been correlated with the grade of colon cancer tumor and high expression levels of FGFR-1 have been correlated with distant liver metastasis [145, 146]. Not only have these components of the FGF system been implicated, but FGF-binding protein (FGF-BP), a chaperone to FGF, has been found to increase in expression in tandem with CRC progression [147]. Ligand binding results in kinase activity and autophosphorylation that leads to recruitment of several factors including PLCγ for signaling through DAG and IP_3, as well as FRS2 or 3 resulting in Ras signaling [148]. Not only do these pathways get activated, but the PI3K/Akt pathway is activated, resulting in EMT mediated through Snail [148]. In CRC, silencing of the fibroblast growth factor receptor-4 (FGFR4) results in a decrease in EMT inducer TGF-β in addition to Snail1 and Twist [149].

PDGF is a strong chemoattractant for mesenchymal cell types and is also mitogenic for many cell types [150]. Its signaling has been implicated in influencing other EMT related signaling pathways, including TGF-β and Wnt [150]. Activation of the PDGF receptor (PDGFR) results in activation of PI3K, the Ras/Raf pathway, and the JAK/STAT pathway, all of which may induce components or effectors of EMT. PDGFR has been noted to be differentially regulated during the process of EMT, being upregulated in the mesenchymal cell type and causing cell dispersion and migration [151]. Accordingly, the EMT related transcription factor Twist1 was found to induce expression of PDGFR and this is required for invadopodia formation, which is a critical determinant of cell motility. Not only does the expression of PDGFR induce EMT behavior in vitro, Twist1-induced metastasis requires PDGFR in vivo [152]. In particular to CRC, PDGFR expression correlates with TGF-β signaling and EMT and ultimately reduced patients disease-free survival and overall survival [153]. In vivo work indicates that knockdown of PDGFR in the C26 cell line results in decreased liver metastasis in a syngeneic mouse model [153].

HGF, also known as scatter factor (SF), was originally identified in the 1980s and is now known to be a glycosylated heterodimer that serves as a ligand for the Met receptor. This factor was initially noted to induce cell scattering and eventual loss of desmosomes [154]. HGF is the only known ligand for the Met receptor tyrosine kinase [155]. HGF is usually produced by fibroblasts and acts in a paracrine manner on cells that express the Met receptor [155]. The interaction of HGF receptor is unique and a detailed review of structure, signaling, and regulation may be found here [155]. Overall, activation of Met by HGF induces proliferation and cell scattering, dependent upon the context in which it is activated. The effects of Met activation can be mediated by several pathways, including those implicated in EMT, especially PI3K/Akt pathways and Ras/Raf pathways [155]. Met receptor was found to be overexpressed in CRC and importantly was later found to be a target of the Wnt pathway, a crucial pathway that is frequently dysregulated in CRC cases [156–158]. Furthermore, HGF treatment of CRC cell lines HCT116 and COLO741 induces increases in β-catenin protein, a critical member of the Wnt pathway; taken together these data indicate that HGF and Wnt signaling pathway can form a feedback loop that ultimately contributes to EMT and poor patient prognosis [159].

TNF-α is cytokine that is normally produced by macrophages, but it is also expressed by tumor cells [160]. TNF-α is found in high levels in CRC samples and level of expression correlates to CRC stage and the likelihood of recurrence in those patients with lymph node metastases [160–162]. CRC patients demonstrate a higher level of circulating TNF-α and this correlates with progressive disease [163]. One mechanism by which this cytokine may play a role in EMT is in sensitizing cells to TGF-β influence via $p38^{MAPK}$ [164]. TNF-α expression also upregulates NF-κB signaling via activation of the IKK complex, which is responsible for phosphorylating the inhibitor of NF-κB [165]. Once the p65 subunit of NFκB is available, it translocates to the nucleus for transcription of the EMT effector, Twist 1, and results in EMT [165]. Considering that CRC development is heavily linked to inflammation and TNF-α is a major cytokine related to an inflammatory response, it is interesting to note that lipopolysaccharide treated animals incur increased rates of metastasis when challenged with a tail-vein injection of breast cancer cells [165]. Furthermore, TNF-α mediated Akt/GSK3β activity is required for the stabilization of another EMT effector, namely Snail, which contributes to the characteristic changeover to a mesenchymal phenotype in CRC cell lines [166].

The signaling pathways that control EMT are dynamic, overlapping, and therefore complex. For example, the TGF-β and Wnt pathways have been documented to depend to some extent on one another [167–169]. TGF-β has also been found to be synergistic with Ras signaling, involving both MAPK and PI3K/Akt pathways [47, 170]. The Snail protein has been associated with expression of Wnt5a and Ror2, activating the non-canonical signaling pathway of the Wnt system and has also been found to be affected by FGFR signaling [97, 148]. The overview presented here is a synopsis of these pathways and further work is needed to elucidate these critical pathways, as it is currently unclear the relationship between all of them [for a review on the EMT network, see [103]. Clarification of the signals that induce and regulate EMT will be valuable for identifying novel strategies and targets for cancer treatment, especially later stage cancers, including CRC.

3.3 EMT-Associated Transcription Factors

The effectors of EMT are transcription factors induced by EMT associated signaling pathways. These transcription factors include Snail (Snail1), Slug (Snail2), Smuc (Snail3), Zeb1 and Zeb2 (SIP1), Twist1 and Twist2, E12/E47 (TCF3), E2–2

(TCF4), Goosecoid, FOXC2, KLF8, Brachyury (Tbx1), Tbx2, Tbx3, Lbx1, Six1, and Prrx1. They each influence the transition by repressing or activating genes responsible for maintenance of the epithelial phenotype or induction of the mesenchymal phenotype. These transcription factors are critical to EMT, as overexpression of transcriptional repressors induces mesenchymal phenotypes in epithelial cells and, importantly, increase the migratory and invasive ability of those cells [12].

Snail family members are zinc-finger type transcription factors with similar structure and bind the consensus sequence CAGGTG [171]. This family of transcription factors has been found to recruit histone deacetylaces (HDACs) to the Ecad promoter, stunting the expression of Ecad [172]. Snail1 has been associated with the repression of several cell adhesion molecules, including Ecad, claudins, cytokeratin-18, desmoplakin, and occludins and its overexpression has been related to dedifferentiated carcinomas, as opposed to non-invasive ones and has been postulated to be largely responsible for EMT initiation in response to certain signals [171, 173–177]. Additionally, Snail correlates to increases in MMP-2, RhoA, and Notch, other factors that play a role in invasive behaviors of cells [178]. Snail overexpression has been documented in several types of cancer, including colon cancer; Snail not only downregulates Ecad but also contributes to drug resistance in colon cancer by downregulating the vitamin D receptor (VDR) in co-operation with Slug (Snail2) (review [179]) [180–182]. Not only does the expression of Snail induce EMT in CRC, but also causes acquisition of a stem cell-like phenotype and the associated chemotherapeutic resistance [183].

The Snail related zinc-finger protein, Slug, was identified to be responsible for desmosome disassembly and induction of cell spreading after a separate study identifying Slug in normal cell migration during neural crest development in the chicken embryo, in addition to Ecad reduction was reported [184–186]. The presence of Slug in colorectal primary tumors tends to be associated with poor prognosis, however not necessarily because of its Ecad repression [187–189]. Slug is also associated with increased expression of the EMT associated transcription factor Zeb1, promoting EMT even further [190]. The final member of the Snail family, Smuc (Snail3), functions in a similar fashion to its family members. It is capable of binding similar consensus sequences and has been found in embryonic tissue, but has not been shown to be required for proper development [191, 192]. More research will elucidate whether this Snail family member is critical to the EMT process, but recent evidence indicates that nuclear Smuc expression correlates with squamous cell carcinoma metastasis [193].

Zeb1 and Zeb2 are highly homologous zinc finger/homeodomain proteins that play a critical role in EMT that bind the consensus sequence CACCT(G) [194, 195]. Like the Snail family of transcription factors, these proteins have been found to recruit HDACs and mediate repression of genes, including Ecad [172]. Zeb1 has been found to be expressed in breast cancer, amongst others, and is associated with repression of several epithelial genes, inducing a mesenchymal phenotype; knockdown of Zeb1 correlates with re-epithelialization and reduction in migratory behavior [196, 197]. Zeb2 was identified as a repressor of Ecad and inducer of a mesenchymal phenotype, even in cell types that express other transcriptional repressors such as Snail [49]. Similar to Zeb1, overexpression of Zeb2 induces loss of epithelial markers, such as Ecad and α- and β-catenin [195]. These transcription factors have been associated with repression of several cadherins, occludins, connexins, and desmosomes, similar to the Snail family [49, 195–197]. Mesenchymal markers are also upregulated by these particular factors, including vimentin [198]. Functionally, knocking down Zeb2 results in decreased cellular migration and invasion in vitro [199].

Twist is another transcription factor that was first identified as an embryonic patterning gene in *Drosophila* and was later found to play a role in EMT and cancer metastasis by reducing epithelial markers and increasing mesenchymal markers such as vimentin and N-cadherin, amongst others [5, 200]. Twist1 and Twist2 are basic-helix-loop-helix (bHLH) proteins that are implicated in EMT by their ability to repress epithelial genes and activation of mesenchymal genes (i.e. N-cadherin, fibronectin, and MMP-2 and -14) [19, 201–203]. Twist has been documented to contribute to EMT by inducing the expression of Snail, perpetuating the transition, as Twist1 can induce Slug within hours and Snail1 days later [204, 205].

Like the other transcriptional repressors, the bHLH protein E12/E47 (TCF3) has been shown to repress Ecad and overexpression of E12/E47 in the epithelial MDCK cell line results in a mesenchymal phenotype, with expression of vimentin and fibronectin [206]. While E12/E47 does accomplish some of the same functions as the other EMT associated transcription factors, it has been shown that there is a unique gene profile induced by E12/E47 that translates to different in vivo behavior of E12/E47 expressing cells as compared to Snail expressing cells [207, 208]. The related E2-2 transcription factors (E2-2A, E2-2B) control a host of EMT related genes similar to E12/E47, but does influence several genes in a distinct fashion as opposed to E12/E47 [209].

Goosecoid is a homeobox factor that can influence EMT by reducing Ecad, amongst other epithelial markers, and activate mesenchymal markers, similar to the several other families of EMT transcription factors [210]. Interestingly, FOXC2 has been implicated in decreased Ecad at the membrane by redirecting nascent Ecad to the cytoplasm instead of the plasma membrane, as opposed to transcriptional repression and has been documented to be induced by other EMT associated transcription factors such as Twist, Snail, and Goosecoid [211]. Krüppel-like factor 8 (KLF8) is a transcription factor that includes a conserved zinc finger domain that binds to the consensus site CACCC and acts as a repressor by its interaction with C-terminal binding protein (CtBP) co-repressors [212]. It has been demonstrated to suppress Ecad production and increase mesenchymal markers and matrix proteases, in addition to reducing expression of KLF4, which may act as a tumor suppressor in CRC [213–216].

Brachyury (Tbx1) is a member of the T-box gene family responsible for repression of Ecad in concert with Slug that ultimately leads to a mesenchymal phenotype [217]. It has been reported that Brachyury expression correlates with poor prognosis of CRC and loss of Brachyury reduces the stemness marker Nanog that is associated with cancer stem cells [218, 219]. Other members of the T-box family have been associated with transition from epithelial phenotype to mesenchymal. Tbx2 has been studied in a breast cancer model and shown to be induced upon TGF-β treatment and induces a mesenchymal phenotype in vitro; conversely, when Tbx2 is knocked down, the metastatic capacity of MDA-MB-435 cells were abrogated [220]. Tbx3 is yet another member of this family of transcription factors associated with EMT. It functions to reduce Ecad and contribute to metastatic potential, in addition to its ability to regulate the expression of Tbx2 [221].

Originally identified in murine embryos' developing nervous system and muscle in 1995, Ladybird homeobox 1 (Lbx1) is a highly conserved homeobox transcription factor [222]. Not only is this transcription factor required for normal muscle development in the embryo, it has been associated with initiation of EMT in breast cancer [223]. It can impart the same effects regarding epithelial and mesenchymal markers that other EMT associated transcription factors can, but it has also been show to be a direct regulator of Zeb1, Zeb2, Snail1, and TGF-β [223]. There is a correlation between Lbx1 expression and poor prognosis in the case of breast cancer [223].

Sine oculis-related homeobox 1 homolog, or Six1, has been implicated in EMT as well. It normally serves as a master regulator in early embryogenesis, controlling normal invasion and migration during development [224]. Its expression in has been found to be upregulated in several types of cancers and specifically capable of inducing expression of mesenchymal markers, loss of epithelial markers, Wnt target genes, and certain stem-like characteristics [224]. It is purported that Six1 induces TGF-β signaling and is a key regulator of the switch from the anti-tumoral to the pro-tumoral effects of the TGF-β signaling pathway by induction of TGF-β type I receptor (TβRI) induction [225, 226]. Recent evidence indicates the related transcription factor Six2 is overexpressed in nephroblastomas and its overexpression leads to increased proliferation and migration of these cancer cells [227].

Paired related homeobox 1 (PRRX1) has been reported to be involved in EMT in colon cancer, inducing invasion and mesenchymal markers, with a decrease in Ecad [228]. Despite the association of PRRX1 overexpression with poor prognosis in colorectal cancer, a separate report concluded that loss of PRRX1 results in MET in a breast cancer model [229]. Further studies will be necessary to define the function of PRRX1 in EMT-MET in separate cancer types.

Several of these transcription factors have been directly linked to CRC [19]. Twist1 is upregulated in colon cancer and is associated with lymph node metastases [230]. Interestingly, in the correlation study, Twist1 overexpression is higher in men as compared to women [230]. Snail and Slug are both overexpressed in human colon cancers and correlate with reduced Ecad and poor prognosis [180, 188]. Zeb1 expression was found highly upregulated in the tumor-host interface area, which is in accordance with EMT in that this invasive front would express more EMT associated factors as compared to the central tumor region and normal mucosa [196]. Furthermore, this expression is related to Wnt signaling pathway activity, as CRC that has intact APC does not overexpress Zeb1 [231]. This is significant in that Wnt signaling is frequently aberrant in CRC due to *APC* mutations. Zeb2 expression follows a similar pattern in both primary and metastatic CRC lesions and the localization of Zeb2 to the invasive edges of the tumor serves as a negative prognostic marker for patients [199]. Six1 is associated with poor survival rates in CRC patients and at the cellular level promotes EMT in the SW480 CRC cell line [232]. Interestingly, while the expression of E12/47 has been correlated to metastasis, a study recently revealed that E12/47 is actually suppressive in the case of colon cancer cell lines but its effects may be marginal in the context of several other EMT encouraging pathways [233]. Brachyury expression has also been correlated to CRC staging and can provide an idea of prognosis [218].

3.4 microRNAs and EMT

MicroRNAs (miRs) are small, non-coding RNAs that primarily target the 5′ end or 3′ UTR of various genes at their miRNA recognition elements (MREs) to silence or signal for degradation of the mRNA [234]. They are generally 22 nucleotides in length and are derived from pri-miRs, which are longer transcribed RNAs that code for several miRs (or a miR cluster) [234]. The miRNAome has been repeatedly documented to play a major role in various steps of cancer progression [19, 235]. While they have been demonstrated to invoke effects upon apoptosis and angiogenesis, amongst others, this section will address those that are relevant to EMT.

MiRs that regulate EMT in both normal embryonic tissues and cancerous tissues are collectively termed "metastamiRs" [235]. For example, the miR-200 family members (miR-200a, -200b, -200c, -141, -429) have been implicated in EMT by their relationship with the Zeb transcription factors; there is a negative feedback loop in which these miRs can repress the expression of Zeb factors to maintain the epithelial state and Zeb transcription factors, upon expression, can repress miR expression [19, 234, 236–238]. The TGF-β and Zeb1 systems are the primary targets of the miR-200 family [21]. Snail1 expression is inhibited by miR-9, let-7d, miR-29b, miRNA-30a, miRNA-30b, and miR-34 in normal cells, while cancer cells tend to be reduced in their expression of these miRs leading to the overexpression of Snail1 [239–243]. Slug also represses miR-200 family members, while miR-200 can inhibit Slug expression, demonstrating negative feedback loops similar to the Zeb/miR-200 family relationship [244]. Epigenetic silencing of miR-203 has been specifically correlated with an upregulation of Slug and miR-1 and miR-200b expression have been shown to prevent EMT via Slug reductions [245]. Twist1 is another transcriptional repressor controlled by let-7d and miRNA-29b, but it is also regulated by miR-580 and miR-214, however it is not clear whether or not it is a direct relationship between miR-214 and Twist1 downregulation [240, 241, 246, 247]. Twist1 directly induces miR-10b, which results in induction of RhoC and that can stimulate invasion, migration, and ultimately metastasis in vivo; further, silencing of the miR-10b via "antagomir" reduces metastasis in a mammary tumor model [248, 249]. The high expression of miR-10b correlates with poor breast cancer patient prognosis [248]. Six1 induces expression of the miR-106b-25 cluster of miRs and this miR cluster (particularly miR-106b and mir-93) is responsible for inhibition of Smad7, which results in increased TGF-β signaling [250]. Ultimately, this leads to the classic changes in phenotype that include the reduction of Ecad and cell membrane associated β-catenin and loss of epithelial functional properties, such as matrix adhesion [250]. Like the EMT associated signaling pathways, the miRs form a complex network of communication regarding the expression levels of EMT associated effectors (and MET associated effectors).

In addition to targeting EMT associated transcription factors, miRs have been found to directly target other key constituents that have been implicated in cell polarity and EMT. miR-9 has been demonstrated to directly target Ecad, reducing the amount of Ecad at the cell surface, releasing β-catenin, and leading to transcription of its target genes and inducing a mesenchymal-like phenotype in certain cell lines [251]. mir-29b targets Snail and Twist, but also induces an increase in Ecad and a decrease in N-cad and MMP-2; loss of this particular miR has been associated with prostate cancer and metastatic potential [241]. miR-34a is a direct repressor of CD44 and loss of miR-34a results in a cancer stem-like phenotype and the metastatic ability of prostate cancer is postulated to be mediated via the control of CD44 expression by this particular miR [252]. Direct targeting of vimentin, a protein implicated in the mesenchymal phenotype, is accomplished by miR-138; reduced expression of this miR has been associated with certain cancer cell types [253]. Not only does miR-138 directly target vimentin, it is also a regulator of Zeb2 and RhoC. The loss of miR-138 results in increased expression of these proteins and contributes to EMT and cancer cell invasion by inducing changes in the actin cytoskeleton, respectively [253]. Likewise, the loss of miR-194 results in a loss of N-cadherin control, leading to increased expression of the mesenchymal marker and increased invasive behavior in gastric cancers [254]. miR-155 expression is responsive to TGF-β signaling and its expression causes a dissolution of cell junctions and an increase in cell migration and invasion by targeting RhoA mRNA and stopping translation of this prototypical Rho GTPase [255]. miR-143 and miR-145 have been implicated in controlling EMT in a prostate model. The expression of either of these miRs induces Ecad expression and reduces mesenchymal markers such as fibronectin and vimentin [256]. Importantly, introduction of miR-143 and -145 to PC-3 cells reduced their ability to invade bone in an intratibial mouse model and negatively correlates to bone metastasis in clinical sampling [256].

miR-204/211 is downregulated in tumor tissues and its loss has been associated with an increase in TβRII expression (which induces EMT transcription factor expression, including Slug) and a decrease in claudins (leading to tight junction loss), likely by transcriptional repression by those upregulated transcription factors [257]. miRs -373 and -520c both directly target CD44 and their expression has been correlated to a migratory and invasive phenotype in vitro and an increase in metastatic potential of non-metastatic cell in vivo [258]. miR-491–5p is a miR that has been associated with maintenance of epithelial polarity. It is upregulated upon TGF-β stimulation and serves to target the protein Par3, which is a crucial modulator of tight junctions and maintenance of apical-basal polarity [259]. miR-661 is required for cell invasion and migration and the mechanism behind this necessity was determined to be caused by its direct targeting of StarD10 and Nectin-1, proteins that are involved in cell polarity and cell-cell adhesion [260].

There are a host of EMT suppressive and promotional miRs [for recent reviews, [see 234, 261], but several have been specifically linked to colorectal cancer progression. Functional and profiling studies have been accumulating regarding perturbations in the global expression of miRs in cancer, including CRC. Several miRs have been correlated with CRC pathogenesis, as the targets of these miRs include several signaling pathways and tumor suppressors, including the Wnt signaling pathway and p53 [262]. Profile analysis of 16 CRC cell lines identified miR-31, miR-96, miR-133b, miR-135b, miR-145, and miR-183 as common miRs amongst the cell lines that are significantly dysregulated [263]. miR-133b and miR-145 were downregulated, while the others were upregulated; miR-31 levels were able to be correlated to pathological stage of cancer patient tissues [263]. An earlier profiling report also identified miR-145 as downregulated, in addition to miR-143 [264]. Expression of miR -135a and -135b are associated with a reduction in the adenomatous polyposis coli (*APC*) gene product, leading to an induction of β-catenin signaling and potentially pro-EMT transcription [265].

Another assay of the changes in miR levels in clinical CRC samples found miR-31 significantly upregulated, along with the miR-17–92 cluster including miR-17–5p, miR-18a, miR-20a, and miR-92 [266]. This work also demonstrated agreement with previous reports of lower levels of miR-143 and miR-145 in cancerous tissues, as compared to non-tumor tissues [266]. Interestingly, an assay of the embryonic microRNAome in comparison to the microRNAome of colonic carcinoma identified a partially shared expression profile, with the miR-17–92 cluster being identified in both samples [267].

In a LIM-1863 model of colon carcinoma, induction of EMT by use of a combination of TGF-β and TNF-α treatment resulted in a marked increase in both miR-21 and miR-31 at the transcriptional and processing level [268]. This effectively propagates the EMT response in the CRC line, inducing a motile phenotype, as miR-21 and -31 induce expression of mesenchymal genes such as fibronectin, laminin-γ-2, and MMP-7, in addition to re-distributing Ecad from the surface to the cytoplasm, which is a primary characteristic of EMT [268]. Despite what appears to be a role in promoting EMT and considering EMT's crucial role in the metastatic process, circulating levels of miR-21 in patient serum is not yet a clearly defined prognostic factor [269, 270].

miR-451 and -338–3p have been associated with epithelial polarity in colon cancer cell line models. The T84 colon cancer cell line can differentiate to a more epithelial phenotype, including tight junction formation and basolateral polarity potentially by facilitating integrin localization [271]. Further studies indicated that miR-338–3p was specifically associated with migration and invasion and ultimately, CRC clinical aggressiveness. One target of miR-338–3p was shown to be smoothened (SMO), which is a key element of the Hedgehog (Hh) signaling pathway, a pathway that has been previously implicated in EMT [272]. Appropriately, reduced levels of miR-338–3p have been associated with advanced stages of CRC [272].

Overall, there are several aspects of EMT that can be controlled by miR activity. The signaling pathways, transcription factors, and components of cellular architecture are all targets of specific miRs. Interference with normal miR balance can lead to EMT and potentially CSCs, which ultimately leads to cancer progression. Recent years have seen an increase in interest in the microRNAome as a target for inhibiting cancer progression. The work that has been accomplished so far in defining particular miR signatures may identify cellular programs to target as well as provide an opportunity to develop novel treatment technologies and therapies, ones that may include directly targeting miRs.

3.5 Other Epigenetic Mechanisms of EMT Control

Epigenetic mechanisms outside of miRs and certain EMT transcriptional repressors also control the transition to a mesenchymal phenotype. Several populations of breast epithelial cells, including stem-like and carcinoma cells, have been analyzed for a cadre of DNA methylation patterns involved in defining breast epithelial epigenetic programs. Particular DNA methylation patterns were associated with individual phenotypes [273]. The Ecad promoter has been shown to be hypermethylated in some tumor cells, which would reduce the transcription of Ecad and result in the characteristic loss of this protein that contributes to EMT [274]. It is worth noting that in diffuse gastric carcinomas, hypermethylation of the Ecad promoter is prevalent [275, 276]. Treatment of cell lines with 5′-aza-2′-deoxycytidine induces demethylation and results in an increase in Ecad production and a switch back to epithelial behavior [172]. Other potential regulators of EMT in CRC are the sFRP proteins, which serve to control Wnt signaling. It has been found that expression of sFRP1 can be reduced in CRC and analysis of patient samples indicates that the promoter of sFRP1 is frequently hypermethylated and sFRP1 transcripts are reduced [277]. The authors propose that this particular epigenetic control cannot necessarily induce EMT, but those cells that have acquired hypermethylation of the sFRP promoter may be more sensitive to Wnt signals and EMT, especially in cases of CRC [278].

Hepatocyte nuclear factor-3 (HNF3) has been associated with enhanced Ecad expression and HNFs are known coactivators for histone acetylation which relaxes chromatin to allow for transcription and also serves as a recruitment factor for other transcriptional machinery [172, 279]. Not only does acetylation potentially play a role in EMT, histone methylation has been documented to play a role in Ecad suppression. Enhancer of zeste homolog 2 (EZH2), a component of the histone methylation complex PRC2, overexpression has been found in several cancers and is known to trimethylate histone 3 at lysine 27 (H3K27), reduce the expression of Ecad, and induce an invasive phenotype [280]. Furthermore, expression of histone demethylases such as lysine-specific demethylase 1 (LSD1) can result in inhibition of EMT phenotype and metastatic potential [172].

4 EMT and Cancer Stem Cells

Tumor heterogeneity has long been a bane to the identification, classification, and treatment of cancer [281]. It has become clear that there is a population of cells that can replicate themselves but also give rise to the diverse cell types found in most tumors and have the ability to participate in establishing tumors that are similar to those of its origin, known as cancer stem-like cells (CSCs) [282]. Albini et al. clarify the definition by raising the points that the staminality of CSCs is not the phenotype of the cell, that these cells are best observed in vivo, and that they are malignant and argue that a CSC may be a transformed normal stem cell or a cancer cell that dedifferentiates to a stem-like state [282]. Originally, CSCs were identified in the context of acute myeloid leukemia (AML) and multiple myelomas [283, 284]. Mammary stem cell lines were identified in 1978 from solid rat tumors, leading the authors to postulate that since these cell types had never been identified before, it was likely these were a rare set of pluripotent cells [285]. Al-Hajj et al. identified a subpopulation of tumorigenic cells from solid, human breast cancers grown in mice that were characterized by a high expression of CD44 and low expression of CD24, both cell surface glycoproteins [286]. These cells were able to generate a diverse lineage of cells and had self-renewal potential, reminiscent of normal stem cells [286]. CSCs are often characterized by their specific cell surface markers and their ability to be cultured in low serum without attachment to tissue culture dishes [287]. Relatedly, they form tumorspheres in these low adhesion conditions and tend to be resistant to therapies and can be highly migratory [287, 288].

Current theories regarding cancer progression have evolved to include the definition of a stem-like cell that has self-renewal capacity reliant on signaling pathways normally associated with development, such as the Wnt/β-catenin pathway [287]. Not only are CSCs able to self-renew, they also establish the heterogenous cell lineages found in most solid tumors and are thus thought to be responsible for the sustaining the tumor, in addition to their ability to seed secondary sites. The capacity of these cell-types to be invasive and their drug resistance potentially explain why there are no cancer treatments that effectively eradicate solid tumors [283, 287]. Specifically, a current aspect of this theory proposes that EMT is the origin for CSCs, as these cell types express similar signaling networks to cells that go through EMT [282, 289]. Stem cell markers were able to be induced by overexpression of EMT associated transcription factors or exposing cells to TGF-β [12, 290, 291]. The EMT program induces invasive behavior and ultimately the ability to seed distant sites and CSC properties have been documented in circulating cancer cells in the case of pancreatic cancer [292, 293]. Currently, circulating tumor cells (CTCs) have been identified in CRC patients, but the molecular background of these cells have not been studied in depth [294]. However, a report identified plastin-3 (PLS3), an actin remodeling protein, as a marker for CTCs in CRC patients [295]. PLS3 overexpression can be induced by TGF-β treatment and was found to correlate to reduced overall survival when detected in peripheral blood of CRC patients [295]. It remains to be seen whether the other CTC markers can be identified and correlated to stemness, metastases, and overall patient survival.

In addition to the transcription factors already discussed, several other factors that are crucial for normal stem cells and developmental processes are likely to be involved in the maintenance of the CSC phenotype, including Oct4, Sox2, Klf4, c-Myc, and Nanog [296]. Oct4 is an embryonic transcription factor that was identified to be re-expressed in a variety of cancers [297]. It belongs to a class of transcription factors in the Pit-Oct-Unc transcription factor family and it notorious for being reserved for expression in pluripotent cell types [298]. Oct4 is responsible for expression of several targets and is usually exclusively expressed in embryonic stem cells and is responsible for maintaining pluripotency [299]. Oct4 is responsible for inducing the expression of Sox2, Oct4 itself, and creatine kinase B, amongst others [298]. Sox2 is a transcription factor that is a member of the SRY (sex determining region Y)-related HMB box gene family that is found overexpressed in several cancer types and is overexpression is sufficient to induce a stem-like phenotype in cancer cells; it has been suggested to be used as a marker of CSCs, as opposed to the traditional surface markers such as CD44 [300–302]. Klf4 is a zinc-finger transcription factor that is generally expressed in differentiated cells of the intestines, lungs, and testes, but has been found to be required for stemness maintenance in conjunction with c-myc and Oct4 [303]. As a note, Klf4's function as either an oncogene or tumor suppressor is highly dependent upon context, as Klf4 can be a tumor suppressor and induce expression of p21 or can be oncogenic and contribute to disease progression by suppression of p53 [304]. Loss of Klf4 expression in mammary epithelial cells leads to an increase in EMT markers and metastasis in vivo [303]. In the case of CRC however, Klf4 has been show to act in an anti-tumoral fashion [213]. C-myc is a bHLH/LZ (leucine zipper) transcription factor that is required for stem cell maintenance and is upregulated in response to pathways associated with EMT, such as Wnt and Shh. Overexpression of c-myc, along with Oct4, Sox2, and Klf4, results in transformation of non-tumorigenic MCF-10A cells into tumorigenic CSCs, however in the case of colorectal CSCs (CCSCs) c-myc has not been documented to be upregulated [296, 305]. Nanog is another known stemness associated gene that has been found in several cancer types. It has been identified to be upregulated in the CCSC and interference with Nanog expression in these cells reduced their tumorigenicity, including having reduced invasiveness [305]. β-catenin can control the expression of Nanog and this correlates with CRC considering the crucial role that dysregulation of the Wnt pathway plays in CRC development and progression [306]. Further, Nanog was identified as a marker for a subpopulation of CRC cells with a stem-like phenotype and correlates to poor prognosis [306, 307]. Nanog expression is increased upon TGF-β stimulation and induces an increase in invasion and migration, functional hallmarks of EMT [307].

Colorectal CSCs (CCSCs) have been identified by several groups based on markers including CD133, CD24, CD29, CD44, CD166 (Activated Leukocyte Cell Adhesion Molecule [ALCAM]), leucine-rich repeat-containing G-protein coupled receptor 4 (Lgr5), and aldehyde dehydrogenase 1 (ALDH1) [20, 30, 308–311]. However, the use of these particular markers are not always definitive, especially if considered singly rather than in conjunction with one another and the fact that these markers are expressed in certain regions of normal colonic mucosa make it significantly difficult to assign these markers solely to CCSCs [20, 30, 288, 312]. Cancer stem-like populations of cells have been identified from cell lines, such as HT-29 [288]. HT-29 spheroid cells were more aggressive than adherent HT-29 in that they were insensitive to chemotherapeutics and increasingly migratory, making them a useful tool for studying the metastatic potential of colon CSCs [288]. CCSCs derived from cell lines HT-29 and HCT-116 are more tumorigenic and metastatic in vivo [313]. The derived CCSCs express decreased Ecad and increased mesenchymal markers, including vimentin, Slug, and Snail, in addition to β-catenin and functionally speaking had reduced adhesion to extracellular matrix components [313]. Lgr5 silencing in colon cancer cell lines results in decreased migration of these cells (SW480 and HT-29) [314]. Relatedly, human CRC samples that demonstrate high expression of stem cell markers, including Lgr5, is predictive of patient relapse and shortened disease-free survival [315].

5 Reversibility of EMT: Mesenchymal to Epithelial Transition (MET)

EMT, and its reverse process, mesenchymal-epithelial transition (MET) play essential roles in normal physiological events including embryogenesis, organ development, and tissue regeneration. However, these events have also been identified as contributing to cancer progression and metastasis [7]. CRC is the third most common cancer in the world, but it is of note that most deaths resulting from CRC are due to metastatic lesions [316]. In 50–60 % of patients, CRC cells metastasize to the liver and in 10–30 % of patients CRC cells metastasize to the lungs [317]. Tumor initiation at a secondary site is generally the rate-limiting step in metastasis which is directed by cell intrinsic factors as well as those from the distal site's microenvironment [318].

In cancer, the metastatic process is thought to consist of the following steps: (I) invasion; (II) intravasation; (III) systemic transport; (IV) extravasation; (V) colonization [7, 319]. While EMT provides an explanation for the first four steps of the metastatic process, MET provides insight for the final step, *i.e.* the colonization required in order to establish a secondary tumor. To form metastases at distant sites from the primary tumor, an epithelial phenotype may be necessary for the metastasizing cells to become established and to form the appropriate microenvironment to support cell proliferation.

Maintaining cellular plasticity in order for cancer cells to undergo EMT and then MET in the appropriate environment is crucial for successful metastasis [292]. While EMT is important to tumor progression, it is not a unidirectional phenomenon, however this concept has thus far been inconsistent with observations that secondary tumors that are derived from migratory cells of the primary tumor share epithelial morphology and some gene expression patterns with the primary tumor. Occurrence of MET is most frequently described by re-expression of Ecad and relocalization of β-catenin to the cell membrane, as opposed to the cytoplasm and nucleus [15]. Other common characteristics of MET include changes in cell-cell junctions, actin cytoskeletal rearrangement, downregulation of the transcription factors Snail1 and Zeb2, coupled with the reversal of mesenchymal cell markers such as vimentin and fibronectin [320].

Mesenchymal characteristics are suggested to be non-essential, if not undesirable, for establishment of a secondary tumor. These cells do not adhere well to each other or surrounding tissues and their proliferation is suppressed [321]. Reversion to an epithelial-like state is thus favorable in order for a new solid tumor mass to become established. Once cells are at a secondary site, the re-expression of Ecad functions to initiate canonical signaling pathways which promote cell survival [322]. Although occasionally described as simply a direct reversal of EMT, MET is far more complex, dynamic, and obscure. Both intrinsic and extrinsic signals contribute to the process, as evidenced by in vitro and in vivo experiments which indicate mesenchymal-type cells that normally have repressed Ecad can be induced to re-express Ecad when in the proper microenvironment [323]. Indeed, distal organs typically provide a microenvironment quite different than that provided in the primary site. The ability of a disseminated cancer cell to adapt to the new signaling molecules, oxygen tension, extracellular milieu, etc. in order to survive, proliferate, and avoid provoking an immune response would be hugely beneficial [14, 322].

Cadherin expression in cancer progression is an important aspect of EMT and MET. Tumor cells with loss of Ecad expression lose their epithelial phenotype during EMT and gain N-cadherin expression, which is associated with motility and invasion [324]. During the course of secondary tumor formation, metastasizing cancer cells discard their mesenchymal phenotype during MET and regain a normal membranous Ecad expression, which allows the cells to aggregate and form new solid islands, which provides a growth advantage in the new tissue [292, 325, 326]. Some findings suggest that MET allows the invaded tumor cells at the secondary site to build connections with resident normal cells by Ecad re-expression [327]. Accordingly, upregulation of Ecad expression and acquisition of epithelial cell features was demonstrated in vitro when prostate cancer cells were cultured with normal hepatocytes [328]. Not only has it been proposed that MET is induced by the mesenchymal-like cells taking up residence amongst normal cells, but it may result not from a gain of signaling but rather from a loss of EMT inducing signals that were found at the primary tumor site [319]. While it may be true that the loss of EMT induction signals influences the MET process, some studies show that the induction of EMT can create a heritable state long after the EMT-inducing signals are removed [329]. The use of Ecad as a hallmark of MET is appropriate as this particular molecule is crucial to formation of cell-cell junctions, but Ecad alone can induce MET. Expression of Ecad in the S180 mouse sarcoma cell line, the MIA PaCa-2 pancreatic cell line, and MDA-MB-231 breast cancer cell line results in induction of MET [322, 323, 330, 331].

The process of nuclear accumulation of β-catenin is reversed in MET, allowing tumor cells to regain epithelial characteristics of the primary tumor. However, because even metastatic lesions are heterogeneous, nuclear accumulation of β-catenin and other EMT-associated characteristics are observable at the invasive front of the secondary tumor. This is an important reminder that EMT and MET are transitions and as such are contextual and progressive; tumor cells manipulate these programs to promote their growth and progression dependent upon their current environmental settings [332]. Studies in human colon carcinoma describe primary tumors and their corresponding liver metastases as having mixed epithelial and mesenchymal populations: the cells toward the center of the tumor retain more epithelial characteristics, while the cells at the invasive front acquire mesenchymal characteristics [332].

Tumor cells expressing high amounts of nuclear β-catenin were found to reduce proliferative activity [321]. Overexpression of p16, a cell cycle inhibitor is initiated by β-catenin in tumor cells at the invasive front, which explains the growth arrest even when *Cyclin D1*, which is associated with proliferation is overexpressed [333]. This explains the loss of epithelial features in well-differentiated CRC is coupled with the cessation of proliferation. In addition, disseminated, dedifferentiated tumor cells of well-differentiated carcinomas must regain their epithelial function in order to expand metastatic growth [334].

FGFs and their receptors FGFRs regulate cell proliferation, differentiation, migration, and inhibition of apoptosis in normal physiological states. These attributes are frequently hijacked by cancer cells to promote cancer cell survival and metastasis. Specifically, changes in expression levels or isoforms of FGFRs have been shown to have a role in MET. FGFR2 transcripts can be alternatively spliced to yield either FGFR2b or FGFR2c, receptors which are normally expressed in a cell type-specific manner. The FGFR2b isoform is normally expressed exclusively in epithelial cells whereas FGFR2c is expressed in mesenchymal cells. Expression of the isoform FGR2b in mesenchymal cell has been shown to be sufficient to induce MET in a model of Dunning prostate cancer [335, 336]. In some cancers, expression of the epithelial-exclusive FGF protein Fox-2 in mesenchymal cells induces an alteration in the splicing pattern of FGFR2. This switch leads to expression of the epithelial FGFR2b protein in the mesenchymal cell, initiating molecular and morphological changes consistent with MET [337]. Two epithelial cell-type specific RNA binding proteins, Rbm35a and Rbm35b, have also been implicated in regulating splicing of FGFR. Expression of either Rbm35a or Rbm35b in mesenchymal-type cells normally expressing FGFR2c caused a switch in splicing of the receptor to the FGFR2b isoform [338].

Inhibition of canonical Wnt-Fzd interactions has been shown to reverse EMT in cancer cells by decreasing EMT-associated transcription factors Slug and Twist [111]. Non-canonical Wnt signaling through Fzd7 has also been identified as sufficient for initiating MET events. It is unsurprising that because this particular receptor isoform is involved in gastrulation, it is also implicated in the morphogenetic changes associated with many CRCs [316]. In a novel in vitro CRC model, depletion of Fzd7 prevented MET-associated events from occurring and cells remained in a mesenchymal state. This same study showed that overexpression of Fzd7 in solid tumors is predominantly expressed in the central areas of well-differentiated cells but rarely or not expressed in poorly differentiated cells at the invasive front [339].

Bone morphogenetic proteins (BMPs) are included in the TGF-β superfamily and are required for a diverse set of signaling pathways involved in development.

In a CRC model, sustained exposure to BMP-2 led to loss of response to growth inhibition signals and to morphological changes associated with EMT. Inhibition of BMP2 signaling reduced motility and invasiveness of metastatic CRC cells and were able to induce MET, potentially through repression of Akt signaling [340]. BMP7 is another related signaling molecule that has been linked to promoting MET. In many different cancer cell lines, BMP7 has been found to be positively correlated to the expression of Ecad and vimentin and has been shown to directly counteract TGF-β1-induced EMT [341, 342]. Furthermore, BMP7 induced MET in melanoma cells by attenuating expression of EMT-associated transcription factors Snail, Slug, Twist and smooth muscle actin (SMA) [343].

Inhibition of epidermal growth factor receptor (EGFR or ErbB1) can increase the expression of Ecad, demonstrating the existence of an inverse relationship between EGFR activity and Ecad expression in de novo human tumors [328]. Fittingly, analyses of prostate cancer metastases in the liver confirmed that expression of Ecad and catenins increased in the metastatic lesion, while EGFR expression was reduced [328]. Src homology phosphotyrosine phosphatase 2 (SHP2) is a non-receptor tyrosine phosphatase involved in signal transduction initiated by the binding of growth factors to EGFR. Aberrant EGFR signaling pathways have been implicated in inducing cancer cell transformation. In vitro inhibition of SHP2 was shown to suppress activation of Ras pathways. This includes suppression of extracellular signal-regulated kinase (ERK) and PI3K-Akt signaling pathways with a corresponding induction of epithelial cell morphology. This was associated with an increase in Ecad expression and attenuation of vimentin and fibronectin levels, all events necessary to be defined an MET [344].

MiRs have been found to be critical regulators of MET. The miR-200 family maintains the epithelial phenotype through direct targeting of the transcriptional repressors of Ecad, thus blocking a mesenchymal phenotype [345]. Furthermore, miR-200 family members inhibit the expression of WAVE3, a protein involved in actin cytoskeleton remodeling to allow for mesenchymal tumor cell migration and invasion. In addition, ectopic expression of miR-200c in lung and breast cancer cells reduced levels of Zeb1, restored Ecad expression, and induced a change in cell morphology to a more epithelial-like phenotype, implicating the miR-200 family in epithelial maintenance [346].

While MET is traditionally associated with the last step in metastases, disruptions in MET activities have been implicated in the generation of several cancers. The pediatric Wilms' tumor is an example in which mutations in genes necessary for MET during development give rise to kidney malignancy [347]. In some human synovial sarcomas, an oncogenic gene fusion in mesenchymal cells in soft tissue leads to morphologic epithelial differentiation, the result of which leads to formation of nests of glandular epithelium [348]. In some ovarian tumors, the epithelial marker Ecad has been shown to be highly expressed. Ecad is not normally expressed at all in the organ and its expression is correlated with malignant transformation and formation of metastases [349]. These examples are exceptions to the general outline of cancer progression and metastatic processes; despite that, it is important to note that these examples demonstrate the requirement for MET in developing a tumor derived from a mesenchymal-like cell.

While EMT and MET have been accepted as critical for embryogenesis, their role in cancer is still under debate from a clinicopathological point of view [350, 351]. Several clinicopathologic studies have attempted to apply the EMT-MET observations in cell culture experiments to how human cancers progress; the associations however did not coincide well. Pathologists expected to find hybrid secondary tumors with some mesenchymal attributes remaining in cells that underwent MET at regions of the tumor near vessels or the invasive front, similar to those observed in cell culture and mouse models. However, in clinical samples such findings have yet to be reported [350]. Most often, in cases of CRC, distant metastases appear strikingly like the primary lesion [321]. Likely this is due to the transient nature of such transitions in cancer which are dynamically regulated during the metastatic process [352].

6 Developing New Treatment Strategies Targeting EMT and Cancer Stem Cells

Clinically, patients with cancer (including CRC) often become resistant to chemotherapy. Although the mechanism of drug resistance is not always well understood, emerging evidence has linked EMT and CSCs to drug resistance. A number of mechanisms are involved in the therapeutic response of cancer cells to anticancer agents. Therapeutic failure can be due to resistance to a single agent or multiple drugs with different structural or mechanistic properties, which is known as multidrug resistance (MDR). Cancer drug resistance accounts for 90 % of treatment failure and it exists against almost every chemotherapeutic agent [353, 354]. Evidence supporting EMT as an important player in drug resistance includes (1) cancer cells with mesenchymal traits are more resistant to chemotherapy as compared to those with epithelial phenotypes, (2) the tumor cells refractory to chemotherapy express higher levels of EMT-associated proteins, (3) the EMT process leads to diminished responsiveness of cancer cells to conventional therapeutic regimens, and (4) reversal of EMT abrogates chemoresistance and re-sensitizes cancer cells to chemotherapy [319, 355, 356].

A wide range of cytokines and membrane proteins are involved in EMT, including TGF-β, TNF-α, IL-18, FGF, EGF, HGF and E-cadherin, N-cadherin, vimentin, and fibronectin. The regulation of EMT is controlled by various signaling pathways including Wnt/β-catenin, Hh, Ras-MAPK, APC, MMP-2 and -9 and their downstream effectors including transcription factors Twist, Snail, Slug, and Zeb1 and Zeb2 [357]. Interestingly, the remodeling process that transforms epithelial cells to mesenchymal cells, in essence, is analogous to stem cells in an embryo. CSCs, which refers to a subpopulation of neoplastic cells with stem cell-like properties within a tumor, have been extensively studied in parallel with EMT for their roles in carcinogenesis [12, 283, 358, 359].

The EMT/CSC phenotype has been implicated in cancer drug resistance. EMT was implicated in cancer progression and metastasis, as Bcl-2, Akt/mTOR, NF-κB, hypoxia-inducible factors (HIFs) [360], macrophage inhibitory cytokine-1 (MIC-1), and ATP-binding cassette (ABC) multidrug transporters [353, 361] are often overexpressed in progressive, metastatic disease. Metastatic cancer cells are more resistant to conventional chemotherapy, possibly due to molecular and morphological changes associated with EMT, therefore it is postulated that drug resistance is partially attributed to EMT [362].

The role of EMT in drug resistance has been demonstrated in cell lines and animal models. In a study using 17 bladder cancer cells lines, resistance of cancer cells to gefitinib (an EGFR inhibitor) was associated with EMT status characterized by lower levels of surface EGFR and Ecad and higher levels of vimentin in the resistant tumor cells [363]. In a recent study, Wang et al. identified differential expression of EMT/CSC markers, such as vimentin, Snail, Twist, MMP-2, and MMP-9 in 2 distinct cell lines derived from a single tumor foci, suggesting that the heterogeneous expression levels were associated with aggressiveness of tumor cells and drug resistance [364]. Another study found that gain of EMT markers correlates with resistance to paclitaxel or cisplatin in the ovarian carcinoma epithelial cell lines, NOS-2, TAOV, SKOV-3, and EOC [365–367]. Therapeutic reversal of EMT by epimorphin, also known as syntaxin-2, a pro-epithelial morphogenic protein, resulted in enhanced sensitivity to carboplatin in ovarian cancer cells [368]. Likewise, Snail and Slug also confer resistance to paclitaxel, adriamycin, and radiotherapy by increasing P-gp expression and/or inhibiting p53-mediated apoptosis shown in breast cancer cells and ovarian cancer cells [177, 369]. Similarly, the EMT phenotype linked to chemoresistance and cancer recurrence has been demonstrated in transgenic or xenograft mouse models. In transgenic CombitTA-Snail mice where Snail expression can be manipulated, Perez-Mancera et al. reported that overexpression of the *Snail* gene induced cancer development and radioprotective potential of cancer cells [370]. In a conditional transgenic mouse model of recurrent neu-induced mammary tumors, Moody et al. showed that recurrent tumors display characteristics of EMT and expressed elevated levels of Snail [371]. The same phenomenon is demonstrated in immunodeficient athymic mice transplanted with xenografts of chemoresistant cancer cells.

In human tumor tissues, microarray data suggests Snail expression is a prognostic predictor of relapse-free survival in women with breast cancer [371]. Expression of Snail in human CRC specimens is associated with EMT/CSC phenotype and mediates chemoresistance to oxaliplatin [183]. However, direct linkage between drug resistance and EMT/CSC phenotype in humans remains to be elucidated, partially due to lack of robust in vivo quantitative markers of EMT/CSC activities.

In summary, drug resistance is a major challenge in cancer therapy. In vitro and in vivo data from animal models have linked EMT/CSC with cancer drug resistance, however direct evidence in humans is yet needed.

6.1 Mechanisms Underlying EMT-Associated Drug Resistance

Marked intratumoral heterogeneity of sensitivity to chemotherapy has been identified relating to a number of chemotherapeutic agents. EMT/CSC population is a source of chemoresistance or radioresistance within tumors. The tumor cells that undergo EMT express higher levels of components associated with drug resistance. Multiple mechanisms have been proposed to explain how EMT leads to drug resistance, especially with respect to MDR. In general, EMT-associated drug resistance can be divided into two types based on mechanism: (1) intrinsic resistance of EMT/CSC cells to chemotherapy and/or (2) acquired drug resistance mediated by the tumor microenvironment [359, 372, 373]. The two mechanisms are not necessarily isolated but can be active in tandem.

Firstly, EMT/CSC cells carry intrinsic resistance to conventional chemotherapy by preserved protective mechanisms against exogenous cytotoxic agents. Recent studies suggest that tumor heterogeneity in genetic or epigenetic status, cell plasticity, and microenvironment, plays a major role in drug resistance [374]. EMT/CSC cells facilitate tumor progression by producing bulk populations of non-tumorigenic cancer cell progeny through differentiation. The larger subpopulation of EMT/CSCs within a tumor, the higher degree of therapeutic failure. For instance, when EMT cells were spiked into non-CSC enriched human mammary epithelial cells (HMLEs) and then treated with paclitaxel, the EMT/CSCs selectively survived treatment [12]. The mechanistic explanations of the intrinsic chemoresistant property of EMT/CSC tumor cells include change in the stage of cell cycle, up-regulation of efflux transporters, and alterations of pharmacological targets [373–375].

Traditional chemotherapeutic agents act by killing cells that divide rapidly; EMT/CSC tend to be more quiescent and divide infrequently, only transiently amplifying cells in order to populate the tumor. Therefore, the quiescent stem-like population tends to be exempt from the cytotoxic effects of chemotherapy. Overexpression of efflux transporters is also a common and important mechanism of drug resistance, resulting in altered pharmacodynamics of drugs. CSCs appear to express higher levels of efflux transporters of the ABC family that exclude cytotoxic agents from the intracellular compartment [354, 376]. Altering of pharmacological targets is another cause of chemoresistance that can exacerbate EMT induced resistance; a common mechanism by which cells can resist drug treatment is by acquiring mutations to molecular pockets that may be targeted by currently available therapeutics.

Secondly, EMT-associated drug resistance could also be mediated by tumor microenvironment (TME), which refers to the dynamic cellular and extracellular components surrounding tumor cells at each stage of carcinogenesis. TME is an important factor that determines survival of CSCs. The host stromal cells in the TME provide soluble and matrix-linked factors, including cytokines and regulatory factors, which in turn inhibit cell division and apoptosis thereby leading to chemoresistance [377]. The exposure to cytotoxic chemotherapy could also enhance release of these factors to alter TME, such as extracellular matrix components, which can induce EMT and drug resistance [373, 378].

Drug resistance can be acquired from the interactions between tumor cells and TME in an EMT-independent manner. It is widely accepted that TME plays a critical role in therapeutic responses and cancer progression. The communication between tumor cells and microenvironment is an essential driver for adaptive mechanisms of tumor cells developing drug resistance, also known as host-derived therapeutic escape. Recruitment and activation of host fibroblasts provide survival signals that allow minor populations of cancer cells to evade therapy. Cancer associated fibroblasts in the TME secrete fibronectin, collagen, laminin, proteoglycans, tenascin, and osteopontin, which transforms malignant cells to highly aggressive cancers [379].

It is a technical challenge to differentiate EMT-dependent and EMT-independent pathways involved in drug resistance, largely due to lack of EMT-specific antagonists. However, it may not be clinically important to distinguish one from the other because they often overlap and cause similar end effects. Drug resistance has become a challenge in cancer therapeutics. Chemoresistant cancer cells exhibit features of EMT; EMT contributes to drug resistance, therapeutic failure, and ultimately disease progression by intrinsic and acquired mechanisms. Under pressure from the TME, tumor cells can generate drug-resistant cells following or in addition to EMT.

6.2 New Treatment Strategies Targeting EMT and CSCs

EMT has been increasingly recognized as a critical therapeutic target of anticancer drugs, as its reversal may inhibit tumor progression, reverse drug resistance, and prevent metastasis. Improvement of current cancer therapies can be achieved by targeting the EMT signal-transduction pathways derived from receptor tyrosine kinases such as Met, IGFR1 and the ErbB family, or non-receptor tyrosine kinases, especially Src. Blocking the TME associated with EMT has also been studied as a new therapeutic strategy. Another approach to developing EMT-targeting drugs is to screen existing drugs for activities against EMT/CSC cells. Nanoparticles are being evaluated as effective drug delivery systems, but it is still a relatively new area of exploration. Development of novel EMT-blocking drugs is an area of active research, although still in its infancy. To date, few strategies have made it to clinical trials. One main challenge is that a feasible and efficient biomarker of EMT is yet lacking. Currently clinical studies on circulating tumor biomarkers based on EMT biology are underway with an ultimate goal to develop a blood-based EMT signature. Undoubtedly such a noninvasive biomarker of EMT once developed will facilitate drug development targeting EMT, as a solid readout for response will be available.

Several anticancer agents have been found to suppress or reverse EMT in vitro or in vivo, including microRNA targeting EMT, epithelial morphogenic protein, angiogenesis inhibitors, adhesion kinase inhibitors, tumor-associated fibroblast inhibitors and human monoclonal antibodies. One example is the miR-200 family that exhibits inhibitory effects on EMT. The EMT process is associated with downregulation of Ecad and upregulation of vimentin expression. Park et al. reported that miR-200 up-regulated Ecad and downregulated vimentin transcription in primary ovarian cancer cells and reduced motility via targeting the Ecad transcriptional repressors, Zeb1 and Zeb2 [236]. Overexpression of miR-429, another member of the miR-200 family, could lead to mesenchymal-epithelial transition (MET) in metastatic ovarian cancer cells [380]. miRs are emerging as a new class of pharmacological targets; it has been suggested that miRs are targetable in several disease, including fibrosis and cancer [381, 382]. They may prove to be robust prognostic markers, as particular miR profiles have been associated with certain cancers. Since individual diseases have particular repression or expression profiles of miRs, specific plans for either inhibition of miR function or re-introduction of a lost miR will be necessary to design successful therapeutic approaches. Efficacious methods for miR delivery and methods of reducing miR expression (adenoviral based miR delivery and antisense miR oligonucleotides or inhibition of the miR processing complex [RISC], respectively) are being developed [382]. In regard to EMT and CSC signatures, several miRs have been implicated not only in colon cancer, but in a host of others; this suggests that modulation of the miR profile for EMT and CSCs could deplete these tumor sustaining cell populations.

Epimorphin, also known as syntaxin-2, functions as a pro-epithelial factor controlling epithelial patterning and morphogenesis in various tissues, such as pancreatic ducts, mammary glands, lungs, and small intestines. Treatment with optimal doses of exogenous epimorphin in ovarian cancer cells induced MET and reversed resistance to platinum, possibly through the αV integrin receptor and the transcription factors C/EBPβ and KLF4 [368]. Akt is an oncogene common in many cancers that can drive EMT. The idea of inhibiting EMT-driving proteins to induce MET events in an effort to reduce or prevent the occurrence of metastases has recently begun to be investigated. In vitro testing of Akt inhibitor phosphatidylinositol ether lipid analogs (PIA) led to a decrease in NF-κB signaling and downregulation of Snail and Twist expression and the mesenchymal cell-specific filament vimentin. Moreover, PIA treatment was able to induce the expression of Ecad and β-catenin. Taken together, in these experiments PIA was able to successfully restore epithelial morphology in mesenchymal cancer cells and thus may provide a novel strategy to reversion of EMT before the cells can metastasize [383].

TME itself has also emerged as a drug target. It represents a dynamic interaction and network between fibroblasts, macrophages, extracellular matrices, and signaling molecules (cytokines, growth factors, proteinases, etc.) secreted by cellular components. TME is not only a paracrine system; it also exerts mechanical effects on multiple steps of tumorigenesis. Physically, epithelial cancers have regions of hypoxia. Mechanical/physical and hypoxic stresses can activate HIFs, GLUT-1, carbonic anhydrase IX (CAIX) and subsequently promote angiogenesis, EMT, and drug resistance. These aspects of TME contribute to in vivo chemoresistance and are clinically important for cancer therapeutics [362]. Silibinin, a hepatoprotective and chemopreventive agent that targets TME constituents (endothelial cells, fibroblasts and immune cells) as well as their interaction with cancer cells, was found to inhibit EMT, tumor growth, angiogenesis, inflammation, and metastasis [384]. Drugs that target both EMT and TME are of particular interest in cancer therapy.

6.3 Screening and Drug Development for Anti-EMT/CSC Drugs

The main obstacle to discovery of EMT/CSC-targeting therapy is lack of powerful and effective tools to distinguish CSCs from normal stem cells, and to differentiate EMT/CSC cells and other tumor cells. The activity of CSC drugs is pivotal to the eventual success of drug development. Although a number of markers of EMT/CSC have been identified, a sensitive and selective surrogate of EMT/CSC activity is still lacking for large-scale drug screening both in vivo and in vitro. Nonetheless, a recent discovery of a cell-based platform using fluorescent reporter-tagged transcriptional factors Oct4 and Sox2 has shown great potential for high throughput screening of small molecules against EMT/CSCs [385]. With the novel platform, the authors screened chemical libraries found that thioridazine, an antipsychotic drug and dopamine receptor antagonist, selectively targets CSCs [385]. Interestingly, dopamine receptors were expressed on CSCs and breast tumors, but not on normal stem cells. Consistently, an epidemiologic study indicated that a lower incidence of cancer occurred in schizophrenic patients who received dopamine antagonist medication [386]. Based on these findings, it is postulated that dopamine receptors can be used as a marker of CSC status, and dopamine receptor antagonists are a potential anti-CSC therapy [385].

As mentioned, a common barrier to clinical efficacy of most targeted therapeutics is overexpression of efflux transporters, such as P-gp. Nanoparticle technology, a multicomponent system, has the advantage of being able to co-deliver anticancer drugs with inhibitors of efflux transporters or to use formulations that include excipients or surfactants that inhibit drug efflux to counteract the ability of cancer cells to expel anti-cancer therapies. Therefore, nanoparticles can effectively deliver drugs to the intracellular compartment of cancer cells and overcome efflux-mediated drug resistance [387].

The advance in nanotechnology has facilitated studies of nanoparticle-based drug delivery systems. Nanoparticles as a targeted delivery system have been evaluated in a wide disease spectrum [388, 389]. Overall, delivering anti-cancer drugs to the site of disease and overcoming drug-resistance have been the first and foremost goals of these efforts. Nanoparticles utilize various forms of "envelopes" including polymer conjugates, liposomes, or dendrimers to encapsulate pharmacologically active drug molecules. The optimal size and molecular weight of nanoparticles give rise to a favorable feature, which is their enhanced permeability and retention effect (EPR) [390]. EPR protects drugs from degradation by enzymes and nucleases in the bloodstream. The surface of nanoparticles can also be selectively conjugated with antibodies or peptides that bind to specific tumor receptors in the extracellular domain. Because nanoparticles are too large for diffusion-mediated transport, when reaching targeted tissues they are taken up by receptor-dependent endocytosis and can then release the encapsulated drugs intracellularly. The intracellular drug release can be engineered to be pH-dependent, hydrophobicity dependent, or osmosis-dependent. Nanoparticles can also be designed to target surface markers that are expressed commonly in CSCs/EMT cells with drug resistant properties, but not in normal cells [389]. Multifunctional nanoparticles integrating drugs, genes, imaging agents, and surface targeting ligands are also under development. They are appealing to scientists and the pharmaceutical industry as a more efficient delivery system. However, poor penetration of nanoparticles into perfusion limited regions of a tumor restricts their potential. Nanoparticles often can only reach the tumor site adjacent to the blood vessels, which is a phenomenon known as transport resistance.

Hence, improvement of intratumoral distribution of nanoparticles is pivotal to the success of this novel technology as a cancer treatment. In addition, the effectiveness of the multi-tasking nanomedicine will also depend on cancer types, selectivity, and minimum toxicity of nanoparticles [387]. Overall, the nanoparticle-based drug delivery system provides a useful tool for efficient targeted anti-cancer drug therapy. This rapidly emerging, cutting-edge technology holds future promise and is making its way to be tested in clinical settings in the near future.

7 Controversy and Conclusions

EMT is a crucial process in cancer progression. It is associated with invasion of cancer from the primary tumor to surrounding tissues and secondary organs. It is also a process that generates CSCs that are responsible for tumorigenicity and resistance to currently available therapeutics, and therefore is a critical determinant of patient outcome. Despite the progress that has been made in understanding EMT from embryonic development to cancer, there remains concern and controversy over certain aspects. Firstly, EMT and MET are transient processes and as such are inherently difficult to observe in vivo, which sometimes brings doubt to the relevance of EMT in cancer pathology. Secondly, the association of EMT with CSCs is controversial based on the fact that CSCs are widely debated in the cancer field. Embryonic stem cells, along with CSCs, are highly dependent on their microenvironment to determine their phenotype and fate; conflicting reports generated regarding CSCs in different cancers and under different conditions confers a sense of ambiguity to CSC biology. It can be reconciled that secondary lesions are well differentiated because of MET, which is the reversion of mesenchymal-like cells to an epithelial phenotype that resembles the primary tumor. The reasons for this reversion are likely due to the microenvironmental signals that are found at the secondary site differing from those of the initial tumor site. Overall, EMT is a crucial determinant of the metastatic ability of tumor cells; as such, EMT and resultant CSCs are currently under investigation as targetable aspects of cancer biology, as they certainly should be. These potential therapeutics will provide ammunition against the most malignant of cancer cells.

References

1. Hanahan D, Weinberg RA (2011) Hallmarks of cancer: the next generation. Cell 144(5):646–674
2. Helling TS, Martin M (2013) Cause of death from liver metastases in colorectal cancer. Ann Surg Oncol
3. Deryugina EI, Quigley J (2006) Matrix metalloproteinases and tumor metastasis. Cancer Metastasis Rev 25(1):9–34
4. Wolf K et al (2003) Compensation mechanism in tumor cell migration: mesenchymal-amoeboid transition after blocking of pericellular proteolysis. J Cell Biol 160(2):267–277

5. Yang J et al (2004) Twist, a master regulator of morphogenesis, plays an essential role in tumor metastasis. Cell 117(7):927–939
6. Micalizzi DS, Farabaugh SM, Ford HL (2010) Epithelial-mesenchymal transition in cancer: parallels between normal development and tumor progression. J Mammary Gland Biol Neoplasia 15(2):117–134
7. Kalluri R, Weinberg RA (2009) The basics of epithelial-mesenchymal transition. J Clin Invest 119(6):1420–1428
8. Thiery J, Sleeman J (2006) Complex networks orchestrate epithelial-mesenchymal transitions. Nat Rev Mol Cell Biol 7(2):131–142
9. Chaffer CL, Weinberg RA (2011) A perspective on cancer cell metastasis. Science 331(6024):1559–1564
10. Kang Y, Massague J (2004) Epithelial-mesenchymal transitions: twist in development and metastasis. Cell 118(3):277–279
11. De Craene B, Berx G (2013) Regulatory networks defining EMT during cancer initiation and progression. Nat Rev Cancer 13(2):97–110
12. Mani SA et al (2008) The epithelial-mesenchymal transition generates cells with properties of stem cells. Cell 133(4):704–715
13. Tsuji T, Ibaragi S, Hu GF (2009) Epithelial-mesenchymal transition and cell cooperativity in metastasis. Cancer Res 69(18):7135–7139
14. Jo M et al (2009) Reversibility of epithelial-mesenchymal transition (EMT) induced in breast cancer cells by activation of urokinase receptor-dependent cell signaling. J Biol Chem 284(34):22825–22833
15. Chaffer CL et al (2006) Mesenchymal-to-epithelial transition facilitates bladder cancer metastasis: role of fibroblast growth factor receptor-2. Cancer Res 66(23):11271–11278
16. Larue L, Bellacosa A (2005) Epithelial-mesenchymal transition in development and cancer: role of phosphatidylinositol 3[prime] kinase//AKT pathways. Oncogene 24(50):7443–7454
17. Leopold L, Vincent J, Wang H, (2012) A comparison of epithelial-to-mesenchymal transition and re-epithelialization. Semin Cancer Biol 22(5–6):471–483
18. Singh A, Settleman J (2010) EMT, cancer stem cells and drug resistance: an emerging axis of evil in the war on cancer. Oncogene 29(34):4741–4751
19. Sanchez-Tillo E et al (2012) EMT-activating transcription factors in cancer: beyond EMT and tumor invasiveness. Cell Mol Life Sci 69(20):3429–3456
20. Ren F, Sheng WQ, Du X (2013) CD133: a cancer stem cells marker, is used in colorectal cancers. World J Gastroenterol 19(17):2603–2611
21. Natalwala A, Spychal R, Tselepis C (2008) Epithelial-mesenchymal transition mediated tumourigenesis in the gastrointestinal tract. World J Gastroenterol 14(24):3792–3797
22. Gregorieff A, Clevers H (2005) Wnt signaling in the intestinal epithelium: from endoderm to cancer. Genes Dev 19(8):877–890
23. Fearon ER, Vogelstein B (1990) A genetic model for colorectal tumorigenesis. Cell 61(5):759–767
24. Medema J, Vermeulen L (2011) Microenvironmental regulation of stem cells in intestinal homeostasis and cancer. Nature 474(7351):318–326
25. Barker N et al (2009) Crypt stem cells as the cells-of-origin of intestinal cancer. Nature 457(7229):608–611
26. Vaiopoulos AG et al (2012) Colorectal cancer stem cells. Stem Cells 30(3):363–371
27. Brabletz S, Schmalhofer O, Brabletz T (2009) Gastrointestinal stem cells in development and cancer. J Pathol 217(2):307–317
28. Barker N, van de Wetering M, Clevers H (2008) The intestinal stem cell. Genes Dev 22(14):1856–1864
29. Ricci-Vitiani L et al (2009) Colon cancer stem cells. J Mol Med 87(11):1097–1104
30. Huang EH et al (2009) Aldehyde dehydrogenase 1 is a marker for normal and malignant human colonic stem cells (SC) tracks SC overpopulation during colon tumorigenesis. Cancer Res 69(8):3382–3389

31. Sipos F, Galamb O (2012) Epithelial-to-mesenchymal and mesenchymal-to-epithelial transitions in the colon. World J Gastroenterol 18(7):601–608
32. Polette M et al (2004) Tumour invasion and matrix metalloproteinases. Crit Rev Oncol Hematol 49(3):179–186
33. Noë V et al (2001) Release of an invasion promoter E-cadherin fragment by matrilysin and stromelysin-1. J Cell Sci 114(Pt 1):111–118
34. Béatrice N-R et al (2003) Upregulation of MMPs by soluble E-cadherin in human lung tumor cells. Int J Cancer 105(6):790–795
35. Whitelock JM et al (1996) The degradation of human endothelial cell-derived perlecan and release of bound basic fibroblast growth factor by stromelysin, collagenase, plasmin, heparanases. J Biol Chem 271(17):10079–10086
36. Fowlkes JL et al (1995) Matrix metalloproteinases as insulin-like growth factor binding protein-degrading proteinases. Prog Growth Factor Res 6(2–4):255–263
37. Yu Q, Stamenkovic I (2000) Cell surface-localized matrix metalloproteinase-9 proteolytically activates TGF-beta and promotes tumor invasion and angiogenesis. Genes Dev 14(2):163–176
38. Radisky DC et al (2005) Rac1b and reactive oxygen species mediate MMP-3-induced EMT and genomic instability. Nature 436(7047):123–127
39. Przybylo JA, Radisky DC (2007) Matrix metalloproteinase-induced epithelial–mesenchymal transition: tumor progression at Snail's pace. Int J Biochem Cell Biol 39(6):1082–1088
40. Mook ORF, Frederiks WM, Van Noorden CJF (2004) The role of gelatinases in colorectal cancer progression and metastasis. Biochim Biophys Acta 1705(2):69–89
41. Lemieux É et al (2009) Constitutively active MEK1 is sufficient to induce epithelial-to-mesenchymal transition in intestinal epithelial cells and to promote tumor invasion and metastasis. Int J Cancer 125(7):1575–1586
42. Zeng Z-S et al (2002) Matrix metalloproteinase-7 expression in colorectal cancer liver metastases: evidence for involvement of mmp-7 activation in human cancer metastases. Clin Cancer Res 8(1):144–148
43. Miettinen J et al (1994) TGF-beta induced transdifferentiation of mammary epithelial cells to mesenchymal cells: involvement of type I receptors. J Cell Biol 127(6 Pt 2):2021–2036
44. Akhurst RJ, Derynck R (2001) TGF-β signaling in cancer—a double-edged sword. Trends Cell Biol 11(11):S44–S51
45. Shipitsin M et al (2007) Molecular definition of breast tumor heterogeneity. Cancer Cell 11(3):259–273
46. Spagnoli FM et al (2000) Inhibition of MMH (Met murine hepatocyte) cell differentiation by TGF(beta) is abrogated by pre-treatment with the heritable differentiation effector FGF1. J Cell Sci 113(Pt 20):3639–3647
47. Peinado H, Quintanilla M, Cano A (2003) Transforming growth factor beta-1 induces snail transcription factor in epithelial cell lines: mechanisms for epithelial mesenchymal transitions. J Biol Chem 278(23):21113–21123
48. Romano LA, Runyan RB (2000) Slug is an essential target of TGFβ2 signaling in the developing chicken heart. Dev Biol 223(1):91–102
49. Comijn J et al (2001) The two-handed E box binding zinc finger protein SIP1 downregulates E-cadherin induces invasion. Mol Cell 7(6):1267–1278
50. Wendt MK, Tian M, Schiemann W (2012) Deconstructing the mechanisms and consequences of TGF-beta-induced EMT during cancer progression. Cell Tissue Res 347(1):85–101
51. Garg M (2013) Epithelial-mesenchymal transition—activating transcription factors—multifunctional regulators in cancer. World J Stem Cells 5(4):188–195
52. Thuault S et al (2006) Transforming growth factor-beta employs HMGA2 to elicit epithelial-mesenchymal transition. J Cell Biol 174(2):175–183
53. Thuault S et al (2008) HMGA2 and Smads co-regulate SNAIL1 expression during induction of epithelial-to-mesenchymal transition. J Biol Chem 283(48):33437–33446
54. Vincent T et al (2009) A SNAIL1-SMAD3/4 transcriptional repressor complex promotes TGF-beta mediated epithelial-mesenchymal transition. Nat Cell Biol 11(8):943–950

55. Verschueren K et al (1999) SIP1, a novel zinc finger/homeodomain repressor, interacts with smad proteins and binds to 5′-CACCT sequences in candidate target genes. J Biol Chem 274(29):20489–20498
56. Postigo AA et al (2003) Regulation of Smad signaling through a differential recruitment of coactivators corepressors by ZEB proteins. EMBO J 22(10):2453–2462
57. Postigo AA (2003) Opposing functions of ZEB proteins in the regulation of the TGFbeta/ BMP signaling pathway. EMBO J 22(10):2443–2452
58. Shirakihara T, Saitoh M, Miyazono K (2007) Differential regulation of epithelial and mesenchymal markers by deltaEF1 proteins in epithelial mesenchymal transition induced by TGF-beta. Mol Biol Cell 18(9):3533–3544
59. Wendt MK, Smith JA, Schiemann W (2009) p130Cas is required for mammary tumor growth and transforming growth factor-beta-mediated metastasis through regulation of Smad2/3 activity. J Biol Chem 284(49):34145–34156
60. Bakin AV et al (2000) phosphatidylinositol 3-kinase function is required for transforming growth factor β-mediated epithelial to mesenchymal transition and cell migration. J Biol Chem 275(47):36803–36810
61. Lamouille S, Derynck R (2011) Emergence of the phosphoinositide 3-kinase-Akt-mammalian target of rapamycin axis in transforming growth factor-beta-induced epithelial-mesenchymal transition. Cells Tissues Organs 193(1–2):8–22
62. Wilkes MC et al (2005) Transforming growth factor-β activation of phosphatidylinositol 3-kinase is independent of Smad2 and Smad3 and regulates fibroblast responses via p21-activated kinase-2. Cancer Res 65(22):10431–10440
63. Bhowmick NA et al (2001) Transforming growth factor-beta1 mediates epithelial to mesenchymal transdifferentiation through a RhoA-dependent mechanism. Mol Biol Cell 12(1):27–36
64. Edlund S et al (2002) Transforming growth factor-β–induced mobilization of actin cytoskeleton requires signaling by small GTPases Cdc42 and RhoA. Mol Biol Cell 13(3):902–914
65. Sorrentino A et al (2008) The type I TGF-[beta] receptor engages TRAF6 to activate TAK1 in a receptor kinase-independent manner. Nat Cell Biol 10(10):1199–1207
66. Yamashita M et al (2008) TRAF6 mediates Smad-independent activation of JNK and p38 by TGF-beta. Mol Cell 31(6):918–924
67. Ozdamar B et al (2005) Regulation of the polarity protein Par6 by TGFbeta receptors controls epithelial cell plasticity. Science 307(5715):1603–1609
68. Hocevar BA, Brown TL, Howe H (1999) TGF-[beta] induces fibronectin synthesis through a c-Jun N-terminal kinase-dependent, Smad4-independent pathway. EMBO J 18(5):1345–1356
69. Engel ME et al (1999) Interdependent SMAD and JNK signaling in transforming growth factor-β-mediated transcription. J Biol Chem 274(52):37413–37420
70. Siegel M, Massague J (2003) Cytostatic and apoptotic actions of TGF-[beta] in homeostasis and cancer. Nat Rev Cancer 3(11):807–820
71. Lee MK et al (2007) TGF-beta activates Erk MAP kinase signalling through direct phosphorylation of ShcA. EMBO J 26(17):3957–3967
72. Janda E et al (2002) Ras and TGFβ cooperatively regulate epithelial cell plasticity and metastasis: dissection of Ras signaling pathways. J Cell Biol 156(2):299–314
73. Vogelmann R et al (2005) TGFβ-induced downregulation of E-cadherin-based cell-cell adhesion depends on PI3-kinase and PTEN. J Cell Sci 118(20):4901–4912
74. Picon A et al (1998) A subset of metastatic human colon cancers expresses elevated levels of transforming growth factor beta1. Cancer Epidemiol Biomarkers Prev 7(6):497–504
75. Tsushima H et al (2001) Circulating transforming growth factor β1 as a predictor of liver metastasis after resection in colorectal cancer. Clin Cancer Res 7(5):1258–1262
76. Kim YH et al (2014) TWIST1 and SNAI1 as markers of poor prognosis in human colorectal cancer are associated with the expression of ALDH1 and TGF-beta1. Oncol Rep 31(3):1380–1388

77. Oft M, Heider KH, Beug H (1998) TGFβ signaling is necessary for carcinoma cell invasiveness metastasis. Curr Biol 8(23):1243–1252
78. Zubeldia IG et al (2013) Epithelial to mesenchymal transition and cancer stem cell phenotypes leading to liver metastasis are abrogated by the novel TGFβ1-targeting peptides P17 and P144. Exp Cell Res 319(3):12–22
79. Tamm I et al (1989) Interleukin 6 decreases cell-cell association and increases motility of ductal breast carcinoma cells. J Exp Med 170(5):1649–1669
80. Tabibzadeh SS et al (1989) Interleukin-6 immunoreactivity in human tumors. Am J Pathol 135(3):427–433
81. Ujiie H et al (2012) Serum hepatocyte growth factor and interleukin-6 are effective prognostic markers for non-small cell lung cancer. Anticancer Res 32(8):3251–3258
82. Tamm I et al (1994) E-cadherin distribution in interleukin 6-induced cell-cell separation of ductal breast carcinoma cells. Proc Natl Acad Sci U S A 91(10):4338–4342
83. Yadav A et al (2011) IL-6 promotes head neck tumor metastasis by inducing epithelial-mesenchymal transition via the JAK-STAT3-SNAIL signaling pathway. Mol Cancer Res 9(12):1658–1667
84. Lyons JG et al (2008) Snail up-regulates proinflammatory mediators and inhibits differentiation in oral keratinocytes. Cancer Res 68(12):4525–4530
85. Chang Q et al (2013) The IL-6/JAK/Stat3 feed-forward loop drives tumorigenesis and metastasis. Neoplasia 15(7):848–862
86. Lo HW et al (2007) Epidermal growth factor receptor cooperates with signal transducer and activator of transcription 3 to induce epithelial-mesenchymal transition in cancer cells via up-regulation of TWIST gene expression. Cancer Res 67(19):9066–9076
87. Xiong H et al (2012) Roles of STAT3 and ZEB1 proteins in E-cadherin down-regulation and human colorectal cancer epithelial-mesenchymal transition. J Biol Chem 287(8):5819–5832
88. Xie G et al (2012) IL-6-induced epithelial-mesenchymal transition promotes the generation of breast cancer stem-like cells analogous to mammosphere cultures. Int J Oncol 40(4):1171–1179
89. Komiya Y, Habas R (2008) Wnt signal transduction pathways. Organogenesis 4(2):68–75
90. Son H, Moon A (2010) Epithelial-mesenchymal transition and cell invasion. Toxicol Res 26(4):245–252
91. Takahashi-Yanaga F, Kahn M (2010) Targeting Wnt signaling: can we safely eradicate cancer stem cells? Clin Cancer Res 16(12):3153–3162
92. Mann B et al (1999) Target genes of beta-catenin-T cell-factor/lymphoid-enhancer-factor signaling in human colorectal carcinomas. Proc Natl Acad Sci U S A 96(4):1603–1608
93. Gradl D, Kuhl M, Wedlich D (1999) The Wnt/Wg signal transducer beta-catenin controls fibronectin expression. Mol Cell Biol 19(8):5576–5587
94. Conacci-Sorrell M et al (2003) Autoregulation of E-cadherin expression by cadherin-cadherin interactions: the roles of beta-catenin signaling, Slug, and MAPK. J Cell Biol 163(4):847–857
95. Stemmer V et al (2008) Snail promotes Wnt target gene expression and interacts with [beta]-catenin. Oncogene 27(37):5075–5080
96. Vincan E, Barker N (2008) The upstream components of the Wnt signalling pathway in the dynamic EMT and MET associated with colorectal cancer progression. Clin Exp Metast 25(6):657–663
97. Ren D, Minami Y, Nishita M, Critical role of Wnt5a–Ror2 signaling in motility and invasiveness of carcinoma cells following Snail-mediated epithelial–mesenchymal transition. Genes Cells 16(3):304–315
98. Dissanayake SK et al (2007) The Wnt5A/protein kinase C pathway mediates motility in melanoma cells via the inhibition of metastasis suppressors and initiation of an epithelial to mesenchymal transition. J Biol Chem 282(23):17259–17271
99. Wright TM et al (2009) Ror2, a developmentally regulated kinase, promotes tumor growth potential in renal cell carcinoma. Oncogene 28(27):2513–2523

100. Enomoto M et al (2009) Autonomous regulation of osteosarcoma cell invasiveness by Wnt5a/Ror2 signaling. Oncogene 28(36):3197–3208
101. Yamamoto H et al (2010) Wnt5a signaling is involved in the aggressiveness of prostate cancer and expression of metalloproteinase. Oncogene 29(14):2036–46
102. Li L et al (2013) Epigenetic identification of receptor tyrosine kinase-like orphan receptor 2 as a functional tumor suppressor inhibiting beta-catenin and AKT signaling but frequently methylated in common carcinomas. Cell Mol Life Sci
103. Katoh M (2011) Network of WNT and other regulatory signaling cascades in pluripotent stem cells and cancer stem cells. Curr Pharm Biotechnol 12(2):160–170
104. Cheng R et al (2014) Wnt5a suppresses colon cancer by inhibiting cell proliferation and epithelial–mesenchymal transition. J Cell Physiol. doi:10.1002/jcp.24566
105. Yook JI et al (2005) Wnt-dependent regulation of the E-cadherin repressor snail. J Biol Chem 280(12):11740–11748
106. Arnold SJ et al (2000) Brachyury is a target gene of the Wnt/β-catenin signaling pathway. Mech Dev 91(1–2):249–258
107. Bo H et al (2013) Upregulation of Wnt5a promotes epithelial-to-mesenchymal transition and metastasis of pancreatic cancer cells. BMC Cancer 13(1):496
108. Krol M et al (2014) Macrophages mediate a switch between canonical and non-canonical wnt pathways in canine mammary tumors. PLoS ONE 9(1):e83995
109. Brabletz T et al (1998) Nuclear overexpression of the oncoprotein beta-catenin in colorectal cancer is localized predominantly at the invasion front. Pathol Res Pract 194(10):701–704
110. Han J et al (2012) Small interfering RNA-mediated downregulation of beta-catenin inhibits invasion and migration of colon cancer cells in vitro. Med Sci Monit 18(7): BR273–BR280
111. Zi X et al (2005) Expression of Frzb/secreted Frizzled-related protein 3, a secreted Wnt antagonist, in humanrogen-independent prostate cancer PC-3 cells suppresses tumor growth and cellular invasiveness. Cancer Res 65(21):9762–9770
112. Bailey JM, Singh K, Hollingsworth MA (2007) Cancer metastasis facilitated by developmental pathways: Sonic hedgehog, Notch, and bone morphogenic proteins. J Cell Biochem 102(4):829–839
113. Kawahira H et al (2005) Hedgehog signaling regulates expansion of pancreatic epithelial cells. Dev Biol 280(1):111–121
114. Karhadkar SS et al (2004) Hedgehog signalling in prostate regeneration, neoplasia and metastasis. Nature 431(7009):707–712
115. Li X et al (2006) Snail induction is an early response to Gli1 that determines the efficiency of epithelial transformation. Oncogene 25(4):609–621
116. Varnat F et al (2009) Human colon cancer epithelial cells harbour active HEDGEHOG-GLI signalling that is essential for tumour growth, recurrence, metastasis and stem cell survival and expansion. EMBO Mol Med 1(6–7):338–351
117. Katoh Y, Katoh M (2008) Hedgehog signaling, epithelial-to-mesenchymal transition and miRNA (review). Int J Mol Med 22(3):271–275
118. Timmerman LA et al Notch promotes epithelial-mesenchymal transition during cardiac development and oncogenic transformation. Genes Dev 18(1):99–115
119. Espinoza I, Miele L (2013) Deadly crosstalk: notch signaling at the intersection of EMT and cancer stem cells. Cancer Lett 341(1):41–45
120. Wang Z et al (2010) The role of Notch signaling pathway in epithelial-mesenchymal transition (EMT) during development and tumor aggressiveness. Curr Drug Targets 11(6):745–751
121. Bao B et al (2011) Notch-1 induces epithelial–mesenchymal transition consistent with cancer stem cell phenotype in pancreatic cancer cells. Cancer Lett 307(1):26–36
122. Niessen K et al (2008) Slug is a direct Notch target required for initiation of cardiac cushion cellularization. J Cell Biol 182(2):315–325
123. Grudzien P et al (2010) Inhibition of notch signaling reduces the stem-like population of breast cancer cells and prevents mammosphere formation. Anticancer Res 30(10):3853–3867

124. Suman S, Das T, Damodaran C (2013) Silencing NOTCH signaling causes growth arrest in both breast cancer stem cells and breast cancer cells. Br J Cancer 109(10):2587–2596
125. Katoh M (2007) Notch signaling in gastrointestinal tract (review). Int J Oncol 30(1):247–251
126. Veenendaal LM et al (2008) Differential Notch and TGFbeta signaling in primary colorectal tumors and their corresponding metastases. Cell Oncol 30(1):1–11
127. Reedijk M et al (2008) Activation of Notch signaling in human colon adenocarcinoma. Int J Oncol 33(6):1223–1229
128. Takebe N, Nguyen D, Yang SX (2014) Targeting Notch signaling pathway in cancer: clinical development advances and challenges. Pharmacol Therap 141(2):140–149
129. Akiyoshi T et al (2008) γ-secretase inhibitors enhance taxane-induced mitotic arrest and apoptosis in colon cancer cells. Gastroenterology. 134(1):131–144
130. Ghaleb AM et al (2008) Notch inhibits expression of the Kruppel-like factor 4 tumor suppressor in the intestinal epithelium. Mol Cancer Res 6(12):1920–1927
131. Moon CM et al (2014) Nonsteroidal anti-inflammatory drugs suppress cancer stem cells via inhibiting PTGS2 (cyclooxygenase 2) and NOTCH/HES1 and activating PPARG in colorectal cancer. Int J Cancer 134(3):519–529
132. Morali OG et al (2001) IGF-II induces rapid beta-catenin relocation to the nucleus during epithelium to mesenchyme transition. Oncogene 20(36):4942–4950
133. Gilles C et al (2003) Transactivation of vimentin by beta-catenin in human breast cancer cells. Cancer Res 63(10):2658–2664
134. Patel GS, Karapetis CS (2013) Personalized treatment for advanced colorectal cancer: KRAS beyond. Cancer Manag Res 5:387–400
135. Okada Y et al (2014) Biomarkers for predicting the efficacy of anti-epidermal growth factor receptor antibody in the treatment of colorectal cancer. Digestion 89(1):18–23
136. Barr S et al (2008) Bypassing cellular EGF receptor dependence through epithelial-to-mesenchymal-like transitions. Clin Exp Metast 25(6):685–693
137. Price JT, Wilson HM, Haites NE (1996) Epidermal growth factor (EGF) Increases the in vitro invasion, motility and adhesion interactions of the primary renal carcinoma cell line, A704. Euro J Cancer 32(11):1977–1982
138. Shibata T et al (1996) Enhancing effects of epidermal growth factor on human squamous cell carcinoma motility and matrix degradation but not growth. Tumour Biol 17(3):168–175
139. Lu Z et al (2001) Epidermal growth factor-induced tumor cell invasion metastasis initiated by dephosphorylation and downregulation of focal adhesion kinase. Mol Cell Biol 21(12):4016–4031
140. Lu Z et al (2003) Downregulation of caveolin-1 function by EGF leads to the loss of E-cadherin, increased transcriptional activity of β-catenin, and enhanced tumor cell invasion. Cancer Cell 4(6):499–515
141. Lal A et al (2002) Mutant epidermal growth factor receptor up-regulates molecular effectors of tumor invasion. Cancer Res 62(12):3335–3339
142. Ciardiello F et al (1991) Differential expression of epidermal growth factor-related proteins in human colorectal tumors. Proc Natl Acad Sci U S A 88(17):7792–7796
143. Molaei M et al (2009) Human epidermal growth factor receptor-2 family in colorectal adenocarcinoma: correlation with survival and clinicopathological findings. Eur J Gastroenterol Hepatol 21(3):289–293
144. Rokita M et al (2013) Overexpression of epidermal growth factor receptor as a prognostic factor in colorectal cancer on the basis of the Allred scoring system. Onco Targets Ther 6:967–976
145. Gagandeep S et al (1996) Expression of basic fibroblast growth factor in colorectal carcinoma. Int J Oncol 8(1):191–195
146. Sato T et al (2009) Overexpression of the fibroblast growth factor receptor-1 gene correlates with liver metastasis in colorectal cancer. Oncol Rep 21(1):211–216
147. Tassi E et al Expression of a fibroblast growth factor–binding protein during the development of adenocarcinoma of the pancreas and colon. Cancer Res 66(2):1191–1198

148. Katoh M (2006) Cross-talk of WNT and FGF signaling pathways at GSK3beta to regulate beta-catenin and SNAIL signaling cascades. Cancer Biol Ther 5(9):1059–1064
149. Pelaez-Garcia A et al (2013) FGFR4 role in epithelial-mesenchymal transition and its therapeutic value in colorectal cancer. PLoS One 8(5):e63695
150. Fischer ANM et al (2006) PDGF essentially links TGF-[beta] signaling to nuclear [beta]-catenin accumulation in hepatocellular carcinoma progression. Oncogene 26(23):3395–3405
151. Jechlinger M et al (2003) Expression profiling of epithelial plasticity in tumor progression. Oncogene 22(46):7155–7169
152. Eckert MA et al (2011) Twist1-induced invadopodia formation promotes tumor metastasis. Cancer Cell 19(3):372–386
153. Steller EJ et al (2013) PDGFRB promotes liver metastasis formation of mesenchymal-like colorectal tumor cells. Neoplasia 15(2):204–217
154. Stoker M, Perryman M (1985) An epithelial scatter factor released by embryo fibroblasts. J Cell Sci 77(1):209–223
155. Graveel CR, Tolbert D, Vande Woude GF (2013) MET: a critical player in tumorigenesis and therapeutic target. Cold Spring Harb Perspect Biol 5(7)
156. Liu C, Park M, Tsao MS (1992) Overexpression of c-met proto-oncogene but not epidermal growth factor receptor or c-erbB-2 in primary human colorectal carcinomas. Oncogene 7(1):181–185
157. Di Renzo MF et al (1995) Overexpression amplification of the met/HGF receptor gene during the progression of colorectal cancer. Clin Cancer Res 1(2):147–154
158. Boon EMJ et al (2002) Wnt signaling regulates expression of the receptor tyrosine kinase met in colorectal cancer. Cancer Res 62(18):5126–5128
159. Rasola A et al (2006) A positive feedback loop between hepatocyte growth factor receptor and [beta]-catenin sustains colorectal cancer cell invasive growth. Oncogene 26(7):1078–1087
160. Grimm M et al (2011) Tumor necrosis factor-alpha is associated with positive lymph node status in patients with recurrence of colorectal cancer-indications for anti-TNF-alpha agents in cancer treatment. Cell Oncol (Dordr) 34(4):315–326
161. Baier K et al (2005) Cytokine expression in colon carcinoma. Anticancer Res 25(3B):2135–2139
162. Csiszar A et al (2001) Characterisation of cytokine mRNA expression in tumour-infiltrating mononuclear cells and tumour cells freshly isolated from human colorectal carcinomas. Eur Cytokine Netw 12(1):87–96
163. Zaloudik J et al (1999) Significance of pre-treatment immunological parameters in colorectal cancer patients with unresectable metastases to the liver. Hepatogastroenterology 46(25):220–227
164. Bates RC, Mercurio AM (2003) Tumor necrosis factor-alpha stimulates the epithelial-to-mesenchymal transition of human colonic organoids. Mol Biol Cell 14(5):1790–1800
165. Li CW et al (2012) Epithelial-mesenchymal transition induced by TNF-alpha requires NF-kappaB-mediated transcriptional upregulation of Twist1. Cancer Res 72(5):1290–1300
166. Wang H et al (2013) Epithelial-mesenchymal transition (EMT) induced by TNF-alpha requires AKT/GSK-3beta-mediated stabilization of snail in colorectal cancer. PLoS ONE 8(2):e56664
167. Eger A et al (2004) [beta]-Catenin TGF[beta] signalling cooperate to maintain a mesenchymal phenotype after FosER-induced epithelial to mesenchymal transition. Oncogene 23(15):2672–2680
168. Tian YC, Phillips AO (2002) Interaction between the transforming growth factor-beta type II receptor/Smad pathway beta-catenin during transforming growth factor-and beta1-mediated adherens junction disassembly. Am J Pathol, 160(5):1619–1628
169. Serra R et al (2011) Wnt5a as an effector of TGFbeta in mammary development and cancer. J Mammary Gland Biol Neoplasia 16(2):157–167

170. Horiguchi K et al (2009) Role of Ras signaling in the induction of snail by transforming growth factor-beta. J Biol Chem 284(1):245–253
171. Nieto MA (2002) The snail superfamily of zinc-finger transcription factors. Nat Rev Mol Cell Biol 3(3):155–166
172. Wang Y, Shang Y (2013) Epigenetic control of epithelial-to-mesenchymal transition and cancer metastasis. Exp Cell Res 319(2):160–169
173. Cano A et al (2000) The transcription factor snail controls epithelial-mesenchymal transitions by repressing E-cadherin expression. Nat Cell Biol 2(2):76–83
174. Batlle E et al (2000) The transcription factor snail is a repressor of E-cadherin gene expression in epithelial tumour cells. Nat Cell Biol 2(2):84–89
175. Ikenouchi J et al (2003) Regulation of tight junctions during the epithelium-mesenchyme transition: direct repression of the gene expression of claudins/occludin by Snail. J Cell Sci 116(Pt 10):1959–1967
176. Dave N et al (2011) Functional cooperation between Snail1 and twist in the regulation of ZEB1 expression during epithelial to mesenchymal transition. J Biol Chem 286(14):12024–12032
177. Tran DD et al (2011) Temporal and spatial cooperation of Snail1 Twist1 during epithelial–mesenchymal transition predicts for human breast cancer recurrence. Mol Cancer Res 9(12):1644–1657
178. Kuphal S et al (2005) Snail-regulated genes in malignant melanoma. Melanoma Res 15(4):305–313
179. Becker KF et al (2007) Analysis of the E-cadherin repressor Snail in primary human cancers. Cells Tissues Organs 185(1–3):204–212
180. Roy HK et al (2005) The transcriptional repressor SNAIL is overexpressed in human colon cancer. Dig Dis Sci 50(1):42–6
181. Pálmer HG et al (2004) The transcription factor SNAIL represses vitamin D receptor expression and responsiveness in human colon cancer. Nat Med 10(9):917–919
182. Larriba MJ et al (2009) Snail2 cooperates with Snail1 in the repression of vitamin D receptor in colon cancer. Carcinogenesis 30(8):1459–1468
183. Fan F et al (2012) Overexpression of snail induces epithelial-mesenchymal transition and a cancer stem cell-like phenotype in human colorectal cancer cells. Cancer Med 1(1):5–16
184. Savagner P, Yamada KM, Thiery J (1997) The zinc-finger protein slug causes desmosome dissociation, an initial and necessary step for growth factor-induced epithelial-mesenchymal transition. J Cell Biol 137(6):1403–1419
185. Nieto MA et al (1994) Control of cell behavior during vertebrate development by Slug, a zinc finger gene. Science 264(5160):835–839
186. Hajra KM, Chen DY, Fearon ER (2002) The SLUG zinc-finger protein represses E-cadherin in breast cancer. Cancer Res 62(6):1613–1618
187. Alves CC et al (2009) Role of the epithelial-mesenchymal transition regulator Slug in primary human cancers. Front Biosci (Landmark Ed) 14:3035–3050
188. Shioiri M et al (2006) Slug expression is an independent prognostic parameter for poor survival in colorectal carcinoma patients. Br J Cancer 94(12):1816–1822
189. Garinis GA et al (2002) Hypermethylation-associated transcriptional silencing of E-cadherin in primary sporadic colorectal carcinomas. J Pathol 198(4):442–449
190. Wels C et al (2011) Transcriptional activation of ZEB1 by Slug leads to cooperative regulation of the epithelial-mesenchymal transition-like phenotype in melanoma. J Invest Dermatol 131(9):1877–1885
191. Kataoka H et al (2000) A novel snail-related transcription factor Smuc regulates basic helix-loop-helix transcription factor activities via specific E-box motifs. Nucleic Acids Res 28(2):626–633
192. Bradley CK et al (2013) The snail family gene snai3 is not essential for embryogenesis in mice. PLoS ONE 8(6):e65344
193. Zhao W et al (2013) Snail family proteins in cervical squamous carcinoma: expression and significance. Clin Invest Med 36(4): E223–E233

194. Gheldof A et al (2012) Evolutionary functional analysis and molecular regulation of the ZEB transcription factors. Cell Mol Life Sci 69(15):2527–2541
195. Vandewalle C et al (2005) SIP1/ZEB2 induces EMT by repressing genes of different epithelial cell-cell junctions. Nucleic Acids Res 33(20):6566–6578
196. Aigner K et al (2007) The transcription factor ZEB1 (deltaEF1) promotes tumour cell dedifferentiation by repressing master regulators of epithelial polarity. Oncogene 26(49):6979–6988
197. Grooteclaes ML, Frisch SM (2000) Evidence for a function of CtBP in epithelial gene regulation and anoikis. Oncogene 19(33):3823–3828
198. Nam E-H et al (2012) ZEB2 upregulates integrin α5 expression through cooperation with Sp1 to induce invasion during epithelial–mesenchymal transition of human cancer cells. Carcinogenesis 33(3):563–571
199. Kahlert C et al (2011) Overexpression of ZEB2 at the invasion front of colorectal cancer is an independent prognostic marker and regulates tumor invasion in vitro. Clin Cancer Res 17(24):7654–7663
200. Thisse B, el Messal M, Perrin-Schmitt F (1987) The twist gene: isolation of a *Drosophila zygotic* gene necessary for the establishment of dorsoventral pattern. Nucleic Acids Res 15(8):3439–3453
201. Alexander NR et al (2006) N-cadherin gene expression in prostate carcinoma is modulated by integrin-dependent nuclear translocation of Twist1. Cancer Res 66(7):3365–3369
202. Terauchi M et al (2007) Possible involvement of TWIST in enhanced peritoneal metastasis of epithelial ovarian carcinoma. Clin Exp Metast 24(5):329–339
203. Yang Z et al (2007) Up-regulation of gastric cancer cell invasion by Twist is accompanied by N-cadherin and fibronectin expression. Biochem Biophys Res Commun 358(3):925–930
204. Smit MA et al (2009) A Twist-Snail axis critical for TrkB-induced epithelial-mesenchymal transition-like transformation, anoikis resistance, and metastasis. Mol Cell Biol 29(13):3722–3737
205. Casas E et al (2011) Snail2 is an essential mediator of Twist1-induced epithelial mesenchymal transition metastasis. Cancer Res 71(1):245–254
206. Pérez-Moreno MA et al (2011) A new role for E12/E47 in the repression ofe-cadherin expression and epithelial-mesenchymal transitions. J Biol Chem 276(29):27424–27431
207. Peinado H et al (2004) Snail and E47 repressors of E-cadherin induce distinct invasive and angiogenic properties in vivo. J Cell Sci 117(13):2827–2839
208. Moreno-Bueno G et al (2006) Genetic profiling of epithelial cells expressing E-cadherin repressors reveals a distinct role for Snail, Slug, and E47 factors in epithelial-mesenchymal transition. Cancer Res 66(19):9543–9556
209. Sobrado VR et al (2009) The class I bHLH factors E2-2A and E2-2B regulate EMT. J Cell Sci 122(Pt 7):1014–1024
210. Hartwell KA et al (2006) The Spemann organizer gene, Goosecoid, promotes tumor metastasis. Proc Natl Acad Sci U S A 103(50):18969–18974
211. Mani SA et al (2007) Mesenchyme Forkhead 1 (FOXC2) plays a key role in metastasis and is associated with aggressive basal-like breast cancers. Proc Natl Acad Sci U S A 104(24):10069–10074
212. van Vliet J, Turner J, Crossley M (2000) Human Kruppel-like factor 8: a CACCC-box binding protein that associates with CtBP and represses transcription. Nucleic Acids Res 28(9):1955–1962
213. Zhao W et al (2004) Identification of Kruppel-like factor 4 as a potential tumor suppressor gene in colorectal cancer. Oncogene 23(2):395–402
214. Wang X et al (2007) Kruppel-like factor 8 induces epithelial to mesenchymal transition epithelial and cell invasion. Cancer Res 67(15):7184–7193
215. Li JC et al (2010) Up-regulation of Kruppel-like factor 8 promotes tumor invasion and indicates poor prognosis for hepatocellular carcinoma. Gastroenterology 139(6):2146–2157 e12

216. Wang X et al (2011) KLF8 promotes human breast cancer cell invasion and metastasis by transcriptional activation of MMP9. Oncogene 30(16):1901–1911
217. Fernando RI et al (2010) The T-box transcription factor Brachyury promotes epithelial-mesenchymal transition in human tumor cells. J Clin Invest 120(2):533–544
218. Kilic N et al (2011) Brachyury expression predicts poor prognosis at early stages of colorectal cancer. Eur J Cancer 47(7):1080–1085
219. Sarkar D et al (2012) BRACHYURY confers cancer stem cell characteristics on colorectal cancer cells. Int J Cancer 130(2):328–337
220. Wang B et al (2012) The T box transcription factor TBX2 promotes epithelial-mesenchymal transition and invasion of normal and malignant breast epithelial cells. PLoS ONE 7(7):e41355
221. Rodriguez M et al (2008) Tbx3 represses E-Cadherin expression enhances melanoma invasiveness. Cancer Res 68(19):7872–7881
222. Jagla K et al (1995) Mouse Lbx1 and human LBX1 define a novel mammalian homeo gene family related to the Drosophila lady bird genes. Mech Dev 53(3):345–356
223. Yu M et al (2009) A developmentally regulated inducer of EMT, LBX1, contributes to breast cancer progression. Genes Dev 23(15):1737–1742
224. McCoy EL et al (2009) Six1 expands the mouse mammary epithelial stem/progenitor cell pool and induces mammary tumors that undergo epithelial-mesenchymal transition. J Clin Invest 119(9):2663–2677
225. Micalizzi DS et al (2009) The Six1 homeoprotein induces human mammary carcinoma cells to undergo epithelial-mesenchymal transition and metastasis in mice through increasing TGF-beta signaling. J Clin Invest 119(9):2678–2690
226. Micalizzi DS et al (2010) Homeoprotein six1 increases TGF-beta type I receptor and converts TGF-beta signaling from suppressive to supportive for tumor growth. Cancer Res 70(24):10371–10380
227. Senanayake U et al (2013) The pluripotent renal stem cell regulator SIX2 is activated in renal neoplasms and influences cellular proliferation and migration. Human Pathol 44(3):336–345
228. Takahashi Y et al (2013) Paired related homoeobox 1, a new EMT inducer, is involved in metastasis and poor prognosis in colorectal cancer. Br J Cancer 109(2):307–311
229. Ocana OH et al (2012) Metastatic colonization requires the repression of the epithelial-mesenchymal transition inducer Prrx1. Cancer Cell 22(6):709–724
230. Valdes-Mora F et al (2009) TWIST1 overexpression is associated with nodal invasion and male sex in primary colorectal cancer. Ann Surg Oncol 16(1):78–87
231. Sanchez-Tillo E et al (2011) Beta-catenin/TCF4 complex induces the epithelial-to-mesenchymal transition (EMT)-activator ZEB1 to regulate tumor invasiveness. Proc Natl Acad Sci U S A 108(48):19204–19209
232. Ono H et al (2012) SIX1 promotes epithelial-mesenchymal transition in colorectal cancer through ZEB1 activation. Oncogene 31(47):4923–4934
233. Zhao H et al (2013) E2A suppresses invasion and migration by targeting YAP in colorectal cancer cells. J Transl Med 11:317
234. Lamouille S et al (2013) Regulation of epithelial–mesenchymal and mesenchymal–epithelial transitions by microRNAs. Current Opinion in Cell Biology 25(2):200–207
235. Franco-Chuaire ML, Magda Carolina SC, Chuaire-Noack L (2013) Epithelial-mesenchymal transition (EMT): principles and clinical impact in cancer therapy. Invest Clin 54(2):186–205
236. Park SM et al (2008) The miR-200 family determines the epithelial phenotype of cancer cells by targeting the E-cadherin repressors ZEB1 ZEB2. Genes Dev 22(7):894–907
237. Burk U et al (2008) A reciprocal repression between ZEB1 and members of the miR-200 family promotes EMT and invasion in cancer cells. EMBO Rep 9(6):582–589
238. Bracken C et al (2008) A double-negative feedback loop between ZEB1-SIP1 and the microRNA-200 family regulates epithelial-mesenchymal transition. Cancer Res 68(19):7846–7854

239. Liu S et al (2012) MicroRNA-9 up-regulates E-cadherin through inhibition of NF-kappaB1-Snail1 pathway in melanoma. J Pathol 226(1):61–72
240. Chang CJ et al (2011) Let-7d functions as novel regulator of epithelial-mesenchymal transition and chemoresistant property in oral cancer. Oncol Rep 26(4):1003–1010
241. Ru P et al miRNA-29b suppresses prostate cancer metastasis by regulating epithelial-mesenchymal transition signaling. Mol Cancer Ther 11(5):1166–1173
242. Kumarswamy R et al (2012) MicroRNA-30a inhibits epithelial-to-mesenchymal transition by targeting Snai1 and is downregulated in non-small cell lung cancer. Int J Cancer 130(9):2044–2053
243. Siemens H et al (2011) miR-34 and SNAIL form a double-negative feedback loop to regulate epithelial-mesenchymal transitions. Cell Cycle 10(24):4256–4571
244. Liu YN et al (2013) MiR-1 and miR-200 inhibit EMT via Slug-dependent and tumorigenesis via Slug-independent mechanisms. Oncogene 32(3):296–306
245. Zhang Z et al (2011) Epigenetic silencing of miR-203 upregulates SNAI2 and contributes to the invasiveness of malignant breast cancer cells. Genes Cancer 2(8):782–791
246. Nairismagi ML et al (2012) Translational control of TWIST1 expression in MCF-10A cell lines recapitulating breast cancer progression. Oncogene 31(47):4960–4966
247. Li B et al (2012) Down-regulation of miR-214 contributes to intrahepatic cholangiocarcinoma metastasis by targeting twist. FEBS J 279(13):2393–2398
248. Ma L, Teruya-Feldstein J, Weinberg RA (2007) Tumour invasion and metastasis initiated by microRNA-10b in breast cancer. Nature 449(7163):682–688
249. Ma L et al (2010) Therapeutic silencing of miR-10b inhibits metastasis in a mouse mammary tumor model. Nat Biotech 28(4):341–347
250. Smith AL et al (2012) The miR-106b-25 cluster targets Smad7, activates TGF-beta signaling, and induces EMT and tumor initiating cell characteristics downstream of Six1 in human breast cancer. Oncogene 31(50):5162–5171
251. Ma L et al (2010) miR-9, a MYC/MYCN-activated microRNA, regulates E-cadherin and cancer metastasis. Nat Cell Biol 12(3):247–256
252. Liu C et al (2011) The microRNA miR-34a inhibits prostate cancer stem cells and metastasis by directly repressing CD44. Nat Med 17(2):211–215
253. Liu X et al (2011) MicroRNA-138 suppresses epithelial-mesenchymal transition in squamous cell carcinoma cell lines. Biochem J 440(1):23–31
254. Song Y et al (2012) Inverse association between miR-194 expression and tumor invasion in gastric cancer. Ann Surg Oncol 19(Suppl 3):S509–S517
255. Kong W et al (2008) MicroRNA-155 is regulated by the transforming growth factor β/smad pathway and contributes to epithelial cell plasticity by targeting RhoA. Mol Cell Biol 28(22):6773–6784
256. Peng X et al (2011) Identification of miRs-143 and -145 that is associated with bone metastasis of prostate cancer and involved in the regulation of EMT. PLoS One 6(5):e20341
257. Wang FE et al (2010) MicroRNA-204/211 alters epithelial physiology. FASEB J 24(5):1552–1571
258. Huang Q et al (2008) The microRNAs miR-373 and miR-520c promote tumour invasion metastasis. Nat Cell Biol 10(2):202–210
259. Zhou Q et al (2010) TGF-{beta}-induced MiR-491–5p expression promotes Par-3 degradation in rat proximal tubular epithelial cells. J Biol Chem 285(51):40019–40027
260. Vetter G et al (2010) miR-661 expression in SNAI1-induced epithelial to mesenchymal transition contributes to breast cancer cell invasion by targeting Nectin-1 and StarD10 messengers. Oncogene 29(31):4436–4448
261. Zhang J, Ma L, (2012) MicroRNA control of epithelial-mesenchymal transition and metastasis. Cancer Metastasis Rev 31(3–4):653–662
262. Slaby O et al (2009) MicroRNAs in colorectal cancer: translation of molecular biology into clinical application. Mol Cancer 8:102
263. Bandres E et al (2006) Identification by Real-time PCR of 13 mature microRNAs differentially expressed in colorectal cancer and non-tumoral tissues. Mol Cancer 5:29

264. Michael MZ et al (2003) Reduced accumulation of specific microRNAs in colorectal neoplasia. Mol Cancer Res 1(12):882–891
265. Nagel R et al (2008) Regulation of the adenomatous polyposis coli gene by the mir-135 family in colorectal cancer. Cancer Res 68(14):5795–5802
266. Motoyama K (2009) et al Over- and under-expressed microRNAs in human colorectal cancer. Int J Oncol 34(4):1069–1075
267. Monzo M et al (2008) Overlapping expression of microRNAs in human embryonic colon and colorectal cancer. Cell Res 18(8):823–833
268. Cottonham CL, Kaneko S, Xu L (2010) miR-21 and miR-31 converge on TIAM1 to regulate migration and invasion of colon carcinoma cells. J Biol Chem 285(46):35293–35302
269. Toiyama Y et al (2013) Serum miR-21 as a diagnostic and prognostic biomarker in colorectal cancer. J Natl Cancer Inst 105(12):849–859
270. Menendez P et al (2013) Prognostic implications of serum microRNA-21 in colorectal cancer. J Surg Oncol 108(6):369–373
271. Tsuchiya S et al (2009) MicroRNA-338–3p and microRNA-451 contribute to the formation of basolateral polarity in epithelial cells. Nucleic Acids Res 37(11):3821–3827
272. Sun K et al (2013) miRNA-338–3p suppresses cell growth of human colorectal carcinoma by targeting smoothened. World J Gastroenterol 19(14):2197–2207
273. Bloushtain-Qimron N et al (2008) Cell type-specific DNA methylation patterns in the human breast. Proc Natl Acad Sci U S A 105(37):14076–14081
274. Graff JR et al (1995) E-cadherin expression is silenced by DNA hypermethylation in human breast and prostate carcinomas. Cancer Res 55(22):5195–5199
275. Tamura G et al (2000) E-Cadherin gene promoter hypermethylation in primary human gastric carcinomas. J Natl Cancer Inst 92(7):569–573
276. Grady WM et al (2000) Methylation of the CDH1 promoter as the second genetic hit in hereditary diffuse gastric cancer. Nat Genet 26(1):16–17
277. Caldwell GM et al (2004) The Wnt antagonist sFRP1 in colorectal tumorigenesis. Cancer Res 64(3):883–888
278. Suzuki H et al (2004) Epigenetic inactivation of SFRP genes allows constitutive WNT signaling in colorectal cancer. Nat Genet 36(4):417–422
279. Soutoglou E et al (2000) Transcriptional activation by hepatocyte nuclear factor-1 requires synergism between multiple coactivator proteins. J Biol Chem 275(17):12515–12520
280. Cao Q et al (2008) Repression of E-cadherin by the polycomb group protein EZH2 in cancer. Oncogene 27(58):7274–7284
281. Dexter DL et al (1978) Heterogeneity of tumor cells from a single mouse mammary tumor. Cancer Res 38(10):3174–3181
282. Albini A, Cesana E, Noonan DM (2011) Cancer stem cells and the tumor microenvironment: soloists or choral singers. Curr Pharm Biotechnol 12(2):171–181
283. Reya T et al (2001) Stem cells, cancer, and cancer stem cells. Nature 414(6859):105–111
284. Lapidot T et al (1994) A cell initiating human acute myeloid leukaemia after transplantation into SCID mice. Nature 367(6464):645–648
285. Bennett DC et al (1978) A possible mammary stem cell line. Cell 15(1):283–298
286. Al-Hajj, M. et al (2003) Prospective identification of tumorigenic breast cancer cells. Proc Natl Acad Sci U S A 100(7):3983–3988
287. Wang K et al (2013) Cancer stem cell theory: therapeutic implications for nanomedicine. Int J Nanomedicine 8:899–908
288. Fan X et al (2011) Isolation and characterization of spheroid cells from the HT29 colon cancer cell line. Int J Colorectal Dis 26(10):1279–1285
289. Castellanos JA, Merchant NB, Nagathihalli NS (2013) Emerging targets in pancreatic cancer: epithelial-mesenchymal transition and cancer stem cells. Onco Targets Ther 6:1261–1267
290. Masui T et al (2013) Snail-induced epithelial-mesenchymal transition promotes cancer stem cell-like phenotype in head and neck cancer cells. Int J Oncol

291. Zhou W et al (2014) Snail contributes to the maintenance of stem cell-like phenotype cells in human pancreatic cancer. PLoS ONE 9(1):e87409
292. Hugo H et al (2007) Epithelial—mesenchymal and mesenchymal—epithelial transitions in carcinoma progression. J Cell Physiol 213(2):374–383
293. Rhim AD et al (2012) EMT and dissemination precede pancreatic tumor formation. Cell 148(1–2):349–361
294. Torino F et al (2013) Circulating tumor cells in colorectal cancer patients. Cancer Treat Rev 39(7):759–772
295. Yokobori T et al (2013) Plastin3 is a novel marker for circulating tumor cells undergoing the epithelial-mesenchymal transition and is associated with colorectal cancer prognosis. Cancer Res 73(7):2059–2069
296. Nishi M et al (2014) Induction of cells with cancer stem cell properties from nontumorigenic human mammary epithelial cells by defined reprogramming factors. Oncogene 33(5):643–652
297. Monk M, Holding C (2001) Human embryonic genes re-expressed in cancer cells. Oncogene 20(56):8085–8091
298. Du Z-W., Cong H-C, Yao Z (2001) Identification of putative downstream genes of oct-4 by suppression-subtractive hybridization. Biochem Biophy Res Comm 282(3):701–706
299. Jin G et al (2002) Stem cell pluripotency and transcription factor Oct4. Cell Res 12(5–6):321–329
300. Avilion AA et al (2003) Multipotent cell lineages in early mouse development depend on SOX2 function. Genes Dev 17(1):126–140
301. Liu XF et al (2014) Cervical cancer cells with positive sox2 expression exhibit the properties of cancer stem cells. PLoS ONE 9(1):e87092
302. Liu K et al (2013) The multiple roles for Sox2 in stem cell maintenance and tumorigenesis. Cell Signal 25(5):1264–1271
303. Tiwari N et al (2013) Klf4 is a transcriptional regulator of genes critical for EMT, including Jnk1 (Mapk8). PLoS ONE 8(2):e57329
304. Rowland BD, Bernards R, Peeper DS (2005) The KLF4 tumour suppressor is a transcriptional repressor of p53 that acts as a context-dependent oncogene. Nat Cell Biol 7(11):1074–1082
305. Shi Z et al (2012) Induced pluripotent stem cell-related genes influence biological behavior and 5-fluorouracil sensitivity of colorectal cancer cells. J Zhejiang Univ Sci B 13(1):11–9
306. Ibrahim EE et al (2012) Embryonic NANOG activity defines colorectal cancer stem cells and modulates through ap1- and tcf-dependent mechanisms. Stem Cells 30(10):2076–2087
307. Meng H-M et al (2010) Over-expression of Nanog predicts tumor progression and poor prognosis in colorectal cancer. Cancer Biol Ther 9(4):295–302
308. Ricci-Vitiani L et al (2007) Identification and expansion of human colon-cancer-initiating cells. Nature 445(7123):111–115
309. O'Brien CA et al (2007) A human colon cancer cell capable of initiating tumour growth in immunodeficient mice. Nature 445(7123):106–110
310. Dalerba Pet al (2007) Phenotypic characterization of human colorectal cancer stem cells. Proc Natl Acad Sci U S A 104(24):10158–10163
311. Kemper K, Grandela C, Medema J (2010) Molecular identification and targeting of colorectal cancer stem cells. Oncotarget 1(6):387–395
312. Levin TG et al (2010) Characterization of the intestinal cancer stem cell marker CD166 in the human and mouse gastrointestinal tract. Gastroenterology 139(6):2072–2082 e5
313. Han XY (2013) et al Epithelial-mesenchymal transition associates with maintenance of stemness in spheroid-derived stem-like colon cancer cells. PLoS ONE 8(9):e73341
314. Hirsch D et al (2013) LGR5 positivity defines stem-like cells in colorectal cancer. Carcinogenesis 2013
315. Merlos-Suárez A et al (2011) The intestinal stem cell signature identifies colorectal cancer stem cells and predicts disease relapse. Cell Stem Cell 8(5):511–524

316. Vincan E, Barker N (2008) The upstream components of the Wnt signalling pathway in the dynamic EMT and MET associated with colorectal cancer progression. Clin Exp Metastasis 25(6):657–663
317. Tan KK, Lopes L, Gde Jr Sim R (2009) How uncommon are isolated lung metastases in colorectal cancer? A review from database of 754 patients over 4 years. J Gastrointest Surg 13(4):642–648
318. Vincan E et al (2007) A human three-dimensional cell line model allows the study of dynamic and reversible epithelial- mesenchymal and mesenchymal-epithelial transition that underpins colorectal carcinogenesis. Cells Tissues Organs 185(1–3):20–28
319. Thiery J (2002) Epithelial-mesenchymal transitions in tumour progression. Nat Rev Cancer 2(6):442–454
320. Maeda M, Johnson KR, Wheelock MJ (2005) Cadherin switching: essential for behavioral but not morphological changes during an epithelium-to-mesenchyme transition. J Cell Sci 118(Pt 5):873–887
321. Brabletz T et al (2001) Variable beta-catenin expression in colorectal cancers indicates tumor progression driven by the tumor environment. Proc Natl Acad Sci U S A 98(18):10356–10361
322. Wells A, Yates C, Shepard CR (2008) E-cadherin as an indicator of mesenchymal to epithelial reverting transitions during the metastatic seeding of disseminated carcinomas. Clin Exp Metastasis 25(6):621–628
323. Chao YL, Shepard CR, Wells A (2010) Breast carcinoma cells re-express E-cadherin during mesenchymal to epithelial reverting transition. Mol Cancer 9:179
324. Chaffer CL, Thompson EW, Williams ED (2007) Mesenchymal to epithelial transition in development disease. Cells Tissues Organs 185(1–3):7–19
325. Zeisberg M, Shah AA, Kalluri R (2005) Bone morphogenic protein-7 induces mesenchymal to epithelial transition in adult renal fibroblasts and facilitates regeneration of injured kidney. J Biol Chem 280(9):8094–8100
326. Kouso H et al (2013) Differences in the expression of epithelial-mesenchymal transition related molecules between primary tumors and pulmonary metastatic tumors in colorectal cancer. Surg Today 43(1):73–80
327. Yao D, Dai C, Peng S (2011) Mechanism of the mesenchymal-epithelial transition its and relationship with metastatic tumor formation. Mol Cancer Res 9(12):1608–1620
328. Yates CC et al (2007) Co-culturing human prostate carcinoma cells with hepatocytes leads to increased expression of E-cadherin. Br J Cancer 96(8):1246–1252
329. Andarawewa KL et al (2007) Ionizing radiation predisposes nonmalignant human mammary epithelial cells to undergo transforming growth factor beta induced epithelial to mesenchymal transition. Cancer Res 67(18):8662–8670
330. Mege RM et al (1998) Construction of epithelioid sheets by transfection of mouse sarcoma cells with cDNAs for chicken cell adhesion molecules. Proc Natl Acad Sci U S A 85(19):7274–7278
331. Seidel B et al (2004) E- and N-cadherin differ with respect to their associated p120ctn isoforms and their ability to suppress invasive growth in pancreatic cancer cells. Oncogene 23(32):5532–5542
332. Brabletz T et al (2005) Invasion and metastasis in colorectal cancer: epithelial-mesenchymal transition, mesenchymal-epithelial transition, stem cells and beta-catenin. Cells Tissues Organs 179(1–2):56–65
333. Palmqvist R et al (2000) Human colorectal cancers with an intact p16/cyclin D1/pRb pathway have up-regulated p16 expression and decreased proliferation in small invasive tumor clusters. Am J Pathol 157(6):1947–1953
334. Brabletz T et al (2005) Opinion: migrating cancer stem cells—an integrated concept of malignant tumour progression. Nat Rev Cancer 5(9):744–749
335. Oltean S, Febbo G, Garcia-Blanco MA (2008) Dunning rat prostate adenocarcinomas and alternative splicing reporters: powerful tools to study epithelial plasticity in prostate tumors in vivo. Clin Exp Metastasis 25(6):611–619

336. Oltean S et al (2006) Alternative inclusion of fibroblast growth factor receptor 2 exon IIIc in Dunning prostate tumors reveals unexpected epithelial mesenchymal plasticity. Proc Natl Acad Sci U S A 103(38):14116–14121
337. Baraniak A, Chen JR, Garcia-Blanco MA (2006) Fox-2 mediates epithelial cell-specific fibroblast growth factor receptor 2 exon choice. Mol Cell Biol 26(4):1209–1222
338. Warzecha CC et al (2009) ESRP1 and ESRP2 are epithelial cell-type-specific regulators of FGFR2 splicing. Mol Cell 33(5):591–601
339. Vincan E et al (2007) Frizzled-7 dictates three-dimensional organization of colorectal cancer cell carcinoids. Oncogene 26(16):2340–2352
340. Kang MH et al (2009) Inhibition of PI3 kinase/Akt pathway is required for BMP2-induced EMT and invasion. Oncol Rep 22(3):525–534
341. Buijs JT et al (2007) Bone morphogenetic protein 7 in the development treatment of bone metastases from breast cancer. Cancer Res 67(18):8742–8751
342. Buijs JT et al (2007) BMP7, a putative regulator of epithelial homeostasis in the human prostate, is a potent inhibitor of prostate cancer bone metastasis in vivo. Am J Pathol 171(3):1047–1057
343. Na YR et al (2009) Bone morphogenetic protein 7 induces mesenchymal-to-epithelial transition in melanoma cells, leading to inhibition of metastasis. Cancer Sci 100(11):2218–2225
344. Zhou XD, Agazie YM (2008) Inhibition of SHP2 leads to mesenchymal to epithelial transition in breast cancer cells. Cell Death Differ 15(6):988–996
345. Gregory A et al (2008) The miR-200 family and miR-205 regulate epithelial to mesenchymal transition by targeting ZEB1 and SIP1. Nat Cell Biol 10(5):593–601
346. Sossey-Alaoui K, Bialkowska K, Plow EF (2009) The miR200 family of microRNAs regulates WAVE3-dependent cancer cell invasion. J Biol Chem 284(48):33019–33029
347. Li CM et al (2002) Gene expression in Wilms' tumor mimics the earliest committed stage in the metanephric mesenchymal-epithelial transition. Am J Pathol 160(6):2181–2190
348. Saito T, Nagai M, Ladanyi M (2006) SYT-SSX1 and SYT-SSX2 interfere with repression of E-cadherin by snail and slug: a potential mechanism for aberrant mesenchymal to epithelial transition in human synovial sarcoma. Cancer Res 66(14):6919–6927
349. Auersperg N et al (2011) E-cadherin induces mesenchymal-to-epithelial transition in human ovarian surface epithelium. Proc Natl Acad Sci U S A 96(11):6249–6254
350. Chui MH (2013) Insights into cancer metastasis from a clinicopathologic perspective: Epithelial-Mesenchymal Transition is not a necessary ste Int J Cancer 132(7):1487–1495
351. Tarin D, Thompson EW, Newgreen DF (2005) The fallacy of epithelial mesenchymal transition in neoplasia. Cancer Res 65(14):5996–6000; discussion 6000–1
352. Tsai JH, Yang J (2013) Epithelial-mesenchymal plasticity in carcinoma metastasis. Genes Dev 27(20):2192–2206
353. Pluchino KM et al (2012) Collateral sensitivity as a strategy against cancer multidrug resistance. Drug Resist Updat 15(1–2):98–105
354. Gottesman MM, Fojo T, Bates SE (2002) Multidrug resistance in cancer: role of ATP-dependent transporters. Nat Rev Cancer 2(1):48–58
355. Bhangu A et al (2012) Epithelial mesenchymal transition in colorectal cancer: seminal role in promoting disease progression and resistance to neoadjuvant therapy. Surg Oncol 21(4):316–323
356. Shang Y, Cai X, Fan D (2013) Roles of epithelial-mesenchymal transition in cancer drug resistance. Curr Cancer Drug Targets
357. Thiery J (2003) Epithelial-mesenchymal transitions in development and pathologies. Curr Opin Cell Biol 15(6):740–746
358. Radisky DC, LaBarge MA (2008) Epithelial-mesenchymal transition and the stem cell phenotype. Cell Stem Cell 2(6):511–512
359. Visvader JE, Lindeman GJ (2008) Cancer stem cells in solid tumours: accumulating evidence and unresolved questions. Nat Rev Cancer 8(10):755–768

360. Bao B et al (2012) Hypoxia induced aggressiveness of prostate cancer cells is linked with deregulated expression of VEGF, IL-6 and miRNAs that are attenuated by CDF. PLoS ONE 7(8):e43726
361. Meng F, Wu G (2012) The rejuvenated scenario of epithelial-mesenchymal transition (EMT) and cancer metastasis. Cancer Metastasis Rev 31(3–4):455–467
362. Teicher BA (2009) Acute and chronic in vivo therapeutic resistance. Biochem Pharmacol 77(11):1665–1673
363. Shrader M et al (2007) Molecular correlates of gefitinib responsiveness in human bladder cancer cells. Mol Cancer Ther 6(1):277–285
364. Wang J et al (2013) Characterization of two novel cell lines with distinct heterogeneity derived from a single human bile duct carcinoma. PLoS ONE 8(1):e54377
365. Kajiyama H et al (2007) Chemoresistance to paclitaxel induces epithelial-mesenchymal transition and enhances metastatic potential for epithelial ovarian carcinoma cells. Int J Oncol 31(2):277–283
366. Rosano L et al (2011) Acquisition of chemoresistance and EMT phenotype is linked with activation of the endothelin A receptor pathway in ovarian carcinoma cells. Clin Cancer Res 17(8):2350–2360
367. Jia L et al (2012) Paclitaxel inhibits ovarian tumor growth by inducing epithelial cancer cells to benign fibroblast-like cells. Cancer Lett 326(2):176–182
368. Yew KH et al (2013) Epimorphin-induced met sensitizes ovarian cancer cells to platinum. PLoS ONE 8(9):e72637
369. Kurrey NK et al (2009) Snail and slug mediate radioresistance and chemoresistance by antagonizing p53-mediated apoptosis and acquiring a stem-like phenotype in ovarian cancer cells. Stem Cells 27(9):2059–2068
370. Perez-Mancera A et al (2005) Cancer development induced by graded expression of Snail in mice. Hum Mol Genet 14(22):3449–3461
371. Moody SE et al (2005) The transcriptional repressor Snail promotes mammary tumor recurrence. Cancer Cell 8(3):197–209
372. Boman BM, Wicha MS (2008) Cancer stem cells: a step toward the cure. J Clin Oncol 26(17):2795–2799
373. Eyler CE, Rich JN (2008) Survival of the fittest: cancer stem cells in therapeutic resistance and angiogenesis. J Clin Oncol 26(17):2839–2845
374. Baguley BC (2010) Multidrug resistance in cancer. Methods Mol Biol 596:1–14
375. Mallini P et al (2013) Epithelial-to-mesenchymal transition: what is the impact on breast cancer stem cells and drug resistance. Cancer Treat Rev
376. Borst P (2012) Cancer drug pan-resistance: pumps, cancer stem cells, quiescence, epithelial to mesenchymal transition, blocked cell death pathways, persisters or what? Open Biol 2(5):120066
377. Vincent-Salomon A, Thiery J (2003) Host microenvironment in breast cancer development: epithelial-mesenchymal transition in breast cancer development. Breast Cancer Res 5(2):101–106
378. Carbone C et al (2011) Anti-VEGF treatment-resistant pancreatic cancers secrete proinflammatory factors that contribute to malignant progression by inducing an EMT cell phenotype. Clin Cancer Res 17(17):5822–5832
379. Paraiso KH, Smalley KS (2013) Fibroblast-mediated drug resistance in cancer. Biochem Pharmacol 85(8):1033–1041
380. Chen J et al (2011) Overexpression of miR-429 induces mesenchymal-to-epithelial transition (MET) in metastatic ovarian cancer cells. Gynecol Oncol 121(1):200–205
381. Vettori S, Gay S, Distler O (2012) Role of MicroRNAs in Fibrosis. Open Rheumatol J 6:130–139
382. Wang V, Wu W (2009) MicroRNA-based therapeutics for cancer. BioDrugs 23(1):15–23
383. Hong KO et al (2009) Inhibition of Akt activity induces the mesenchymal-to-epithelial reverting transition with restoring E-cadherin expression in KB KOSCC-25B oral squamous cell carcinoma cells. J Exp Clin Cancer Res 28:28

384. Deep G, Agarwal R (2013) Targeting tumor microenvironment with silibinin: promise and potential for a translational cancer chemopreventive strategy. Curr Cancer Drug Targets 13(5):486–499
385. Sachlos E et al (2012) Identification of drugs including a dopamine receptor antagonist that selectively target cancer stem cells. Cell 149(6):1284–1297
386. Dalton SO et al (2005) Risk for cancer in a cohort of patients hospitalized for schizophrenia in Denmark, 1969–1993. Schizophr Res 75(2–3):315–324
387. Kirtane AR, Kalscheuer SM, Panyam J (2013) Exploiting nanotechnology to overcome tumor drug resistance: challenges and opportunities. Adv Drug Deliv Rev
388. Gao Z, Zhang L, Sun Y (2012) Nanotechnology applied to overcome tumor drug resistance. J Control Release 162(1):45–55
389. Iyer AK et al (2013) Role of integrated cancer nanomedicine in overcoming drug resistance. Adv Drug Deliv Rev
390. Matsumura Y, Maeda H (1986) A new concept for macromolecular therapeutics in cancer chemotherapy: mechanism of tumoritropic accumulation of proteins and the antitumor agent smancs. Cancer Res 46(12 Pt 1):6387–6392

Omics and Biomarkers Development for Intestinal Tumorigenesis

Mehmet Koyuturk and Rod K. Nibbe

Abstract While considerable research into colorectal cancer (CRC) has implicated many genetic alterations that trigger the disease and sustain its progression, there are few well-validated, clinically useful molecular biomarkers of CRC. The observation that cancer is highly diverse across individual tumors is manifested at the molecular level by concomitantly diverse patterns of gene expression. However, while analysis of gene expression has been used to identify candidate biomarkers of cancer, such biomarkers frequently do not cross validate well on independent datasets and this has raised legitimate concerns regarding the usefulness of gene expression based markers. It is has been postulated that by integrating the *functional* information of gene products into the approach, networks of mechanistically related gene products may be identified and used to develop more robust biomarkers. Many such approaches focus on established signaling pathways for this purpose; however, pathways consisting of a few proteins interacting in a serial fashion oversimplify, and provide inadequate models for, a complex phenotype (e.g. CRC) mediated by a constellation of interacting gene products. Here, we discuss several integrative techniques based on cellular networks (protein–protein interactions) and incorporation of lower-coverage, but functionally relevant proteomic data, and show the power these techniques hold for prioritizing disease genes for biomarker discovery and biological verification of function.

Keywords Colorectal cancer biomarkers · Proteomics interactome · Computational-biology · Integrative-omics · Information theory · Protein–protein interactions

---

M. Koyuturk (✉)
Department of Electrical Engineering & Computer Science, Case Western Reserve University, Cleveland, OH, USA
e-mail: mehmet.koyuturk@case.edu

R. K. Nibbe
Systems Biology NeoProteomics, Inc., Cleveland, OH, USA
e-mail: rkn2660@gmail.com

V. W. Yang, A. B. Bialkowska (eds.), *Intestinal Tumorigenesis*,
DOI 10.1007/978-3-319-19986-3_12

1 Introduction

One expectation of the *–omics* revolution is that high-dimensional data sets from proteomic and genomic experiments should be integrated with *functional* information to provide a more complete account of the molecular changes underlying the etiology of human disease. In the past decade there has been a rapid increase in the number and size of databases which catalog specific gene annotations and experimental protein or gene expression data. Network modeling of protein–protein interactions provides an ideal context for integrating these data [1–5]. Network modeling approaches enable the discovery of disease-associated networks by leveraging protein–protein interaction (PPI) databases which have been created from literature curation, inference by homology, co-expression or co-evolution, high-throughput data, or a combination of these [6]. These PPI databases capture tens of thousands of interactions—both direct (physical) and indirect (functional)—among thousands of gene products involved in mediating a normal cellular state.

Network Biology It has been hypothesized that the dysregulation of certain network modules, or "subnetworks," located within the global PPI may have a causative role in certain disease states [4, 5]. Network modeling, or "network biology," is a new generation of integrative, systems biology approaches well suited to identifying these subnetworks, and in turn elucidating the molecular mechanisms of disease initiation or progression, and providing more robust models of disease heterogeneity consistent with the goals of personalized medicine. With the additional cellular information provided by these network models and accompanying search algorithms, it is expected these models will lead to the discovery of more clinically relevant biomarkers, or drive the identification of companion diagnostics in drug discovery pipelines.

In the following section, we give a brief overview of the rationale and goals related to three separate network based techniques for prioritizing candidate disease genes, not only for the aforementioned reason of biomarker discovery, but also to prioritize experiments to verify the function of these genes in the pathophysiology of disease. In Sect. 3, we provide a more detailed discussion of the mathematical and computational basis for each approach, and summarize key results of previously published studies where each one was applied in a specific context of CRC.

1.1 Approach Details

1.1.1 Approach 1: *S*ubnetwork *A*nalysis and *S*coring *Sy*stem: *SASSy*

SASSy (see Fig. 1) is a high-throughput method to identify a compact set of functionally related proteins discriminative of a specific disease phenotype. The first step in the process involves the discovery of subnetworks based on a "seed" of a priori disease genes, such as those found by proteomic profiling of a set of test and control samples. We call a network discovered using this seed of disease genes a "disease-associated sub-network". Such a sub-network is a small module derived from the search of a much larger PPI network, one that represents all the known proteins and interactions among them in a given species. The task of creating disease-associated subnetworks is commonly carried out using a commercial software product, e.g. Ingenuity (*IPA*) or Thompson Reuters (*MetaCore*). However, these subnetworks contain many nodes and interactions (typically ≧ 50), some of which are likely to be false positives. Therefore, for these networks to be useful for biomarker discovery or biological verification of function, a method is needed to prune these subnetworks of those proteins/interactions which, although they may be "topologically close" to the seed genes, may in fact have no functional role in the disease phenotype. To prune the subnetworks, we first create a model of subnetwork *activity* using gene

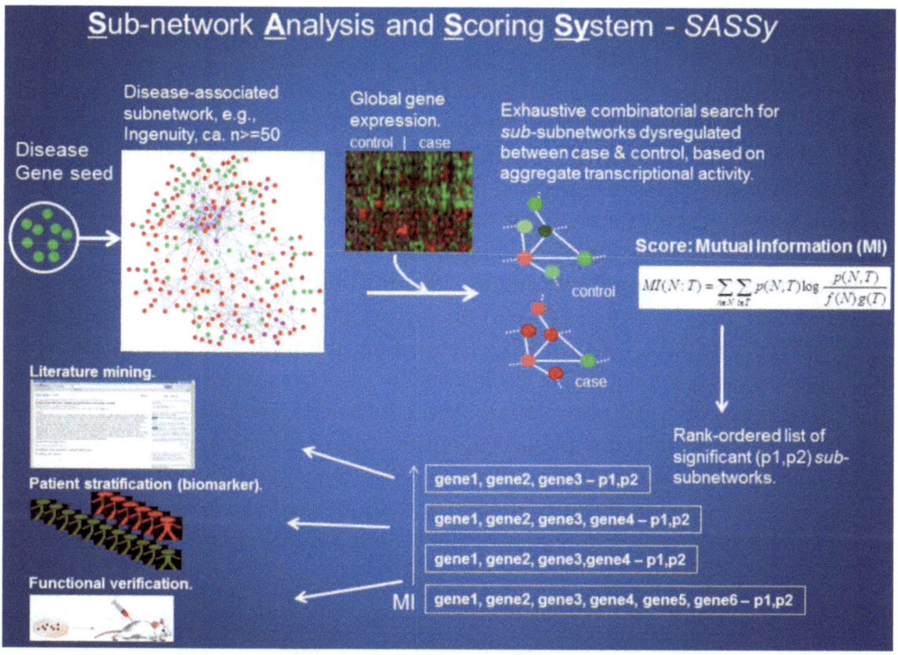

Fig. 1 A "seed" of disease genes, for example one found by proteomic profiling between case and control, is used to search a global protein–protein interaction (PPI) network for disease-associated subnetworks of interacting proteins. A disease-associated subnetwork will contain proteins from the seed (*green*) but also other proteins inferred to be relevant by virtue of their proximity to the seed proteins. To prune a subnetwork of false positives, gene expression data is integrated and used as a model of subnetwork activity. The disease-associated subnetwork is exhaustively searched for *sub*-subnetworks whose activity is dysregulated between case and control as measured by mutual information (MI). The output is a list of *sub*-subnetworks (typically $n \leq 8$) rank-ordered by MI, each having two independent measures of statistical significance (p1 & p2). The genes in the high-ranking sub-subnetworks are used to mine the literature for novel disease associations; train classifiers for biomarker discovery and patient stratification; or to propose testable hypotheses by traditional low-throughput experiments to verify function in a biologically relevant model (e.g. cell culture or animal)

expression data. Next, using an information-theoretic approach, we exhaustively search the subnetwork for small (typically $n \leqq 8$) combinations of genes whose transcriptional activity is most significantly dysregulated between test and control. We postulate that the gene products in these *sub*-subnetworks are the most likely candidates to have a functional role in causing or sustaining the disease phenotype, and as such are useful for prioritizing low-throughput verification experiments or driving a biomarker discovery pipeline (e.g., developing robust classifiers).

1.1.2 Approach 2: A Disease-Associated Subnetwork Discovery Method Based on Protein Crosstalk: *X-talker*

X-talker (see Fig. 2) also integrates proteomic and gene expression data in the context of PPI networks to identify small subnetworks significantly dysregulated between test and control, but differs in two important computational ways: (1) it leverages publically available PPI data and gene expression sets alleviating the need for an expensive commercial product to generate the disease-associated subnetworks, and (2) it incorporates a novel method for generating these subnetworks based on a model of protein "crosstalk." Protein crosstalk refers to the way a given protein affects one or more biological pathways (e.g., the EGFR signaling pathway) to regulate the pathway's activity, or when the protein's expression is dysregulated, to interfere with the pathway's normal activity.

We ground *X-talker* on the hypothesis that the observed fold change of the proteomic targets may be associated with the synergistic dysregulation of their interacting partners at the level of mRNA. Therefore, the proteomic targets present a biologically relevant location in the interactome to bootstrap a search for disease associated "hotspots", the observation of which has considerable support in the literature [4, 7–10]. From a computational perspective, we postulate that subnetworks which exhibit significant association with the proteomic seed targets (crosstalk) should also show a significant change in activity between control and phenotype, where change in activity is modeled as it was in the *SASSy* approach (see previous

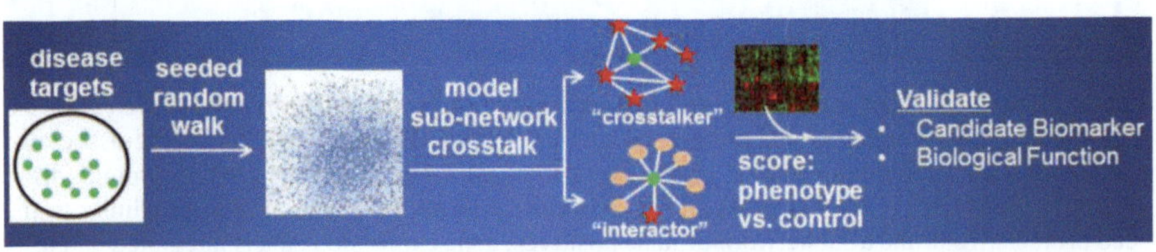

Fig. 2 A schematic of the Crosstalker approach. A "seed" of disease genes, for example one found by proteomic profiling between case and control, is used to "walk" (search) a global protein–protein interaction (PPI) network for disease hotspots, i.e. subnetworks of proteins significantly associated with one or more seed genes. Such subnetworks we define as "crosstalkers" (non-significant subnetworks are termed "interactors"), specific locations in the interactome predicted to be significantly dysregulated in the disease phenotype. If appropriate gene expression data is available, the crosstalker subnetworks may subsequently scored for their dysregulation using *SASSy*. Crosstalker was flexibly designed to work with any suitably formatted public PPI and gene expression data set

section). Unlike comparable approaches, *X-talker* does not rely on canonical disease-related genes or proteins [11]; rather, it is a global, experimentally driven discovery method that identifies network interactions statistically significant with respect to all targets in an experimentally-derived set.

Beyond the aforementioned bias and noise *X-talker* overcomes, it also overcomes the limitations of certain proprietary search algorithms, for instance the one used by the commercial software used to generate the disease-associated subnetworks in the *SASSy* approach. *X-talker* substantially improves on the overall approach based on a plausible model of protein association and activity, and is flexibly adaptable to a wide variety of publicly available proteomic, genomic and PPI datasets. It takes as input a set of proteins identified via proteomic screening, a comprehensive set of human protein–protein interactions, gene expression data for phenotype and control samples for all genes in the network, and returns a set of dysregulated subnetworks in close proximity of the proteomic targets.

1.1.3 Approach 3: *C*ombinatorially Dys*R*egul*A*ted Sub*NE*tworks: *CRANE*

SASSy and *X-talker* use "subnetwork activity" as the expression profile of a protein subnetwork. Here, the activity of a subnetwork in a sample is defined as the *aggregate* expression of the genes in the subnetwork in that sample. While quite useful, this additive scheme captures the coordination between the dysregulation of interacting gene products to a limited extent, since interacting genes may not be additive in their functions. For example, consider two genes whose coordinate function is disrupted in a certain phenotype. If the disruption causes up-regulation of one of these genes and the down-regulation of the other, then this effect will be canceled out by additive measures like subnetwork activity [12].

The third approach we discuss in this chapter is based on the observation that coordinated changes in the mRNA-level expression of interacting proteins can exhibit combinatorial patterns as well (see Fig. 3) [13]. For example, in the context of predicting metastasis of CRC, one can search for rules such as "if gene x is up-regulated, gene *y* is down-regulated, and gene *z* is up-regulated in a patient, then the tumor is more likely to metastasize into the liver". To identify such patterns, we quantize gene expression into discrete expression levels and represent the expression profile of a subnetwork as a multi-dimensional random variable that represents the combination of expression states of the genes in the subnetwork. We then quantify the dysregulation of each subnetwork based on the mutual information between this multi-dimensional variable and phenotype, i.e., the information on phenotype gained by knowledge of the expression state of the subnetwork. This formulation allows development of efficient search algorithms for significant subnetworks, thereby enabling unbiased search of the human PPI network for protein subnetworks dysregulated in metastatic CRC. The resulting algorithm is named *CRANE*.

In the context of predicting metastasis of CRC, *CRANE* overs two benefits over other approaches: (1) *CRANE* captures tumor heterogeneity more elaborately by modeling the interactions between gene products in a subnetwork combinatorially,

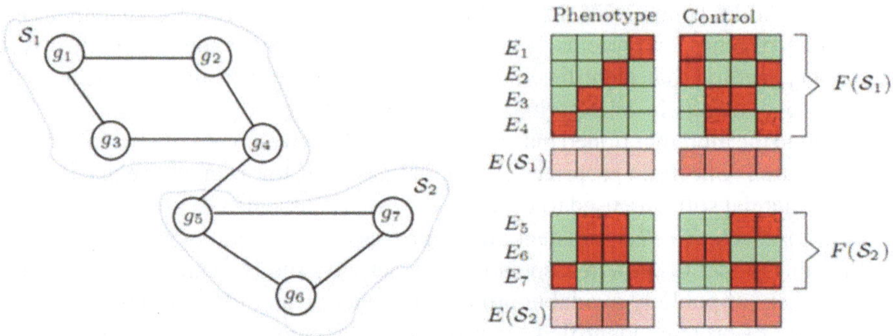

Fig. 3 A combinatorial approach to discover dysregulated subnetworks. Consider two subnetworks, S1 and S2, composed of four and three genes, respectively, and the corresponding gene expression pattern of these genes measured in four phenotype and four control samples. Red represents an up-regulated gene, green a down-regulated gene. Using an aggregate (additive) measure of gene expression to "score" dysregulation between test and control, where the activity of the subnetwork in each sample is represented as the average expression of the four genes, control and phenotype are clearly separable in S1. That is, the subnetwork activity in each phenotype sample is light red, and dark red in each control sample. However, in S2, an aggregate measure of activity cannot discriminate phenotype and control. To account for the kind heterogeneity frequently observed in complex diseases like cancer, CRANE first binarizes the gene expression data, rendering every gene in every sample as "on or off", and then searches the PPI for closely connected subnetworks that maximize the *combinatorial information* between phenotype and control [13]

as opposed to additive modeling of subnetwork activity. (2) Through efficient search of the subnetwork space, CRANE lifts the need to seed the search using proteomic data and enables an unbiased search using only gene expression and protein interaction data.

2 Materials

2.1 SASSy

Microarray mRNA expression was measured by a cDNA microarray on 171 human colon tissue samples of various stages (normal 16, stage B 41, stage C 25, stage D 50, metastatic 39) using the Affymetrix Human EXON 1.0 ST chip. Expression values for normal and stage D samples for select genes in our networks, plus a decoy database of 1000 genes for hypothesis testing, were generously provided to us by the Case Comprehensive Cancer Center.

Proteomic Seed Genes We performed 2D in-gel (2D-DIGE system, GE Healthcare) electrophoresis using digests obtained from 12 Duke's D tissue biopsies (with 12 matched controls), which had been snap frozen at the time of surgery. Quantitative image processing of the gels resolved 150 spots significantly, differently expressed at two-fold or greater between test and control. The spots were excised from the gels and digested by trypsin, and subsequently analyzed by LC-MS/MS for protein identification. Sixty seven proteins were identified using the SEAQUEST software based on a strict measure of statistical significance for peptide detection.

PPI Network Database We used the MetaCore PPI database and associated "closest neighbor" search algorithm from Thompson Reuters (formerly GeneGo) to generate the four disease-associated subnetworks. The gene symbol for each gene in a subnetwork was exported to a text file and used in the scoring procedure.

Software The prototype implementation of *SASSy* was developed at the Center for Proteomics and Bioinformatics at Case Western Reserve University using the Matlab software, and subsequently licensed by the university to NEO Proteomics, where it was re-implemented in JAVA to commercial grade software.

2.2 X-talker

Proteomic targets (seed) Two proteomic seeds were compared: Nibbe et al. [14] ($n=67$) and Friedman et al. [15] ($n=55$). The Nibbe seed was the proteomic seed used in the testing of SASSy. The Freidman seed was similarly obtained from profiling 12 samples obtained from six patient tumors (w/matched controls) of mixed pathological stage.

PPI Database The Human Reference Database (HPRD) [16] was downloaded from the HPRD website and contained 35023 binary interactions between 9299 proteins, as well as 1060 protein complexes consisting of 2146 proteins. We integrated the binary interactions and protein complexes using a matrix model (e.g., each complex is represented as a clique between the proteins in the complex), to obtain a PPI network composed of 42781 binary interactions among 9442 proteins. Sixty of the proteomic seeds from the data of Nibbe et al. had at least one interaction in HPRD, while 37 of the seeds from the data of Friedman et al. had at least one interaction in HPRD. Fourteen of the proteins in the two seed sets were common.

Microarrays Two microarrays were downloaded from the Gene Expression Omnibus at NCBI. *GSE8671*: represents 64 experiments using mRNA isolated from tissue biopsies obtained from 32 patients (matched CRC tumor and adjacent normal mucosa) performed on an Affymetrix GeneChip (Human U133 Plus 2.0).

GSE10942: represents 48 experiments on matched tissue biopsies (24 patients) performed on an Illumina array (Human ref-8, v2.0).

Software A prototype of the methods described here was implemented with the MATLAB software (v2007b). For classification we used a support vector machine implemented using Matlab's svmtrain() and svmclasify() functions. The Pajek software was used for network visualization.

2.3 *CRANE*

Microarrays In our experiments, we use two CRC-related microarray datasets obtained from GEO (Gene Expression Omnibus; *http://www.ncbi.nlm.nih.gov/geo/index.cgi*). These datasets, referenced by their accession number in the GEO database, include the following relevant data:

GSE6988 contains expression profiles of 17,104 genes across 29 *vs.* 51 colorectal tumor samples with and without liver metastasis, respectively.

GSE3964 contains expression profiles of 5845 genes across 28 *vs.* 18 colorectal tumor samples with and without liver metastasis, respectively.

PPI Network Database The human protein–protein interaction data used in our experiments was obtained from the Human Protein Reference Database (HPRD) [16]. This dataset contained 35023 binary interactions among 9299 proteins, as well as 1060 protein complexes consisting of 2146 proteins. We integrated the binary interactions and protein complexes to obtain a PPI network composed of 42781 binary interactions among 9442 proteins.

Software A prototype of *CRANE* was implemented with the MATLAB software and subsequently licensed by the university to NEO Proteomics, where it was re-implemented in JAVA to commercial grade software.

3 Methods

3.1 *SASSy*

3.1.1 Approach

Consider a relatively large ($\geq$50 nodes) disease-associated subnetwork generated from a "seed" of disease-associated genes, for example, as identified by proteomic profiling (see Fig. 1). While the subnetwork may be significantly "associated" with some or many genes in the seed, such a subnetwork graph entails no quantitative measure of the activity of singles genes or a group of genes with respect to the phenotype it is associated with. We postulate that the activity of *some* genes in the subnetwork is coordinately dysregulated between test and control samples, and hence these genes (proteins) may have a *functional* role in causing or sustaining the phenotype. Conversely, the activity of certain other genes in the subnetwork is *not* dysregulated between test and control, and therefore these genes are false positives with respect to functional relevance.

To find the functionally dysregulated genes we create a model of subnetwork activity based on gene expression, and assign each protein in the subnetwork the transcriptional value of its cognate gene (obtained from appropriately normalized microarray experiments on the same control/phenotype used to generate the seed genes). We do this for every gene in the subnetwork for all samples on the microarray, yielding a G × S array we call the activity matrix, where the rows represent the subnetwork genes {1…i} and the columns the control and test samples {1…j}. We estimate the aggregate activity of the entire subnetwork in each sample by the normalized sum of the gene expression values. As activity is a continuous variable, we discretize it into n equally spaced bins using Sturge's rule, the same method employed by Ideker and co-workers [7]:

$$nBins = log_2(j) + 1$$

This leaves us with a discrete measure of subnetwork activity for each control and test sample. Next, we compute the marginal distribution of the activity for each phenotype, p(N) and p(T), where N = control and T = test, and the joint distribution of activity, p(N, T), and then dysregulation is estimated by the mutual information (MI):

$$MI(N:T) = \sum_{n \in N} \sum_{t \in T} p(N,T) \log \frac{p(N,T)}{f(N)g(T)}$$

Mutual information effectively measures the ability of a subnetwork genotype to predict the phenotype. The benefit of this formulation is that it makes no assumption on the distribution of the phenotype or genotype variables; it works on empirical distributions to characterize the dependency between the two variables. Finally, we exhaustively search the subnetwork for combinations of n genes that maximize MI, the metric we use to rank order the result set. In addition, statistical significance of each result is based on two p-values (p_1, p_2) computed from two separate null distributions of MI, respectively: (1) 10K random n-gene combinations, and (2) 10K random permutations of phenotype.

3.2 Results

We applied *SASSy* to identify candidate, subnetwork based biomarkers of late stage (Duke's D) CRC [14]. Much research has focused on the *early* detection of CRC, but far less attention has been paid to whether or not there are molecular markers specific to a late-stage (Duke's D) phenotype, and whether or not these markers present opportunities for therapeutic intervention to improve the prognosis for long-term survival, which for a late stage diagnosis is less than ten percent [17].

We performed proteomic profiling of twenty-four tissue biopsies obtained from stage D tumors ($n = 12$) with matched controls ($n = 12$). We identified sixty seven proteins significantly, differentially expressed (up- and down-regulated) between test and control. The list of proteomic targets ("seed") was imported into a commercial network discovery software tool (Thompson Rueter's *MetaCore*, formerly

Table 1 Significantly dysregulated genes in Duke's stage D CRC. Bold gene names Indicate the gene was also detected as significant by proteomic profiling

Signature # and Mutual Information (MI) score	Genes dysregulated in Stage D CRC	Statistical significance
1, MI = 0.4981	***CapG***, *HIST1H2AB* *IGFBP3* ,*ITGB1* *TUBA4A*, *TUBB3*	*p*H1 < 0.0004 *p*H2 << 0.0001
2, MI = 0.4628	*CSNK2A2*, *GNA12*, *PDGFRB*, *PLA2G12A* *PLK1*, ***SERPINA1***	*p*H1 < 0.0009 *p*H2 << 0.0001
3, MI = 0.6063	*CCT3*, *CCT5* *CCT7*, *DYNLT3* *PLK1*, *TRAP1*	*p*H1 < 0.0007 *p*H2 << 0.0001
4, MI = 0.5398	*ACTL6A*, ***HSP90AB1*** *IFITM1*, *POLR2D* *SMARC4A*, *RPS2*	*p*H1 < 0.0002 *p*H2 << 0.0001

GeneGo, version 4.6 build 12332) that permits a search of a PPI database comprised of tens of thousands of manually curated protein–protein interactions. The search returned four subnetworks (each with $n \leqq 75$ nodes) significantly associated with the proteomic targets. Each of these disease-associated subnetworks was subsequently "scored" using *SASSy*. The gene expression data were obtained by cDNA microarray (see Sect. 2.1) performed on patient tissue biopsies, 16 controls and 50 stage D tumors. Through scoring each of these subnetworks we resolved four independent signatures composed of six genes each significantly dysregulated in late stage CRC (Table 1). We chose one of the signatures (row 1, Table 1) to verify its dysregulation using an independent cohort of tissue samples, three stage D biopsies with matched controls. We chose this signature because of the involvement two targets have in CRC progression (IGFBP3, [18]) and malignant transformations (ITGB1, [19]). We verified the dysregulation of five signature genes at the level of protein and mRNA (see Fig. 4).

3.3 X-talker

3.3.1 The Power of Freely Available, Public Data

In the past decade a burgeoning volume of well-annotated data has accumulated in public databases and been made freely available to researchers. Notable examples are the gene expression omnibus (GEO) hosted at NCBI [20], as well as several PPI databases, for example the Human Protein Reference Database (HPRD) [16], the database of genetic and protein interactions (BioGRID) [21], the Search Tool for the Retrieval of Interacting Genes/proteins (STRING) [22], and several others. With respect to the quality of the GEO database as a whole, despite the inherent experimental variability of the individual contributions (microarray datasets), GEO

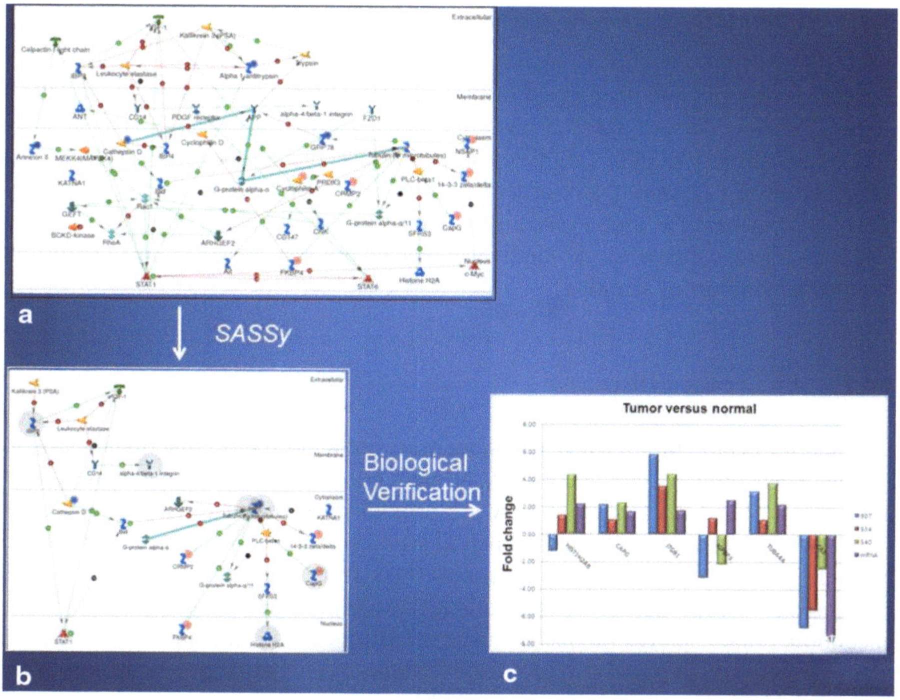

Fig. 4 **a** A large (ca. 50 nodes) disease-associated subnetwork (Thompson Reuters, *MetaCore*) discovered using a seed of sixty-seven genes found by proteomic profiling significantly, differentially expressed in twelve late stage (Duke's D) CRC tumor biopsies with matched controls. A red or blue circle next to the gene name indicates the gene was observed significantly up or down regulated (respectively) at the level of protein, i.e. it was present in the proteomic seed. **b** *SASSy* scored *sub*-subnetwork. Signature genes (*shaded*) were significantly dysregulated at the level on mRNA and were expanded one hop to infer functional relevance. **c** Biological verification of dysregulation in an independent cohort of samples. Shown is the relative expression change of signature proteins and mRNA, tumor *versus* normal, for three patients (507, 534, and 540). mRNA values are the difference between the means measured using the normal samples ($n=16$) and stage D samples ($n=50$) obtained from the microarray. Protein-fold change was determined by label-free mass spectrometry except for IGFBP3, which was determined by Western blot. As most targets were up regulated in the tumor, carbonic anhydrase I (*CA1*) is included as a loading control to indicate that the observed up-regulation of most targets was not merely due to a greater amount of tumor digest on column

is evidently very robust in terms of the pathophysiological concordance across the individual experiments that measured the same disease phenotype [23]. Similarly, when PPI databases composed of variable quality evidence are assembled into one, and each protein interaction is carefully annotated with its underlying evidence [24], researchers can restrict their focus to only high-confidence interactions. When integrated within an appropriate discovery pipeline, one that implements expert algorithms, these databases can be the basis for the discovery of powerful biomarkers.

3.4 Approach

X-talker is described in the context of human CRC. Please see *Materials* for details on the relevant datasets. *X-talker* has three steps whose goal is to (1) *discover* sub-networks within a global PPI significant for a disease phenotype (e.g., colon cancer) versus control; (2) *score* (rank order) the sub-networks by their information content; (3) transform the high scorers into features to train a mathematical classifier (*biomarker*) to validate the predictive power of the sub-networks in (ideally) an independent cohort of samples.

3.4.1 Sub-network Discovery

In order to develop biologically sound measures to quantify the functional association between proteins, we use a random walk based algorithm to compute cross-talk scores. These scores capture network proximity and connectivity to proteomic seeds. In order to account for experimental artifacts and incompleteness of data, we use permutation tests to empirically assess the significance of the crosstalk scores computed by these algorithms. Our statistical evaluation scheme is based on a reference model that captures the basic characteristics of the proteomic seeds, in terms of the number of seeds and their degree distribution. This algorithm and assessment of statistical significance are illustrated in Fig. 5.

After all proteins in the network are scored according to the significance of their proximity to all proteomic seeds, we construct two "candidate subnetworks" for each proteomic seed: (i) subnetwork induced by all interacting partners of the seed protein which we term "*Interactor*" subnetwork, (ii) subnetwork induced by the interacting partners that have significant ($p<0.001$) crosstalk scores, which we term "*Crosstalker*" subnetwork. The concepts of interactor and crosstalker subnetworks are also illustrated in Fig. 5.

3.4.2 Subnetwork Scoring

We evaluate the coordinate dysregulation of each candidate subnetwork with respect to the phenotype of interest, using mRNA expression data for test and control samples. The dysregulation of each candidate subnetwork is evaluated using the information theoretic scheme utilized by SASSy, which is described in Sect. 3.1.

3.4.3 Statistical Significance of Synergistic Dysregulation

We assess the statistical significance of coordinate differential expression for each candidate subnetwork. In order to do so, for a given Q, we generate a null distribution for synergistic differential expression of subnetworks that reflect the topological properties of Q. Since Q is composed of proteins that are connected to each other

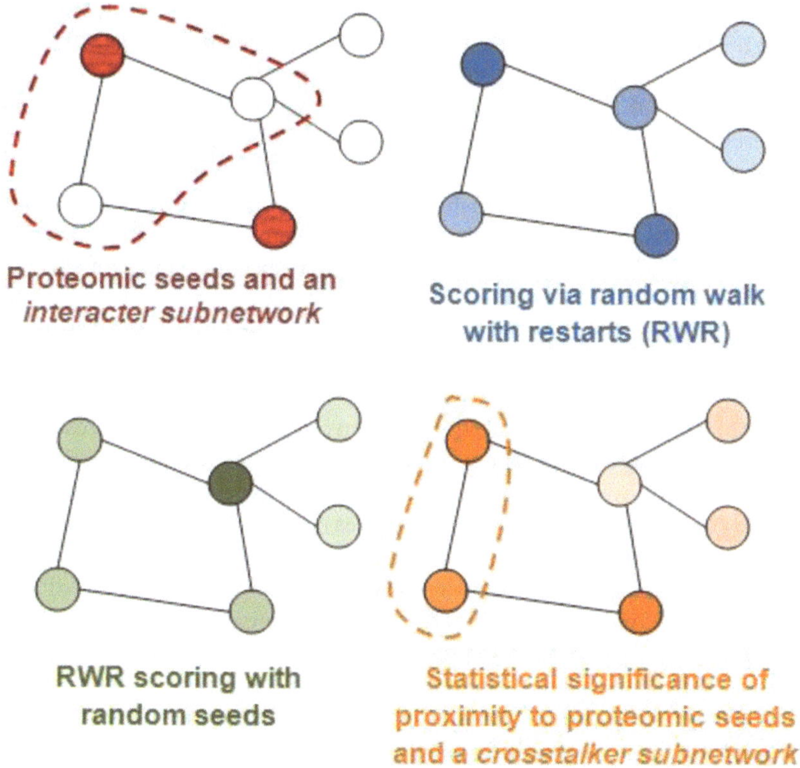

Fig. 5 Random walk based algorithm for identification of crosstalker sub-networks. A hypothetical PPI network and two hypothetical proteomic seeds is shown on the *upper-left panel*. An interactor subnetwork for each proteomic seed is defined as one composed of the interacting partners of that seed in the PPI network, as highlighted in the figure. Using a random walk with restarts (RWR) algorithm, each protein in the network is scored according to their proximity to the proteomic seeds. In the *upper right panel*, the shade of blue represents the RWR-based scores for each protein. In order to correct for ascertainment bias that might be introduced by the number of known interactions of a protein, we also compute the RWR-based score for random seeds multiple times. In the *lower left panel*, the shade of *green* represents the mean RWR-based score for each protein for several random seed sets. In the *lower right panel*, the shade of orange represents the statistical significance of the RWR-based score to the actual proteomic seeds (darker color indicates more significance), with a background distribution of scores with respect to random seeds. The crosstalker subnetwork for a seed protein is composed of the interacting partners of that protein that have significant RWR-based scores, as highlighted in the figure

via a single protein (that is, the corresponding proteomic seed), the null distribution should also be derived from subnetworks that consist of the same number of proteins, which are connected to each other through a single protein in the network. Therefore, we first construct a bag D of proteins in the network with degree at least $|Q|$, i.e, $D=\{v \epsilon V: |N(v)| \geq |Q|\}$, where V denotes the set of all proteins in the network and $N(v)$ denotes the set of interacting partners of protein v. Subsequently, we choose a protein v from D uniformly at random. Finally, we choose $|Q|$ proteins uniformly at random from $N(v)$ to construct a random instance $Q^{(i)}$ representative of Q. Repeating this procedure n times (in our experiments, we use $n = 1000$) and computing $I(Q^{(i)})$, we obtain a null distribution of synergistic differential expression for subnetworks similar to Q. Observe that, only the size of $Q^{(i)}$ depends on Q in this procedure. For this reason, in our experiments, we do not explicitly generate a null distribution for each candidate subnetwork of interest. Rather, we generate a null distribution for subnetworks of size 2, 4, 8, 16, 32, 64. Then we interpolate the mean and standard deviation of coordinate differential expression for these distributions, to obtain a curve that characterizes the behavior of coordinate differential expression with respect to subnetwork size.

3.4.4 Classification (In Silico Cross-Validation)

In order to assess the reproducibility of discovered subnetworks across different data sets and evaluate the potential of our framework for feature selection in classification of CRC, we perform cross-classification experiments. In these experiments, we use the aggregate expression profiles ($e(Q)$) of crosstalker and interactor subnetworks as features for classification. For this purpose, in each experiment, we select the crosstalker (or interactor) subnetworks with coordinate differential expression ($I(Q)$) one standard deviation above the mean of the permuted dataset, according to a specific mRNA expression data set (e.g., GSE8671). Assume that there are K such subnetworks. Then, for each $k \leq K$, we use the k subnetworks with maximum $I(Q)$ to train an SVM classifier on the same data set (GSE8671). Subsequently, we use this classifier to predict the class (tumor vs. normal) of each sample on a different data set (GSE10950). We evaluate the performance of the classifier using the harmonic mean of precision (selectivity) and recall (sensitivity), known as the F-measure, defined as:

$$\text{F-measure} = 2 \text{ X precision X recall}/\left(\text{precision} + \text{recall}\right)$$

Here, precision is the fraction of true positives among all samples classified as tumor and recall is the fraction of tumor samples called accurately by the classifier among all tumor samples.

3.5 Results

Here, we briefly discuss the results obtained using the proteomic seed of Nibbe et al. Interested readers will find a full discussion of the results of this approach using two proteomic seeds, one genomic seed (CRC driver genes), and two microarray datasets for cross-validation in [15]. For each proteomic target in the seed, a candidate subnetwork consisting of its interactors, termed the *interactor* subnetwork, was obtained, resulting in a total of 46 interactor subnetworks. Similarly, a *crosstalker* subnetwork was obtained. Thus, in the discovery step, each target "induces" an interactor subnetwork and a crosstalker subnetwork. The nature of these distinct subnetwork types was discussed above.

Discovery and Scoring For the purpose of discussion we will refer to a subnetwork by the proteomic target that induced the subnetwork (e.g. *TCP1*). Of the 46 candidate subnetworks associated with proteomic targets, 10 unique interactor subnetworks (green squares) exhibited significant MI scores. For five of these subnetworks (CCT2, TCP1, SYNCRIP, HNRPF and HNRPH1) the crosstalker version of the subnetworks was found to have enhanced MI on one or the other microarray datasets. Two crosstalker subnetworks (red diamonds), CCT2 and TCP1, show improvement over their corresponding interactor subnetwork on both arrays. Notably, on GSE10950, the dysregulation score of the TCP1 crosstalker subnetwork is significant, while the corresponding interactor subnetwork failed to show significance (Fig. 6). Figure 7 shows the crosstalker subnetworks for which the synergy was higher than the corresponding interactor subnetwork. As expected, the graph reveals that many proteomic targets reside within or near dense subnetworks of crosstalkers.

Post-Trancriptional Dysregulation of TCP1 Subnetwork: We observed that several of the subnetworks generated using the two proteomic seed sets contained proteins in common. In particular, certain sub-units of the TCP1 complex exhibited marked crosstalk in the subnetwork induced by CCT2. In addition, we had previously shown [14] that certain sub-units of this complex (CCT3, CCT5, and CCT7) were also significant for the late-stage CRC phenotype. TCP1 (or TCPa) is a hetero-oligomeric complex comprised of two stacked ring structures, each composed of eight known subunits and plays a functional role in maintaining the CRC phenotype. Specifically, it was shown [25, 26] to be required for the proper biogenesis of PLK1, a kinase that has a critical role in cytokinesis. However, other than their role as sub-units in the formation of the TCP complex little is known about the independent role, if any, of these sub-units in CRC [27]. Consequently, these targets present an opportunity for follow-on mechanistic studies. For this reason, we verified the protein expression of TCP1, CCT3, CCT5, CCT7, and PLK1 by western blot in a separate cohort of three patient sample pairs not used in the screening phase, and compared this to the average expression at the level of mRNA (Fig. 8). Consistent with our hypothesis, the data indicate co-regulation at the level of mRNA and protein, but also reveal the wide variability of expression of these targets among individual patients. CCT3 and CCT7 were dramatically overexpressed in two patients (507 and 534), but less so in patient 540, which was similar to the pattern for PLK1.

Classification: Cross-Validation We evaluated the quality of the crosstalker versus interactor subnetworks in terms of their ability to classify tumor versus control on the microarrays, using an SVM-based classifier in a cross-validation approach. The significant subnetworks in each group were first ranked by MI, and the features were valued by superposing the mRNA expression values of each gene in the subnetwork. We also compared the classification performance of proteomics-driven subnetworks with that of genomics-driven subnetworks, by replacing the set of seed proteins with a set of proteins coded by candidate driver genes of CRC [28]—referred to here as CAN (candidate CRC driver) genes. The results of the classification experiments are shown in Fig. 9. When trained on GSE10950 and validated on GSE8671, proteomic crosstalkers outperformed the interactor subnetworks(both proteomic and genomic) when the number of features used to

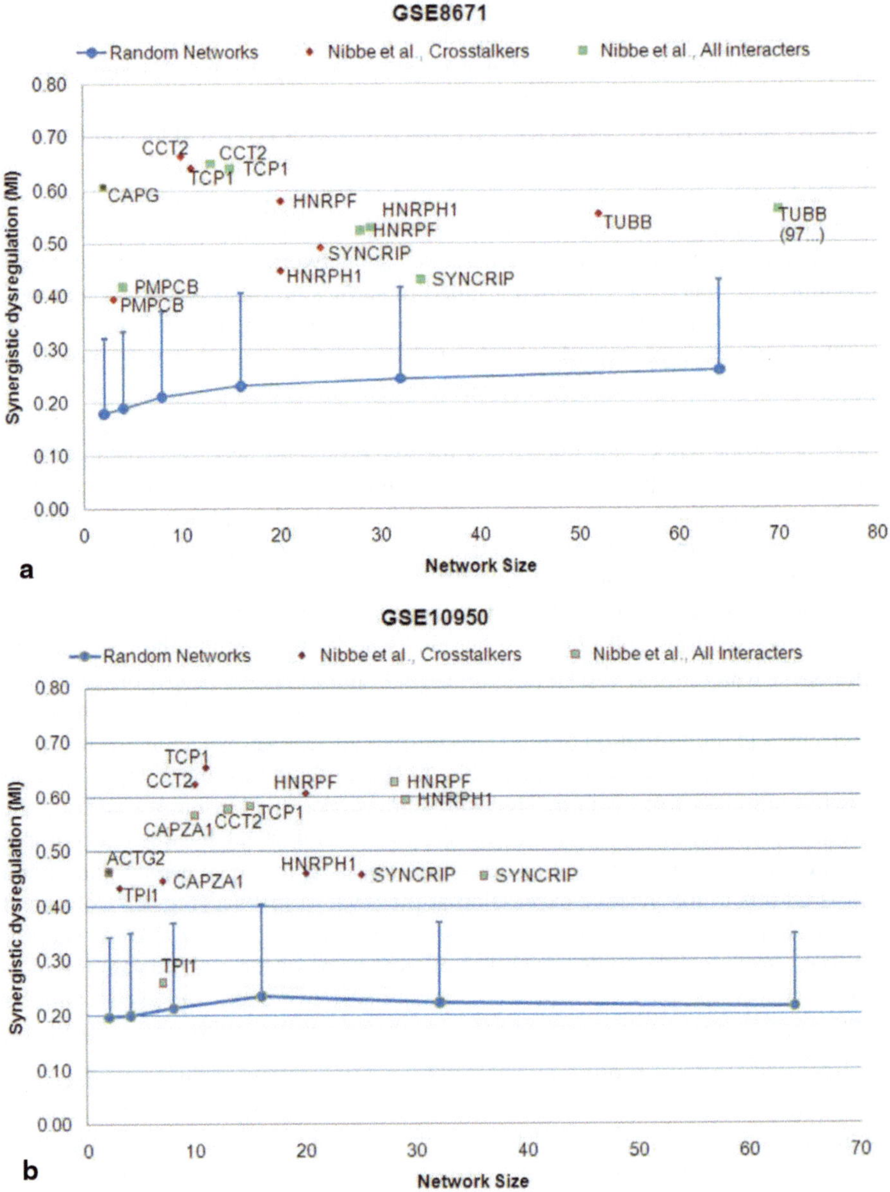

Fig. 6 Synergistic dysregulation versus network size for candidate sub-networks associated with proteomic seeds obtained from Nibbe et al. Sub-network dysregulation (i.e. mutual information of sub-network mRNA expression profile with phenotype class) versus network size for candidate sub-networks. All interactors (green squares) and crosstalkers (red diamonds) were scored using (a) GSE10950 and (b) GSE8671. The blue lines represent the linear interpolation of the means of the estimated null distributions computed for random candidate subnetworks of size 2, 4, 8, 16, 32, and 64, using the respective arrays. Vertical bars represent one standard deviation from the mean

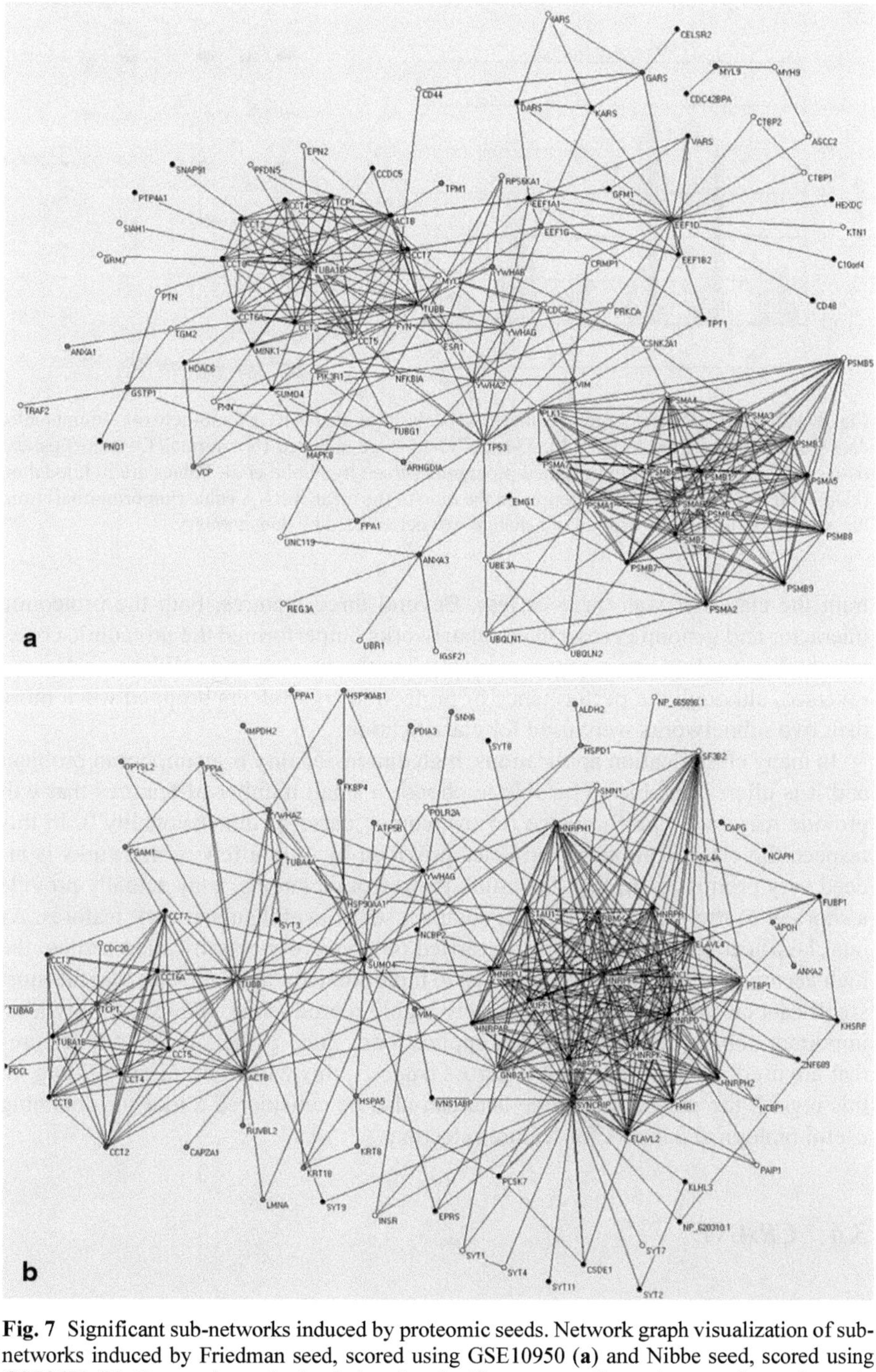

Fig. 7 Significant sub-networks induced by proteomic seeds. Network graph visualization of sub-networks induced by Friedman seed, scored using GSE10950 (**a**) and Nibbe seed, scored using GSE8671 (**b**). Proteomic seeds that induced a significant crosstalker sub-network are shown in red, other proteomic seeds are shown in orange, crosstalkers are black and interactors are white

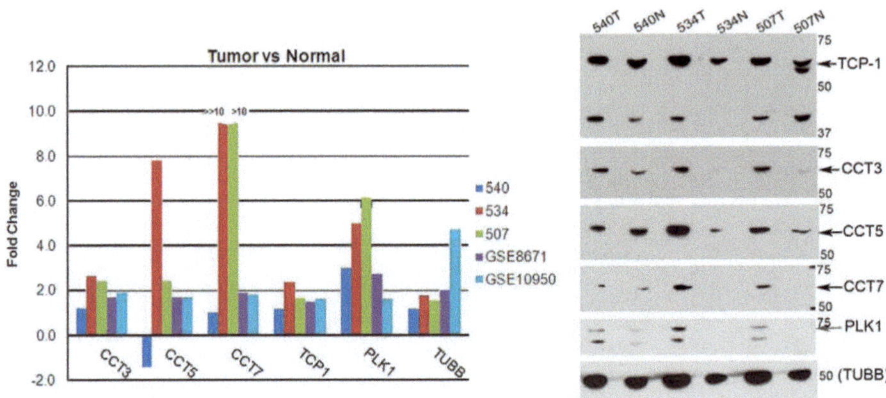

Fig. 8 Validation of select targets predicted to be dysregulated in TCP1 sub-network. Immunoblot data were obtained from three (540, 534, 507) late-stage matched (N=normal/T=tumor) patient tissue biopsies not used in the original proteomic screen by Nibbe et al. Values are in kilodalton (kDa). GSE8671 and GSE10950 represent the ratio of the mean mRNA value (tumor/normal) from the respective microarray array. Fold change was determined by densitometry

train the classifier was three or less. Beyond three features, both the proteomic interactor and genomic crosstalker subnetworks outperformed the proteomic crosstalkers (Fig. 9). Performance was similar when the training and validation sets were reversed, although the performance of proteomic crosstalkers dropped when more than two subnetworks were used for classification.

In many classification applications, high dimensionality is an important problem and it is often desirable to be able to choose a small number of features that will provide reasonable performance (to overcome "curse of dimensionality"). In this respect, the classification performance provided by only a few subnetworks is indeed very promising, in that "crosstalk to proteomic targets" may actually provide a shortcut to the identification of a compact set of useful subnetwork features. As our classification experiments were carried out in a cross-classification setting, the high accuracy of classification using up to three subnetworks indicates that the most significant crosstalker subnetworks were highly reproducible. Reproducibility is an important concern in classification applications, since if the subnetwork features that are used are not reproducible across datasets, this will result in over-fitting. In this regard, the use of proteomic data can also be considered a tool for obtaining useful biological insights for feature selection.

3.6 *CRANE*

3.6.1 Approach

We define the expression state of a set S of k interacting gene products as FS= {ê1, ê2 ,…, êk } = $\{L, H\}^k$, where êi represents the binary expression state of the ith gene

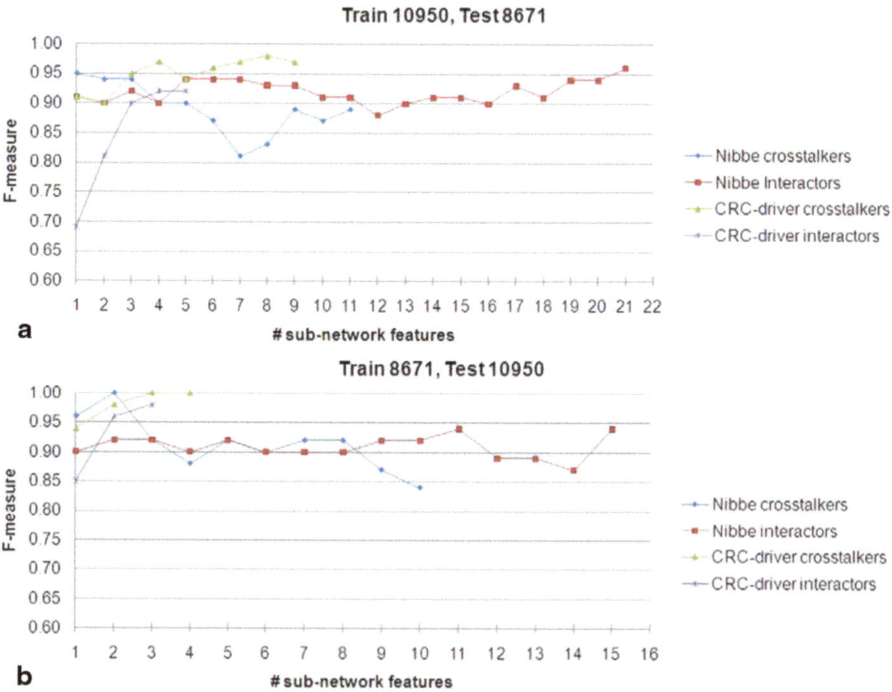

Fig. 9 Cross-validation performance comparison of sub-network based classifiers. The sub-networks induced by proteomic and genomic seeds were first ranked by mutual information with phenotype (MI). Then the normalized mRNA expression values for the genes were aggregated to compute a feature for each sub-network with significant MI. These features were used to train an SVM-based classifier to distinguish normal from tumor using GSE10950, and then cross-validated on GSE8671 (**a**), and vice-versa (**b**)

in the network. In other words, FS is a random variable that represents the combination of binary expression states of the genes in S. We binarize the expression levels of the genes by first normalizing the gene expression matrix so that each column of the gene expression matrix has equal sum and each row has equal mean and standard deviation. Subsequently, we set the largest 25 % of the numbers in this matrix to H (high expression) and the remaining 75 % to L (low expression).

We then define the *combinatorial coordinate dysregulation* of a given subnetwork S as follows:

$$\Delta_{\text{combinatorial}}(S) = I(F_S; C) = H(C) - H(C|\hat{e}_1, \hat{e}_2, \ldots, \hat{e}_k) \quad (12.1)$$

While the computation of combinatorial coordinate dysregulation is straightforward, the problem of identifying subnetworks with high combinatorial coordinate dysregulation is intractable. Motivated by this consideration, we decompose the combinatorial coordinate dysregulation of a subnetwork into components that are associated with individual subnetwork states. Namely, they define

$$J(f_s;C) = p(f_s)\sum\nolimits_{c\in\{0,1\}} p(c\,|f_s)\log(p(c\,|f_s)/p(c)), \tag{12.2}$$

where

$$I(\mathrm{Fs};\mathrm{C}) = \sum\nolimits_{fs\in\{H,L\}^k} J(fs;C). \tag{12.3}$$

Here, fS ∈ $\{H, L\}^k$ denotes an observation of the random variable F_S, *i.e.*, a specific combination of the expression states of the genes in S, c denotes an observation of the random variable C, *i.e.*, a specific phenotype, and p(x) denotes P (X=x), *i.e.*, the probability that random variable X is equal to x (similarly, p(x|y) denotes P (X=x|Y=y)). In biological terms, J (fS; C) can be considered a measure of the information provided by *subnetwork state* fS on phenotype C. Based on this definition, we derive a useful bound on J(fS; C) that can be computed using statistics of smaller subnetworks. Using this bound, we develop an efficient branch-and- bound algorithm, *CRANE,* for the identification of subnetworks and associated state functions informative of phenotype (*i.e.*, with high J (fS; C)).

Once we identify significant subnetworks that are combinatorially dysregulated in metastatic CRC, we train neural networks to use these subnetworks for predicting colon cancer metastasis on the two independent datasets obtained from GEO. The neural network is composed of two layers. In the first layer, the gene products in each subnetwork are integrated to compute a signature for each subnetwork. In the second layer, the signatures of the subnetworks are integrated to predict the likelihood of metastasis. In all experiments, one dataset is used for subnetwork discovery and model building; the other dataset is used for testing.

3.6.2 Results

For each of the datasets mentioned above, we discover informative state functions (in terms of discriminating tumor samples with or without metastasis) using *CRANE.* While state functions that are indicative of either metastatic or non-metastatic phenotype can have high *J*(.) values, we use only those that are indicative of (i.e., knowledge of which increases the likelihood of) metastatic phenotype for classification and further analyses, since such state functions are directly interpretable in terms of their association with metastasis. we use subnetworks discovered on the *GSE6988* dataset to train classifiers and we test the resulting classifiers on all the samples of *GSE3964*. Similarly, we use subnetworks discovered on *GSE3964* to train classifiers using the same dataset and perform testing of these classifiers on 28 metastatic and 20 randomly selected non-metastatic samples of *GSE6988*. The cross-classification performance of subnetworks discovered by an algorithm is not only indicative of the power of the algorithm in discovering subnetworks that are descriptive of phenotype, but also the reproducibility of these subnetworks across different datasets.

The results of these cross-classification experiments are shown in Fig. 10. As seen in the figure, the subnetworks identified by *CRANE* outperform the subnetworks

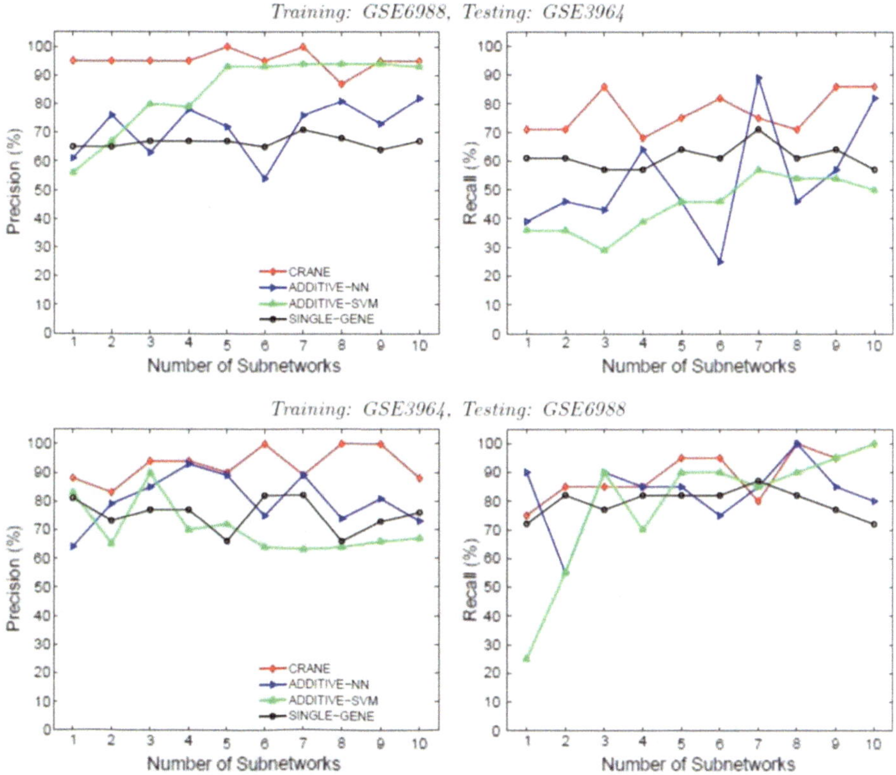

Fig. 10 Classification performance of subnetworks identified by *CRANE* in predicting colon cancer metastasis, as compared to single gene markers and subnetworks identified by algorithms that aim to maximize additive coordinate dysregulation. Subnetworks identified by *CRANE* are used to train neural networks (NNs), while those identified by the additive algorithm are used to train NNs, as well as support vector machines (SVMs). In the graphs, horizontal axes show the number of disjoint subnetwork features (with maximum combinatorial or additive coordinate dysregulation) used in classification, and vertical axes show the precision and recall achieved by the single gene or subnetwork features discovered by each algorithm

identified by other algorithms in predicting metastasis of colorectal cancer. In fact, in both cases, *CRANE* has the potential to deliver very high accuracy using very few subnetworks (maximum precision of 100 % on both *GSE6988* and *GSE3964*, maximum recall of and 95 % and 86 % for classification of samples in *GSE6988* and *GSE3964*, respectively). While we use a simple feature selection method here for purposes of illustration, the performance of *CRANE* subnetworks are quite consistent, suggesting that these performance figures can indeed be achieved by developing elegant methods for selection of subnetwork features. These results are rather impressive, given that the best performance that can be achieved by the additive algorithm is 82 %/93 % precision and 89 %/100 % recall for the classification of *GSE3964* and *GSE6988*, respectively. Note that, while the performance of other algorithms is improved by increasing number of subnetwork features, the performance of *CRANE* appears to decline. This is likely because *CRANE* represents subnetwork features as multi-dimensional state functions. Therefore, while a few subnetworks each containing a few genes provide sufficient information for accurate classification, the accuracy declines as more subnetworks are incorporated because of the growth in dimensionality.

Figure 11 shows a sample subnetwork that illustrates how the dysregulated subnetworks identified by integrative algorithms can generate novel hypotheses that can drive experimental biology of colorectal cancer. This subnetwork contains membrane-bound proteins TNFSF11, MMP1, BCAN, MMP2, TBSH1, and SPP1. For this subnetwork, the state function LLLLLH (in respective order) indicates metastatic phenotype with J-value 0.33. The combinatorial coordinate dysregulation of this subnetwork is 0.72, while its additive coordinate dysregulation is 0.37, i.e.,

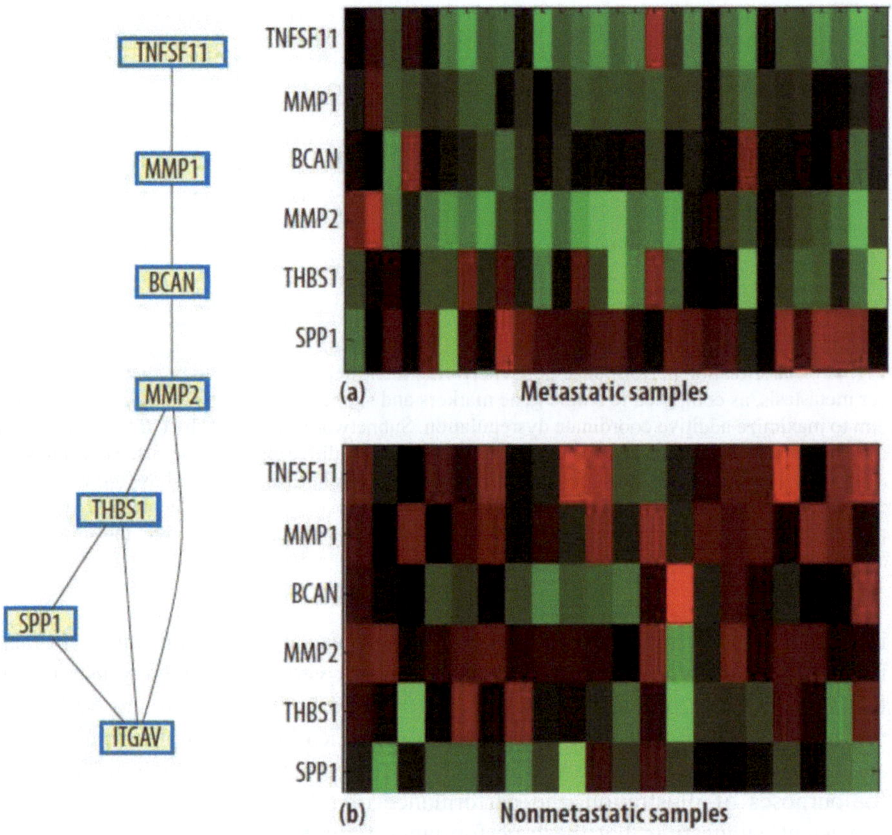

Fig. 11 Hypothesis-driver subnetwork: interaction diagram illustrating key interactions with gene products from a subnetwork identified by CRANE as indicative of CRC metastasis. *Left panel*: A combinatorially dysregulated subnetwork identified by CRANE [3] in the context of the metastasis of colorectal cancer and the interactions between this subnetwork and integrin alpha-chain V (ITGAV). *Right panel*: Heatmap of gene expression for these proteins in metastatic and nonmetastatic colon tumor samples; tone of green and red respectively show down- and up-regulation

this subnetwork would likely escape detection by the additive algorithm. As seen, SPP1, TBSH1, and MMP2 interact with integrin alpha-chain V (ITGAV). Integrins are heterodimeric integral membrane proteins that are composed of an alpha chain and a beta chain. They are known to play a major role in mediating cell adhesion and cell motility, processes known to be involved in metastasis of colorectal cancer. However, alpha-v/beta-5 integrin does not exhibit significant differential expression at the mRNA-level, suggesting that the subnetwork identified by CRANE may be a signature of its post-translational dysregulation in metastatic cells. Therefore this subnetwork suggests a number of follow-up experiments, including investigation of the post-translational dysregulation of integrin heterodimers in metastatic samples, and pharmacological inhibition or siRNA interference of the integrin dimers, to evaluate the role of these proteins in maintaining the metastatic phenotype.

4 Notes and Caveats

Integration of a PPI network with microarray data comes with a number of pre-processing considerations. First, our method requires a one-to-one mapping of the genes represented on the array with the genes in the PPI. This means that some of the data on the array will go unused in the integration because coverage in the PPI will usually be only a fraction of that represented on the array, and vice-versa—sometimes genes in the PPI will not be represented on the array. In the prototype, a check is performed when the PPI and microarray are imported to the software, and any genes not represented on both datasets are noted. While a gene that is present in the network but absent in the array is used (since it provides information on the indirect interactions among other proteins), it is not considered while scoring subnetwork dysregulation. Additionally, many microarrays have a many-to-one relationship with respect to their probes and the corresponding gene, respectively. Consequently, a pre-processing step is required to reduce this relation to a one-to-one basis. A simple averaging of the probe expressions on a sample by sample basis may be appropriate; in other cases taking the probe with the highest mean expression value across samples may be justified. Normalization of the gene expression values is also advised, in a row-wise (across samples) and (sometimes) column-wise (across genes) fashion, in order to remove any possibility of gene bias in the scoring scheme. Finally, since PPI networks suffer from large amounts of missing data and ascertainment bias, biased network properties have to be taken into account in the null model while assessing statistical significance.

Acknowledgments The research presented here was supported, in part, by National Institutes of Health Grants UL1-RR024989 from the National Center for Research Resources (Clinical and Translational Science Awards), P30-CA043703 from the Case Western Reserve University Cancer Center Proteomics Core, and T32-GM008803 from the NIGMS (Institutional National Research Service Award). This work was also supported, in part, by NSF CAREER Award CCF-0953195.

All inquiries related to *SASSy*, *X-TALKER*, or *CRANE* should be directed to john.schenkel@neoproteomics.net.

References

1. Vidal M (2005) Interactome modeling. FEBS Lett 579:1834–1838
2. Edelman EJ, Guinney J et al (2008) Modeling cancer progression via pathway dependencies. PLoS Comput Biol 4:e28
3. Auffray C (2007) Protein subnetwork markers improve prediction of cancer outcome. Mol Syst Biol 3:141–142
4. Ideker T, Sharan R (2008) Protein networks in disease. Genome Res 18(4):644–652
5. Nibbe RK, Chowdhury SA et al (2011) Protein–protein interaction networks and subnetworks in the biology of disease. Wiley Interdiscip Rev Syst Biol Med 3(3):357–367
6. Shoemaker BA, Panchenko AR (2007) Deciphering protein-protein interactions. Part I. Experimental techniques and databases. PLoS Comput Biol 3:e42
7. Chuang HY, Lee E, Liu YT, Lee D, Ideker T (2007) Network based classification of breast cancer metastasis. Mol Syst Biol 3:140
8. Goh KI, Cusick ME et al (2007) The human disease network. Proc Natl Acad Sci U S A 104:8685–8690
9. Barrenas F, Chavali S et al. (2009) Network properties of complex human disease genes identified through genome-wide association studies. PLoS One 4:e8090
10. Oti M, Brunner HG (2007) The modular nature of genetic diseases. Clin Genet 71(1):1–11
11. Vanunu O, Magger O et al (2010) Associating genes and protein complexes with disease via network propagation. PLoS Comput Biol 6(1):e1000641
12. Liu Y, Koyutürk M et al (2012) Gene interaction enrichment and network analysis to identify dysregulated pathways and their interactions in complex diseases. BMC Syst Biol 6(1):65
13. Chowdhury SA, Nibbe RK et al (2011) Subnetwork state functions define dysregulated subnetworks in cancer. J Comput Biol 18(3):263–281
14. Nibbe RK, Markowitz S et al (2009) Discovery and scoring of protein interaction subnetworks discriminative of late stage human colon cancer. Mol Cell Proteomics 8(4):827–845
15. Friedman DB, Hill S et al (2004) Proteome analysis of human colon cancer by two-dimensional difference electrophoresis and mass spectrometry. Proteomics 4:793–811
16. Prasad TSK, Goel R et al (2009) Human protein reference database—2009 update. Nucleic Acids Res 37:D767–D772
17. Ries LAG, Melbert D et al (2007) SEER cancer statistics review, 1975–2004, National Cancer Institute, National Institutes of Health Bethesda, MD
18. Slattery ML, Samowitz et al (2004) Associations among IRS1, IRS2, IGF1, and IGFBP3 genetic polymorphisms and colorectal cancer. Cancer Epidemiol Biomark Prev 13:1206–1214
19. Stallmach A, von Lampe B et al (1992) Diminished expression of integrin adhesion molecules on human colonic epithelial cells during the benign to malign tumour transformation. Gut 33:342–346
20. Barrett T, Wilhite SE et al (2013) NCBI GEO: archive for functional genomics data sets–update. Nucleic Acids Res 41(Database issue):D991–D995
21. Chatr-Aryamontri A, Breitkreutz BJ et al (2012) The BioGRID interaction database: 2013 update. Nucleic Acids Res 37(Database issue):D412–D416
22. Jensen et al (2009) STRING 8—a global view on proteins and their functional interactions in 630 organisms. Nucleic Acids Res 37(Database issue):D412–D416
23. Dudley JT, Tibshirani R (2009) Disease signatures are robust across tissues and experiments. Mol Syst Biol 5:307
24. Turner B, Razick S (2010) iRefWeb: interactive analysis of consolidated protein interaction data and their supporting evidence. Database (Oxford) 2010:baq023
25. Nibbe RK, Koyutürk M et al (2010) An integrative -omics approach to identify functional sub-networks in human colorectal cancer. PLoS Comput Biol 6(1):e1000639

26. Liu X, Lin CY et al (2005) CCT chaperonin complex is required for the biogenesis of functional Plk1. Mol Cell Biol 25:4993–5010
27. Coghlin C, Carpenter B et al. (2006) Characterization and over-expression of chaperonin t-complex proteins in colorectal cancer. J Pathol 210:351–357
28. Sjöblom T, Jones S et al. (2006) The consensus coding sequences of human breast and colorectal cancers. Science 314:268–274

Targeted Therapies For Intestinal Tumorigenesis

E. Gabriela Chiorean, Andrew Coveler, Jon Grim and William M. Grady

Abstract Novel therapies for colorectal cancer based on discoveries made over the last three decades are moving into clinical practice. These therapies have extended the lives of patients with advanced colorectal cancer and are anticipated to ultimately lead to improved cure rates. The majority of the new therapies are based on targeting key signaling pathways that are deregulated in colorectal cancer, including the MAPK signaling pathway, PI3K signaling pathway, VEGF signaling pathway, and mTOR signaling pathway. The best results to date have come from targeting the EGF-MAPK and VEGF signaling pathways, although many other promising targeted therapies are in early phase clinical studies. An emerging area in the management of colorectal cancer is immunotherapy. In this chapter, we highlight some of the areas of novel therapy development, notably signal pathway targeted therapies and tumor immunotherapy.

Keywords Colorectal cancer · Targeted therapy · Signaling pathway · Receptor tyrosine kinase · Immunology

1 Introduction

Novel therapies for colorectal cancer based on discoveries made over the last three decades are moving into clinical practice. These therapies are primarily based on the discovery of deregulated signaling pathways that are central to colorectal cancer

W. M. Grady (✉) · E. G. Chiorean · A. Coveler · J. Grim
Clinical Research Division, Fred Hutchinson Cancer Research Center,
1100 Fairview Ave. N. D4-100, Seattle, WA 98109, USA
e-mail: wgrady@fhcrc.org

E. G. Chiorean · A. Coveler · J. Grim · W. M. Grady
Department of Medicine, University of Washington School of Medicine, Seattle, WA, USA
e-mail: wgrady@fhcrc.org

J. Grim
Research and Development Service, VA Puget Sound Health Care System, Seattle, WA, USA

V. W. Yang, A. B. Bialkowska (eds.), *Intestinal Tumorigenesis*,
DOI 10.1007/978-3-319-19986-3_13

and are directed at mutant or aberrantly expressed receptor tyrosine kinases. Therapies directed at the EGF-MAPK and VEGF signaling pathways have been shown to be effective in the clinic and many other therapies are in development. These therapies have extended the lives of patients with colorectal cancer and are likely to lead to improved cure rates. Another emerging and promising line of therapy for colorectal cancer is aimed at enhancing the anti-tumor immune response. In this chapter, we highlight some of the areas of novel therapy development, notably signal pathway targeted therapies and tumor immunotherapy.

2 Targeted Therapy—Receptor Tyrosine Kinases

2.1 General Paradigm of Signaling Pathways

Receptor tyrosine kinases (RTKs) are cell surface molecules that serve as “on/off switches” that initiate signaling cascades that ultimately result in cellular phenotypes. Although differences exist between these families of ligands and receptors, they share a general paradigm of signal transduction that starts with extracellular ligands (either present in the extracellular space or attached to the surface of neighboring cells) binding to RTKs. Ligand binding transmits a signal across the cell membrane through conformational changes in RTKs that leads to the recruitment and phosphorylation of intracellular molecules. This triggers further downstream signaling that is transmitted by a number of intertwined signaling networks, including the RAS/RAF/MEK/ERK pathway and the PI3K/AKT/mTor pathway. These signals are ultimately transmitted to the nucleus, where they result in transcriptional programs that influence a wide array of mitogenic processes, including proliferation, apoptosis, and cell motility.

2.2 Abnormal RTK Signaling in Malignant Cells

RTK pathways play a central role in the precise regulation of cellular growth and survival, thus it is not surprising that these pathways are often altered in a wide array of human cancers. The RTKs that are most relevant to colorectal cancer biology and therapy are Epidermal Growth Factor Receptor/ERBB1 (EGFR) [1], Vascular Endothelium Growth Factor (VEGFR) [2], MET (the receptor for Hepatocyte Growth Factor (HGF) [3], and Insulin Growth Factor Receptor 1 (IGFR1) [4]. Over-active signaling by these RTKs can occur due to overexpression of RTKs and/or their ligands and to RTK constitutive activation secondary to genetic alterations. In addition to the RTKs themselves, intracellular post-receptor kinases that transmit signals downstream are also frequently altered in human cancers, including RAS/RAF/MEK/ERK and PI3K/AKT/mTor [5]. While upregulation or activation of these pathways occurs via diverse mechanisms in different tumor types, careful study of hundreds of human colorectal cancer samples have revealed that positive regulators are most commonly activated by point mutations or gene amplification, while negative regulators are inactivated by detrimental mutations or complete gene deletions [6]. Together, these genetic changes result in multiple phenotypes associated with cancer, including increased proliferation, invasion, metastasis, and angiogenesis; and decreased programmed cell death. Notably, cancer cells often appear to develop a critical dependence on these overactive pathways (known as oncogene addiction), such that "addicted" tumors are exquisitely sensitive to pathway inhibition, at least in model systems [7–10]. This creates a therapeutic opportunity and forms the basis for many targeted anti-cancer therapies [11, 12]. In fact, most RTK pathways implicated in cancer can now be targeted in some fashion. Common strategies include using antibodies or other molecules to disrupt the interaction between ligand and receptor or using specific inhibitors that shut down constitutively active versions of cell surface or intracellular kinases. Each of these approaches is discussed in more detail below.

3 Cell Surface RTKS Relevant to CRC Biology and Therapy

3.1 EGF/R and VEGF/R

These RTKs and pathways are discussed in detail in the following section.

3.2 Insulin Growth Factor (IGF)

The Insulin/Insulin Growth Factor signaling pathway is best known for its regulation of blood sugar and metabolism in mammals. However, its effects in both normal and cancer cells extends beyond these roles and include mitogenic and anti-apoptotic functions. The principal components of this pathway include three cell surface RTKs (the insulin receptor, the insulin-like growth factor receptor 1 (IGF1R), and the insulin-like growth factor receptor 2 (IGF2R),) and three ligands (insulin, IGF1, and IGF2). While insulin is a central regulator of metabolism, IGF1/IGF2 are induced by Growth Hormone and are mitogenic. Intracellular events after ligand binding and RTK activation include phosphorylation of the IRS proteins, which then activate the PI3K and Ras-MAPK pathways. Significant positive and negative crosstalk between these various components leads to complex signaling outputs. Two excellent reviews summarize the biology of IGF and highlight the importance of IGF in cancer [13, 14].

Additional molecular interactions of components in the IGF pathway with proteins in other signaling pathways have implications for cancer treatment strategies. Of particular relevance to currently available targeted therapies, the EGF and IGF families of RTKs have been shown to crosstalk. This is significant for breast cancer, as IGF pathways appear to mediate resistance to ERBB2 targeted therapies [15]. More relevant to CRC is the finding that EGF and IGF crosstalk can influence the response and mediate resistance to EGF inhibitors in many tumor types [16–19].

Activation of the IGF pathway is seen in many cancers, including CRC. Cell lines and primary tumors can overexpress receptors (IGF1R) or ligands (IGF1 or IGF2), which may lead to more aggressive behavior of the cancer and affect prognosis [4, 20, 21, 22]. Mutations in these components, on the other hand, are rarely found in human cancers. Preclinical models of IGF hyperactive tumors demonstrate that IGF activity induces tumor growth and that IGF blockade/inhibition can cause tumor regression [23, 24, 13, 14, 25]. Furthermore, epidemiologic studies show an increased cancer risk in people with overactive IGF pathways, such as people with adult-onset diabetes [26]. Finally, the association of obesity and a sedentary lifestyle with cancer incidence and outcomes may be related to abnormal insulin levels [27, 26, 28]. Thus, the IGF pathway is an attractive candidate for targeted anticancer therapeutics.

3.3 Strategies to Target the IGF Pathway in Patients with CRC

Two distinct approaches have been used to inhibit the IGF pathway in patients with CRC. The first class of agents is monoclonal antibodies against either IGF1R or IGF ligands (including Figitumumab (CP-751,871), Ganitumab (AMG-479), Cixutumumab (IMCA12), and Dalotuzumab (MK-0646). The second class of agents are small molecule tyrosine kinase inhibitors that target IGF1R selectively, or that have more promiscuous activity (including OSI-906, AXL1717, XL228, and BMS-754807). These agents have been extensively tested in unselected patients with CRC, generally in combination with standard chemotherapeutics or with other targeted therapies (EGFR inhibitors). These trials have failed to show a benefit for these compounds, regardless of whether these were given as single agents or in combination with other approved therapies (see, for example [29–31], (J Clin Oncol 29: 2011 (suppl; abstr 3501)). Because pre-clinical studies have implicated Akt/mTor pathways in resistance to these agents, multiple follow-up trials have been performed testing IGF1R mAbs or TKIs in combination with Akt/mTor inhibitors. While some of these are still ongoing and results are not yet known (NCT01122199), available data have again show a lack of anti-tumor activity. For example, a trial of OSI-906 in combination with the mTor inhibitor everolimus showed no benefit for heavily pretreated patients with metastatic CRC (J Clin Oncol 29: 2011 (suppl; abstr e14054)). Finally, ongoing efforts are combining IGF inhibitors with MAPK pathway inhibitors (e.g., MEK162); results of these studies are not yet known (NCT01562899). A molecular explanation for these broadly negative results reported to date has not yet emerged, but could be related to compensatory upregulation of IGF pathway components, including insulin or IGF1 or IGF2. This lack of anti-tumor activity has also been observed in other tumor types, causing many pharmaceutical companies to abandon IGF targeted agents [32]. However, subsequent to these studies, potential biomarkers that may predict response or resistance to IGF targeted compounds have been identified, which could improve patient selection in future trials [33, 34]. In addition, ongoing pre-clinical studies may identify additional mechanisms of resistance that will guide future combination approaches.

3.4 MET

The c-MET RTK and its ligand, hepatocyte growth factor (HGF), are components of an important signaling pathway that regulates cell proliferation, survival, angiogenesis and invasion/metastasis. It is activated in many human cancers, including lung cancer, medullary thyroid cancer, prostate cancer, and CRC (Recently reviewed in [35]). Studies of human cell lines and primary tumor samples have shown that approximately 50 % of CRCs overexpress c-MET. Tumor c-MET levels have also been associated with liver metastasis and with poor prognosis [36–39], supporting the role of this pathway in invasion and metastasis. Likewise, serum HGF levels have been associated with poor prognosis in CRC [40]. These data make the HGF/c-MET pathway an attractive therapeutic target.

Activation of the pathway is initiated when HGF binds to c-MET. This triggers receptor autophosphorylation, which, in turn, recruits intracellular signaling molecules that ultimately result in the strong mitogenic phenotypes outlined above. Across human cancers, c-MET is rarely mutated but instead is overexpressed due to increased gene copy number and/or increased gene transcription. The HGF ligand can also be aberrantly overexpressed in cancer, leading to autocrine growth signaling. These signals are transmitted, in part, via the PI3K pathway and the RAS pathway, which are also frequently deregulated in human CRC.

Additional clinically relevant crosstalk occurs between c-MET and other RTKs, including EGFR and VEGFR, resulting in synergistic responses to growth factors or to unanticipated resistance to targeted therapies. For example, in certain situations, EGFR or ERBB3 can induce activation of signaling pathways downstream of MET leading to escape from the anti-tumor effects of therapies that inhibit HGF/MET. Furthermore, there is evidence that c-Met overexpression mediates resistance to EGFR inhibitors across tumor types, including CRC [41, 42]. This crosstalk has important implications for therapies targeting these pathways in CRC, as combined blockade of multiple pathways will likely be necessary for these agents to be effective in the treatment of patients with CRC.

In vitro and in vivo models have been used to validate the c-MET/HGF axis as a causally important cancer pathway and as a legitimate therapeutic target. Mice overexpressing HGF or c-MET develop a wide array of cancers, including hepatocellular cancer, skin cancer, breast cancer, and lung cancer [43–46]. Furthermore, HGF signaling promotes cancer cell growth and spread; Zhang et al. engrafted human cell lines into a genetically engineered mouse expressing high levels of human

HGF. Using this model, they showed HGF significantly enhanced proliferation and metastasis of human cancer cell lines, including CRC cell lines. This model is an ideal translational platform, and it has been used to test HGF/c-Met targeted therapeutics, demonstrating the efficacy of a number of c-Met directed therapies [47].

3.5 Strategies to Target the c-MET Pathway in Patients with CRC

Drug development has focused on two areas: monoclonal antibodies that bind to c-MET or to HGF and small molecule inhibitors of c-MET that counteract the intracellular signaling cascade triggered by c-MET activation. Because of the extensive crosstalk between c-MET and other pathways, the majority of data generated to date have been in combination with other targeted agents, including EGFR inhibitors and angiogenesis inhibitors.

A number of monoclonal antibodies targeting c-MET pathway components have been developed as therapeutics for patients with c-MET overexpressing cancers, including onartuzumab, rilotumumab, and ficlatuzumab. Onartuzumab (MetMab) is a monoclonal antibody targeting the extracellular portion of c-Met. It is similar to other monoclonal antibodies in that it blocks binding of HGF to c-Met, but it is unique in that it does not induce dimerization of c-Met, thus it should have minimal agonist activity. Notably, a single patient with metastatic gastric cancer had a prolonged response to Onartuzumab as a single agent in a phase I trial [35]. However, this success appears uncommon, as results with this drug as a single agent in CRC show little clinical benefit, even when patients are selected according to ovexpression of c-MET [48]. Given the crosstalk between c-MET and other CRC pathways, especially VEGF, the greater benefit of c-MET inhibitors may be in combination with other agents. Recent phase Ib/II clinical trial results show a significant increase in response rate for patients with *RAS* wild type CRC treated with panitumumab and rilotumumab [49]. There is an ongoing randomized Phase II clinical trial of Onartuzumab in combination with standard chemotherapy (FOLFOX+Avastin) as first line therapy for patients with metastatic colorectal cancer [50]. The results of this study are not yet known.

c-MET specific kinase inhibitors are being tested across multiple tumor types. Tivantinib has the most mature clinical data. In c-MET overexpressing hepatocellular cancer, tivantinib as a single agent was beneficial in the second line setting [51]. This approach is now being tested in a phase III trial. In NSCLC, tivantinib failed to improve survival when combined with the EGFR inhibitor erlotinib in an unselected population of patients with this disease, but there may be a benefit in the subgroup with high c-MET expression [52]. In CRC, unselected patients treated with tivantinib in combination with irinotecan and cetuximab showed a trend towards benefit, but this was not statistically significant (J Clin Oncol 31, 2013 (suppl; abstr 3508)). This study did not include sufficient patients to determine if high c-MET expression was a biomarker of response to tivantinib. A large number of other specific inhibitors are being tested in phase I trials, but results from these trials are not yet available.

Because of the high degree of similarity between kinase domains in RTKs, the majority of available c-MET inhibitors are not specific for c-MET. Instead, these agents inhibit multiple kinases, including RON, VEGF, and ALK. Because of the extensive crosstalk between c-MET and other pathways, these less selective inhibitors may be more effective than the specific inhibitors listed above, albeit with the potential for increased side effects. Many of these agents are being tested in phase I studies for refractory solid tumors. Included in this group are crizotinib, foretinib and cabozantinib. XL-184 (Cabozantinib) is a multitargeted kinase inhibitor that inhibits both c-MET and VEGFR2. Because c-MET is a possible escape pathway after inhibition of VEGFR2, dual inhibition of these pathways is a rationale drug target [53]. It has proven activity in medullary thyroid cancer, which often overexpresses these pathways [54]. It has shown activity against human CRC in a xenograft model, where 12 of 12 tumors showed either stable disease or response to this treatment (Abstract: Cancer Research August 14, 2013 73:LB-302). It is currently being tested in patients with metastatic CRC in combination with the EGFR inhibitor panitumumab (NCT02008383).

In conclusion, existing data for c-MET inhibitors has not demonstrated significant benefit in the treatment of CRC, but further optimization of both patient selection and combination therapies may yield more effective treatments for patients.

3.6 Mitogen Activated Protein Kinase (MAPK) Pathway

Another signaling pathway that plays an important oncogenic role in cancer, including colorectal cancer, is the MAPK pathway [55, 56]. This pathway is typically activated by an RTK, which results in the serial phosphorylation of a cascade of kinases that that ultimately induce cell proliferation and cell survival. Activating mutations in genes encoding proteins in the MAPK pathway have been found commonly in CRC, such as *KRAS* and *BRAF* [57]. KRAS is a GTPase that is activated by RTKs and that activates the RAF kinase, BRAF. KRAS is constitutively active in 40–60 % of human CRC secondary to oncogenic mutations in codons 12, 13, and 61. *BRAF* has a V600E mutation in 5–10 % of CRC, resulting in hyperactivation of this kinase. Notably, mutant *BRAF* confers a poor prognosis, but has not been shown to predict response to treatment [58]. Activated Raf phosphorylates MEK1 and MEK2, and these kinases activate downstream pathway members via phosphorylation of ERK, which has a range of cytoplasmic and nuclear targets that ultimately result in transcriptional programs resulting in increased cell proliferation and decreased apoptosis. Significant crosstalk links the MAPK pathway to other prominent CRC pathways, including the EGFR pathway and the IGFR pathway.

Pharmacologic inhibition of the MAPK pathway can be achieved by targeting any of the core components, including RAS, RAF, MEK, or ERK. Indeed, each of these components has been targeted in pre-clinical models with demonstration of inhibition of the pathway and suppression of tumor growth. Some compounds have moved on to clinical trials in CRC patients. The broad experience with inhibitors of this pathway is discussed below.

3.7 RAS Inhibitors

Activation of the RAS family of proteins (KRAS, NRAS, HRAS) can initiate signaling via the MAPK pathway in both normal and transformed tissues. Constitutively active mutations in *KRAS* (and to a lesser extent, *NRAS*) make it an attractive target for therapeutic intervention. Historically, a myriad of diverse attempts to target RAS itself have failed (reviewed in [59]), and these negative results have shifted attention to inhibition of downstream kinases. However, recent pre-clinical work from Shokat and colleagues combined structural biochemistry and high throughput screening of compounds to identify selective inhibitors of a specific *KRAS* mutant, *RASG12C* [60]. As predicted, these compounds inhibited the growth of lung cancer cell lines with this mutation, but had no minimal effects on cells with other specific *KRAS* mutations or with wild type *KRAS*. At present, it is not yet clear if this promising approach can be utilized to inhibit other forms of oncogenic *KRAS* resulting from other mutation, including those more commonly associated with CRC, such as G12D and G12V. Other RAS inhibitors have also been developed and some selective activity in studies of *KRAS*-mutant cell lines, including CRC cells [61–63]. Together, these developments are important steps towards targeting RAS in human cancer patients.

3.8 RAF Inhibitors

The RAF proteins include ARAF, BRAF, and CRAF. BRAF has received the most attention, because activating mutations affecting amino acid V600 in this protein are commonly found in cancer, especially melanoma [64]. This has led to the development of RAF inhibitors (vemurafinib (PLX-4032) and dabrafinib) with proven and dramatic clinical activity in patients with melanoma. Other available inhibitors currently in clinical trials include LGX818, XL281, PLX3603 and RAF265. Interestingly, BRAF inhibitors can have a paradoxical activating effect in cancers that express wild type RAF proteins and this may limit their overall utility. The mechanism responsible for this unanticipated activation of BRAF has recently been reviewed [65]. As noted previously, human CRC have *BRAF* mutations in 5–10 % of cases, and these mutations are associated with a poor prognosis. Given its profound activity in *BRAF* mutant melanoma, vemurafinib was tested in CRC patients who also had V600E *BRAF* mutations in their cancers. However, no anti-CRC activity was demonstrated (J. Clin. Oncol. 28 (suppl.), abstr. 3534 (2010)). In order, to investigate the differences in response between tumor types, additional laboratory work was performed and has now implicated both the EGFR and PI3K pathways in mediating vemurafenib resistance in CRC [65–67]. Most ongoing clinical trials include combination therapies to inhibit these resistance pathways (NCT00326495, NCT01791309, and NCT01719380) and results are eagerly awaited. A recent case report suggests dramatic benefits may be achieved with such combined approaches [68].

3.9 MEK Inhibitors

MEK1 and MEK2 are downstream signaling components of numerous cell surface RTKs and are also direct targets of the RAF kinases [69]. The downstream location of these kinases in these signaling cascades make them attractive targets for cancer therapy. They may be particularly useful for tumors with intrinsic or acquired resistance to agents targeting upstream components, including EGFR and BRAF [70]. A number of MEK inhibitors have been developed and tested in preclinical models and in clinical trials [69]. Early results of the MEK inhibitor PD184352 showed no activity in unselected CRC patients [71]. However, later generation MEK inhibitors show efficacy in preclinical cell line-based CRC models [72, 73]. In these studies, mutation of *KRAS* and *BRAF* predicted sensitivity to MEK inhibitors. Given the efficacy of BRAF inhibitors in mutant melanoma, MEK inhibitors have been pursued in this disease as well, and it is now well established that *BRAF* mutations and NRAS activation predict efficacy of MEK inhibitors in melanoma [74]. Furthermore, combining BRAF and MEK inhibition is of proven clinical benefit in metastatic melanoma [75]. This experience has guided clinical trials of MEK inhibitors in CRC. While some current trials are testing new generation MEK inhibitors (including selumetinib (AZD6244), MEK162, Trametinib, and PD0325901) as single agents, many are using these agents in combination with standard chemotherapy or with targeted therapies for patients with metastatic colorectal cancer. For example, preliminary but promising data on the combination of dabrafenib and trametinib for *BRAF* mutant colon cancer was recently reported (J Clin Oncol 31, 2013 (suppl; abstr 3507). Another ongoing trial is testing combined inhibition of EGFR, BRAF, and MEK in patients with *BRAF*-mutant colorectal cancer (NCT01750918). It is hoped that these trials will demonstrate clinical benefits for heavily pretreated or poor risk subgroups of CRC patients. Additional correlative studies will explore mechanisms of resistance to the MEK inhibitors.

3.10 ERK Inhibitors

Immediately downstream of MEK1/2 are the ERK1/2 proteins. Upon activation, the ERK family of kinases can translocate to the nucleus, where they phosphorylate transcription factors that drive mitogenic phenotypes. Although ERKs represent attractive targets of anticancer therapy, few inhibitors are available. However, a recent report describes a novel ERK inhibitor with significant activity in pre-clinical models [76]. As with the MEK inhibitors described above, ERK inhibitors may be useful in tumors resistant to other agents targeting the MAPK pathway [77, 78].

3.11 PI3K/AKT/mTor Pathway

Phosphoinositide 3-kinases (PI3K) are lipid kinases that serve to transmit signals from cell surface receptors (EGFR, IGFR, PDGFR) to intracellular signaling components, such as Akt and mTor, among others. Together, these multiple components constitute the PI3K pathway [79, 80]. Activation of the PI3K pathway leads to decreased apoptosis, increased proliferation, and other hallmark behaviors of cancer [81, 82]. A number of components of this pathway are commonly altered in human cancers, including CRC. The most common mutations are activating mutations in *PIK3CA*, the p110 subunit of PI3K itself, which is mutated in approximately 15–20 % of CRCs [83]. Other components, notably PTEN, which is a phosphatase that inhibits the PI3K pathway, *AKT1, AKT2*, and *PDK1* are mutated in other tumor types, but are infrequently mutated in CRC. PTEN expression is absent in approximately 40 % of CRCs secondary to a variety of mechanisms, including loss of heterozygosity, which may lead to decreased expression of the wild-type allele. However, given that 20 % of human CRCs appear to have activation of the PI3K pathway and microsatellite unstable (MSI) CRCs seem to be particularly sensitive to PI3K inhibitors, this is an attractive target for therapeutic targeting [84].

Drug development has focused on three related pathway components, alone or in combination: PI3K inhibitors, which target the enzyme itself, and Akt inhibitors and mTor inhibitors, which target critical downstream signaling molecules. Many of these compounds have been tested in patients with advanced colorectal cancer. Finally, recent data has suggested that patients with *PIK3CA* mutant CRCs benefit from aspirin therapy in the postoperative (adjuvant) setting.

3.12 PI3K Inhibitors

There is limited experience to date with PI3K inhibitors in patients with CRC. A phase I trial of XL147 included a small number of patients with CRC. In this trial, 43.9 % of patients had stable disease, but only one of these patients had CRC [85]. BKM120 is another PI3K inhibitor that has been tested in Phase I clinical trials that included patients with CRC; some had stable disease while on treatment [86]. BYL719 specifically inhibits the alpha isoform of PI3K and is being tested in *PIK3CA* mutant solid tumors including CRC. CAL-101 (Idelalisib), a specific inhibitor of the delta isoform of PI3K, has shown benefit in hematologic malignancies. The utility of these agents for CRC patients remains undefined, but additional trials, including those testing PI3K inhibitors in combination with other agents, are currently recruiting patients.

3.13 AKT Inhibitors

AKT is downstream of PI3K and other pathways and is often amplified or deregulated in human cancers [79]. In CRC, preclinical studies have linked aberrant Akt activity with increased proliferation and metastasis and decreased cell death. Using colorectal cancer cell lines, Agarwal et al. showed that the AKT inhibitor MK-2206 inhibits growth and stimulates apoptosis both in vitro and in vivo. They also showed a link between the IGF1 pathway and the Akt pathway, which suggests that the combination of IGF inhibitors and AKT inhibitors may be needed to effectively treat CRCs [87]. A related study in breast cancer cells showed that mutant *PIK3CA* or PTEN deletion predicted sensitivity to MK-2206, suggesting a strategy for predicting efficacy in CRC [88]. In patients, MK-2206 has been tested as a single agent in a Phase I study for solid tumors, where it was well tolerated. Correlative data showed it effectively decreased AKT activity in most patients [89]. Single agent MK-2206 is currently being tested in clinical trials for patients with refractory CRC (NCT01186705 and NCT01802320).

Perifosine (aka KRX-0401) is an alkylphospholipid that has multiple targets, including AKT and PI3K. Although this drug has minimal activity as a single agent in CRC, it generated considerable interest due to a phase II study of previously treated CRC patients, where the combination of perifosine and standard chemotherapy significantly improved both time to progression and overall survival [90]. Unfortunately, a larger randomized phase III study unexpectedly showed absolutely no benefit with this combination. While other AKT inhibitors are available, experience in CRC is limited; these agents (GSK2141795, GDC-0068 and AZD5363) are mostly being tested in broad phase I studies and results are pending.

3.14 mTor Inhibitors

mTor is a downstream component of the PI3K pathway and is regulated by a number of other growth promoting pathways. A number of inhibitors of mTor are available, including rapamycin and the related compounds temsirolimus, everolimus, and ridaforolimus [91]. Pre-clinical studies in cell lines and in mouse models show mTor inhibition can effectively slow the growth of PI3K-driven cancers [92, 93]. Completed clinical trials show these agents to be tolerable for patients with colorectal cancer, but evidence of significant clinical efficacy as single agents is lacking. For example, Ng et al. performed sequential phase II studies of everolimus in heavily pre-treated patients with metastatic colorectal cancer. While there was a stable disease rate of 30 %, median survival was 5–6 months. The authors concluded that everolimus did not have meaningful clinical benefit for these patients. Interestingly, molecular analysis of tumor samples showed that mutant *KRAS* CRCs had inferior outcomes compared to wild-type *KRAS* CRCs [94]. Similar findings with temsirolimus have been reported by Spindler et al. [95]. This suggests that dual inhibition of the RAS pathway and mTor pathway may be needed to attain optimal clinical benefit. Current trials are testing mTor inhibitors in combination with other agents.

3.15 PI3K Modulation to Counter Resistance to Other Agents

Studies across multiple tumor types show that the PI3K pathway may be an important determinant of resistance to standard chemotherapy, radiation therapy, and targeted therapies, including EGFR inhibitors [96]. Inhibition of PI3K in cell lines and other model systems can successfully restore sensitivity to these therapies. Based on the findings in these pre-clinical studies, clinical trials are testing this possibility in patients by combining standard chemotherapy regimens with either the pan-PI3K inhibitor PX-866 (NCT01252628), the PI3K/mTor dual inhibitor PF-05212384 (NCT01937715), or with the mTor inhibitor everolimus (NCT00522665 and NCT01047293).

3.16 The Role of Adjuvant Aspirin (ASA) in Patients with PIK3A Mutated Colorectal Cancer

While the agents discussed above have insufficient data to recommend their use outside of a clinical trial at this time, recently published retrospective data suggest a benefit of post-operative ASA use in patients with *PIK3CA* mutant colorectal cancers. Liao et al. reviewed data from the Nurses Health Study and identified 161 patients with *PIK3CA* mutant colorectal cancers [97]. Of these patients, those who reported regular ASA use in the post-operative setting had an increase in overall survival over those who did not take ASA. This benefit of ASA was not seen in those patients with wild-type *PIK3CA* CRCs. In a similar retrospective analysis, Domingo et al. reviewed data on approximately 100 patients who reported use of ASA after curative intent surgery for stage II or III colorectal cancer [98]. When stratified according to *PIK3CA* status, the benefit of ASA in reducing the risk of recurrence was confined to the patients with CRCs with mutant *PIK3CA*. The implications of these two studies are potentially far-reaching and suggest CRCs from patients eligible for adjuvant therapy should be tested for mutation of *PIK3CA*, and those with mutant PIK3CA should receive ASA post-operatively, unless there is a clear contraindication to its use.

3.17 Summary

In this chapter, we have reviewed state of the art of targeted therapies for colon cancer. The obvious success stories of VEGF and EGFR targeted therapies should be contrasted with the disappointing results, to date, with other targeted agents. A common theme is that targeting single components is often ineffective or only transiently effective, and the extensive crosstalk between signaling pathways plays a large part in mediating resistance to targeted agents. Much progress has been made in understanding these mechanisms of resistance, and this has translated to rational combination treatments that have met with success in preclinical models. Continued efforts to define mechanisms of resistance to various agents will improve selection of combinations with higher chances of sustained anti-cancer activity.

4 Current Targeted Therapies for Growth Factor Pathways: EGF and VEGF Signaling

4.1 Current Targeted Therapies for Growth Factor Pathways

Colorectal cancer (CRC) is molecularly heterogenous. Recently, a number of genetic and epigenetic aberrations that associate with prognosis and that may also be suitable for therapeutic targeting have been identified. Of the approximately 150,000 patients diagnosed annually with colorectal cancer in the United States, almost 25% will also have synchronous metastatic disease [99]. An additional 35–40% of patients with no evidence of distal metastatic disease will develop metastases to distant organs after their initial presentation. Targeted therapy directed against the epidermal growth factor receptor (EGFR), and the vascular endothelial growth factor (VEGF) and its receptors (VEGFRs) have proven to be effective for the treatment of metastatic CRC [100, 101]. These molecularly based agents are active as monotherapy against colorectal tumors (anti-EGFR agents cetuximab and panitumumab), as well as in combination with chemotherapy (VEGF targeted antibodies bevacizumab and aflibercept, and the anti-EGFR antibodies), or as single agents in the refractory setting (VEGFR-tyrosine kinase inhibitor regorafenib). Both EGFR and VEGF targeted agents in combination with chemotherapy (5-fluorouracil and oxaliplatin, FOLFOX) have been tested in the adjuvant setting, after resection of colon primary tumors, but no benefit in either disease-free or overall survival was observed [NCCTG 0147, and NSABP C-08 trials] [102, 103]. The reason for the lack of benefit with cetuximab added to FOLFOX had been postulated to be toxicity and treatment intensity related. The lack of survival benefit with bevacizumab after colorectal cancer resection has been explained by the cytostatic rather than cytotoxic effect of bevacizumab, with drug exposure causing growth arrest rather than cell death. Furthermore, the potential for enhanced tumor aggressiveness upon cessation of VEGF blockade has been documented in preclinical models, but that was unlikely the case, as the patterns of tumor progression observed in the NSABP-C08 randomized trial did not reveal any differences in the treatment groups [103]. Below we will discuss these targeted therapies, and will further describe the role of biomarkers for precision medicine in CRC.

4.2 EGFR

The EGFR family is composed of four related trans-membrane receptors that dimerize (EGFR/ERBB1, HER2/ERBB2, HER3/ERBB3 and HER4/ERBB4) to bind a variety of ligands including EGF, HB-EGF (heparin-bound EGF), NRG, TGFα, amphiregulin, and epiregulin [104]. EGFR is over-expressed in up to 80 % of CRC, and is frequently associated with poor prognosis [105]. EGFR signals through major downstream effector pathways including the RAS/RAF/MAPK pathway, which regulates cell proliferation; phosphatidylinositol-3 kinase (PI3K)/PTEN/ Akt/ mTOR, which inhibits apoptosis and induces proliferation, cell invasion and angiogenesis [106]; and the JAK/STAT pathway, which regulates apoptosis and cell survival [107]. Despite the common protumorigenic effects of EGFR signaling on colorectal cancer cells, treatment with EGFR targeted monoclonal antibodies, cetuximab (Erbitux; Lilly) and panitumumab (Vectibix; Amgen), does not uniformly result in clinical benefit, and response rates are modest in the overall colorectal cancer population [108, 109]. In addition, interestingly, small molecule tyrosine kinase inhibitors of the EGFR have shown little activity in CRC [110, 111]. This discrepancy between antibody therapy and tyrosine kinase inhibitors may stem from the fact that monoclonal antibodies directed at EGFR can induce antibody dependent cytotoxicity and can inhibit EGFR function through a variety of mechanisms including competitive inhibition of ligand binding to the extracellular portion of the receptor, prevention of receptor dimerization and induction of receptor downregulation. The next generation of anti-EGFR therapies currently in clinical trials are being developed to optimize EGFR receptor targeting and non-specific engagement of the partner human epidermal growth factor receptor (HER) families, and to maximize antibody dependent cytotoxicity (ADCC) [112].

4.3 EGFR Monoclonal Antibodies

4.3.1 Cetuximab and Panitumumab as Single Agents

Cetuximab is an IgG1 human-mouse chimeric monoclonal antibody, which acts both by direct binding to the epidermal growth factor receptor (EGFR) thereby blocking ligand interaction and subsequent EGFR tyrosine kinase activation, as well as by binding of its Fcγ receptor to immune effector cells (such as NK cells, macrophages and polymorphonuclear cells), which then trigger antibody-dependent cytotoxicity (ADCC) [104]. Panitumumab is an $IgG2_k$ subtype fully human monoclonal antibody whose mechanism of action is based on direct EGFR blockade [113].

Initial clinical trials with anti-EGFR antibodies demonstrated response rates in only 8–10 % of patients with refractory metastatic colorectal cancer, but due to improved survival, and delayed cancer progression (increased progression-free survival, PFS), the FDA approved cetuximab in 2004 and panitumumab in 2006 for patients with advanced CRC [100, 113].

In 2008 a major advance in the use of anti-EGFR antibodies in the treatment of advanced CRC occurred with the discovery of the predictive value of the *KRAS* gene status. Clinical reports demonstrated striking differences in benefit from anti-EGFR agents in CRC based on the presence or absence of activating *KRAS* gene mutations in exon 2 [114]. KRAS, the human homolog of the Kirsten rat sarcoma-2 virus oncogene, is a small G-protein that is a component of the EGFR-RAS-

MAPK signaling pathway. Oncogenic mutations in *KRAS* occur with an incidence of 28.7% in all human cancers [115, 116], and in about 40% of colorectal tumors [115]. KRAS contributes to colorectal carcinogenesis and cancer progression, and may confer a worse prognosis [117].

Ultimately, even with an intact wild-type (WT) *KRAS* gene in exon 2 less than 20% of colorectal cancer patients respond to anti-EGFR therapies. In the randomized phase III CO.17 trial lead by the National Cancer Institute of Canada (NCIC CTG) for patients refractory to chemotherapy, cetuximab was compared to best supportive care, and demonstrated significant survival benefit (6.1 vs 4.6 months, $p=0.0046$) [118]. Furthermore, patients with WT *KRAS* exon 2 tumors had improved response rates (RR) (13 vs 1%), PFS (3.7 vs 1.9 months), and an overall survival (OS) increase by almost 5 months (9.5 vs 4.8 months, $p<0.001$) [118, 119]. In a randomized phase III study versus best supportive care, panitumumab resulted in RR of 10% vs 0, and increased the PFS (HR 0.54, $p<0.0001$), but survival rates were similar, likely due to patients' pre-planned cross-over to the treatment arm upon disease progression [113]. A follow-up analysis demonstrated that patients with WT *KRAS* tumors had even higher response rates (17%), and median PFS (12.3 vs 7.3 weeks, $p<0.0001$), then initially reported, but no difference in OS (Table 1) [120]. These studies lead to the approval by the FDA for cetuximab (2004), and panitumumab (2006) in mCRC refractory to prior chemotherapies: 5-FU, irinotecan and oxaliplatin. Given the predictive value of *KRAS* exon 2 (codons 12 or 13) mutational status for benefit from anti-EGFR antibodies in mCRC, in 2009 both the American Society of Clinical Oncology and the FDA issued recommendations for using both cetuximab and panitumumab exclusively in patients with WT *KRAS* exon 2 [121].

4.4 Anti-EGFR Antibodies with Chemotherapy in Metastatic CRC

EGFR targeted therapies have demonstrated additive benefit to several conventional cytotoxic chemotherapy backbones. A decade ago, phase II trials assessing cetuximab with irinotecan versus cetuximab alone in patients with irinotecan refractory CRC showed improved responses (23% vs 11%, $p=0.007$), and time to progression (4.1 vs 1.5 months, $p<0.001$) [100]. Several mechanisms of cetuximab-mediated irinotecan resensitization have been postulated: decreased drug efflux via cross-talk between EGFR and the multidrug resistance protein P-Gp [138, 139], restoration of apoptosis [140], and inhibition of DNA damage-repair [141]. For second line therapy in mCRC (unselected for *KRAS* WT status), the randomized phase III EPIC trial of cetuximab plus irinotecan versus irinotecan alone noted response rates of 16 vs 4%, median PFS of 4 vs 2.6 months, as well as improved quality of life, but survival rates were similar in the two treatment groups, possibly due to cross-over to the cetuximab arm upon disease progression [109]. In support of a beneficial effect of anti-EGFR therapy for second-line treatment of mCRC patients, in the 181 trial, panitumumab also showed significantly increased PFS (5.9 vs 3.9 months) and response rates (35 vs 10 %), with a trend for OS benefit (14.5 vs 12.5 months, $p=0.12$) when combined with FOLFIRI, versus FOLFIRI alone [137].

More recently, CRYSTAL was a randomized phase III trial for first line treatment of mCRC that used cetuximab combined with 5-fluorouracil and irinotecan (FOLFIRI). This trial demonstrated increased PFS in the overall population compared to FOLFIRI alone, (9.9 vs 8.7 months) [122]. An updated analysis of the WT *KRAS* exon 2 patients noted a significant overall survival benefit with the addition of cetuximab (OS 23.5 vs 20 months, $p=0.009$), as well as superior response rates (57.3 vs 39.7 %, $p<0.01$). In the same clinical trial, the tumors' *BRAF* mutational status was not predictive of response, but it was prognostic [123]. Based on these data, on July 2012, the FDA approved cetuximab combined with FOLFIRI as first-line treatment for patients with WT *KRAS* and EGFR expressing mCRC.

Cetuximab has also been studied with oxaliplatin-based chemotherapy (5-fluoro-uracil/leucovorin plus oxaliplatin–FOLFOX, or capecitabine plus oxaliplatin -CAPOX) in mCRC. The phase III COIN study randomized patients with mCRC to first-line FOLFOX or CAPOX with or without cetuximab [124]. Among patients with CRCs with WT *KRAS*, there was an increased response rate in the arm that included cetuximab (64 vs 57 %, $p=0.049$), however, no benefit was observed in OS or PFS (Table 1). Similarly, in the NORDIC VII study of bolus 5-FU/leucovorin plus oxaliplatin (FLOX) with or without cetuximab, there was no benefit associated with the addition of anti-EGFR antibody therapy (PFS 8.7 vs 7.9 months) [125]. Although smaller phase II trials showed benefit when cetuximab was combined with FOLFOX [126–130] overall the cetuximab with oxaliplatin-based therapy studies has generated conflicting results.

Panitumumab was studied in the randomized phase III PRIME study and it significantly improved PFS when added to FOLFOX in the first line treatment of WT *KRAS* exon 2 mCRC (PFS 9.6 vs 8 months ($p=0.02$), with a trend for increased OS 23.9 vs 19.7 months ($p=0.072$) [132]. Less frequent than mutations in *KRAS* exon 2, other *RAS* family genes have been documented in mCRC [142, 143]. Among patients with WT *KRAS* exon 2 CRC in PRIME, 17 % had other mutations in *KRAS* exons 3–4 or *NRAS* exons 2–4. In a subsequent analysis, patients with CRCs with WT *KRAS* exons 2–4 and *NRAS* exons 2–4 had significant improvement in overall survival with FOLFOX plus panitumumab compared to FOLFOX alone (26 vs 20.2 months, $p=0.04$), [131] leading to its approval by the FDA in the first-line setting for metastatic CRC (mCRC) in 2014.

In an extension of treatment strategies that employ anti-EGFR and cytotoxic agents for the treatment of patients with CRC, studies have been conducted in which anti-EGFR agents have been compared head-to-head to bevacizumab. Cetuximab has been studied vis-a-vis bevacizumab in combination with FOLFIRI, as first-line therapy for patients with WT *KRAS* exon 2 mCRC in the FIRE-3 German AIO study [133]. Cetuximab was superior to bevacizumab with respect to overall survival (28.7 vs 25 months, $p=0.017$), albeit OS was not the primary study endpoint. The response rate (primary endpoint) had a trend of improvement with cetuximab (62 vs 58 %, $p=0.183$), and the PFS rates were similar. Retrospective review of the "extended" RAS status: *KRAS* exons 3 and 4, and *NRAS* exons 2–4, in the

Table 1 Clinical trials with anti-EGFR antibodies in colorectal cancer

Trial	Nr. Patients	Treatment	RR (%)	*p*	PFS (mo)	*p*	OS (mo)	*p*
First line chemotherapy+/−cetuximab								
CRYSTAL [122, 123]	599 vs 599 350 vs 316	FOLFIRI+/−cetuximab Overall *KRAS* WT	47/39 57/40	0.004 <0.001	8.9/8.0 9.9/8.4	0.048 0.001	19.9/18.6 23.5/20.0	0.31 0.009
COIN [124]	367 vs 362	CAPOX/FOLFOX+/−cetuximab	64/57	0.049	8.6/8.6	0.60	17.0/17.9	0.67
Nordic VII [125]	185 vs 194 97 vs 97	FLOX+/−cetuximab Overall *KRAS* WT	49/41 46/47	0.15 0.89	8.3/7.9 7.9/8.7	0.31 0.66	19.7/20.4 20.1/22.0	0.67 0.48
OPUS [126, 127]	168 vs 169 97 vs 82	FOLFOX+/−cetuximab Overall *KRAS* WT	46/36 57/34	0.064 0.002	7.2/7.2 8.3/7.2	0.62 0.006	18.3/18.0 22.8/18.5	0.91 0.39
CELIM [128, 129]	56 vs 55	FOLFOX+cetuximab vs FOLFIRI+cetuximab	68/57	0.23	11.2/10.5	0.40	35.8/29.0	0.90
CECOG [130]	74 vs 77	FOLFOX+cetuximab vs FOLFIRI+cetuximab	43/45	NS	8.6/8.3	0.73	17.4/18.9	0.92
First line chemotherapy+/−panitumumab								
PRIME [131, 132]	593 vs 590 325 vs 331 259 vs 253	FOLFOX+/−panitumumab *KRAS* exon 2 WT, *KRAS* exons 2–4 WT and *NRAS* exons 2–4 WT	55/48 NA	0.068 NA	9.6/8 10.1/7.9	0.02 0.004	23.8/19.4 25.8/20.2	0.03 0.009
First line chemotherapy with cetuximab/panitumumab vs bevacizumab								
FIRE-3 [133, 134]	295 vs 297 171 vs 171	FOLFIRI+cetuximab vs FOLFIRI+bevacizumab *KRAS* exon2 WT, *KRAS* exons 2–4 WT and *NRAS* exons 2–4 WT	62/58 66/60	0.183 0.32	10.0/10.3 10.4/10.2	0.547 0.540	28.7/25.0 33.1/25.6	0.017 0.011
PEAK [135]	142 vs 143 88 vs 82	FOLFOX+panitumumab vs FOLFOX+bevacizumab *KRAS* exon 2 WT, *KRAS* exons 2–4 WT and *NRAS* exons 2–4 WT	58/54 64/61	NS NS	10.9/10.1 13.0/9.5	0.353 0.029	34.2/24.3 41.3/28.9	0.009 0.058
CALGB 80405 [136]	578 vs 559	FOLFOX/FOLFIRI+cetuximab FOLFOX/FOLFIRI+bevacizumab	NA	NA	10.4/10.8	0.55	29.9/29.0	0.34

Table 1 (continued)

Trial	Nr. Patients	Treatment	RR (%)	*p*	PFS (mo)	*p*	OS (mo)	*p*
Second line chemotherapy+/−cetuximab or panitumumab								
Cunningham [100]	218/111	Cetuximab ± irinotecan	23/11	0.007	4.1/1.5	< 0.001	8.6/6.9	0.48
EPIC [109]	648/650	Irinotecan ± cetuximab	16/4	< 0.0001	4/2.6	< 0.0001	10.7/10	0.71
181 [137]	303/294	FOLFIRI ± panitumumab *KRAS* WT exon 2	35/10	< 0.001	5.9/3.9	0.004	14.5/12.5	0.12
Single agent cetuximab or panitumumab								
CO.17 [118, 119]	287/285 117/113	Cetuximab/best supportive care *KRAS* WT exon 2	8/0 13/1	< 0.001 < 0.001	n/a 3.7/1.9	< 0.001	6.1/4.6 9.5/4.8	0.005 < 0.001
Van Cutsem [113] Amado [120]	231/232 124/119	Panitumumab/Best supportive care *KRAS* WT exon 2	10/0 17/0	< 0.001 < 0.001	1.9/1.7 2.9/1.7	< 0.0001 < 0.0001	7/7 8.1/7.6	0.81 ns

FIRE-3 study noted marked improvement in OS (33.1 vs 25.6 months, $p=0.011$) for patients treated with FOLFIRI and cetuximab vs bevacizumab if no mutations in *KRAS* or *NRAS* were present (Table 1), while any mutations in *KRAS* exons 2–4 or in *NRAS* exons 2–4 conferred no benefit from cetuximab versus bevacizumab (RR 38 vs 51.2%, $p=0.097$; PFS 7.5 vs 10.1 months, $p=0.085$; OS 20.3 vs 20.6 months, $p=0.06$) [134]. Results from the FIRE-3 trial need to be reproduced in other randomized phase III trials to confirm the optimal biological treatment to be used in combination with FOLFIRI for first-line WT *KRAS* CRC.

Panitumumab was also compared to bevacizumab; both in combination with FOLFOX for first-line treatment of WT *KRAS* CRC in the randomized phase II trial, PEAK [135]. For patients with WT *KRAS* exon 2 tumors and for patients with WT *KRAS* exons 2–4 and *NRAS* exons 2–4 tumors both progression-free and/or overall survival were markedly improved with panitumumab (OS 34.2 vs 24.3 months, $p=0.009$) (Table 1). The phase III trial CALGB 80405 among 1137 patients with WT *KRAS* exon 2 CRC who were randomized to FOLFOX/FOLFIRI with either cetuximab or bevacizumab, was anticipated to confirm the superiority of EGFR-targeting compared to bevacizumab [136]. Nevertheless, while response rates were not yet reported, the survival outcomes were equivalent in all treatment arms (29+months). Analyses of the extended *KRAS* and *NRAS* genes status will be up-coming and may further help elucidate the optimal first line strategy for WT *KRAS*, *NRAS* mCRC patients.

Clinical trials (CAIRO2 and PACCE) have also assessed conventional cytotoxic chemotherapy (both oxaliplatin- and irinotecan-based) with dual anti-VEGF plus anti-EGFR antibodies. The addition of anti-VEGF plus anti-EGFR therapy to irinotecan or oxaliplatin based regimens increased toxicity and diminished efficacy [144, 145]. Thus, it is not recommended for bevacizumab to be administered concurrently with either cetuximab, or panitumumab.

4.5 Anti-EGFR Antibodies and Radiotherapy

EGFR modulates cellular radiation response and DNA repair and its overexpression in CRCs has been linked to radiation resistance [146]. EGFR blockade induces radiosensitivity in several preclinical models. Cetuximab is FDA approved in combination with radiotherapy for treatment of squamous cell head and neck carcinoma, emphasizing its effective radiosensitizing properties. Several studies, including the randomized phase II EXPERT-C trial, tested cetuximab with chemoradiotherapy in rectal adenocarcinomas, and the results showed good tolerability, pathologic complete responses (pCR) no better than historical controls (10–15%), but overall survival rates (5-year OS 80–90%) appear to be encouraging [147–150]. Similarly, panitumumab was combined with chemoradiotherapy in rectal cancer and showed higher risk of severe diarrhea (40% grade 3/4) and pCR of 10–20% [151, 152]. Larger randomized studies with long-term follow up are needed to confirm a definite benefit with cetuximab or panitumumab and chemoradiotherapy in rectal cancer patients.

4.6 EGFR Biomarkers of Response and Mechanisms of Resistance-Precision Medicine

EGFR is overexpressed in many CRCs, which led to the use of anti-EGFR therapy for CRC. However, it has been recognized that increased expression is not required for EGFR targeted therapies to be effective [153].

EGFR gene amplifications may affect the response to anti-EGFR agents, as CRCs with mutant *KRAS* may occasionally respond to cetuximab in the presence of overexpressed EGFR [154–156]. Another component of the ERBB family, HER2, is amplified in a subset of CRC and may contribute to anti-EGFR resistance, even in patients with WT *KRAS/NRAS/BRAF/PIK3CA* mCRC [157]. EGFR somatic gene mutations in exons 18–21, which are major predictive markers for the efficacy of EGFR tyrosine kinase inhibitors (TKI) in non-small cell lung cancer, occur rarely in CRC (incidence 0.4–3 %), [158, 159] and could explain the lack of benefit with EGFR TKI in CRC. Moreover, *EGFR* gene mutations do not correlate with response to anti-EGFR antibodies in CRC. Interestingly, vertical EGFR blockade with monoclonal antibody therapy and EGFR-TKI may be synergistic in vitro [160], and preliminary reports show clinical activity [161]. Randomized trials may help define a role for EGFR-TKI with or without anti-EGFR monoclonal antibodies in CRC. Additional molecular components of the EGFR pathway are the EGFR ligands, such as amphiregulin and epiregulin, and high RNA transcript levels or gene overexpression, have been shown to predict higher response rates to anti-EGFR antibodies, and improvements in PFS or OS [162–164]. EGFR posttranslational modification, i.e., ubiquitination, with dysregulated receptor internalization and degradation, has been implicated in acquired cetuximab resistance [165, 166]. This process leads to activation of HER3, as well as the ability to bypass EGFR dependency and activate Src kinase pathway signaling instead [165].

Although EGFR is not an effective predictive marker for anti-EGFR therapy, mutant *KRAS* is the most significant negative predictive factor for response to anti-EGFR antibodies in patients with metastatic CRC. Nevertheless, the presence of *KRAS* exon 2 mutations in codons 12 and 13 (40 % of mCRC patients), and even extended RAS mutations in *KRAS* exons 3, 4 and neuroblastoma-ras (*NRAS*) exons 2–4 (another 20 % of mCRC patients), do not solely explain the *de novo* resistance to EGFR blockade, as only 30 % of patients respond to this class of therapy. In addition, some patients with *KRAS* mutations may respond to anti-EGFR antibodies; *KRAS* codon 13 mutations (G13D) were associated with improved OS and PFS after cetuximab treatment, compared to *KRAS* mutations in codon 12 [167].

V-raf murine sarcoma viral oncogene homologue B1 (*BRAF*) is the main downstream effector from *KRAS*, and several studies showed limited, or no benefit from either cetuximab or panitumumab in patients with CRCs with *BRAF* mutations (incidence < 10 %), [168] while others observed mostly a negative effect on survival [123]. Thus, it is currently believed that mutant *BRAF* is primarily a prognostic factor in mCRC and not a predictive marker for response to anti-EGFR therapy. Interestingly, for some patients with WT *BRAF* colorectal cancers, expression array analysis identified a *BRAF*-mutant like signature associated with poor outcomes, including in patients treated with cetuximab, which suggested that the poor prognosis associated with mutant *BRAF* is a consequence of the activation of a specific set of BRAF target genes [169].

The global activation of the phosphatidylinositol 3-kinase (PI3K)/protein kinase B (AKT)/mammalian target of rapamycin (mTOR) pathway in CRC may also contribute to resistance to EGFR targeted therapies although the data supporting this are not consistent [170]. Alterations in the PI3K pathway include mutations in *PIK-3CA* (10–20 % of CRC) [57, 171–173], and loss of or mutations in the phosphatase and tensin homologue (*PTEN*) (loss of expression in approximately 40 % of CRC and mutation in approximately 10 % of CRC) [174–176].

Another mechanism of resistance to monoclonal antibody therapy is related to antibody dependent cytotoxicity (ADCC). Cetuximab is able to initiate immune responses via the interaction of the Fc portion of the IgG1 molecule with receptors on NK cells and other immune cells. Possibly affecting the ADCC induced by cetuximab, polymorphisms in the Fcγ receptors (FcγRIIa-131H/H and FcγRIIIa-V158F) on immune cells may correlate with improved PFS after cetuximab treatment [177].

Epithelial—mesenchymal transition, which can be associated with AKT and STAT3 activation, has also been involved in both cetuximab and EGFR-TKI resistance [178, 179]. Other biological interactions mediating resistance may occur between EGFR and the insulin growth factor receptor (IGF1R) or MET, both capable of activating downstream pathways. The IGF1R has not demonstrated a major role in the primary resistance to anti-EGFR therapies in CRC, but when overexpressed it may be prognostic for longer survival [180]. MET, the hepatocyte growth factor receptor can activate both the PI3K/AKT/mTOR and the RAF/MEK/MAPK pathways in a RAS-independent manner. *MET* may be amplified in CRC, including in *KRAS* mutant subtypes.

The clinical effectiveness of anti-EGFR antibodies, even when initially successful, is limited by the emergence of resistance, usually after 3–15 months of treatment. *KRAS* mutations may develop while on anti-EGFR treatment in up to 50 % of patients [181]. In addition, polyclonal mutations in *NRAS* and *BRAF* have been identified after anti-EGFR therapy and likely contribute to acquired treatment resistance [182]. In all, these mutations can result in the overall activation of the mitogen-activated protein kinase kinase (MEK), and extracellular signal-regulated kinase (ERK) signaling. Interestingly, combinatorial blockade of EGFR and MEK, but not MEK alone, has been shown to induce antitumor efficacy in resistant CRCs in preclinical studies. Therapeutic MEK blockade may represent a new strategy for restoring sensitivity to anti-EGFR therapies. Among patients who do not develop *KRAS* mutations while on anti-EGFR treatment, *MET* amplification has been identified as a putative mechanism of acquired resistance [41]. While an infrequent event (< 1 %), *KRAS* gene amplifications have also been associated with both primary and secondary resistance to anti-EGFR therapies in CRC [183].

The complexity of the EGFR signaling network explains the difficulty of predicting clinical responsiveness and also creates a number of potential mechanisms responsible for primary and secondary resistance to anti-EGFR agents. It is likely that identifying a specific profile of biomarker alterations will be important to select synergistic combinations of therapeutic inhibitors that will be less susceptible to the development of treatment resistant CRCs. New hypotheses suggest that testing downstream terminal phosphoprotein effectors of pathways affected by EGFR, such as pMEK1 or p70S6 kinase, (higher levels correlating with worse PFS and OS) could be better predictors of outcomes after anti-EGFR treatment in CRC [184].

4.7 VEGF

Angiogenesis, or neovascularization contributes to CRC growth and metastasis, and it requires the stimulation of the vascular endothelial cells through the release of angiogenic peptides, of which the vascular endothelial growth factor (VEGF) is the most potent. High VEGF expression correlates with CRC stage and is a prognostic factor for metastatic disease recurrence for colon cancer patients [185]. Stimulated by hypoxia, growth factors (e.g., EGF, PDGF, FGF), and oxidative stress, VEGF induces endothelial cells proliferation, increases vascular permeability, and contributes to the formation of the stromal fibrin matrix which appears to promote tumor cell invasion [2].

Five VEGF ligands (VEGF-A through VEGF-E) can be secreted by tumors, but the critical one for neovascularization is VEGF-A. VEGFs bind to the VEGF receptors (VEGFRs)—1, 2 and 3 on the surface of endothelial cells, with VEGFR-2 being critical for angiogenesis [2]. The VEGF axis can be inhibited by targeting the VEGF ligand (bevacizumab, zev-aflibercept), by blocking the binding of the VEGF ligand to its receptors with antibodies to the extracellular domain of VEGFR-2 (IMC-1121B, ramucirumab), or by small molecules tyrosine kinase inhibitors of the intracellular kinase domain of the VEGFRs (regorafenib, sorafenib, sunitinib, and others).

Angiogenesis blockade may impact cancer outcomes and improve response to systemic therapies via several mechanisms: (1) normalizing and decreasing the permeability of tumor vasculature thus increasing the oxygenation and delivery of therapeutics to the tumor tissue; [186] (2) decreasing new blood vessel formation which may decrease or stabilize tumor growth; (3) enhancing tumor cells apoptosis; [187] (4) improving the function of endogenous dendritic cells and other immune effector cells, which may augment the anti-tumor immune response; [188–190]. Anti-VEGF therapies may prevent chemo- and radiotherapy resistance, in spite of the increased VEGF expression induced by these therapies [191].

4.8 Bevacizumab

Bevacizumab (Avastin; Genentech) is a fully humanized monoclonal antibody that inhibits all human VEGF-A isoforms from binding to VEGFR-1 and -2. Bevacizumab is the first biological agent approved for metastatic colorectal cancer, and is indicated in both first and second-line treatment in combination with chemotherapy. Contrary to anti-EGFR monoclonal antibodies, which have single agent activity in mCRC, bevacizumab seems to only benefit when combined with chemotherapy.

4.9 First Line Therapy in mCRC

The pivotal phase III trial, AVF2107g, demonstrated an approximately 5 month survival improvement (20.3 vs 15.6 months), 4 month increase in PFS (10.6 vs 6.2 months), and 10% increase in response rates (45 vs 35%) when bevacizumab was added to irinotecan and 5-FU/leucovorin (IFL) chemotherapy (Table 2) [101]. The same results were not reproduced when bevacizumab was combined with oxaliplatin-based chemotherapy (FOLFOX or CAPOX) in the NO16966 trial, where response rates (38% in both groups) and overall survival (21.3 vs 19.9 months) were not increased by bevacizumab, but the PFS rates were statistically higher (9 vs 8.4 months) [192]. While these results suggest that bevacizumab is more effective when combined with irinotecan compared to oxaliplatin-based regimens, a prospective observational community based registry study (ARIES) suggested that outcomes were equivalent with either FOLFOX or FOLFIRI combined with bevacizumab for first-line therapy of mCRC (OS 23.7 vs 25.5 months) [193]. The recently reported phase III TRIBE study described high rates of response (65 vs 53%) and improved PFS (12.1 vs 9.7 months) when bevacizumab was combined with a regimen of 5-FU/leucovorin/oxaliplatin/irinotecan (FOLFOXIRI) compared to FOLFIRI; nevertheless, survival data is not yet mature to make definitive recommendations [194]. Interestingly, in the TRIBE study as well as in others using FOLFOXIRI plus bevacizumab, particular benefit was noted for patients with mutant *BRAF* mCRC, who otherwise have an unusually poor prognosis (OS 24 months with FOLFOXIRI/bevacizumab vs 14–16 months with standard regimens) [195]. Bevacizumab is also beneficial when added to fluoropyrimidine therapy alone, either 5-FU/LV or capecitabine, for elderly patients or for those who are not candidates for more intense regimens (Table 2) [196, 197].

The initial report on the randomized phase III CALGB 80405 trial at the 2014 American Society of Clinical Oncology (ASCO) for patients treated with either FOLFOX or FOLFIRI and bevacizumab, described lower survival rates of 26.9 months among 409 patients treated with FOLFOX/bevacizumab, compared to 33.4 months among 81 patients treated with FOLFIRI/bevacizumab. These results build on the synergism between irinotecan-based therapy and anti-VEGF agents that has been documented in other clinical trials.

Ongoing studies that are assessing the clinical effects of FOLFOX vs FOLFIRI chemotherapy backbones with bevacizumab (MAVERICC), and FOLFOXIRI vs FOLFOX with bevacizumab (STEAM, VISNU-1) are in progress.

Table 2 Clinical trials with anti-VEGF/VEGFR therapies in colorectal cancer

Trial	Nr. Patients	Treatment	RR %		PFS (mo)	*p*	OS (mo)	*p*
First line chemotherapy+/−bevacizumab								
AVF2107g [101]	411/402	IFL+/−bevacizumab	45/35	0.004	10.6/6.2	<0.001	20.3/15.6	<0.001
NO16966 [192]	701/699	FOLFOX/ CAPOX+/−bevacizumab	38/38	0.99	9.4/8	0.002	21.3/19.9	0.077
AVEX [196]	140/140	Capecitabine+/−bevacizumab	n/a		9.1/5.1	<0.0001	n/a	
ARIES [193]	968/243	FOLFOX+bevacizumab vs FOLFIRI+bevacizumab	56/52	NS	10.3/10.2	0.688	23.7/25.5	0.625
TRIBE [194]	252/256	FOLFOXIRI+bevacizumab vs FOLFIRI+bevacizumab	65/53	0.006	12.1/9.7	0.006	31/25.8	0.125
CALGB 80405 [136]	409 81	FOLFOX+bevacizumab FOLFIRI+bevacizumab	n/a n/a				26.9 33.4	
Second line chemotherapy+/−antiVEGF therapy								
ECOG3200 [198]	291/286	FOLFOX+/−bevacizumab	23/9	<0.0001	7.3/4.7	<0.0001	12.9/10.8	0.0011
ML18147 [199]	411/409	Crossover chemotherapy+/−bevacizumab	4/5	0.31	5.7/4.1	<0.0001	11.2/9.8	0.0062
VELOUR [200]	612/614	FOLFIRI+/−aflibercept	20/11	0.0001	6.9/4.7	<0.0001	13.5/12.0	0.0032
Anti-VEGF therapy for refractory disease								
CORRECT	500/253	Regorafenib vs placebo	1/0	0.19	1.9/1.7	<0.0001	6.4/5.0	0.0052

4.10 Second Line Therapy in mCRC

Several studies addressed the question of whether following disease progression on first line therapy there is a role for anti-VEGF treatment in the second line setting. Initially, bevacizumab has been tested in the randomized phase III study ECOG3200 for the second-line treatment of patients naïve to prior anti-VEGF therapy. In combination with FOLFOX versus FOLFOX alone, bevacizumab improved overall survival by 2 months (12.9 vs 10.8 months), as well as PFS (2.6 months increase), and overall response rates (23 vs 9 %) [198].

For patients with prior bevacizumab exposure during first line therapy, the ML18146 trial tested bevacizumab with cross-over chemotherapy in the second-line setting. Continuing VEGF blockade in the second line treatment of mCRC while changing the chemotherapy backbones from irinotecan-based to oxaliplatin-based or the reverse upon transition to the second line therapy, showed mild benefit compared to conventional cytotoxic chemotherapy alone: 1.4 months increase in OS (11.2 vs 9.8 months, $p=0.0062$), but response rates were low in both arms (4 vs 5 %) [199].

4.11 Ziv-Aflibercept

Ziv-aflibercept (Zaltrap; Sanofi-Aventis) is a fully humanized recombinant fusion protein that consists of portions of the extracellular domains of both VEGF-R1 and VEGF-R2 proteins fused to the Fc portion of human immunoglobulin G1 capable of binding all forms of VEGF-A, VEGF-B, and the placental growth factors (PIGF) 1 and 2.

Ziv-aflibercept has been studied in a large randomized international phase III trial (VELOUR) in the second-line treatment of mCRC in combination with FOLFIRI vs FOLFIRI/placebo for patients who were previously treated with oxaliplatin, and with bevacizumab (30 % of the study subjects). [200] Ziv-aflibercept conferred a slight increase in overall survival by 1.5 months when added to FOLFIRI (13.5 vs 12 months, $p=0.0032$), and improved PFS (6.9 vs 4.7 months, $p<0.001$), and response rates (20 vs 11 %, $p<0.001$), compared with the control arm, further confirming the benefit of angiogenic blockade in second-line therapy for mCRC.

While the FDA granted approval in 2011 for ziv-aflibercept in second-line treatment of mCRC after progression on prior oxaliplatin-based therapy, clinical practitioners are now faced with the opportunity of using VEGF blockade in both first and second-line therapy for CRC patients. Nevertheless, the choice between continuing bevacizumab, based on the ML18147 trial (OS 11.2 vs 9.8 months) vs using ziv-aflibercept based on the VELOUR trial (OS 13.5 vs 12 months) where only 30 % of patients had prior bevacizumab exposure, has proven difficult.

Ziv-aflibercept has also been studied for the first-line therapy of mCRC in combination with FOLFOX, in the randomized phase II trial AFFIRM [201]. Among 236 patients, ziv-aflibercept/FOLFOX compared to FOLFOX alone did not increase either response rates (49.1 vs 45.9 %) or median PFS (8.48 vs 8.77 months). Randomized studies are being planned for testing ziv-aflibercept versus regorafenib in the refractory setting for mCRC patients.

4.12 *Regorafenib*

Regorafenib (BAY 73–4506) is a novel oral multikinase inhibitor of VEGFR-2, as well as of VEGFR-1, -3, TIE2, the platelet-derived growth factor receptor (PDGFR)-β, the fibroblast growth factor receptor (FGFR)-1, and the mutant oncogenic kinases KIT, RET, RAF1 and BRAF. Regorafenib demonstrated antitumor activity in preclinical models including colorectal cancer, [202] and in a phase I study of patients with refractory mCRC. In that study, 74 % of patients experienced disease control, and pharmacodynamic evidence of angiogenesis inhibition [203].

Following these encouraging data, the phase III international trial CORRECT randomized refractory mCRC patients to regorafenib or placebo, with the primary endpoint of overall survival. At a planned interim analysis the study reached its efficacy boundary, with an overall survival benefit of 23 % for regorafenib. Regorafenib conferred a small but significant 1.4 months survival improvement (6.4 vs 5 months, $p = 0.0052$), but the disease control rate was 40 vs 15 % for the placebo group, and the hazard ratio for PFS was 51 % better with regorafenib [204]. Adverse events were significantly higher with regorafenib, especially the hand foot skin reaction (17 % grade 3), fatigue, diarrhea, anorexia, mucositis, hypertension and voice changes (> 20 %, all grade), and the overall quality of life was not improved compared to placebo. Regorafenib is currently FDA approved to be administered orally at 160 mg daily × 3 weeks every 4 weeks, however to prevent patient drug discontinuation due to toxicity, most physicians in the U.S utilize a lower starting dose (120 or 80 mg) with careful upward tapering of the dose, providing good tolerability. Of note, although definitive correlative biomarkers for outcomes are still in progress, regorafenib seems to benefit mCRC patient irrespective of their *KRAS* mutation status.

An ongoing study is testing regorafenib with FOLFIRI in second-line mCRC patients. The activity seen with regorafenib single-agent in mCRC is notable, as most other VEGFR TKI tested so far have not resulted in any significant survival advantage in any line of therapy. It is likely that the effects on multiple oncogenic pathways and on the tumor stroma, in addition to the antiangiogenic effect may be contributing to the activity seen with regorafenib.

4.13 *VEGF and Biomarkers—Precision Medicine*

Several anti-VEGF therapies have been developed, but so far we do not have conclusive evidence of a predictive biomarker in clinical practice. Several groups observed conflicting results regarding the correlation between bevacizumab-induced hypertension and colorectal cancer prognosis [205, 206, 207]. In colorectal cancer, mutant *KRAS* and *BRAF*, or *TP53* do not appear to predict treatment response to bevacizumab [208, 209]. Molecular markers that accurately predict response to anti-VEGF therapy are still needed.

With regards to angiogenesis pathway biomarkers, several have been tested as prognostic or predictive markers of benefit from antiVEGF agents, however, none have been shown to be clinically useful to date. In various tumors, studies of plasma VEGF-A levels, the target of bevacizumab, yielded conflicting results regarding the correlation with response and PFS with bevacizumab-based therapy, [210, 211] but several studies across tumor types, including colorectal cancer, suggest that high plasma VEGF-A levels confer worse survival [212]. Furthermore, it has been well described that there is not a clear association between a response as measured by VEGF-A expression changes and a survival benefit with bevacizumab-containing regimens in CRC [213]. The pivotal study AVF2107g which led to the FDA approval of bevacizumab with chemotherapy in metastatic CRC, found that plasma VEGF-A levels higher vs lower than the median (44 pg/mL) conferred worse survival both in the control group (13 vs 18 months, $p=0.0005$), as well as in the bevacizumab group (16 vs 25 months), albeit without significant correlation with response or progression-free survival [212]. In addition, there was no correlation between the plasma and the tumor VEGF-A levels. Several retrospective analyses found no correlation between tumor VEGF-A levels and outcomes in CRC [214, 215]. More recently though, the MAX trial of the Australasian Gastrointestinal Trials Group reported that low tumor VEGF-A levels conferred improved response rates to bevacizumab containing regimen (53 vs 38 %, $p=0.03$), as well as superior PFS (9.3 vs 7.2 months, $p=0.008$) [216].

Other angiogenic biomarkers such as soluble plasmaVEGFR-1 and -2 levels have been extensively studied. VEGF induces ligand-dependent downregulation of VEGFR-2, thus there is an inverse relationship between plasma VEGF-A and VEGFR-2 levels. Circulating VEGFR-2 levels may be considered a marker of tumor "VEGF dependency" [217]. High VEGFR-2 levels conferred superior survival in mCRC patients treated with bevacizumab in the phase III BEBYP trial, [218] as well as prolonged responses in the NO16966 trial, [219] but other studies were inconclusive [220]. Small studies showed converse results for plasma VEGFR-1: low levels associated with improved responses and overall survival [220]. For patients treated with VEGFR tyrosine kinase inhibitors, low VEGFR-2 levels were associated with trends for improved survival, but not with response [221].

Certain Single-nucleotide polymorphisms (SNPs) in angiogenic genes affect gene transcription and some have been linked with clinical benefit from bevacizumab-containing therapy [210, 222, 223]. Potential VEGF and VEGFR SNPs have been identified retrospectively in clinical trials in solid tumors, and several were studied prospectively. Most allelic variants in *VEGF-A*, *VEGFR-1,* and *VEGFR-2* genes have not demonstrated a definitive correlation with treatment response, but the *VEGFR-2* rs12505758C/T variant showed a trend of correlation with worse PFS and OS [224]. Recently, the *VEGF-A* rs25648C/C polymorphism was associated with improved prognosis for CRC patients treated with bevacizumab therapy in the phase III MAX trial [225]. This may be due to the C/C genotype correlation with lower VEGF-A expression levels.

Multiple challenges exist not only in improving our understanding of the complexity of the angiogenic pathways, but also in determining the optimal way to use anti-angiogenic therapy given the fact that the intrinsic benefit of anti-VEGF agents is limited, and only realized in the context of concurrent cytotoxic chemotherapy. This complexity is likely part of the reason for our lack of success in identifying predictive biomarkers for improved outcomes with VEGF blockade, which suggests novel approaches are critically needed to identify biomarkers for anti-angiogenic therapies.

5 Immunotherapy

5.1 Introduction

The treatment of cancer is complex and the response of a cancer to a particular therapy varies depending on the stage of the cancer. More advanced cancer is typically more difficult to cure than early stage cancer. This situation can be likened to chess in which the opening moves, when all options are available, determine the course and eventual outcome of the game; whereas the strategic options at the end of the game are limited [226]. Immunologic therapies are often used in patients with advanced cancer and though recent successes in the clinic exist, an understanding of the early steps in tumor development is required to understand how we can optimally harness the immune system to treat cancer. Immune surveillance is the mechanism by which the immune system targets and destroys developing malignancies. Malignant cells evade immunosuppression by downregulation of intrinsic immunogenicity [227]. In order to appreciate potential therapeutic options that rely on anti-tumor activities of the immune system or that can inhibit tumor-mediated suppression of the immune response, we will first provide an overview of relevant immunology; immune mechanisms that mediate immune responses against cancer cells; and mechanisms through which tumors evade the immune system.

5.2 Immunology Overview

The immune system's main function is to eliminate anything that is recognized as "not self", be it a virus, bacterium, cells in a transplanted organ or tumor cells. Immunosurveillance is a monitoring process of the immune system that detects and destroys neoplastic and pre-neoplastic cells. The immune system has two main arms-the innate immune system and the adaptive immune system. The innate immune response is mainly made of neutrophils, monocytes/macrophages, dendritic cells, mast cells, and natural killer cells [228]. The innate immune system consists of two principal reactions, which are inflammation and antiviral defense. The innate immune system has a fixed repertoire of responses, which means it does not remember and modify its response based on prior encounters with the non-self entity, which is in contrast to the adaptive immune response (discussed below). The innate immune system has evolved to recognize structures of microbes that are essential to the survival of these microbes. Consequently, the structures are recurrent over a broad spectrum of microbes. These structures are called pathogen associated molecular patterns (PAMPs) and are recognized by Toll-like receptors (TLRs). An example is the bacterial lipopolysaccharide (LPS), which is recognized by TLR4. Of note, these structures can be used to stimulate inflammation and help generate an immune response when used as an adjuvant to a vaccine. The receptors of the innate immune response are not capable of participating in anti-host cell immune responses.

In contrast, the adaptive immune response is more flexible and can recognize non microbial antigens because the receptors on the cells in the adaptive immune system are encoded by genes produced by recombination of genes segments that can produce a near-endless number of unique receptors. Receptors that recognize self are eliminated through an editing process against self-reactive lymphocytes. The adaptive immune response is initiated by the activation of antigen receptors on lymphocytes. B lymphocytes can recognize macromolecules such as proteins and lipids. Subsequently, B cells generate humoral immune responses yielding antibodies. Most T lymphocytes sense peptide fragments of protein and only in the context of presentation by other cells called antigen presenting cells. In fact the immune system has developed a highly specialized system for capturing and displaying antigens to lymphocytes via the major histocompatibility complex (MHC).

The innate and adaptive immune systems interact with each other as the innate immune response consists of cells such as dendritic cells, which present antigens to these T cells [228]. The innate immune system can eliminate newly transformed cancer cells via both an immediate and longer-term response [229]. The immediate, or innate, immune response includes cytotoxic T lymphocytes (CD8), which directly destroy cells. These destroyed cells are seen as foreign substances that induce an immune response through the release of foreign proteins (aka antigens). Antigen presenting cells such as monocytes, macrophages and dendritic cells remove the recently destroyed tumor cells and present antigens from the tumor cells to helper-T cells (CD4). T helper cells that recognize the tumor antigens then proliferate and mature into effector T cells and memory T cells in response to the antigen presenting cells. Effector T cells secrete cytokines that stimulate and interact with other white blood cells. Memory T cells retain the antigen affinity of the originally activated T cells for rapid activations at a later time. The adaptive immune system then expands in a clonal manner, that is they proliferate within 2 days resulting in millions of specific clones to the same antigen [228].

Obviously, in order for tumors to develop they must escape the immune system. Immunoediting is a process that leads to the modification of tumor cells so that they do not elicit an immune response. Just as many tumors exploit embryologic programs to drive growth, they use innate mechanisms to suppress the immune system. The immune system has an extensive regulatory system to prevent autoimmunity and destruction of host cells. Using the same mechanisms that prevent auto-immunity, tumors suppress the immune response creating anergy towards the tumor [230]. This can be done by various methods including the induction of T regulatory cells and the destruction of T cells. Tumor mediated destruction of T cells (T memory, killer and/or helper T cells) through feedback mechanisms like the Programmed Death 1 Pathway (PD) and Cytotoxic T-Lymphocyte Antigen 4 (CTLA-4) are evident in many tumor. Therapies directed at this mechanism have been developed and are in clinical trials [228]. These feedback mechanisms exist to control autoimmunity, but tumors overexpress ligands activating these pathways and thus suppress the immune response. The exploitation of immune mechanisms and evasion of immune surveillance are abilities cancer cells acquire during tumor progression. Three immune hallmarks are required for a tumor to develop and these are the (1) ability to thrive in a chronically inflamed microenvironment, (2) ability to evade immune recognition and the (3) ability to suppress immune reactivity [230].

5.3 The Immune System and Colorectal Cancer

5.3.1 The Adaptive Immune System and Colorectal Cancer

Studies of primary tumor tissues, mouse models of cancer, and in vitro cell line systems have all demonstrated that the immune response in a tumor affects the behavior of many cancers, including colorectal cancer, and may in fact predict clinical outcome better than the traditional TNM staging [231]. Although there is evidence the immune response in a tumor can affect tumor behavior, the best way to evaluate the immune response remains unclear. T cell populations in the tumor can be determined and measured using standard immunohistochemistry (IHC), which allows both the location and quantity of the T cells in the tumor to be assessed. Immune cells can be found surrounding colorectal cancers, intermingled with the cancer cells throughout the tumor, or deep inside the tumors, and the location of the T cells appears to be an indicator of anti-tumor activity of the immune cells [232]. The tumor-infiltrating lymphocytes (TIL) appear to be important prognostic factors [233, 234] that positively associate with disease-free, disease specific and overall survival. High densities of TILs correlate with improved survival in patients with colorectal adenocarcinoma (CRC) [235]. Galon has shown T cell infiltration at the center of the tumor (CT) and/or at the invasive margin (IM) yields significant prognostic information. To identify lymphocytes, the cell surface marker CD3 is identified using standard immunohistochemistry (IHC) techniques. The IHC results can be assessed for staining intensity (high vs. low) and the location of the immune cells can be determined (center (CT) vs margin (IM)). Using these scoring parameters, Galon and colleagues showed that tumors with $CD3_{CT}^{Hi}$ $CD3_{IM}^{Hi}$, $CD3_{CTIM}^{Het}$, and $CD3_{CT}^{Lo}$ $CD3_{IM}^{Lo}$ patterns had a favorable prognosis, with the median overall survival (mOS) of these three categories not being reached (79 and 60 months, respectively).

The immune system evaluation was predictive of survival regardless of stage. TIL appears to be predictive of outcome in both early stage disease and late stage disease, i.e. incurable metastatic CRC [235]. Using a validated immunostaining assay, Halama and colleagues assessed immune cell infiltrates at the invasive margin of the CRC liver metastases and correlated the results with response to chemotherapy ($N = 101$) [236]. The invasive margin was defined as a 500 micrometer width on each side of the border between malignant cells and liver tissue. Tissue specimens were immunohistochemically analyzed for their infiltration with CD3 (T cell marker), CD8 (cytotoxic T cell marker), granzyme B (cytotoxic T cell marker) and FOXP3 (regulatory T cell marker). A strong association between immune cell densities at the invasive margin of CRC liver metastases was found and in fact, TIL densities at the invasive margin of liver metastases predicted a response to chemotherapy with a sensitivity of 79 % and specificity of 100 %. Patients with metastatic CRCs having a CD3 cell count above 572 cells/mm^2 and either a CD8+ cell density higher than 170 cells/mm^2 or granzyme B+ cell density higher than 23 cells/mm^2 have a high likelihood of responding to chemotherapy. For practical reasons the derived rule rounded the required CD3 cell count to above 600 cells/mm^2 and either a CD8+ cell density of higher than 200 cells/mm^2 or a granzyme B+ cell density of higher than 25 cell/mm^2. The score designated a double weight to CD3 and a single weight to CD8 and granzyme B ($2+1+1$). CRCs with a score of 0–2 have a low density of T cells and are predicted to be non-responders. As an example, if a biopsy was obtained and the invasive margin was evaluated for CD3, CD8 and granzyme B with a resulting cell count of 650, 210 and 20 cells/mm^3 for CD3, CD8 and granzyme B the resulting score would $2+1+0=3$. This would be scored as high with a predicted PFS of 13.7 months. If the score was low (0–2) the PFS would be predicted to be 5.1 months. Though the score ended up binary, high and low, the two high scores had different PFS with a score of 3 and 4 having PFS of 10.3 and 16.6 months respectively [236].

5.3.2 The Innate Immune System and Colorectal Cancer

The innate immune system is composed of cells that generate an immune response via phagocytosis, generation of cytokines, and induction of antigen presenting cells. Member of the innate immune system include neutrophils, Natural Killer (NK) cells, macrophages, and others. We will now focus on the macrophages because of recent studies that suggest that they are a key component of the innate immune response in colorectal cancer. The two types of circulating phagocytes, neutrophils and monocytes, are blood cells that are recruited to sites of infection, where they recognize and ingest microbes for intracellular killing [228]. Monocytes enter extravascular tissues and differentiate into cells called macrophages, which, unlike neutrophils, survive in these sites for long periods. Macrophages can be characterized as having one of two major phenotypes: M1 and M2. M1 activity inhibits cell proliferation, causes tissue damage, and promotes a Th1 immune response; while M2 activity promotes cell proliferation, tissue repair, and promotes a Th2 immune response [237]. The Th1 response is triggered by IL-2 and IL-12 and yields IFN-gamma as its main cytokine. Th1 responses activate [238] macrophages to phagocytose and digest intracellular bacteria. The Th2 response is triggered by IL-4 and yields effector cytokines of IL-4, -5 and -13. The main effector cells are eosinophils, basophils and mast cells. Notably, the innate immune system appears to have the potential to have either pro- or anti-tumorigenic effects, which is in contrast to the anti-tumor adaptive immune response. Tumor Associated Macrophages (TAMs), especially M2, are frequently associated with progression of several types of cancer, including colon cancer [239]. Colorectal cancers appear to have found ways to promote M2 activity, e.g. by secreting IL-4, which in turn promotes cell proliferation and tissue repair. Intratumoral macrophages that express the macrophage markers CD68 and CD163 are enriched in colon carcinoma. The cytokine profile of the tumor microenvironment is extremely important to the phenotype of the local mononuclear phagocytes, which include TAMs. Macrophages are educated by the tumor microenvironment, so they adopt a trophic role that facilitates angiogenesis, matrix breakdown and tumor-cell egress and may increase the metastatic potential of tumors [240]. An experiment in macrophage-deficient mice demonstrated the role macrophages play in metastatic progression. In these macrophage deficient mice the incidence and initial rates of growth of primary tumors were not different from those seen in normal mice, but the rate of tumor progression was slowed and their metastatic ability was almost completely abrogated when compared with mice containing normal numbers of macrophages. In addition to matrix breakdown yielding a potential for tumor cell egress, migratory TAMs are also equipped to enter areas of the tumor where vascularization is needed., where they can synthesize angiogenic regulators, resulting in the formation of new blood vessels allowing tumor growth and yielding access of tumor cells to the vasculature for escape into the circulation [238].

5.4 Tumor Mediated Immune Suppression or Escape

5.4.1 Tumor Intrinsic Factors That Mediate Immune Escape

Tumors acquire a variety of escape mechanisms in order to evade the anti-tumor effects of the immune system. Mutations and dysregulation in the major histocompatibility complex (MHC) class I antigen presentation system are reasonably frequent [241]. The deregulated MHC is a potential mechanism for inducing a "stealth" phenotype via loss of the MHC class I antigen expression. If a tumor cell does not present antigens in the context of the MHC, a T cell cannot recognize the tumor cell. In a manner analogous to the selection of drug resistant tumor variants, the immune system could be a driving force for the selection of MHC I deficient tumor types. Another proposed mechanism for immune escape is oncogene interference with antigen presentation. An in vitro murine model of ras transformation demonstrated an inverse correlation between MHC class I surface expression and RAS oncogene expression [242]. Using IHC the expression pattern of six major MHC I antigen presentation machinery components were compared in high-grade dysplasia in colorectal neoplasms that had mutant or wild-type *KRAS*. A stage-related increase in MHC class I antigen loss was observed as well as an association with *KRAS* mutations and downregulation of the antigen presentation machinery. The exact connection of how MHC antigen presentation machinery was down-regulated remains to be fully elucidated.

Another mechanism responsible for tumor cell evasion of anti-tumor immune responses appears to be mediated by cancer initiating cells in tumors. It appears that the immune system is modulated by tumor cells very early in development of tumorigenesis and that cancer-initiating cells (CICs), which are also known as cancer stem-like cells, are responsible for these early tumor-immune cell interactions. CICs are, arguably, the cells believed to be responsible for tumor initiation, progression and resistance. CICs and PBMCs obtained from patients with a diagnosis of CRC have been evaluated for their immunogenicity [243]. Proliferation of T cells was inhibited by co-culture with CICs, but not with the more differentiated tumor cells. The membrane expression of IL-4 and to a lesser extent IL-4R were associated with CIC compared to more differentiated tumor cells and it appears that IL-4 is involved in at least one of the immune escape mechanisms of CICs.

A major factor limiting immune recognition of cancer cells is the fact tumors arise from the organism's own tissue and, therefore, mainly express self-antigens to which the individual's T cells have developed tolerance, either centrally or peripherally. This situation is manifested as tolerance of T cells that display a high avidity for normal self-antigens expressed by the tumor, leaving only T cells with low avidity [244]. Adoptive transfer experiments, where single cell suspensions of T cells from lymph nodes and spleen are harvested in one animal and infused into another, have demonstrated tumor antigen-specific naïve T cells are rapidly anergized when transferred to tumor bearing mice [245].

5.4.2 Immune Escape Mechanisms That Affect Specific Types of Immune Cells

Myeloid cells with immune suppressive functions, such as myeloid-derived suppressor cells (MDSC) or tumor associated macrophages (TAM), have recently received much attention based on their recognition as potent suppressors of T cell- and potentially NK cell-mediated immunity with a wide repertoire of effector mechanisms [241]. The heterogeneity of MDSC is large and the descriptions of human phenotypes are varied making a consolidated assessment of the mechanisms and effects of these cells on tumors challenging. They exhibit an array of suppressive functions including depletion of amino acids essential for T cell function; production of suppressive cytokines; interference with T cell homing, and collaboration with Tregs. They also contribute to tumor angiogenesis and the production of reactive nitrogen and oxygen species.

Regulatory T cells (Tregs) appear to also be mediators of immune escape. They were initially described in the 1970s but assigning them a characteristic phenotype as CD4+CD25+cells was necessary for their recognition as key players in tumor immunology. Since then, increased levels of Tregs have been noted in many types of cancer and have been shown to correlate with a poor disease course and prognosis. Tregs are subdivided into two major groups: the naturally occurring forms derived from the thymus and the adaptive variants generated in the periphery that is outside the thymus or for example near the tumor. Tumor immunosurveillance resembles autoreactivity, and Tregs, a dominant factor in peripheral tolerance, can act as a double-edged sword. Suppressive lymphocyte subsets are being identified and the phenotype as well as mode of action can vary significantly among them. The suppressive repertoire of Tregs is very broad and ranges from contact-dependent mechanisms to the secretion of suppressive molecules by which they efficiently blunt T cell-, NK cell-, and DC-mediated immune responses [241].

Dendritic Cells (DCs) are the most important antigen-presenting cells (APCs). DCs (iDCs) are initially produced in an immature state. They then recognize and capture antigens in peripheral tissues and subsequently they develop into mature (mDCs). These mDCs migrate to lymph nodes where, once activated, they prime CD4 and CD8 specific T cell responses. Antigen presentation by iDCs in an inadequate cytokine milieu leads to the inhibition of T cell functions. Quantitative and qualitative defects of circulating DCs have been observed in several types of cancer, including colon cancer [246]. Colorectal cancer patients exhibit an impaired capacity to generate iDCs from blood monocytes compared to healthy controls as shown by a higher percentage of CD14 expressing cells. The co-stimulatory molecule CD40 is down regulated. At a postoperative time point it was observed there was recovery of the ability to generate DCs from CRC patients, compared to controls, although tumor resection did not improve the phenotypic and functional characteristics of these mature DCs. Tumor cells, especially those in early stages of cancer development construct several immune escape mechanisms that involve DCs, to elude host immune system control [247]. The cytokines implicated include IL-10, IL-6, M-CSF, GM-CSF and VEGF. Gangliosides, prostanoids, NO, hyaluronan and polyamines are other mediators thought to impair phenotypic and functional differentiation. Even some tumor antigens, MUC1, PSA, HER-2/neu, are thought to impair maturation and alter differentiation and/or antigen procession.

In summary, there are a variety of escape mechanisms that tumor cells including colorectal cancer cells have demonstrated. Innately, tumor cells can have defective antigen presenting systems via mutations in the antigen presenting system, but it appears to be more likely that defects in antigen presentation arise from suppressor functions mediated by mutated genes. In addition, cancer cells use the innate tolerance mechanisms, which are designed to prevent autoimmunity to not only dampen the immune response, but to also promote metastases.

5.5 Therapies

5.5.1 Introduction

In 1893 a 31-year-old patient with round celled sarcoma of the neck had five operations within 3 years. At the last operation it was not possible to remove the entire tumor. Two weeks after the operation, a severe erysipelas infection occurred and "the sarcoma entirely disappeared" [248]. Seven years later, the patient remained disease free. This is one of the first reports linking immunity with cancer regression and has served as a demonstration of the potential of the immune system to suppress cancer. Roughly 100 years later, in 1994 [249], biologic therapy with IL-2 demonstrated significant and persistent antitumor activity effects in patients with advanced metastatic melanoma or renal cell cancer. At the time, this was the best available evidence that immunologic therapy could be effective in selected patients with solid tumors. Somewhat not surprising then, ipilimumab, a CTLA-4 antibody, was later shown to improve overall survival in patients with metastatic melanoma [250]. In the same year that the effectiveness of anti-CTLA-4 therapy was shown, the first cancer vaccine (sipuleucel-T) received FDA approval for the treatment of prostate cancer [251]. Available data support the notion that most cancers are vulnerable to immunologically-based therapies [241]. To date, most immunotherapies targeting colorectal cancer have used either a vaccine approach or an adoptive T cell therapy approach.

Given the rapidity at which the field of immunotherapy is changing and the unpredictable nature of vaccine based therapies, the following examples of anti-tumor vaccines were chosen not as specific examples, but as examples of how vaccines are designed and evaluated prior to a phase III randomized clinical trial are performed. The therapies that are highlighted below target specific antigens, but immunotherapeutic approaches are not limited to these antigens. The approach of a cancer vaccine is to strengthen and train the patient's immune system to achieve a focused and specific attack on cancer cells. Cancer vaccines typically are not toxic and are much more tolerable than traditional cytotoxic chemotherapy. Combination approaches with check point inhibitors such as anti-PD1 and Anti-CTLA4 are under investigation given their potential to augment the anti-tumor activity of cancer vaccines.

5.5.2 Dendritic Cell Vaccines

Dendritic cell vaccines are generated by isolating dendritic cells (DCs) from the peripheral blood of patients and then "pulsing" the DCs with either allogenic tumor lysates or manufactured peptides. Dendritic cells are antigen presenting cells and as they are autologous the concept behind this approach is that the host immune system will present the antigens to the adaptive immune cells regardless of whether they are manufactured peptides or tumor cell lysates. Of interest, dendritic cell vaccines are typically administered as an intradermal injection in order to harness specific aspects of the vast immune cell population in the skin. A recent phase II dendritic cell vaccine study in colorectal cancer patients ($N=20$) employed DCs pulsed with MAGE (melanoma associated antigen) [252]. The vaccine was administered every 2 weeks for a total of ten vaccines. The best observed response was stable disease per RECIST criteria. The median survival was only 5.3 months though the median survival from initial diagnosis was 43.1 months; indicating that the subjects in this trial had end-stage disease and that the effects of this therapy may be underestimated. No serious adverse events were noted.

In a Phase I/II study of a CEA loaded DC vaccine eight of ten patients were able to receive at least one series of vaccination [253]. Patients were vaccinated four times every 2 weeks. Two patients had a decrease in CEA levels, and the toxicity was mild in all cases with fevers and some induration. In another Phase I DC vaccine study, 10 patients underwent three leukaphoresis sessions at 4 weekly intervals to generate dendritic cell for three sets of injections [254]. The MUC1 cDNA insert flanked by motifs encouraging integration and promotion of the protein was cultured with the DCs. The patients had disease stabilization and no toxic effects from the therapy. This trial highlights an issue regarding DC based vaccines, which is scalability. Indeed, despite sipuleucel-T demonstrating an overall survival benefit in patients with prostate cancer, the cost and process of making each person his own vaccine has led to limited use of this therapy.

5.5.3 Tumor Antigen Viral Vector Vaccines

One possible way to induce a potent antitumor response is to use a virus to deliver tumor associated antigens. Poxviruses such as vaccinia, fowlpox and canarypox have found widespread use as vaccine vectors for the treatment of infectious diseases and cancer. A modified vaccinia Ankara (MVA) virus is commonly used as it is attenuated. A promising antigen for tumor antigen viral vector vaccines is the human oncofetal antigen 5T4. This antigen is a leucine-rich membrane glycoprotein that is expressed at high levels on placenta and also on a wide range of human carcinomas including colorectal cancer while it has restricted expression on normal tissues [255]. MVA and 5T4 have been combined to create the TroVax vaccine. In a phase I/II study of this vaccine, 17/22 patients enrolled completed the protocol and were evaluable. The patients received three injections at weeks 0, 4 and 8. If a clinical response was detected by week 12 or if an immunologic response was detected between weeks 0 and 12, the patients were offered two more injections at weeks 14 and 20. The clinical responses were encouraging, and the adverse events were limited to local events at the injection site. No vector was detected in the peripheral blood. No patient had detectable 5T4 specific antibody responses at baseline, and 14 of 17 patients showed detectable 5T4 antibody response after two or more vaccinations. Proliferation assays were performed to measure cellular response, i.e. CD4/CD8 responses, and 8 patients developed de novo immune responses.

5.5.4 Peptide Vaccines

Recent efforts related to peptide vaccine development have focused on the use peptides generated from mutated genes. An advantage of peptide vaccines that has made them appealing for clinical use compared to other vaccine types is that they can be produced on a large scale. Mutant *RAS* has been the basis for promising peptide vaccines that are currently under clinical investigation. *RAS* mutations are often found in adenocarcinomas and occur in 40–50 % of colorectal cancers. *RAS* is mutated by a single amino acid substitution in 90 % of cases. A 13-mer mutant

RAS peptide administered with an adjuvant was able to generate an immune response in a phase I study, which led to a phase II trial in patients with colorectal ($N=7$) and pancreatic ($N=5$) cancer with no evidence of disease [256]. All patients enrolled in this study had tumors with mutant *KRAS*. The adjuvant was DETOX™ composed of 250 mcg cell wall skeleton of *Mycobacterium phlei* and 25 mcg of monophosphoryl lipid A (MPL) from *Salmonella Minnesota* R595. The purpose of this adjuvant is to stimulate the innate immune response near the peptides, which is then believed to induce a bystander effect or epitope spreading so that the immune response will include these peptides. The vaccine was administered every 4 weeks up to six vaccinations or until disease progression. The vaccine was well tolerated with no serious systemic side effects. The most common side effects were injection site reaction, fatigue and diarrhea. The mean DFS was 35 months for pancreas cancer patients 27 months for colon cancer. DFS and OS improvement was seen in patients who had positive immune responses.

5.5.5 Cytotoxic Chemotherapy and Tumor Immunotherapy

Chemoimmunotherapy has gained attention recently as a consequence of the recent successes of tumor immunotherapy. Although chemotherapy is usually thought to be immunosuppressive, a regimen of FOLFOX plus gemcitabine plus GM-CSF and low-dose IL2 (GOLFIG) is designed to be immunogenic [257, 258]. In a phase II trial of this regimen, treatment was associated with improved response rates and time to progression (TTP). Interestingly six patients who developed signs of autoimmunity, e.g. arthritis, had a significantly longer TTP (23.8 vs 10 months) and OS (31.8 vs 16.8) months. In a small Phase III trial, the response rates were 63 vs 34% [259]. Gemcitabine is thought to play an important role in this regimen by inducing tumor cell death, which releases antigens. The major barrier to further development of GOLFIG is the lack of interest among pharmaceutical companies for funding development of this regimen because these agents are not under patent protection.

5.5.6 Summary

The fundamental components of the immune system and of the anti-tumor immune response are better understood now, and this understanding has led to renewed hope that the promise of tumor immunotherapy will be realized. In this section, we have reviewed the components of the immune system that participate in anti-tumor responses and have discussed examples of cancer immunotherapy using vaccine based methods. It is widely anticipated that immunotherapy for colorectal cancer will move into the clinic in the near future, however, the specific types of immunotherapies that will prove to be most effective remain to be determined. Finally, in evidence of the anticipated adoption of tumor immunotherapy for the treatment of CRC, an emerging form of immunotherapy based on immune checkpoint inhibition (e.g. anti-PD1 therapy) is rapidly moving into clinical care and is likely to be the first immunotherapy based treatment for colorectal cancer that is available for clinical use.

References

1. Venook AP (2005) Epidermal growth factor receptor-targeted treatment for advanced colorectal carcinoma. Cancer 103:2435–2446
2. Hicklin DJ, Ellis LM (2005) Role of the vascular endothelial growth factor pathway in tumor growth and angiogenesis. J Clin Oncol 23:1011–1027
3. Comoglio PM, Giordano S, Trusolino L (2008) Drug development of MET inhibitors: targeting oncogene addiction and expedience. Nat Rev Drug Discov 7:504–516
4. Ewing GP, Goff LW (2010) The insulin-like growth factor signaling pathway as a target for treatment of colorectal carcinoma. Clin Colorectal Cancer 9:219–223
5. Shaw RJ, Cantley LC (2006) Ras, PI(3)K and mTOR signalling controls tumour cell growth. Nature 441:424–430
6. Cancer Genome Atlas Network (2012) Comprehensive molecular characterization of human colon and rectal cancer. Nature 487:330–337
7. Bertotti A, Burbridge MF, Gastaldi S et al (2009) Only a subset of Met-activated pathways are required to sustain oncogene addiction. Sci Signal 2:er11
8. Chin L, Tam A, Pomerantz J et al (1999) Essential role for oncogenic Ras in tumour maintenance. Nature 400:468–472
9. Dry JR, Pavey S, Pratilas CA et al (2010) Transcriptional pathway signatures predict MEK addiction and response to selumetinib (AZD6244). Cancer Res 70:2264–2273
10. Sharma SV, Gajowniczek P, Way IP et al (2006) A common signaling cascade may underlie "addiction" to the Src, BCR-ABL, and EGF receptor oncogenes. Cancer Cell 10:425–435
11. Torti D, Trusolino L (2011) Oncogene addiction as a foundational rationale for targeted anticancer therapy: promises and perils. EMBO Mol Med 3:623–636
12. Weinstein IB (2002) Cancer. Addiction to oncogenes—the Achilles heal of cancer. Science 297:63–64
13. Pollak M (2012) The insulin and insulin-like growth factor receptor family in neoplasia: an update. Nat Rev Cancer 12:159–169
14. Pollak M (2008) Insulin and insulin-like growth factor signalling in neoplasia. Nat Rev Cancer 8:915–928
15. Lu Y, Zi X, Zhao Y et al (2001) Insulin-like growth factor-I receptor signaling and resistance to trastuzumab (Herceptin). J Natl Cancer Inst 93:1852–1857
16. Buck E, Eyzaguirre A, Rosenfeld-Franklin M et al (2008) Feedback mechanisms promote cooperativity for small molecule inhibitors of epidermal and insulin-like growth factor receptors. Cancer Res 68:8322–8332
17. Hu YP, Patil SB, Panasiewicz M et al (2008) Heterogeneity of receptor function in colon carcinoma cells determined by cross-talk between type I insulin-like growth factor receptor and epidermal growth factor receptor. Can Res 68:8004–8013
18. Huang F, Xu LA, Khambata-Ford S (2012) Correlation between gene expression of IGF-1R pathway markers and cetuximab benefit in metastatic colorectal cancer. Clin Cancer Res 18:1156–1166
19. Jameson MJ, Beckler AD, Taniguchi LE et al (2011) Activation of the insulin-like growth factor-1 receptor induces resistance to epidermal growth factor receptor antagonism in head and neck squamous carcinoma cells. Mol Cancer Ther 10:2124–2134
20. Hakam A, Yeatman TJ, Lu L et al (1999) Expression of insulin-like growth factor-1 receptor in human colorectal cancer. Human Pathol 30:1128–1133
21. Ouban A, Muraca P, Yeatman T et al (2003) Expression and distribution of insulin-like growth factor-1 receptor in human carcinomas. Human Pathol 34:803–808
22. Pollak MN, Perdue JF, Margolese RG et al (1987) Presence of somatomedin receptors on primary human breast and colon carcinomas. Cancer Lett 38:223–230
23. Chang CH, Wang Y, Trisal P et al (2012) Evaluation of a novel hexavalent humanized anti-IGF-1R antibody and its bivalent parental IgG in diverse cancer cell lines. PLoS ONE 7:e44235

24. Hassan AB, Howell JA (2000) Insulin-like growth factor II supply modifies growth of intestinal adenoma in Apc(Min/ +) mice. Cancer Res 60:1070–1076
25. Yang XF, Beamer WG, Huynh H et al (1996) Reduced growth of human breast cancer xenografts in hosts homozygous for the lit mutation. Cancer Res 56:1509–1511
26. Giovannucci E, Michaud D (2007) The role of obesity and related metabolic disturbances in cancers of the colon, prostate, and pancreas. Gastroenterology 132:2208–2225
27. Calle EE, Rodriguez C, Walker-Thurmond K et al (2003) Overweight, obesity, and mortality from cancer in a prospectively studied cohort of U.S. adults. N Engl J Med 348:1625–1638
28. Noto H, Goto A, Tsujimoto T et al (2013) Latest insights into the risk of cancer in diabetes. J Diabetes Investig 4:225–232
29. Becerra CR, Salazar R, Garcia-Carbonero R et al (2014) Figitumumab in patients with refractory metastatic colorectal cancer previously treated with standard therapies: a nonrandomized, open-label, phase II trial. Cancer Chemother Pharmacol 73:695–702
30. Cohn AL, Tabernero J, Maurel J et al (2013) A randomized, placebo-controlled phase 2 study of ganitumab or conatumumab in combination with FOLFIRI for second-line treatment of mutant KRAS metastatic colorectal cancer. Ann Oncol 24:1777–1785
31. Reidy DL, Vakiani E, Fakih MG et al (2010) Randomized, phase II study of the insulin-like growth factor-1 receptor inhibitor IMC-A12, with or without cetuximab, in patients with cetuximab- or panitumumab-refractory metastatic colorectal cancer. J Clin Oncol 28:4240–4246
32. Guha M (2013) Anticancer IGF1R classes take more knocks. Nat Rev Drug Discov 12:250
33. Carden CP, Molife LR, De Bono JS (2009) Predictive biomarkers for targeting insulin-like growth factor-I (IGF-I) receptor. Mol Cancer Ther 8:2077–2078
34. Zha J, O'brien C, Savage H et al (2009) Molecular predictors of response to a humanized anti-insulin-like growth factor-I receptor monoclonal antibody in breast and colorectal cancer. Mol Cancer Ther 8:2110–2121
35. Blumenschein GR Jr, Mills GB, Gonzalez-Angulo AM (2012) Targeting the hepatocyte growth factor-cMET axis in cancer therapy. J Clin Oncol 30:3287–3296
36. Di Renzo MF, Olivero M, Giacomini A et al (1995) Overexpression and amplification of the Met/HGF receptor gene during the progression of colorectal cancer. Clin Cancer Res 1:147–154
37. Samame Perez-Vargas JC, Biondani P, Maggi C et al (2013) Role of cMET in the development and progression of colorectal cancer. Int J Molecular Sci 14:18056–18077
38. Takeuchi H, Bilchik A, Saha S et al (2003) c-MET expression level in primary colon cancer: a predictor of tumor invasion and lymph node metastases. Clin Cancer Res 9:1480–1488
39. Zeng ZS, Weiser MR, Kuntz E et al (2008) c-Met gene amplification is associated with advanced stage colorectal cancer and liver metastases. Cancer Lett 265:258–269
40. Toiyama Y, Miki C, Inoue Y et al (2009) Serum hepatocyte growth factor as a prognostic marker for stage II or III colorectal cancer patients. Int J Cancer 125:1657–1662
41. Bardelli A, Corso S, Bertotti A et al (2013) Amplification of the MET receptor drives resistance to anti-EGFR therapies in colorectal cancer. Cancer Discov 3:658–673
42. Engelman JA, Zejnullahu K, Mitsudomi T et al (2007) MET amplification leads to gefitinib resistance in lung cancer by activating ERBB3 signaling. Science 316:1039–1043
43. Ritchie ME, Trask RV, Fontanet HL et al (1991) Multiple positive and negative elements regulate human brain creatine kinase gene expression. Nucl Acids Res 19:6231–6240
44. Sakata H, Takayama H, Sharp R et al (1996) Hepatocyte growth factor/scatter factor overexpression induces growth, abnormal development, and tumor formation in transgenic mouse livers. Cell Growth Differ 7:1513–1523
45. Stabile LP, Lyker JS, L and SR et al (2006) Transgenic mice overexpressing hepatocyte growth factor in the airways show increased susceptibility to lung cancer. Carcinogenesis 27:1547–1555
46. Takayama H, Larochelle WJ, Sharp R et al (1997) Diverse tumorigenesis associated with aberrant development in mice overexpressing hepatocyte growth factor/scatter factor. Proc Natl Acad Sci U S A 94:701–706

47. Zhang YW, Staal B, Essenburg C et al (2010) MET kinase inhibitor SGX523 synergizes with epidermal growth factor receptor inhibitor erlotinib in a hepatocyte growth factor-dependent fashion to suppress carcinoma growth. Cancer Res 70:6880–6890
48. Dienstmann R, Serpico D, Rodon J et al (2012) Molecular profiling of patients with colorectal cancer and matched targeted therapy in phase I clinical trials. Mol Cancer Ther 11:2062–2071
49. Van Cutsem E, Eng C, Nowara E et al (2014) Randomized phase Ib/II trial of rilotumumab or ganitumab with panitumumab versus panitumumab alone in patients with wild-type KRAS metastatic colorectal cancer. Clin Cancer Res 20:4240–4250
50. Bendell JC, Ervin TJ, Gallinson D et al (2013) Treatment rationale and study design for a randomized, double-blind, placebo-controlled phase II study evaluating onartuzumab (MetMAb) in combination with bevacizumab plus mFOLFOX-6 in patients with previously untreated metastatic colorectal cancer. Clin Colorectal Cancer 12:218–222
51. Santoro A, Rimassa L, Borbath I et al (2013) Tivantinib for second-line treatment of advanced hepatocellular carcinoma: a randomised, placebo-controlled phase 2 study. Lancet Oncol 14:55–63
52. Scagliotti GV, Novello S, Schiller JH et al (2012) Rationale and design of MARQUEE: a phase III, randomized, double-blind study of tivantinib plus erlotinib versus placebo plus erlotinib in previously treated patients with locally advanced or metastatic, nonsquamous, non-small-cell lung cancer. Clin Lung Cancer 13:391–395
53. Yakes FM, Chen J, Tan J et al (2011) Cabozantinib (XL184), a novel MET and VEGFR2 inhibitor, simultaneously suppresses metastasis, angiogenesis, and tumor growth. Mol Cancer Ther 10:2298–2308
54. Elisei R, Schlumberger MJ, Muller SP et al (2013) Cabozantinib in progressive medullary thyroid cancer. J Clin Oncol 31:3639–3646
55. Fang JY, Richardson BC (2005) The MAPK signalling pathways and colorectal cancer. Lancet Oncol 6:322–327
56. Roberts PJ, Der CJ (2007) Targeting the Raf-MEK-ERK mitogen-activated protein kinase cascade for the treatment of cancer. Oncogene 26:3291–3310
57. De Roock W, Claes B, Bernasconi D et al (2010) Effects of KRAS, BRAF, NRAS, and PIK3CA mutations on the efficacy of cetuximab plus chemotherapy in chemotherapy-refractory metastatic colorectal cancer: a retrospective consortium analysis. Lancet Oncol 11:753–762
58. Chen D, Huang JF, Liu K et al (2014) BRAFV600E mutation and its association with clinicopathological features of colorectal cancer: a systematic review and meta-analysis. PLoS ONE 9:e90607
59. Baines AT, Xu D, Der CJ (2011) Inhibition of Ras for cancer treatment: the search continues. Future Med Chem 3:1787–1808
60. Ostrem JM, Peters U, Sos ML et al (2013) K-Ras(G12C) inhibitors allosterically control GTP affinity and effector interactions. Nature 503:548–551
61. Burns MC, Sun Q, Daniels RN et al (2014) Approach for targeting Ras with small molecules that activate SOS-mediated nucleotide exchange. Proc Natl Acad Sci U S A 111:3401–3406
62. Maurer T, Garrenton LS, Oh A et al (2012) Small-molecule ligands bind to a distinct pocket in Ras and inhibit SOS-mediated nucleotide exchange activity. Proc Natl Acad Sci U S A 109:5299–5304
63. Shima F, Yoshikawa Y, Ye M et al (2013) In silico discovery of small-molecule Ras inhibitors that display antitumor activity by blocking the Ras-effector interaction. Proc Natl Acad Sci U S A 110:8182–8187
64. Turajlic S, Ali Z, Yousaf N et al (2013) Phase I/II RAF kinase inhibitors in cancer therapy. Expert Opin Investig Drugs 22:739–749
65. Lito P, Rosen N, Solit DB (2013) Tumor adaptation and resistance to RAF inhibitors. Nat Med 19:1401–1409
66. Corcoran RB, Ebi H, Turke AB et al (2012) EGFR-mediated re-activation of MAPK signaling contributes to insensitivity of BRAF mutant colorectal cancers to RAF inhibition with vemurafenib. Cancer Discov 2:227–235

67. Prahallad A, Sun C, Huang S et al (2012) Unresponsiveness of colon cancer to BRAF(V600E) inhibition through feedback activation of EGFR. Nature 483:100–103
68. Connolly K, Brungs D, Szeto E et al (2014) Anticancer activity of combination targeted therapy using cetuximab plus vemurafenib for refractory BRAF (V600E)-mutant metastatic colorectal carcinoma. Curr Oncol 21:e151–e154
69. Akinleye A, Furqan M, Mukhi N et al (2013) MEK and the inhibitors: from bench to bedside. J Hematol Oncol 6:27
70. Poulikakos PI, Solit DB (2011) Resistance to MEK inhibitors: should we co-target upstream? Sci Signal 4:pe16
71. Rinehart J, Adjei AA, Lorusso PM et al (2004) Multicenter phase II study of the oral MEK inhibitor, CI-1040, in patients with advanced non-small-cell lung, breast, colon, and pancreatic cancer. J Clin Oncol 22:4456–4462
72. Little AS, Balmanno K, Sale MJ et al (2011) A correction to the research article titled: "Amplification of the driving oncogene, KRAS or BRAF, underpins acquired resistance to MEK1/2 inhibitors in colorectal cancer cells" by A. S. Little, K. Balmanno, M. J. Sale, S. Newman, J. R. Dry, M. Hampson, P. A. W. Edwards, P. D. Smith, S. J. Cook. Sci Signal 4:er2
73. Yamaguchi T, Kakefuda R, Tajima N et al (2011) Antitumor activities of JTP-74057 (GSK1120212), a novel MEK1/2 inhibitor, on colorectal cancer cell lines in vitro and in vivo. Int J Oncol 39:23–31
74. Ascierto PA, Schadendorf D, Berking C et al (2013) MEK162 for patients with advanced melanoma harbouring NRAS or Val600 BRAF mutations: a non-randomised, open-label phase 2 study. Lancet Oncol 14:249–256
75. Flaherty KT, Infante JR, Daud A et al (2012) Combined BRAF and MEK inhibition in melanoma with BRAF V600 mutations. N Engl J Med 367:1694–1703
76. Morris EJ, Jha S, Restaino CR et al (2013) Discovery of a novel ERK inhibitor with activity in models of acquired resistance to BRAF and MEK inhibitors. Cancer Discov 3:742–750
77. Hatzivassiliou G, Liu B, O'brien C et al (2012) ERK inhibition overcomes acquired resistance to MEK inhibitors. Mol Cancer Ther 11:1143–1154
78. Nissan MH, Rosen N, Solit DB (2013) ERK pathway inhibitors: how low should we go? Cancer Discov 3:719–721
79. Yuan TL, Cantley LC (2008) PI3K pathway alterations in cancer: variations on a theme. Oncogene 27:5497–5510
80. Zhang J, Roberts TM, Shivdasani RA (2011) Targeting PI3K signaling as a therapeutic approach for colorectal cancer. Gastroenterology 141:50–61
81. Hanahan D, Weinberg RA (2000) The hallmarks of cancer. Cell 100:57–70
82. Hanahan D, Weinberg RA (2011) Hallmarks of cancer: the next generation. Cell 144:646–674
83. Kandoth C, Mclellan MD, Vandin F et al (2013) Mutational landscape and significance across 12 major cancer types. Nature 502:333–339
84. Vilar E, Mukherjee B, Kuick R et al (2009) Gene expression patterns in mismatch repair-deficient colorectal cancers highlight the potential therapeutic role of inhibitors of the phosphatidylinositol 3-kinase-AKT-mammalian target of rapamycin pathway. Clin Cancer Res 15:2829–2839
85. Shapiro GI, Rodon J, Bedell C et al (2014) Phase I safety, pharmacokinetic, and pharmacodynamic study of SAR245408 (XL147), an oral pan-class I PI3K inhibitor, in patients with advanced solid tumors. Clin Cancer Res 20:233–245
86. Rodon J, Brana I, Siu LL et al (2014) Phase I dose-escalation and -expansion study of buparlisib (BKM120), an oral pan-Class I PI3K inhibitor, in patients with advanced solid tumors. Investig New Drugs 32(4):670–81
87. Agarwal E, Chaudhuri A, Leiphrakpam PD et al (2014) Akt inhibitor MK-2206 promotes anti-tumor activity and cell death by modulation of AIF and Ezrin in colorectal cancer. BMC Cancer 14:145
88. Sangai T, Akcakanat A, Chen H et al (2012) Biomarkers of response to Akt inhibitor MK-2206 in breast cancer. Clin Cancer Res 18:5816–5828

89. Yap TA, Yan L, Patnaik A et al (2011) First-in-man clinical trial of the oral pan-AKT inhibitor MK-2206 in patients with advanced solid tumors. J Clin Oncol 29:4688–4695
90. Bendell JC, Nemunaitis J, Vukelja SJ et al (2011) Randomized placebo-controlled phase II trial of perifosine plus capecitabine as second- or third-line therapy in patients with metastatic colorectal cancer. J Clin Oncol 29:4394–4400
91. Kim DD, Eng C (2012) The promise of mTOR inhibitors in the treatment of colorectal cancer. Expert Opin Investig Drugs 21:1775–1788
92. Blaser B, Waselle L, Dormond-Meuwly A et al (2012) Antitumor activities of ATP-competitive inhibitors of mTOR in colon cancer cells. BMC Cancer 12:86
93. Deming DA, Leystra AA, Farhoud M et al (2013) mTOR inhibition elicits a dramatic response in PI3K-dependent colon cancers. PLoS ONE 8:e60709
94. Ng K, Tabernero J, Hwang J et al (2013) Phase II study of everolimus in patients with metastatic colorectal adenocarcinoma previously treated with bevacizumab-, fluoropyrimidine-, oxaliplatin-, and irinotecan-based regimens. Clin Cancer Res 19:3987–3995
95. Spindler KL, Sorensen MM, Pallisgaard N et al (2013) Phase II trial of temsirolimus alone and in combination with irinotecan for KRAS mutant metastatic colorectal cancer: outcome and results of KRAS mutational analysis in plasma. Acta Oncol 52:963–970
96. Burris HA 3rd (2013) Overcoming acquired resistance to anticancer therapy: focus on the PI3K/AKT/mTOR pathway. Cancer Chemother Pharmacol 71:829–842
97. Liao X, Lochhead P, Nishihara R et al (2012) Aspirin use, tumor PIK3CA mutation, and colorectal-cancer survival. New Eng J Med 367:1596–1606
98. Domingo E, Church DN, Sieber O et al (2013) Evaluation of PIK3CA mutation as a predictor of benefit from nonsteroidal anti-inflammatory drug therapy in colorectal cancer. J Clin Oncol 31:4297–4305
99. Siegel R, Desantis C, Jemal A (2014) Colorectal cancer statistics, 2014. CA Cancer J Clin 64:104–117
100. Cunningham D, Humblet Y, Siena S et al (2004) Cetuximab monotherapy and cetuximab plus irinotecan in irinotecan-refractory metastatic colorectal cancer. N Engl J Med 351:337–345
101. Hurwitz H, Fehrenbacher L, Novotny W et al (2004) Bevacizumab plus irinotecan, fluorouracil, and leucovorin for metastatic colorectal cancer. New Eng J Med 350:2335–2342
102. Alberts SR, Sargent DJ, Nair S et al (2012) Effect of oxaliplatin, fluorouracil, and leucovorin with or without cetuximab on survival among patients with resected stage III colon cancer: a randomized trial. JAMA 307:1383–1393
103. Allegra CJ, Yothers G, O'connell MJ et al (2011) Phase III trial assessing bevacizumab in stages II and III carcinoma of the colon: results of NSABP protocol C-08. J Clin Oncol 29:11–16
104. Ciardiello F, Tortora G (2008) EGFR antagonists in cancer treatment. N Engl J Med 358:1160–1174
105. Mendelsohn J (1992) Epidermal growth factor receptor as a target for therapy with antireceptor monoclonal antibodies. J Natl Cancer Inst Monogr 125–131
106. Mendelsohn J, Baselga J (2006) Epidermal growth factor receptor targeting in cancer. Semin Oncol 33:369–385
107. Bournazou E, Bromberg J (2013) Targeting the tumor microenvironment: JAK-STAT3 signaling. Jak-Stat 2:e23828
108. Saltz LB, Meropol NJ, Loehrer PJ Sr et al (2004) Phase II trial of cetuximab in patients with refractory colorectal cancer that expresses the epidermal growth factor receptor. J Clin Oncol 22:1201–1208
109. Sobrero AF, Maurel J, Fehrenbacher L et al (2008) EPIC: phase III trial of cetuximab plus irinotecan after fluoropyrimidine and oxaliplatin failure in patients with metastatic colorectal cancer. J Clin Oncol 26:2311–2319
110. Meyerhardt JA, Zhu AX, Enzinger PC et al (2006) Phase II study of capecitabine, oxaliplatin, and erlotinib in previously treated patients with metastastic colorectal cancer. J Clin Oncol 24:1892–1897

111. Rothenberg ML, Lafleur B, Levy DE et al (2005) Randomized phase II trial of the clinical and biological effects of two dose levels of gefitinib in patients with recurrent colorectal adenocarcinoma. J Clin Oncol 23:9265–9274
112. Berge E, Thompson C, Messersmith W (2011) Development of novel targeted agents in the treatment of metastatic colorectal cancer. Clin Colorectal Cancer 10:266–278
113. Van Cutsem E, Peeters M, Siena S et al (2007) Open-label phase III trial of panitumumab plus best supportive care compared with best supportive care alone in patients with chemotherapy-refractory metastatic colorectal cancer. J Clin Oncol 25:1658–1664
114. Lievre A, Bachet JB, Boige V et al (2008) KRAS mutations as an independent prognostic factor in patients with advanced colorectal cancer treated with cetuximab. J Clin Oncol 26:374–379
115. Forbes S, Clements J, Dawson E et al (2006) Cosmic 2005. Brit J Cancer 94:318–322
116. Malumbres M, Barbacid M (2003) RAS oncogenes: the first 30 years. Nat Rev Cancer 3:459–465
117. Andreyev HJ, Norman AR, Cunningham D et al (1998) Kirsten ras mutations in patients with colorectal cancer: the multicenter "RASCAL" study. J Nat Cancer Inst 90:675–684
118. Jonker DJ, O'callaghan CJ, Karapetis CS et al (2007) Cetuximab for the treatment of colorectal cancer. New Eng J Med 357:2040–2048
119. Karapetis CS, Khambata-Ford S, Jonker DJ et al (2008) K-ras mutations and benefit from cetuximab in advanced colorectal cancer. New Eng J Med 359:1757–1765
120. Amado RG, Wolf M, Peeters M et al (2008) Wild-type KRAS is required for panitumumab efficacy in patients with metastatic colorectal cancer. J Clin Oncol 26:1626–1634
121. Allegra CJ, Jessup JM, Somerfield MR et al (2009) American Society of Clinical Oncology provisional clinical opinion: testing for KRAS gene mutations in patients with metastatic colorectal carcinoma to predict response to anti-epidermal growth factor receptor monoclonal antibody therapy. J Clin Oncol 27:2091–2096
122. Van Cutsem E, Kohne CH, Hitre E et al (2009) Cetuximab and chemotherapy as initial treatment for metastatic colorectal cancer. New Eng J Med 360:1408–1417
123. Van Cutsem E, Kohne CH, Lang I et al (2011) Cetuximab plus irinotecan, fluorouracil, and leucovorin as first-line treatment for metastatic colorectal cancer: updated analysis of overall survival according to tumor KRAS and BRAF mutation status. J Clin Oncol 29:2011–2019
124. Maughan TS, Adams RA, Smith CG et al (2011) Addition of cetuximab to oxaliplatin-based first-line combination chemotherapy for treatment of advanced colorectal cancer: results of the randomised phase 3 MRC COIN trial. Lancet 377:2103–2114
125. Tveit KM, Guren T, Glimelius B et al (2012) Phase III trial of cetuximab with continuous or intermittent fluorouracil, leucovorin, and oxaliplatin (Nordic FLOX) versus FLOX alone in first-line treatment of metastatic colorectal cancer: the NORDIC-VII study. J Clin Oncol 30:1755–1762
126. Bokemeyer C, Bondarenko I, Hartmann JT et al (2011) Efficacy according to biomarker status of cetuximab plus FOLFOX-4 as first-line treatment for metastatic colorectal cancer: the OPUS study. Ann Oncol 22:1535–1546
127. Bokemeyer C, Bondarenko I, Makhson A et al (2009) Fluorouracil, leucovorin, and oxaliplatin with and without cetuximab in the first-line treatment of metastatic colorectal cancer. J Clin Oncol 27:663–671
128. Folprecht G, Gruenberger T, Bechstein W et al (2014) Survival of patients with initially unresectable colorectal liver metastases treated with FOLFOX/cetuximab or FOLFIRI/cetuximab in a multidisciplinary concept (CELIM-study). Ann Oncol 25(5):1018–25
129. Folprecht G, Gruenberger T, Bechstein WO et al (2010) Tumour response and secondary resectability of colorectal liver metastases following neoadjuvant chemotherapy with cetuximab: the CELIM randomised phase 2 trial. Lancet Oncol 11:38–47
130. Ocvirk J, Brodowicz T, Wrba F et al (2010) Cetuximab plus FOLFOX6 or FOLFIRI in metastatic colorectal cancer: CECOG trial. World J Gastroenterol 16:3133–3143

131. Douillard JY, Oliner KS, Siena S et al (2013) Panitumumab-FOLFOX4 treatment and RAS mutations in colorectal cancer. N Engl J Med 369:1023–1034
132. Douillard JY, Siena S, Cassidy J et al (2010) Randomized, phase III trial of panitumumab with infusional fluorouracil, leucovorin, and oxaliplatin (FOLFOX4) versus FOLFOX4 alone as first-line treatment in patients with previously untreated metastatic colorectal cancer: the PRIME study. J Clin Oncol 28:4697–4705
133. Heinemann V, Fischer Von Weikersthal L, Decker T et al (2013) Randomized comparison of FOLFIRI plus cetuximab versus FOLFIRI plus bevacizumab as first-line treatment of KRAS wild-type metastatic colorectal cancer: German AIO study KRK-0306 (FIRE-3). J Clin Oncol LBA3506
134. Stintzing S, Jung A, Rossius L et al (2013) Analysis of KRAS/NRAS and BRAF mutations in FIRE-3: a randomized phase III study of FOLFIRI plus cetuximab or bevacizumab as first-line treatment for wild-type KRAS (exon 2) metastatic colorectal cancer patients. Eur J Cancer 49 (Suppl) 3, abstract 17
135. Schwartzberg LS, Rivera F, Karthaus M et al (2014) PEAK: a randomized, multicenter phase II study of panitumumab plus modified fluorouracil, leucovorin, and oxaliplatin (mFOLFOX6) or bevacizumab plus mFOLFOX6 in patients with previously untreated, unresectable, wild-type KRAS exon 2 metastatic colorectal cancer. J Clin Oncol 32 (21):2240-7
136. Venook AP, Niedzwiecki D, Lenz AJ et al (2014) CALGB/SWOG 80405: phase III trial of irinotecan/5-FU/leucovorin (FOLFIRI) or oxaliplatin/5-FU/leucovorin (mFOLFOX6) with bevacizumab or cetuximab for patients with KRAS wild-type untreated metastatic adenocarcinoma of the colon or rectum. ASCO J Clin Oncol 32:5s (suppl; abstr LBA3), Chicago
137. Peeters M, Price TJ, Cervantes A et al (2010) Randomized phase III study of panitumumab with fluorouracil, leucovorin, and irinotecan (FOLFIRI) compared with FOLFIRI alone as second-line treatment in patients with metastatic colorectal cancer. J Clin Oncol 28:4706–4713
138. Meyers MB, Yu P, Mendelsohn J (1993) Crosstalk between epidermal growth factor receptor and P-glycoprotein in actinomycin D-resistant Chinese hamster lung cells. Biochem Pharmacol 46:1841–1848
139. Xu Y, Villalona-Calero MA (2002) Irinotecan: mechanisms of tumor resistance and novel strategies for modulating its activity. Annals Oncol 13:1841–1851
140. Sclabas GM, Fujioka S, Schmidt C et al (2003) Restoring apoptosis in pancreatic cancer cells by targeting the nuclear factor-kappaB signaling pathway with the anti-epidermal growth factor antibody IMC-C225. J Gastrointest Surg 7:37–43; discussion 43
141. Bandyopadhyay D, Mandal M, Adam L et al (1998) Physical interaction between epidermal growth factor receptor and DNA-dependent protein kinase in mammalian cells. J Biol Chem 273:1568–1573
142. Peeters M, Oliner KS, Parker A et al (2013) Massively parallel tumor multigene sequencing to evaluate response to panitumumab in a randomized phase III study of metastatic colorectal cancer. Clin Cancer Res 19:1902–1912
143. Vaughn CP, Zobell SD, Furtado LV et al (2011) Frequency of KRAS, BRAF, and NRAS mutations in colorectal cancer. Genes Chromosomes Cancer 50:307–312
144. Hecht JR, Mitchell E, Chidiac T et al (2009) A randomized phase IIIB trial of chemotherapy, bevacizumab, and panitumumab compared with chemotherapy and bevacizumab alone for metastatic colorectal cancer. J Clin Oncol 27:672–680
145. Tol J, Koopman M, Cats A et al (2009) Chemotherapy, bevacizumab, and cetuximab in metastatic colorectal cancer. New Eng J Med 360:563–572
146. Rodemann HP, Dittmann K, Toulany M (2007) Radiation-induced EGFR-signaling and control of DNA-damage repair. Int J Rad Biol 83:781–791
147. Dewdney A, Cunningham D, Tabernero J et al (2012) Multicenter randomized phase II clinical trial comparing neoadjuvant oxaliplatin, capecitabine, and preoperative radiotherapy with or without cetuximab followed by total mesorectal excision in patients with high-risk rectal cancer (EXPERT-C). J Clin Oncol 30:1620–1627

148. Fokas E, Conradi L, Weiss C et al (2013) Preoperative chemoradiation therapy with capecitabine/oxaliplatin and cetuximab in rectal cancer: long-term results of a prospective phase 1/2 study. Int J Radiat Oncol Biol Phys 87:992–999
149. Velenik V, Ocvirk J, Oblak I et al (2012) Cetuximab in preoperative treatment of rectal cancer—term outcome of the XERT trial. Radiol Oncol 46:252–257
150. Weiss C, Arnold D, Dellas K et al (2010) Preoperative radiotherapy of advanced rectal cancer with capecitabine and oxaliplatin with or without cetuximab: a pooled analysis of three prospective phase I-II trials. Int J Rad Oncol Biol Phys 78:472–478
151. Helbling D, Bodoky G, Gautschi O et al (2013) Neoadjuvant chemoradiotherapy with or without panitumumab in patients with wild-type KRAS, locally advanced rectal cancer (LARC): a randomized, multicenter, phase II trial SAKK 41/07. Ann Oncol 24:718–725
152. Pinto C, Di Fabio F, Maiello E et al (2011) Phase II study of panitumumab, oxaliplatin, 5-fluorouracil, and concurrent radiotherapy as preoperative treatment in high-risk locally advanced rectal cancer patients (StarPan/STAR-02 Study). Ann Oncol 22:2424–2430
153. Chung KY, Shia J, Kemeny NE et al (2005) Cetuximab shows activity in colorectal cancer patients with tumors that do not express the epidermal growth factor receptor by immunohistochemistry. J Clin Oncol 23:1803–1810
154. Cappuzzo F, Finocchiaro G, Rossi E et al (2008) EGFR FISH assay predicts for response to cetuximab in chemotherapy refractory colorectal cancer patients. Ann Oncol 19:717–723
155. Moroni M, Veronese S, Benvenuti S et al (2005) Gene copy number for epidermal growth factor receptor (EGFR) and clinical response to antiEGFR treatment in colorectal cancer: a cohort study. Lancet Oncol 6:279–286
156. Sartore-Bianchi A, Moroni M, Veronese S et al (2007) Epidermal growth factor receptor gene copy number and clinical outcome of metastatic colorectal cancer treated with panitumumab. J Clin Oncol 25:3238–3245
157. Bertotti A, Migliardi G, Galimi F et al (2011) A molecularly annotated platform of patient-derived xenografts ("xenopatients") identifies HER2 as an effective therapeutic target in cetuximab-resistant colorectal cancer. Cancer Discov 1:508–523
158. Barber TD, Vogelstein B, Kinzler KW et al (2004) Somatic mutations of EGFR in colorectal cancers and glioblastomas. N Engl J Med 351:2883
159. Metzger B, Chambeau L, Begon DY et al (2011) The human epidermal growth factor receptor (EGFR) gene in European patients with advanced colorectal cancer harbors infrequent mutations in its tyrosine kinase domain. BMC Med Genet 12:144
160. Matar P, Rojo F, Cassia R et al (2004) Combined epidermal growth factor receptor targeting with the tyrosine kinase inhibitor gefitinib (ZD1839) and the monoclonal antibody cetuximab (IMC-C225): superiority over single-agent receptor targeting. Clin Cancer Res 10:6487–6501
161. Weickhardt AJ, Price TJ, Chong G et al (2012) Dual targeting of the epidermal growth factor receptor using the combination of cetuximab and erlotinib: preclinical evaluation and results of the phase II DUX study in chemotherapy-refractory, advanced colorectal cancer. J Clin Oncol 30:1505–1512
162. Jacobs B, De Roock W, Piessevaux H et al (2009) Amphiregulin and epiregulin mRNA expression in primary tumors predicts outcome in metastatic colorectal cancer treated with cetuximab. J Clin Oncol 27:5068–5074
163. Jonker DJ, Karapetis CS, Harbison C et al (2014) Epiregulin gene expression as a biomarker of benefit from cetuximab in the treatment of advanced colorectal cancer. Br J Cancer 110:648–655
164. Khambata-Ford S, Garrett CR, Meropol NJ et al (2007) Expression of epiregulin and amphiregulin and K-ras mutation status predict disease control in metastatic colorectal cancer patients treated with cetuximab. J Clin Oncol 25:3230–3237
165. Lu Y, Li X, Liang K et al (2007) Epidermal growth factor receptor (EGFR) ubiquitination as a mechanism of acquired resistance escaping treatment by the anti-EGFR monoclonal antibody cetuximab. Cancer Res 67:8240–8247

166. Wheeler DL, Huang S, Kruser TJ et al (2008) Mechanisms of acquired resistance to cetuximab: role of HER (ErbB) family members. Oncogene 27:3944–3956
167. Tejpar S, Celik I, Schlichting M et al (2012) Association of KRAS G13D tumor mutations with outcome in patients with metastatic colorectal cancer treated with first-line chemotherapy with or without cetuximab. J Clin Oncol 30:3570–3577
168. Di Nicolantonio F, Martini M, Molinari F et al (2008) Wild-type BRAF is required for response to panitumumab or cetuximab in metastatic colorectal cancer. J Clin Oncol 26:5705–5712
169. Popovici V, Budinska E, Tejpar S et al (2012) Identification of a poor-prognosis BRAF-mutant-like population of patients with colon cancer. J Clin Oncol 30:1288–1295
170. Sartore-Bianchi A, Martini M, Molinari F et al (2009) PIK3CA mutations in colorectal cancer are associated with clinical resistance to EGFR-targeted monoclonal antibodies. Cancer Res 69:1851–1857
171. Frattini M, Signoroni S, Pilotti S et al (2005) Phosphatase protein homologue to tensin expression and phosphatidylinositol-3 phosphate kinase mutations in colorectal cancer. Cancer Res 65:11227
172. Ogino S, Nosho K, Kirkner GJ et al (2009) PIK3CA mutation is associated with poor prognosis among patients with curatively resected colon cancer. J Clin Oncol 27:1477–1484
173. Samuels Y, Velculescu VE (2004) Oncogenic mutations of PIK3CA in human cancers. Cell Cycle 3:1221–1224
174. Jhawer M, Goel S, Wilson AJ et al (2008) PIK3CA mutation/PTEN expression status predicts response of colon cancer cells to the epidermal growth factor receptor inhibitor cetuximab. Cancer Res 68:1953–1961
175. Laurent-Puig P, Cayre A, Manceau G et al (2009) Analysis of PTEN, BRAF, and EGFR status in determining benefit from cetuximab therapy in wild-type KRAS metastatic colon cancer. J Clin Oncol 27:5924–5930
176. Perrone F, Lampis A, Orsenigo M et al (2009) PI3KCA/PTEN deregulation contributes to impaired responses to cetuximab in metastatic colorectal cancer patients. Ann Oncol 20:84–90
177. Bibeau F, Lopez-Crapez E, Di Fiore F et al (2009) Impact of FcγRIIa-FcγRIIIa polymorphisms and KRAS mutations on the clinical outcome of patients with metastatic colorectal cancer treated with cetuximab plus irinotecan. J Clin Oncol 27:1122–1129
178. Fuchs BC, Fujii T, Dorfman JD et al (2008) Epithelial-to-mesenchymal transition and integrin-linked kinase mediate sensitivity to epidermal growth factor receptor inhibition in human hepatoma cells. Cancer Res 68:2391–2399
179. Oliveras-Ferraros C, Vazquez-Martin A, Cufi S et al (2011) Stem cell property epithelial-to-mesenchymal transition is a core transcriptional network for predicting cetuximab (Erbitux) efficacy in KRAS wild-type tumor cells. J Cell Biochem 112:10–29
180. Cappuzzo F, Varella-Garcia M, Finocchiaro G et al (2008) Primary resistance to cetuximab therapy in EGFR FISH-positive colorectal cancer patients. Brit J Cancer 99:83–89
181. Misale S, Yaeger R, Hobor S et al (2012) Emergence of KRAS mutations and acquired resistance to anti-EGFR therapy in colorectal cancer. Nature 486:532–536
182. Misale S, Arena S, Lamba S et al (2014) Blockade of EGFR and MEK intercepts heterogeneous mechanisms of acquired resistance to anti-EGFR therapies in colorectal cancer. Sci Transl Med 6:224ra226
183. Valtorta E, Misale S, Sartore-Bianchi A et al (2013) KRAS gene amplification in colorectal cancer and impact on response to EGFR-targeted therapy. Int J Cancer 133:1259–1265
184. Perkins G, Lievre A, Ramacci C et al (2010) Additional value of EGFR downstream signaling phosphoprotein expression to KRAS status for response to anti-EGFR antibodies in colorectal cancer. Int J Cancer 127:1321–1331
185. Takahashi Y, Kitadai Y, Bucana CD et al (1995) Expression of vascular endothelial growth factor and its receptor, KDR, correlates with vascularity, metastasis, and proliferation of human colon cancer. Cancer Res 55:3964–3968

186. Jain RK (2001) Normalizing tumor vasculature with anti-angiogenic therapy: a new paradigm for combination therapy. Nat Med 7:987–989
187. Pidgeon GP, Barr MP, Harmey JH et al (2001) Vascular endothelial growth factor (VEGF) upregulates BCL-2 and inhibits apoptosis in human and murine mammary adenocarcinoma cells. Br J Cancer 85:273–278
188. Gabrilovich DI, Ishida T, Nadaf S et al (1999) Antibodies to vascular endothelial growth factor enhance the efficacy of cancer immunotherapy by improving endogenous dendritic cell function. Clin Cancer Res 5:2963–2970
189. Griffioen AW (2008) Anti-angiogenesis: making the tumor vulnerable to the immune system. Cancer Immunol Immunother 57:1553–1558
190. Kandalaft LE, Motz GT, Busch J et al (2011) Angiogenesis and the tumor vasculature as antitumor immune modulators: the role of vascular endothelial growth factor and endothelin. Curr Top Microbiol Immunol 344:129–148
191. Beinert T, Binder D, Oehm C et al (2000) Further evidence for oxidant-induced vascular endothelial growth factor up-regulation in the bronchoalveolar lavage fluid of lung cancer patients undergoing radio-chemotherapy. J Cancer Res Clin Oncol 126:352–356
192. Saltz LB, Clarke S, Diaz-Rubio E et al (2008) Bevacizumab in combination with oxaliplatin-based chemotherapy as first-line therapy in metastatic colorectal cancer: a randomized phase III study. J Clin Oncol 26:2013–2019
193. Bendell JC, Bekaii-Saab TS, Cohn AL et al (2012) Treatment patterns and clinical outcomes in patients with metastatic colorectal cancer initially treated with FOLFOX-bevacizumab or FOLFIRI-bevacizumab: results from ARIES, a bevacizumab observational cohort study. Oncologist 17:1486–1495
194. Falcone A, Cremolini C, Masi G et al (2013) FOLFOXIRI/bevacizumab versus FOLFIRI/bevacizumab as first-line treatment in unresectable metastatic colorectal cancer patients; results of the phase III TRIBE trial by GONO Group. J Clin Oncol 31 Suppl 3505
195. Loupakis F, Cremolini C, Salvatore L et al (2014) FOLFOXIRI plus bevacizumab as first-line treatment in BRAF mutant metastatic colorectal cancer. Eur J Cancer 50:57–63
196. Cunningham D, Lang I, Marcuello E et al (2013) Bevacizumab plus capecitabine versus capecitabine alone in elderly patients with previously untreated metastatic colorectal cancer (AVEX): an open-label, randomised phase 3 trial. Lancet Oncol 14:1077–1085
197. Kabbinavar FF, Schulz J, Mccleod M et al (2005) Addition of bevacizumab to bolus fluorouracil and leucovorin in first-line metastatic colorectal cancer: results of a randomized phase II trial. J Clin Oncol 23:3697–3705
198. Giantonio BJ, Catalano PJ, Meropol NJ et al (2007) Bevacizumab in combination with oxaliplatin, fluorouracil, and leucovorin (FOLFOX4) for previously treated metastatic colorectal cancer: results from the Eastern Cooperative Oncology Group Study E3200. J Clin Oncol 25:1539–1544
199. Bennouna J, Sastre J, Arnold D et al (2013) Continuation of bevacizumab after first progression in metastatic colorectal cancer (ML18147): a randomised phase 3 trial. Lancet Oncol 14:29–37
200. Van Cutsem E, Tabernero J, Lakomy R et al (2012) Addition of aflibercept to fluorouracil, leucovorin, and irinotecan improves survival in a phase III randomized trial in patients with metastatic colorectal cancer previously treated with an oxaliplatin-based regimen. J Clin Oncol 30:3499–3506
201. Pericay C, Folprecht G, Saunders G et al (2012) Phase 2 randomized, non-comparative, open-label study of aflibercept and modified FOLFOX6 in the first-line treatment of metastatic colorectal cancer (AFFIRM). Ann Oncol 23(Suppl 4) (Abstract O-0024)
202. Wilhelm SM, Dumas J, Adnane L et al (2011) Regorafenib (BAY 73-4506): a new oral multikinase inhibitor of angiogenic, stromal and oncogenic receptor tyrosine kinases with potent preclinical antitumor activity. Int J Cancer 129:245–255
203. Strumberg D, Scheulen ME, Schultheis B et al (2012) Regorafenib (BAY 73-4506) in advanced colorectal cancer: a phase I study. Br J Cancer 106:1722–1727

204. Grothey A, Van Cutsem E, Sobrero A et al (2013) Regorafenib monotherapy for previously treated metastatic colorectal cancer (CORRECT): an international, multicentre, randomised, placebo-controlled, phase 3 trial. Lancet 381:303–312
205. Hurwitz HI, Douglas PS, Middleton JP et al (2013) Analysis of early hypertension and clinical outcome with bevacizumab: results from seven phase III studies. Oncologist 18:273–280
206. Osterlund P, Soveri LM, Isoniemi H et al (2011) Hypertension and overall survival in metastatic colorectal cancer patients treated with bevacizumab-containing chemotherapy. Br J Cancer 104:599–604
207. Scartozzi M, Galizia E, Chiorrini S et al (2009) Arterial hypertension correlates with clinical outcome in colorectal cancer patients treated with first-line bevacizumab. Ann Oncol 20:227–230
208. Hurwitz HI, Yi J, Ince W et al (2009) The clinical benefit of bevacizumab in metastatic colorectal cancer is independent of K-ras mutation status: analysis of a phase III study of bevacizumab with chemotherapy in previously untreated metastatic colorectal cancer. Oncologist 14:22–28
209. Ince WL, Jubb AM, Holden SN et al (2005) Association of k-ras, b-raf, and p53 status with the treatment effect of bevacizumab. J Natl Cancer Inst 97:981–989
210. Bates DO, Catalano PJ, Symonds KE et al (2012) Association between VEGF splice isoforms and progression-free survival in metastatic colorectal cancer patients treated with bevacizumab. Clin Cancer Res 18:6384–6391
211. Longo R, Gasparini G (2007) Challenges for patient selection with VEGF inhibitors. Cancer Chemother Pharmacol 60:151–170
212. Hegde PS, Jubb AM, Chen D et al (2013) Predictive impact of circulating vascular endothelial growth factor in four phase III trials evaluating bevacizumab. Clin Cancer Res 19:929–937
213. Grothey A, Hedrick EE, Mass RD et al (2008) Response-independent survival benefit in metastatic colorectal cancer: a comparative analysis of N9741 and AVF2107. J Clin Oncol 26:183–189
214. Jubb AM, Hurwitz HI, Bai W et al (2006) Impact of vascular endothelial growth factor-A expression, thrombospondin-2 expression, and microvessel density on the treatment effect of bevacizumab in metastatic colorectal cancer. J Clin Oncol 24:217–227
215. Jubb AM, Oates AJ, Holden S et al (2006) Predicting benefit from anti-angiogenic agents in malignancy. Nat Rev Cancer 6:626–635
216. Price Tj TA, Bruhn M et al (2014) Assessment of IL-6, IL-8, bFGF, PDGF-BB, and VEGF-A as prognostic and predictive biomarkers for anti-VEGF in metastatic colorectal cancer. Gastrointestinal cancers symposium. J Clin Oncol 32, 2014 (suppl 3; abstr 502)
217. Cremolini C, Loupakis F, Bocci G et al (2014) Biomarkers and response to bevacizumab—letter. Clin Cancer Res 20:1056–1057
218. Masi G, Loupakis F, Salvatore L et al (2012) A randomized phase III study evaluating the continuation of bevacizumab beyond progression in metastatic colorectal cancer patients who received bevacizumab as part of first-line treatment: results of the BEBYP trial by Gruppo Oncologico Nord Ovest (GONO). Ann Oncol 23 supplement 9 (9)
219. Foernzler D, Delmar P, Kockx M et al (2010) Tumor tissue based biomarker analysis in NO16966: a randomized phase III study of first-line bevacizumab in combination with oxaliplatin-based chemotherapy in patients with metastatic colorectal cancer. In: Oncol JC (ed) Gastrointestinal cancers symposium. Abstract 374
220. Aoyagi Y, Iinuma H, Horiuchi A et al (2010) Association of plasma VEGF-A, soluble VEGFR-1 and VEGFR-2 levels and clinical response and survival in advanced colorectal cancer patients receiving bevacizumab with modified FOLFOX6. Oncology Lett 1:253–259
221. Jurgensmeier JM, Schmoll HJ, Robertson JD et al (2013) Prognostic and predictive value of VEGF, sVEGFR-2 and CEA in mCRC studies comparing cediranib, bevacizumab and chemotherapy. Br J Cancer 108:1316–1323

222. Formica V, Palmirotta R, Del Monte G et al (2011) Predictive value of VEGF gene polymorphisms for metastatic colorectal cancer patients receiving first-line treatment including fluorouracil, irinotecan, and bevacizumab. Int J Colorectal Dis 26:143–151
223. Loupakis F, Ruzzo A, Salvatore L et al (2011) Retrospective exploratory analysis of VEGF polymorphisms in the prediction of benefit from first-line FOLFIRI plus bevacizumab in metastatic colorectal cancer. BMC Cancer 11:247
224. Loupakis F, Cremolini C, Yang D et al (2013) Prospective validation of candidate SNPs of VEGF/VEGFR pathway in metastatic colorectal cancer patients treated with first-line FOLFIRI plus bevacizumab. PLoS ONE 8:e66774
225. Chionh Fjm GV, Chueh AC et al (2014) Single nucleotide polymorphisms (SNPs) in vascular endothelial growth factor (VEGF) family genes as predictive or prognostic biomarkers in patients with metastatic colorectal cancer: analysis of the phase III MAX study. J Clin Oncol 32:5s (suppl; abstr 3609)
226. Dhodapkar MV (2013) Personalized immune-interception of cancer and the battle of two adaptive systems—when is the time right? Cancer Prev Res 6:173–176
227. Bissell MJ, Radisky D (2001) Putting tumours in context. Nat Rev Cancer 1:46–54
228. Abbas AK, Lichtman AH, Pillai S (2014) Basic immunology: functions and disorders of the immune system. Elsevier Saunders, Philadelphia
229. Kim R, Emi M, Tanabe K (2007) Cancer immunoediting from immune surveillance to immune escape. Immunology 121:1–14
230. Cavallo F, De Giovanni C, Nanni P et al (2011) 2011: the immune hallmarks of cancer. Cancer Immunol Immunother 60:319–326
231. Galon J, Costes A, Sanchez-Cabo F et al (2006) Type, density, and location of immune cells within human colorectal tumors predict clinical outcome. Science 313:1960–1964
232. Pihl E, Nairn RC, Milne BJ et al (1980) Lymphoid hyperplasia: a major prognostic feature in 519 cases of colorectal carcinoma. Am J Pathol 100:469–480
233. Ohtani H (2007) Focus on TILs: prognostic significance of tumor infiltrating lymphocytes in human colorectal cancer. Cancer Immun 7:4
234. Pages F, Galon J, Dieu-Nosjean MC et al (2010) Immune infiltration in human tumors: a prognostic factor that should not be ignored. Oncogene 29:1093–1102
235. Pages F, Kirilovsky A, Mlecnik B et al (2009) In situ cytotoxic and memory T cells predict outcome in patients with early-stage colorectal cancer. J Clin Oncol 27:5944–5951
236. Halama N, Michel S, Kloor M et al (2011) Localization and density of immune cells in the invasive margin of human colorectal cancer liver metastases are prognostic for response to chemotherapy. Cancer Res 71:5670–5677
237. Mills CD (2012) M1 and M2 macrophages: oracles of health and disease. Crit Rev Immunol 32:463–488
238. Lin EY, Nguyen AV, Russell RG et al (2001) Colony-stimulating factor 1 promotes progression of mammary tumors to malignancy. J Exp Med 193:727–740
239. Zhang Y, Sime W, Juhas M et al (2013) Crosstalk between colon cancer cells and macrophages via inflammatory mediators and CD47 promotes tumour cell migration. Eur J Cancer 49:3320–3334
240. Pollard JW (2004) Tumour-educated macrophages promote tumour progression and metastasis. Nat Rev Cancer 4:71–78
241. Poschke I, Mougiakakos D, Kiessling R (2011) Camouflage and sabotage: tumor escape from the immune system. Cancer Immunol Immunother 60:1161–1171
242. Atkins D, Breuckmann A, Schmahl GE et al (2004) MHC class I antigen processing pathway defects, ras mutations and disease stage in colorectal carcinoma. Int J Cancer 109:265–273
243. Volonte A, Di Tomaso T, Spinelli M et al (2013) Cancer-initiating cells from colorectal cancer patients escape from T cell-mediated immunosurveillance in vitro through membrane-bound IL-4. J Immunol 192:523–532
244. Mapara MY, Sykes M (2004) Tolerance and cancer: mechanisms of tumor evasion and strategies for breaking tolerance. J Clin Oncol 22:1136–1151

245. Staveley-O'carroll K, Sotomayor E, Montgomery J et al (1998) Induction of antigen-specific T cell anergy: an early event in the course of tumor progression. Proc Natl Acad Sci U S A 95:1178–1183
246. Orsini G, Legitimo A, Failli A et al (2013) Defective generation and maturation of dendritic cells from monocytes in colorectal cancer patients during the course of disease. Int J Mol Sci 14:22022–22041
247. Pinzon-Charry A, Maxwell T, Lopez JA (2005) Dendritic cell dysfunction in cancer: a mechanism for immunosuppression. Immunol Cell Biol 83:451–461
248. Coley WB (1893) II. Hawkins on tubercular peritonitis. Ann Surg 17:462–464
249. Rosenberg SA, Yang JC, Topalian SL et al (1994) Treatment of 283 consecutive patients with metastatic melanoma or renal cell cancer using high-dose bolus interleukin 2. JAMA 271:907–913
250. Hodi FS, O'day SJ, Mcdermott DF et al (2010) Improved survival with ipilimumab in patients with metastatic melanoma. New Eng J Med 363:711–723
251. Higano CS, Small EJ, Schellhammer P et al (2010) Sipuleucel-T. Nat Rev Drug Discov 9:513–514
252. Burgdorf SK, Fischer A, Myschetzky PS et al (2008) Clinical responses in patients with advanced colorectal cancer to a dendritic cell based vaccine. Oncol Rep 20:1305–1311
253. Sakakibara M, Kanto T, Hayakawa M et al (2011) Comprehensive immunological analyses of colorectal cancer patients in the phase I/II study of quickly matured dendritic cell vaccine pulsed with carcinoembryonic antigen peptide. Cancer Immunol Immunother 60:1565–1575
254. Loveland BE, Zhao A, White S et al (2006) Mannan-MUC1-pulsed dendritic cell immunotherapy: a phase I trial in patients with adenocarcinoma. Clin Cancer Res 12:869–877
255. Harrop R (2006) Vaccination of colorectal cancer patients with modified vaccinia Ankara delivering the tumor antigen 5T4 (TroVax) induces immune responses which correlate with disease control: a phase I/II trial. Clin Cancer Res 12:3416–3424
256. Toubaji A, Achtar M, Provenzano M et al (2008) Pilot study of mutant ras peptide-based vaccine as an adjuvant treatment in pancreatic and colorectal cancers. Cancer Immunol Immunother 57:1413–1420
257. Correale P, Fulfaro F, Marsili S et al (2005) Gemcitabine (GEM) plus oxaliplatin, folinic acid, and 5-fluorouracil (FOLFOX-4) in patients with advanced gastric cancer. Cancer Chemother Pharmacol 56:563–568
258. Correale P, Montagnani F, Miano S et al (2008) Biweekly triple combination chemotherapy with gemcitabine, oxaliplatin, levofolinic acid and 5-fluorouracil (GOLF) is a safe and active treatment for patients with inoperable pancreatic cancer. J Chemother 20:119–125
259. Carlson RH (2011) Advanced colorectal cancer: GOLFIG shows benefit over FOLFOX-4. Oncol Times UK 8:11

Index

V. W. Yang, A. B. Bialkowska (eds.), *Intestinal Tumorigenesis*,
DOI 10.1007/978-3-319-19986-3

MIX
Papier aus verantwortungsvollen Quellen
Paper from responsible sources
FSC® C105338

If you have any concerns about our products,
you can contact us on
ProductSafety@springernature.com

In case Publisher is established outside the EU,
the EU authorized representative is:
Springer Nature Customer Service Center GmbH
Europaplatz 3, 69115 Heidelberg, Germany

Printed by Libri Plureos GmbH
in Hamburg, Germany